Applying Evolutionary Archaeology

A Systematic Approach

Jim: I hope you find parts of this interesting.

Best,
Mike

Applying Evolutionary Archaeology

A Systematic Approach

Michael J. O'Brien

and

R. Lee Lyman

University of Missouri–Columbia
Columbia, Missouri

Kluwer Academic / Plenum Publishers
New York Boston Dordrecht London Moscow

Library of Congress Cataloging-in-Publication Data

O'Brien, Michael J. (Michael John), 1950–
Applying evolutionary archaeology: a systematic approach/Michael J. O'Brien and R. Lee Lyman
p. cm.
Includes bibliographical references and index.
ISBN 0-306-46253-2 (Hardbound)—ISBN 0-306-46254-0 (Paperbound)
1. Archaeology—Methodology. 2. Archaeology—Philosophy. 3. Evolution. 4. Anthropology, Prehistoric. I. Lyman, R. Lee. II. Title.

CC75.7 .O27 2000
930.1—dc21 99-089696

ISBN 0-306-46253-2 (Hardbound)
ISBN 0-306-46254-0 (Paperback)

233 Spring Street, New York, N.Y. 10013

http://www.wkap.nl/

10 9 8 7 6 5 4 3 2 1

A C.I.P. record for this book is available from the Library of Congress.

Printed in the United States of America

Preface

Anthropology, and by extension archaeology, has had a long-standing interest in evolution in one or several of its various guises. Pick up any lengthy treatise on humankind written in the last quarter of the nineteenth century and the chances are good that the word *evolution* will appear somewhere in the text. If for some reason the word itself is absent, the odds are excellent that at least the concept of change over time will have a central role in the discussion. After one of the preeminent (and often vilified) social scientists of the nineteenth century, Herbert Spencer, popularized the term in the 1850s, evolution became more or less a household word, usually being used synonymously with change, albeit change over extended periods of time. Later, through the writings of Edward Burnett Tylor, Lewis Henry Morgan, and others, the notion of evolution as it applies to stages of social and political development assumed a prominent position in anthropological discussions.

To those with only a passing knowledge of American anthropology, it often appears that evolutionism in the early twentieth century went into a decline at the hands of Franz Boas and those of similar outlook, often termed *particularists*. However, it was not evolutionism that was under attack but rather comparativism—an approach that used the ethnographic present as a key to understanding how and why past peoples lived the way they did (Boas 1896). Anthropology's admiration of the comparative approach began to sour after roughly 1900, but evolutionism itself continued as a strong undercurrent in anthropological work, notably that of archaeologists. They were beginning to discover that the temporal span of human existence on the North American continent, although not on the order seen in Europe, extended back well beyond the recent ethnographic past. If time was fairly deep, then perhaps one could begin to talk about evolutionary change in terms of the archaeological record. Some Americanist archaeologists of the 1920s and 1930s picked up on the notion of evolution, but they made a fundamental mistake—one that undermined any chance of using scientific evolutionism as a means of explaining the record. In short, they transferred a biological theory into archaeology, but they failed to make clear, probably because of a lack of clarity in their own minds, exactly what was evolving. Their colleagues were not persuaded—one went so far in the 1940s as to imply that artifacts do not breed—and scientific, or Darwinian, evolutionism as a legitimate explanatory framework in archaeology became dormant until the 1970s.

Ironically, archaeologists, because of what they study, could have been at the forefront of a move to make evolutionism an integral part of the study of humankind, yet in retrospect it is easy to see how and why they failed. The effort was doomed not because of lack of theory, which had been around since 1859, but because no logical connection was made between that theory and the archaeological record. Note that we did not say that *no* connection was made; rather, we said that no *logical* connection was made. Anthropologists made a logical connection between evolutionism and what they studied—organisms as opposed to the products made, used, and discarded by those organisms—but the connection extended only to the somatic aspects of humans, that is, only to the bodily, or organic, features that humans exhibit. Recently, anthropological attention has begun to shift to evolution as it pertains to human behavior, but only rarely does one see more than passing mention of the possibility that inorganic remains are as much a part of evolution—and hence of the evolutionary record—as are bones, teeth, and skin. Our primary purpose in writing this book is to demonstrate that the archaeological record *is* an evolutionary record in the same way that the organic fossil record is. As such, the archaeological record is completely amenable to the methods used by paleobiologists to reconstruct phylogenetic histories. As we will show, however, archaeology—both theoretically and methodologically—has yet to make this connection in anything approaching robust fashion.

Our position—perhaps *underlying principle* is a better term—is that the year 1859 was a watershed in the study of nature, of which *Homo sapiens* and our phylogenetic cousins and ancestors are only a small part. In that year Charles Darwin published *On the Origin of Species*, which to our minds changed forever the ways in which nature is examined. We in no way suggest that what Darwin had to say has escaped sound revision (as well as scurrilous misinterpretation); rather, we suggest only that it set about a revolution in the way descent and modification of organisms are viewed. Darwin's opinions on what he termed "descent with modification" are, as far as we know, as valid today as they were the day they were penned. His ideas not only on the process of adaptation but also on the features produced by the process, unfortunately also known as "adaptations," spawned a philosophical (and later methodological) revolution that is still ongoing. In essence, Darwin provided the theoretical basis for studying change in organisms—not *all* kinds of change, but certainly all kinds of change that take place over extended (multigenerational) periods of time. It has been the role of scientists ever since to refine that basis and to examine the wonderful particulars spawned by nature, some of which form the archaeological record.

Anthropology for the most part has played a secondary role in providing a coherent picture of the development of the one organism it is best suited to study. Archaeology, often considered the stepchild of anthropology, has accomplished even less. Competition between paradigms—worldviews of how we perceive things to be—has overshadowed the principles and certainly the interesting details of human evolution. The first few chapters in many introductory anthropology textbooks discuss evolution, genetics, and natural selection but end up excluding nonsomatic attributes—culture in a general sense—from consideration, or the authors divide selection into two kinds, natural and cultural, and focus on the latter as the driving force in effecting change in human evolution. We minimize this distinction. To us, organisms, including humans, change in part because of selection, regardless of how that selection is presented to the organism. The source of selection, or of any other evolutionary mechanism, really makes no difference to the organism. It could be the physical environment, the social environment, or both. Or it might be pure chance that causes change within a lineage of organisms. Our job as anthropologists is to understand not only the context in which evolution occurred but also the outcome of the process. And it is the myriad outcomes that underscore the fascinating story of the emergence and development of humankind.

To us, these outcomes are as manifest in such things as stone tools as they are in physical characteristics of organisms. Biologists certainly do not find it strange to talk about extending the phenotype of birds to include nests and nest-building behaviors, given that those behaviors and the products they create—nests—protect the germ line of the organisms. Neither do they find it strange to talk about dams and dam-building behavior as parts of a beaver's phenotype. Many anthropologists might buy into this argument as well but, ironically, they draw the line when it is suggested that tools and tool-making behaviors or the way people dress or build their houses are parts of the human phenotype. Why does this dual view exist? Is it because we have set humans apart from the rest of the natural world? Are humans quantitatively and/or qualitatively *so* different that they warrant not only a whole new set of laws but also a different set of philosophical questions with which to examine themselves? We do not think so, and we would like the opportunity to convince skeptics that the theory necessary for explaining change in the human lineage was set forth 140 years ago. Furthermore, we would like the opportunity to convince archaeologists that the record they study is a historical narrative of the evolutionary process.

Our approach consists of two interrelated strategies. The first deals with examining several ontological and epistemological issues that we view as fundamental to the eventual incorporation of a Darwinian evolutionary approach in archaeology. In so doing, we backtrack a bit and pick up the origins of those issues in both archaeology and biology. We note, however, that this book is not a complete history of evolutionary thought in archaeology, nor do we attempt to detail all of the issues and subissues with which evolutionary biologists have wrestled over the last century. Rather, we introduce those topics that we view as being fundamental to applying Darwinian evolutionary theory to archaeology. The second strategy deals with teasing apart numerous methodological issues that confront anyone using a fossil record—be it an organic record comprising the remains of organisms or an inorganic record comprising the things made, used, and discarded by those organisms—to examine evolutionary pathways. How one confronts these issues stems directly from how epistemological issues are addressed. One's choice of units and methods must be directed by epistemological considerations. Our adoption of the word *systematics* for use in the title of this book underscores our belief not only that theory guides the selection of one set of units and methods over another but also that the strand that connects the wide gap between theory and explanation must be spun carefully and systematically.

Our goal is to produce a book that will serve as an introduction to evolutionary archaeology and that almost anyone with a basic curiosity in things human can pick up and with a little patience understand. Darwinian evolution is not difficult to comprehend, but philosophical and methodological debates that have grown up around the subject over the last half-century often *are* complex and difficult to follow even for those with extensive backgrounds in the subject. Extension of those debates to archaeology, although soundly constructed, has tended to be telegraphic and loaded with polemical statements, with the result that too few archaeologists have bothered to wade through the arguments. This, it seems to us, is one reason why the resurgence in applying Darwinian evolutionism to archaeology that began in the 1970s has not attracted more attention within the discipline. Without question, Americanist archaeology has paid little attention to the epistemological debates that have occurred in evolutionary biology, but also without question is the fact that those archaeologists (ourselves included) who have urged consideration of those debates have not taken the time to construct arguments that can be readily understood. And above all, the units necessary to track change in the archaeological record remain obscure. We hope to remedy the situation and illustrate by dint of example some of the important considerations in incorporating Darwinian evolutionism into archaeology. As a guide to terminology, we include a glossary of some of the more important terms and concepts used. Finally, we point out that we cannot hope in these pages to cover all of the important issues, and critics may well complain that there remain large gaps in our coverage. This is correct; instead of trying to cover all the bases, we tend to emphasize the basic and fundamental issues. There still remains plenty of fertile ground for others to cover.

Acknowledgments

We gratefully acknowledge the advice and assistance of our sponsoring editor at Kluwer/Plenum, Eliot Werner, of our production editor, Herman Makler, and of our copy editor, Jacquelyn Coggin. We thank Elliott Sober, Marc Ereshefsky, and Tom Jones for their detailed comments on the original book proposal and an earlier draft of Chapter 1, and Ray Ethington and Mike Schiffer for comments on an earlier draft of the manuscript. Bob Leonard read the final manuscript and pointed out several lapses in logic and other potentially embarrassing blunders, as did E. J. O'Brien, who also edited the manuscript for both content and style. He also rewrote several sections and made them more intelligible. This is the sixth book of ours on which he has worked, and we greatly appreciate the help and advice he has provided over the years. All figures were created by Dan Glover, who also tracked down various source materials, prepared the references and index, and made numerous other significant contributions to the project. Don Luetjen, Joe Kinker, and Roy Hathcock allowed us to photograph projectile points in their collections for use in the book. Finally, we acknowledge the advice and constructive criticism provided by Bob Dunnell over the last two decades.

Contents

Figures and Tables

FIGURES

TABLES

Applying Evolutionary Archaeology

A Systematic Approach

Chapter 1

Darwinian Theory and Archaeology

> Darwin's theory of evolution through natural selection was undoubtedly the most revolutionary theory of all time. It surpassed even the astronomical revolution ushered in by Copernicus in the significance of its implications for our understanding of the nature of the universe and of our place and role in it.... Moreover, Darwin's masterly marshalling of the evidence ... remain[s] to this day an intellectual monument that is unsurpassed in the history of human thought. (Muller 1949:459)

Over the past decade or so, an increasing number of archaeologists have begun to show an interest in employing evolutionary theory to examine and explain variation in the material record—an interest that is manifest in the growing list of articles and books on the subject. But, knowledgeable students of the discipline might ask, have not archaeologists long been interested in applying evolutionism to what they study? Haven't the examination and explanation of change—the essential tasks of evolutionists—always been at the core of everything that archaeologists do? If so, then why single out the last decade for special treatment? Our answer is that although archaeologists *have* been interested in evolution—an interest that crops up from time to time, only to be lost in the shuffle as new and trendier paradigms come along—there are decided differences between modern archaeological interest in evolution and that of previous decades. One difference is that the evolutionism currently under consideration is of a kind that is not as easily displaced as some of its counterparts. In fact, it *cannot* be displaced in archaeology anymore than it can in other disciplines where the focus is on a historical record.

Of all such disciplines, only anthropology, and by extension archaeology, have failed to understand the explanatory significance of this kind of evolutionism, but there are signs that this is changing.

The kind of evolutionism we have in mind traces its roots directly to 1859 and the publication of Charles Darwin's *On the Origin of Species*, and it bears almost no resemblance to other brands of evolutionism that have interested anthropologists in general and archaeologists in particular over the years. The incorporation of Darwinian evolutionism into archaeology represents a radical departure from the unilinear and progressive cultural evolutionism of Edward B. Tylor (1871) and Lewis Henry Morgan (1877), which figured so prominently in the writings of prehistorians of the late nineteenth century (Dunnell 1988b; O'Brien 1996d). It also is a significant departure from the cultural evolutionism of Leslie White (1945, 1949, 1959a, 1959b), V. Gordon Childe (1951a, 1951b), Julian Steward (1955), and their followers (e.g., Sahlins and Service 1960), which played a comparably significant role in the processual archaeology of the 1960s and 1970s. A major difference between these kinds of evolutionism and Darwinian evolutionism, at least as the difference applies in archaeology, is a shift in focus away from the evolution of culture per se and toward the evolution of particular cultural phenomena. As we shall see, this shift in perspective is not easily made because of a deeply rooted tradition that views culture itself and not cultural phenomena as the appropriate topic of anthropological and archaeological study.

Sandwiched between the decline in archaeology of the Tylor–Morgan brand of evolutionism in the early twentieth century and the rise in the 1960s and 1970s of what for simplicity we refer to as Whitean cultural evolution were several attempts to use Darwinian evolutionism as a means of explaining some of the immense variation seen in the archaeological record. References to biological evolution, or at least to evolutionary metaphors, were fairly common in Americanist[1] archaeology after about 1910 (e.g., Gladwin 1936; Kidder 1915, 1932; Kroeber 1931a). Some historical reviews of evolutionism in archaeology (e.g., Dunnell 1980; O'Brien and Holland 1990) make it sound as if archaeologists all but ignored Darwinian evolutionism until an enlightened few made it accessible in the late 1970s. Most such reviews spend considerable time discussing the differences between Darwinian biological evolutionism and Whitean cultural evolutionism—an emphasis that reflects not only the latter's influence on the processual archaeology of the 1960s and 1970s but also its continued influence, albeit to a lesser degree, on modern archaeological thought. Such discussion is warranted, but usually ignored is the fact that Americanists earlier this century made serious, though flawed, efforts to incorporate elements of Darwinian evolutionism into

[1]We use the term *Americanist* archaeology (as opposed to *American* archaeology) to emphasize archaeology undertaken by American-trained archaeologists, regardless of the location of their research area.

their work (Lyman et al. 1997a, 1997; Lyman and O'Brien 1997; O'Brien and Lyman 1998, 1999a). Note what A. V. Kidder (1932:8), one of the leading culture historians of the period 1910–1940 and a person regularly vilified by archaeologists of the 1960s, had to say on the subject of evolution:

> Archaeologists, noting that modern biology has mounted above the plane of pure taxonomy, have attempted to follow that science into the more alluring fields of philosophic interpretation, forgetting that the conclusions of the biologist are based on the sound foundation of scientifically marshalled facts gathered during the past century by an army of painstaking observers. This groundwork we utterly fail to possess. Nor will it be easy for us to lay, because the products of human hands, being unregulated by the more rigid genetic laws which control the development of animals and plants, are infinitely variable. But that is no reason for evading the attempt. It has got eventually to be done, and the sooner we roll up our sleeves and begin comparative studies of axes and arrowheads and bone tools, make classifications, prepare accurate descriptions, draw distribution maps and, in general, persuade ourselves to do a vast deal of painstaking, unspectacular work, the sooner shall we be in position to approach the problems of cultural evolution, the solving of which is, I take it, our ultimate goal.

Kidder's insightful remarks are important because they indicate how some culture historians in the 1930s conceived of the phenomena they were studying. With a few changes in wording, Kidder's quote could stand up under the scrutiny of the most conservative evolutionary archaeologist: Things in the archaeological record *evolve* over time. And notice the sentence that contains the phrases "begin comparative studies," "make classifications," "prepare accurate descriptions," and "draw distribution maps." These not only are the basic things that an archaeologist interested in evolution does but they also represent what the majority of archaeologists have always done. Interestingly, they also are the everyday tasks of an evolutionary biologist, especially of a paleobiologist. For our purposes here, paleobiology denotes the study and explanation of the paleontological record using evolutionary theory. Paleontology is more a descriptive or biostratigraphic enterprise, similar in many ways to the characterizations of culture history provided by the processual archaeologists of the 1960s (e.g., Flannery 1967).

Robert Dunnell (1985b:732) noted that "archaeology is far more similar to paleontology and evolutionary biology than it is to sociocultural anthropology," though this parallelism has rarely been discussed. In fact, archaeology and paleobiology have a number of interests in common. As we will see throughout this book, they have more in common than simply routine tasks; the natural link that exists between the two disciplines is one of the keys to successfully creating an evolutionary archaeology. Unfortunately, that natural link often is obscured because archaeologists forget that their subject matter is historical in nature (Lyman and O'Brien 1998). As such, it requires that a particular set of analytical strategies be used to document and explain change. Failure on the part of archaeologists to make the critical distinction between different approaches to understanding the natural world has meant that throughout much of its existence, archaeology has been based on the wrong model of science. This went unnoticed by Kidder and his contemporaries, just as, with few exceptions, it has gone unnoticed ever since.

Kidder made his comments a decade before the unifications of the great biological schools of the early twentieth century—first the unification of paleontologists and zoologists into what has been labeled the naturalist school, and second the unification of the naturalists and the geneticists. Julian Huxley (1942) appropriately labeled these collective events the *Modern Synthesis*. That synthesis led to a more elevated status for paleobiologists, who had long been considered the poor cousins of true biologists, who studied living organisms rather than dead ones. But it was those cousins who had long held to the view that the fossil record, a historical sequence, contained an enormously powerful key to understanding not only the evolution of life but also its modern configuration. Prior to the Synthesis, few biologists paid attention to their fossil-loving colleagues, but by the 1940s, helped along especially by publication of George Gaylord Simpson's (1944) *Tempo and Mode in Evolution*, recognition of the fossil record as an important means of understanding such processes as speciation became more commonplace.

THE LONG SEARCH FOR A WORKABLE EVOLUTIONISM

It is clear that archaeologists took little notice of the theoretical and methodological issues that were central to the Synthesis; thus, they never picked up on the significance of the issues for studying the archaeological record. The lack of progress made by archaeologists in recognizing the applicability of Darwinian evolutionism to the material record is evident in the words of William G. Haag, who, twenty-nine years after Kidder (1932:8) had made his point (see quotation), wrote that "when cultural evolution is considered a natural extension of organic evolution, our thinking about the total process of evolution will be greatly clarified" (Haag 1961:440; see also Haag 1959). Although Haag, an archaeologist with extensive training in zoology, never applied evolutionary theory to an examination of the archaeological record, his admonition demonstrates that he made no hard-and-fast division between kinds of evolution. To Haag, evolution was evolution, regardless of the modifier that might be placed in front of it. His comments fell on the eve of the intellectual revolution that archaeology would go through in the 1960s, and they are equally as important as Kidder's in a historical sense. Notice that Haag said, "When cultural evolution is considered a *natural* extension of organic evolution"; he did not say, "When it is considered as a *parallel track* to organic evolution." It is evident that little progress was made in the three decades after Kidder laid out his agenda for extending biological evolution to encompass things in the archaeological record, and the situation did not improve much in the two decades following Haag's comments. Failure on the part of processual archaeologists—literally, archaeologists interested in understanding how cultural processes work in effecting change—to appreciate the message of these two culture historians killed any chances of incorporating Darwinian evolutionism into archaeology. Whitean cultural evolution was still very much alive and served to underpin much of the processual archaeology that was carried out in the 1960s and 1970s.

Kidder probably would have predicted the false starts and sidetracks that Americanist archaeology would take in its long search for a workable evolutionism. He correctly noted that archaeology lacked both the basic data and a theory that could account for cultural phenomena. It is obvious that Kidder, like Haag, did not need a *separate* theory to explain cultural variation; the theory was already available. It was a matter of convincing archaeologists to use it. To those interested in incorporating Darwinian evolutionary theory in archaeology, Kidder's words could not have been better put, because he focused on the issue that underlies any evolutionary study—the documentation of variation. He did not see variation in such things as pottery as being fundamentally different from genetically controlled variation; he just saw more of it being produced as a result of the almost limitless imagination of humans and their enormous capacity for effecting change in their social and physical environment. Others, such as Harold Gladwin (1936), worried that because cultural evolution occurred so much more rapidly than biological evolution, Darwin's ideas might not be applicable to the archaeological record—a view that is commonplace in the discipline today.

Kidder excelled at documenting variation—he did it throughout his life, first in the Southwest and then in Mesoamerica—but there was one thing missing from his archaeological procedures. In the rush to document variation, he and other like-minded culture historians forgot (or never realized) that the choice of units one uses in any scientific analysis must be guided by theory. In Kidder's case, the problem was to figure out how to apply evolutionary theory to the archaeological record. He suspected the theory was powerful in terms of its explanatory ability, but he was unable to arrive at a logical means of connecting artifacts with that theory. Kidder's lack of knowledge of both how evolution worked and the many epistemological issues that are central to Darwinian evolutionism—many of which were not well formulated in Kidder's day—forced him to rely instead on vague metaphors when it came to evolution and the material record. As a result, his efforts were later derogated by Walter Taylor (1948:46) as being "neither historiography nor cultural anthropology" and characterized as a "description of artifacts ... for its own sake and for the sake of comparative study on a purely descriptive level." But recall Kidder's (1932:8) long quote cited earlier, where he indicated that laying the "sound foundation of scientifically marshalled facts" must precede "philosophic interpretation." Kidder was merely following what he perceived to be the research design of the successful (paleo)biologist: Get the facts and then connect them to the theory. But was this the path to success?

If Kidder had truly understood Darwinian evolutionism, he would have had a theoretical base from which to launch the formation of units useful for measuring evolutionary change. In the absence of theory, common sense became the warrant culture historians, Kidder included, used for unit construction (Lyman et al. 1997b). Some of the units routinely employed in archaeology would, ironically, have worked very well in evolutionary studies, but culture historians did not realize this. Without theory, many of the so-called explanations of how and why things change—the epitome of evolutionism—were nothing more than personal interpretations based on common sense and anthropological knowledge. Unfortunately, common sense—because it is contingency bound and ethnocentric—can play no role in science.

It certainly is not surprising that culture historians played no part in the biological debates that led to the Modern Synthesis, since, from the beginning, Americanist archaeology had its own interests—measuring time and building cultural chronologies on the basis of the material record—and its own body of methods and techniques for pursuing those interests (Lyman et al. 1997b; O'Brien and Lyman 1999a; Willey and Sabloff 1993). The intellectually stimulating debates that were going on between the major camps in biology—the naturalists and the geneticists—over the nature of evolution resulted in their deciding in the early 1940s that they were in several respects epistemological allies as opposed to members of separate disciplines. To be sure, each camp was investigating different things, but the resulting alliance produced a powerful means of examining not only the production of variation in nature but also what happened to that variation in a historical sense. Why, when they began showing interest in the one thing that biologists had long taken for granted—evolution—did culture historians not take more notice? Some culture historians—for example, Harold Colton—were even trained as biologists, though long before the Synthesis occurred. However, their attempts to incorporate various aspects of biological evolution into their archaeological research were as naive and ill-fated as those made by archaeologists with no formal biological training.

Extending the Human Phenotype

The general lack of enthusiasm on the part of archaeologists over biological issues is traceable to a deep-rooted belief that despite some superficial similarities, biological evolution and cultural evolution are just too different to treat under the same umbrella. Kidder (1932:8) and other cultural historians who had an interest in evolution evidently believed it was possible to solve "the problems of cultural evolution," but they could not reconcile their rudimentary knowledge of how evolution worked relative to somatic features with how it might work with artifacts. Thus, despite their view that variation was variation, regardless of the "laws" (Kidder 1932:8) that regulated its production, it was too great a leap to make from the evolution of organic features, governed in large part by genetic laws, to the evolution of inorganic features, which were "unregulated" (Kidder 1932:8) by those laws. This is no black mark solely against culture historians because, as we will see later, even some modern evolutionary biologists have problems incorporating human culture under the Darwinian evolutionary umbrella. Skin and bones, it appears, are appropriate subjects under Darwinian evolutionism, but not such things as stone tools and ceramic vessels.

Yet it is exactly such extrasomatic features that we view as being as subject to evolutionary processes as is any somatic feature. Put more properly, the *individuals* who made and used stone tools and ceramic vessels were subject to evolutionary processes. Products of technology are not adaptive reflections but rather active components of the adaptive process. They are active given their variation; the variants represent alternative solutions to adaptive problems and have different reproductive consequences for their makers and users, and thus their own replication. To us, things in the archaeological record are as much parts of human phenotypes as skin and bones, and as such, they are capable of yielding data relevant to understanding both the process of evolution and the specific evolutionary histories of their possessors. This is a profoundly different way of viewing archaeological objects than the traditional one of "material culture," though it should not be unduly troubling. Most biologists routinely view such things as a bird's nest as part of the bird phenotype in the same way that they view its beak and feathers. Why should pots, projectile points, and the like not be thought of in the same way as hair, skin, and other human somatic features? Certainly, we should have no trouble accepting that the behaviors that lead to creation of a ceramic vessel or a stone tool are phenotypic. Accepting the *results* of behaviors as phenotypic, then, requires only another small step—a logical extension that once made opens the door for Darwinian evolutionism to walk through. Hence, we can begin to talk about such things as selection and drift in terms of how they shaped the variation that shows up in the archaeological record.

Biologists have no trouble accepting bird nests, beaver dams, spider webs, and the like as phenotypic because the behaviors that lead to these end products are at least in part genetically controlled. So far as we know, no one has ever found a gene or series of genes that control how a beaver builds a dam or how a spider spins a web, but never having found such genes does not destroy the logical proposition that such activities are controlled or at least conditioned by them. As such, behaviors, or at least the genetic bases for certain behaviors, are inherited. If we accept the rather noncontroversial notion that one of the functions of an organism as an interactor with its environment is to act as a vehicle for the replicative units it carries—the genes—then it stands to reason that the vehicle has to do its job of protecting the germ-line replicators so that information is passed to the succeeding generation. If the vehicle does not do its job, or if it is prevented from doing its job, the germ line dies. Nature has shaped an almost infinite number of vehicles, some of which perform their jobs better than others in a given environment. The evolutionist's job is to figure out how and why, under a given set of circumstances, some individuals are more successful—more fit—than others in passing their genes on to succeeding generations. Lest this sound as if it is the most narrow reading of an evolutionist's job, we point out that the investigation of the myriad pathways to success that organisms, including humans, have taken is anything but narrow and uninteresting. It brings us into direct contact with all of the wonderfully complex features produced by the evolutionary process, whether or not those features are somatic, that potentially affect an organism's fitness.

The point we are making is with respect to the genotype–phenotype distinction. If the genotype dictates the formation of "bodily" portions of the phenotype—the somatic features—and those portions protect the germ-line replicators, then the genes that control nest-building behavior can be considered as producing a further protection for the germ line. The logic is identical. Richard Dawkins (1990:198) makes the same argument:

> The house of a caddis is strictly not a part of its cellular body, but it does fit snugly round the body. If the body is regarded as a gene vehicle, or survival machine, it is easy to see the stone house as a kind of extra protective wall, in a functional sense the outer portion of the vehicle. It just happens to be made of stone rather than chitin. Now consider a spider sitting at the centre of her web. If she is regarded as a gene vehicle, her web is not a part of that vehicle in quite the same obvious sense as a caddis house, since when she turns round the web does not turn with her. But the distinction is clearly a frivolous one. In a very real sense her web is a temporary functional extension of her body, a huge extension of the effective catchment area of her predatory organs.

We can extend this argument by considering the remains of a mud-dauber's nest and a fragment of daub from a Mississippian-period wall-trench house from the southeastern United States—both recovered from identical archaeological contexts (O'Brien and Holland 1995b). Modern biologists would have no difficulty in dealing with both the dauber's nest and the nest-building behaviors within the framework of the extended phenotype. Neither would they have trouble seeing nest-building behaviors as historical features, that is, as features that change over time. Yet many archaeologists will feel compelled to introduce a *behavioral analog*—one perhaps derived from the ethnographic or ethnohistorical literature—to explain the human-constructed house remains. If, however, the step from a genetic basis for morphological development to a genetic basis for behavior is conceptually negligible, then the step from behavior to extended phenotype—here, mud-dauber nests and Mississippian-period wall-trench houses, or caddis stone houses and spider webs—also is negligible.

Evolutionary Archaeology and Reductionism

Modern efforts at sketching out an evolutionary archaeology (for a collection of the seminal literature, see O'Brien 1996b; for more recent titles, see O'Brien et al. 1998; Lyman and O'Brien 1998) either implicitly or explicitly hinge on extending the phenotype to include things found in the archaeological record. These attempts have been based not only on a better understanding of the complexities of evolution than the level of understanding exhibited by culture historians, but concomitantly on understanding the need to avoid charges of reductionism—the distillation of general theoretical principles from one field for use in another—which was the death knell of attempts made by some culture historians to incorporate evolutionism into archaeology. Their naivete in general toward evolution made them easy prey for fellow archaeologists such as J. O. Brew (1946), who pointed out in so many words that artifacts do not breed. This and other shots occasionally fired into the camp of budding evolutionists by colleagues who saw through the reductionist arguments were enough to quell any uprisings that might have been in the planning stage (see the review in Lyman and O'Brien 1997).

We do not need to make reductionist arguments to bring archaeology under the umbrella of Darwinian evolutionism. All we need to do is first, extend the phenotype to include objects in the material record, and second, demonstrate that the material record is as much a *historical* record as is the fossil record. Because it is historical, the archaeological record marks the passage of time. But the objects themselves—flakes of stone, pottery sherds, animal bones—are more than simply chronological markers, just as fossilized organisms are more than simply a means to a chronological end (e.g., Eldredge and Gould 1977). Artifacts carry the imprint of human manufacture, use, discard, loss, and myriad other *activities*, to use a term popularized by Michael Schiffer (1976). In effect, they are capsules of variation that were formed at particular instants in the past, and once strung together, these capsules present us with a historical recording of how certain variants replaced others over time. The historical nature of the archaeological record is what sets archaeology apart from ethnology as a distinct field of study. Once artifacts are placed in proper chronological order and shown to comprise a historical lineage, evolutionary theory is a rich source of explanations for why lineages took the shape they did (Lyman and O'Brien 1998). These lineages, in turn, speak directly to the lineages of humans responsible for producing the artifacts. There is nothing reductionistic in any part of this procedure.

Having said this, we need to remember that the archaeological record is an incomplete historical accounting and is in fact more a record of evolutionary successes than failures (O'Brien and Holland 1990). Large pieces are missing from it, just as they are from the paleontological record. We have to admit that not all products and by-products of human manufacture and use preserve equally, just as soft tissues of organisms do not usually escape the ravages of time and the elements. No archaeological site contains a full accounting of materials made, used, thrown away, or lost by a group of people, just as no fossil locality contains the remains of all organisms that lived together in a community. And even if they did, what would those localities tell us about what was going on in other places inhabited by members of the same species? Thus, archaeologists, as do their paleobiologist colleagues, compile histories using evidence scattered over wide regions and then correlate those histories both to extend the master sequence temporally and to fill in some of the gaps (O'Brien and Lyman 1999a). Regardless of how complete or incomplete a historical sequence is, the bottom line is that what archaeologists and paleobiologists construct are just that—historical sequences. Such sequences can be described, sometimes in great detail, but the end product is simply a statement of *how* things went together sequentially, not a statement of *why* they went together the way they did. As important as sequences are—and one certainly has to get the temporal ordering right before other questions can even be posed—they hardly qualify as the end products of scientific inquiry. They are, as Robert O'Hara (1988) points out, historical chronicles as opposed to historical narratives.

Recognition of this fact led to the creation of a new kind of archaeology in the 1960s, cultural processualism, as archaeologists became dissatisfied with adding another bit or two to the numerous chronological sequences that culture historians had labored to construct. Such efforts were seen as things of the past, to be conducted only by those not particularly concerned with asking the why-type questions—in short, by those disinterested in making archaeology scientific. The really exciting work was in the arena of explanation—asking *why* things were the way they were, not simply *how* they had come to be that way. Not surprisingly, evolution, when it played a role at all in processual archaeology, was viewed as a Whitean pathway and not as a Darwinian process. With one or two exceptions, no one saw Darwinian evolutionism for what it is—a means of explaining those myriad pathways that culture historians had spent so much time constructing.

SORTING THROUGH THE ISSUES

As evolutionists, we are pleased with the renewed attention Darwinian theory currently is receiving in archaeology, but we also realize that there are two related measures of success: how faithfully evolutionary theory is being applied, and how widely it is being applied as an explanatory framework. With regard to the first point, how can we call evolutionary archaeology successful if what is produced in its name does not follow logically from evolutionary theory (Lyman and O'Brien 1998; O'Brien 1996d)? We have to be careful here, because we are not suggesting that Darwinian theory is a cure-all for archaeological problems. In fact, we argue that it is not and cannot be a panacea because it was not written in archaeological terms. Neither is it a cure-all for the myriad problems that confront evolutionary biologists; thus, Darwin's original theory has been refined and extended—it has evolved—as we learn more about the natural world. One could, we suppose, argue that so much extension has been made that Darwin himself would not recognize modern evolutionism, but we would bet that he *would* recognize it. He might have to work through some pretty heavy rhetoric, but he would get there eventually. And when he did, he would find his central tenets pretty much the way he originally constructed them.

With regard to the second point, of what use is theory, regardless of how well it explains something, if no one bothers to employ it in real-world situations because either it is not understood or it appears that the theory is being applied in contrived situations? It should come as no surprise that archaeologists, as with scientists in general, have large stakes in the paradigms under which they operate and hence are not easily convinced, for example, that the manner in which the discipline has traditionally approached the material record is based on an inappropriate model of science. What about the theory being used to explain the nature of the record? Could it be inappropriate as well? Or, worse yet, is it possible that the "theory" really is not theory?

Some archaeologists are so strongly wedded to one paradigm or another that they will never be convinced that Darwinian evolutionism has much to offer (e.g., Boone and Smith 1998). There also are those who are unwilling to accept the view that things in the material record are as phenotypic as somatic features (e.g., Maschner 1998)—the underlying premise of evolutionary archaeology. Although perhaps expressing an interest in Darwinian evolutionism, they are much more at home with the notion that the appearance of culture successfully uncoupled humans from the effects of selection. To this we would say exactly what philosopher Daniel Dennett (1995:365) said: "*We haven't seen the long run yet.* Mother Nature's experiment with culture on this planet is only a few thousand generations old" (emphasis in the original). Finally, there are those archaeologists who might see some value in a Darwinian approach to understanding the archaeological record but who are entirely uncomfortable with adopting biological concepts and methods. This is what one such critic, Bruce Trigger, had to say on the subject:

> I agree that evolutionary archaeology, with its emphasis on selection, has more to offer archaeology than does processual archaeology, with its cultural evolutionary emphasis on progress. What seems essential, however, is to recognize that among human beings selection has ceased to be largely external—it also involves culturally mediated perceptions, agency, and choice.... I believe that evolutionary archaeology would be more widely accepted and better equipped to account for sociocultural phenomena if its sponsors abandoned what seems like a reductionist biological terminology in favor of one that explicitly takes account of the unique, emergent aspects of human behavior. (Trigger 1998:364)

However, we believe there are numerous archaeologists who either are convinced or can be convinced that evolutionism is the most powerful tool available for explaining change. Acceptance by a wide audience, however, will hinge directly on attaining a basic understanding of what evolutionary archaeology is and what it is not. We are going to have to be clear about such issues as what is meant by Darwinian evolution, what it is that evolves, how change is measured, and so on. Even among those with an interest in evolutionary archaeology there appears to be some confusion over basic premises as well as method. For example, the title of a recent book, *Darwinian Archaeologies* (Maschner 1996), suggests that there is more than one kind of evolutionary archaeology based on Darwinian theory. This is not only incorrect but also an impossibility. There is only one Darwinian biology, for example. This does not imply that theories do not have components—Ernst Mayr (1991b), for example, divides Darwinian theory into at least five interrelated theories or subtheories—but this in no sense implies that there are multiple evolutionary biologies. The same applies to evolutionary archaeology. Part of the confusion, we suspect, resides in lack of familiarity on the part of archaeologists with the biological literature, and part of it rests on the fact that classical evolutionary theory does not begin to include all the terms needed to address archaeological phenomena (Dunnell 1995; Teltser 1995b, 1995c). Part of it also resides in conflation of the terms *evolution* and *ecology* into *evolutionary ecology* (sometimes labeled, more appropriately, *behavioral* ecology)—an approach that according to some authors (e..g., Boone and Smith 1998) has considerable overlap with evolutionary archaeology, at least in principle (but see O'Connell 1995; Broughton and O'Connell 1999). Conversely, we see the two as having little in common in terms of analytical goals (Lyman and O'Brien 1998), though this in no way denigrates the analysis of function—the centerpiece of ecological studies.

The principles behind Darwinian evolution—Mayr's (1991b) separate theories—are not difficult to understand; rather, it is the philosophical and methodological debates that surround the subject that are complex and sometimes difficult to follow. Yet without an adequate epistemological grounding—that is, a grounding in the debates over how we have come to know what we think we know about organic evolution—Darwinian evolutionism itself can quickly become little more than a form of reductionism. In this book we attempt to sort through the myriad issues raised in evolutionary biology and to focus on those that we view as being fundamental to any effort to extend evolutionism to the archaeological record. This has required that we set forth in clear language how to take a theory that to this point has seen little use in archaeology and demonstrate its explanatory power relative to the material record. Most important, it has also required that we escape the pitfalls that ensnared earlier attempts by culture historians (see Lyman and O'Brien 1997). We do this by demonstrating that cultural phenomena—artifacts—can be naturally incorporated under the Darwinian umbrella.

It is impossible in this book to cover all the significant issues involved in extending evolutionism to the archaeological record. The topics that we selected, and which are outlined in the remainder of this chapter, comprise three arenas—the ontological, the epistemological, and the methodological—each of which interconnects with the other two in ways that prohibit clear separation. Thus, our arrangement of topics in individual chapters is partly arbitrary, but the order of the chapters themselves reflects our ideas of how best to get across the fact that at its base evolutionary archaeology is about constructing lineages that can be explained in terms of Darwinian theory. The lack of clear separation of topics should come as no surprise. As Darwin described his treatment of descent with modification in *On the Origin of Species*, it was "one long argument" (Darwin 1859:459) as opposed to many shorter, more concise ones.

Essentialism versus Materialism

Fundamentally, Darwinian evolution is not only a different theory than others employed in science but, of more importance, it is also a different *kind* of theory (Mayr 1959a). In Chapter 2, we examine how the kind of science that has grown up around Darwinian evolutionism differs from other kinds of science. Failure to keep the distinction clear robs Darwinian evolutionism of its uniqueness, rooted as it is in a materialist as opposed to a typological, or essentialist, ontology (Lewontin 1974b). Materialism, because it gives equal consideration to time and space, is capable of addressing issues that are historical in nature. Conversely, essentialism merges time and space into a single entity; thus, the kinds of questions asked under the essentialist umbrella are ahistorical in nature. Americanist archaeology has always considered itself scientific, though the label oftentimes was more implicitly than explicitly applied. The boredom with culture history that Americanist archaeology experienced in the late 1950s (Lyman et al. 1997b) paved the way for the professional arrival of Lewis Binford and his clarion call to make archaeology explicitly scientific. Despite great fanfare, the new archaeology, or processualism, eventually fell short of this lofty objective, though its legacy is very much alive today (e.g., Spencer 1997, 1998). The reason for its failure was a built-in error that no one, including philosophers with an interest in archaeology, noticed: In their haste to be scientific, archaeologists selected physical science, with its essentialist ontology, as the model to emulate. In the process, they overlooked the fact that their subject matter—things in the archaeological record—clearly fell in the materialist realm.

Because it was based on an inappropriate model of science, processual archaeology was incapable of constructing a theory that dealt solely with its historical subject matter. The best it could do—and for a while this was seen as perfectly reasonable by its advocates—was to draw its theoretical inspiration from anthropology, which, though it was seldom noticed, had long depended on common sense (a patently essentialist ontology) instead of science as a means of explaining why the world looked the way it did. Processual archaeologists borrowed anthropological theory, especially cultural-evolutionary theory (also patently essentialistic), and applied it to the archaeological record, but this was an inappropriate theory for explaining a historical record (Lyman and O'Brien 1997). In archaeology, in large part because of its close ties to anthropology, there has always been a strong tendency to view evolution in terms of progress and stages, which were the epistemological underpinnings of Whitean cultural evolutionism. The driving force in helping humans up the ladder of progress comprised two basic elements: (1) an "urge" internal to all organisms, including humans, to improve their lot in life, which was the "motive force as well as the means of cultural evolution" (White 1943:339), and (2) culture—that "extrasomatic mechanism employed by a particular animal species in order to make its life secure and continuous" (White 1959b:8). As Kroeber (1946:9) observed, "What White means by evolution is a fixed, necessary, inherent, and predetermined process." This is decidedly not the Darwinian view we adopt here.

For some reason, most anthropologists and archaeologists still hold fiercely to the belief that *culture* is the key not only to human survival and adaptation but also to unlock the mysteries of human survival and adaptation. Thus, analysis is framed around a circular cause-and-effect relationship. As the scenario goes, once hominids attained the status of culture-bearing animals, culture gave us its most cherished offspring—intent—as our means to conquer (usually) the forces of the unknown. Intention, together with its sibling, invention, keeps humanity moving forward (upward, on the Whitean evolutionary ladder) in the midst of constant environmental perturbations. Under such a perspective, human intent is given a causal role. In contrast, our position is that although intent is a powerful mechanism for generating variation, it has no ultimate explanatory power for two reasons. First, it is archaeologically invisible. Second, by including it, one is attributing "cause to the phenomena being studied rather than [locating] cause in the theoretical system" (Dunnell 1989:37); that is, by including human intent, "cause is attributed to the actors, to the data.... Human intentions thus are substituted for theory" (Dunnell 1992a:214). Dunnell indicates the "substantive tautology [of confusing human motivation or intent with scientific cause] precludes empirical testing" (1989:37); in short, "reason-giving ... does not admit falsification" (1992a:215).

But we emphasize that human intent is not excluded from consideration; human "decision making, even informed decision making, is not inconsistent with selectionism" (Rindos 1989:13). Intentionality is, in our view, best seen as a way of "increasing variation within the [cultural] system" (Rindos 1989:15). Charles Spencer (1997:225) suggests evolutionary archaeology "derides the concept of directed variation as one-step adaptation or Lamarckism." He is correct, but the reason for the derision is not to retain selection as the sole mechanism driving change. Rather, we deride it because such a Lamarckian approach precludes understanding the origin of the mechanism of directed variation: "[N]o culturally based method for creating appropriately directed variation could have arisen" by human intent aimed at adaptive need (Rindos 1989:16–17). As Robert Leonard and Tom Jones (1987:216) note, "In setting humans apart from other kinds of organisms in evolutionary studies [and thus calling upon human intent as an explanation], we face considerable difficulties in, among other things, coming to grips with the historical transition from predominantly genetic to predominantly cultural modes of information transmission."

Today, many archaeologists still cling tenaciously to the essentialist model of how science works, some because they firmly believe that it is the proper model, others because of a perception that nothing better exists. Something better does exist, but it comes at considerable intellectual price. Saying that archaeology should be a materialist science is easy; actually carrying out the switch from essentialist to materialist thinking is not. There is no better means of highlighting the difficulties encountered when one adopts a materialist perspective than by focusing on a question that has long been at the center of debate in evolutionary biology: What is a species? Most people would assume this question was answered decisively decades ago, but it most assuredly was not. Evolutionary biologists and philosophers of biology routinely question not only what a species is (e.g., Davis 1996; Ghiselin 1974a, 1974b, 1981, 1987; Hull 1976; Kitcher 1984a, 1984b; Kitts 1984; Kitts and Kitts 1979; Mayr 1987; O'Hara 1993; Schwartz 1981; Sober 1984a, 1984b; Sokal and Crovello 1970; Wilson 1996) and how to define it (e.g., Cracraft 1983, 1987, 1989a, 1989b; Mayden 1997; Mayr 1957, 1963, 1976b, 1982c, 1993; Mishler and Brandon 1987; Mishler and Donoghue 1982; Simpson 1961) but also how a species originates and how and why it changes into something else (e.g., Mayr 1987). Others even question the.usefulness of the species concept (e.g., Ereshefsky 1989; Hull 1998). One issue that crops up from time to time concerns the "reality" of species: Are they real entities, or are they simply convenient categories that we have created in order to help us categorize nature?—a question that we contend should be of utmost concern to anyone interested in Darwinian evolutionism.

It is understandable why the issue of what a species is should play such a central role in biology, but why introduce it in a discussion of archaeology? Why not leave wrangling over such issues to philosophers and biologists? One might argue that when we get right down to it, regardless of whether species are real or not, are we not all members of what we at least claim is the same species? Furthermore, since the greatest part of the archaeological record is less than 50,000 years old—the point at which paleontologists agree that modern *Homo sapiens sapiens* was the only hominid around—why worry about what a species is or about how and why speciation occurs? The answers to these questions are not immediately obvious. Neither the concept of species nor that of speciation has been applied archaeologically except in situations where there is a change in species of the hominid group or groups responsible for specific archaeological signatures. Thus, species and speciation might be considered archaeologically with reference to the evolution of *Homo erectus* from *Homo habilis* or when tools in the Middle East are assigned to either *Homo sapiens* or *Homo heidelbergensis*, but not when the tools date to the more recent prehistoric past. Why, then, should an archaeologist working in the Americas, where the record (so far) appears to date only to the last 13,000 years or so, worry about species and speciation? Our answer is that if one is going to adopt an evolutionary approach in archaeology, that adoption must be based in materialism. There is absolutely no room for equivocation on that point. The manner in which the issues of species and speciation have been handled by biologists and paleontologists is critical not only to understanding how materialism approaches the issue of change but also in demonstrating the intellectual and practical problems that are encountered in so doing.

Extending the Argument to Archaeology

In Chapter 3, we examine how twentieth-century archaeologists have handled the issue of change and how they have attempted to link change in the archaeological record to evolution. Culture historians and processualists alike often thought that what they were doing was based in evolutionism, but the kind of evolutionism that underlay their efforts was anything but Darwinian in nature. Ironically, some of the methods that have been developed in Americanist archaeology, especially those used by culture historians to keep track of time, are quite at home in a materialist framework. There was, however, no scientific theory to guide either the development or the application of those methods, the result being that the units created to do such things as measure time were indiscriminately used for other purposes as well. In the end, archaeology became an interpretive discipline based in large part on commonsensical warrants instead of on theory.

Our discussion in Chapter 3 highlights the fact that archaeology has always acted more as a confirmatory science than as an explanatory one (Dunnell 1982, 1992a)—archaeologists start with established, commonsensical notions of how and why the archaeological record is structured the way it is and then set out to confirm those notions. Under a confirmatory approach, how do we know why the record is structured a certain way? By using ethnographic or ethnohistorical analogs as keys to the past. The end result of this analogical approach is an interpretation of the archaeological record rather than an explanation of it (Dunnell 1994; O'Brien 1996b)—an approach that would work if human behavior were invariant, which it clearly is not. Behavior varies over time and space, and it does so in nonpredictable ways—behavior is historical. Any approach that denies the explanatory significance of the historical component is inappropriate for studying the material record. Analogy has its place in a historical science, but not as a surrogate for a historical chronicle.

Parallels with Paleobiology

The parallels between evolutionary archaeology and paleobiology are striking, as they should be, since each is grounded in Darwinian theory. Evolutionary archaeology is geared toward providing Darwinian explanations of the archaeological record, just as paleobiology explains the paleontological record. This comprises building cultural lineages—what O'Hara (1988) refers to as historical chronicles—and constructing explanations of those lineages—O'Hara's evolutionary narratives (see also Szalay and Bock 1991). Or to put it another way (Jones et al. 1995:29), evolutionary archaeology comprises writing a "description of the historical patterns of differential trait representation and [constructing] arguments as to how evolutionary [processes] acted to create those patterns." Paleobiology entails exactly the same procedures. Both steps, regardless of whether they are taken in archaeology or in paleontology, employ concepts embedded within Darwinian evolutionary theory. Relevant concepts include lineage (a temporal line of change owing its existence to heritability), selection (a mechanism of change), transmission mechanisms, innovation/mutation, diffusion/genetic recombination, and heritability. The last concept ensures that we are examining change within a lineage rather than merely a temporal sequence or convergence; in the latter case, similarity is of the analogous sort and, although potentially important, must be separated from homologous similarity to ensure accurate historical chronicles.

Paleobiologists have been working on evolutionary problems much longer than have archaeologists; thus, we as archaeologists have much to learn from them, especially how to avoid reinventing the proverbial wheel. We examine this issue in more detail in Chapter 4 and contrast the paleobiology of the 1970s and later with what was going on in archaeology during that period. We demonstrate that many of the important issues with which paleobiologists were wrestling, although they should have been of utmost interest to archaeologists, were ignored. Archaeologists instead wrangled over how to make their discipline scientific—which translated into how to make it anthropological. In the process, they overlooked something that paleobiologists took as part of their creed: History is important.

Archaeological Units

Much of the epistemological and methodological confusion in archaeology, and in biology as well, is over units. There is no more fundamental issue in a materialist science than the construction and application of units. What are applicable units, and how does one construct them? Culture historians became mired in the materialist paradox—described in Chapter 2—in part because they used the same units for different kinds of work, missing the critical point that many of the units they were using, though they were ideally suited for one purpose, were not well-suited for other purposes (Lyman and O'Brien 1997; O'Brien and Lyman 1998). Biologists face the same problem with the species concept; in fact, the whole issue of what a species is and how the concept can be applied in understanding the natural world is a classic case of epistemological confusion over kinds of units (Hull 1997). We address these concerns in Chapter 5. We cover such units as kinds, types, groups, and classes, as well as some of the methods (e.g., paradigmatic classification, taxonomic classification, and phenetics) that produce the various units used in biology and traditional archaeology. In several respects, Chapter 5 is the most important of all the chapters, because it is the units that are critical to the success of Darwinian evolutionism. To proceed in orderly fashion from kind of science to theory and finally to explanation requires the use of *systematics*, which is nothing more than a logical means of creating a set of units for a specified purpose. "Logical" as used here means that the units used to partition variation are derived naturally from the kind of science and the theory being employed. The title we selected for this book emphasizes the importance of systematics to evolutionary archaeology.

Because evolutionism is an entirely different kind of approach and contains a different kind of theory than that with which most archaeologists are familiar, many of the data that routinely are generated for use in archaeological analysis are inappropriate for the kinds of historical investigation that derive from a Darwinian perspective—an issue that is as pertinent in archaeology (Dunnell 1980, 1982, 1985a, 1985b, 1986, 1988a; O'Brien and Holland 1990, 1992; O'Brien et al. 1994) as it is in biology (Dobzhansky 1951; Ghiselin 1966, 1974b, 1981; Hull 1992, 1997; Mayden 1997; Mayr 1959a, 1963, 1977, 1987; Sober 1980, 1984a). To compound the problem, many of the *methods* routinely used to generate archaeological data are inappropriate for examining implications derived from evolutionary theory. All archaeologists would admit that they have a fundamental interest in variation as it exists across space and through time. Interest in variation, of course, is not unique to archaeology; biologists and physicists share an interest in variation as it exists in the natural world, though their interests are manifestly different in terms of the nature of the variation they study. Some units—populations, species, demes, and the like—are inappropriate for physics, just as other units—atomic number and atomic weight, for example—are inappropriate for evolutionary biology. Note, however, that the appropriateness of one unit over another is not dictated by whether the thing to be investigated is or ever was alive. Rather, it is based on the kind of science that is being conducted.

The point is that units are not transferrable from one kind of science to another unless the properties under investigation are similar. When time and space are inconsequential—for example, when the subject is physical–chemical interactions within an organism—the same kinds of units might be used by both a biologist and a physicist. However, when time and space are consequential, units are not automatically interchangeable. This should be noncontroversial: Historical entities—those that exhibit limited temporal and spatial distributions—require particular kinds of units for their investigation; ahistorical entities—those that do not exhibit temporal and spatial distributions, or, more appropriately, those that occur in *all* temporal and spatial contexts—require different kinds of units. If the subject matter dictates the kind of science that is appropriate, then it also dictates the kinds of units that are appropriate.

Archaeologists have devised their own battery of units—artifact types, phases, components, and the like—to deal with their subject matter, but if archaeology has traditionally followed the wrong model of science, then there is reason to suspect that some of the units commonly used might be of little or no value in an evolutionary study. Because archaeologists have not questioned the model of science after which the discipline has modeled itself, the debates that have taken place over the use and misuse of one unit or another have not paralleled the penetrating philosophical discussions that have taken place in biology. We are not suggesting that archaeologists simply adopt the units used in biology. Although biologists study phenotypic variation, as do archaeologists, there is no reason to suspect that the former have invented the language and units necessary for incorporating the archaeological record under an evolutionary umbrella (Dunnell 1995; O'Brien 1996c; O'Brien and Holland 1995b). Nor is there any reason to expect that they would have.

The bottom line is that to make evolutionism more than a metaphysical exercise requires real-world expression that can be accessed through the systematic use of appropriate methods and units. Archaeologists are not the only ones faced with this dilemma: Any discipline that derives its subject matter from living organisms faces it, including paleobiology:

> By the very nature of its materials, paleontology can have nothing direct to say about evolutionary mechanisms. Using fossils, we cannot study mutation, recombination, and selection with any degree of practicality, and certainly these and related concepts would never have come from paleontology. Likewise our modern concepts of speciation, while perhaps less difficult to apply to paleontological data, do not emerge as self-evident precepts when one examines data on the fossil record.... [I]t is indeed difficult to identify any coherent segment of evolutionary theory, no matter how broadly construed, as having originated in paleontology. (Eldredge 1979:9)

The Construction of Historical Lineages

The paleontological record, as it provides a historical sequence of variation, played a role in Darwin's formulations, but the notion of descent with modification did not and could not have come from his observations of that record. It came in large part from his observations of generational changes in various kinds of English breeding stock, in part from his earlier observations while on the *Beagle*, and in part from his knowledge of economics. This in no way implies, however, that the fossil record falls outside the scope of Darwinian evolutionism; it simply means that Darwinian theory did not originate in paleontology. Why else would paleontologists regularly be housed in geology departments rather than in biology departments? As Laudan (1992:57) notes, such departmental affiliation is a result of history: "[T]hrough most of the nineteenth and twentieth centuries geologists found [paleontological] results crucial while biologists found them peripheral." This situation changed after the Modern Synthesis in the 1940s.

Things might seem even worse in archaeology, given that Darwin did not even have the archaeological record in mind when he formulated his theory. But having said that, we cannot, as David Rindos (1989:5) once pointed out, blame Darwin for not doing our work for us. It is up to us to devise and systematically use methods and units that allow us to incorporate archaeological materials into evolutionism. In Chapter 6, we first examine how biologists and paleobiologists have constructed historical lineages and then use that examination as the basis for exploring how archaeologists can best construct artifactual lineages using the units discussed in Chapter 5. One key issue in examining lineages is separating the tempo of evolution from the mode of evolution—a dichotomy that is well-rooted in biology and reached its clearest exposition through the work of George Gaylord Simpson in the 1940s. We examine this issue in Chapter 7, focusing specifically on the differences between phyletic gradualism and punctuated equilibrium, and how those differences are manifest in the paleobiological and archaeological records.

Explaining the Lineages

Once we have built lineages and tested the reliability of the construction, we can either sit back and enjoy the end product, or we can begin the truly grueling and often fruitless job of trying to explain why a lineage took the shape that it did. We take up this topic in Chapter 8. In organizing this chapter and trying to decide what to include, we were reminded of a question one of our colleagues once asked: "How is Darwinian evolutionism going to help me do archaeology?" Our answer was that evolutionism was not going to help him *do* archaeology, but it might put him on the road to deriving explanations for some of the things he observed archaeologically—explanations that made sense in terms of the theory from which they were derived. But the explanations have to be framed in archaeological terms, not biological ones. Simply put, our goal in Chapter 8 is to demonstrate that Darwinian evolutionism is capable of solving *some* archaeological problems using archaeological units. In turn, the units are used to slice up variation in the archaeological record in terms that make sense theoretically. The key point is that data generated from real "stuff" are needed to make evolutionary archaeology more than simply a metaphysical exercise. To be able to explain something scientifically requires systematic use of theory, methods, and units, but it also requires observations that are made on objects or events. All science works this way. Einstein's theoretical deductions about the cosmos, for example, made for interesting conversation until it was observed that one or more of them was supported with real-world data. The important point is that physicists looked for and continue to look for the data necessary to support Einstein's equations—data that have very demanding characteristics and thus must be generated using equally demanding standards. Likewise, archaeologists often need specific information to address specific issues, and when those issues involve how-type questions, archaeologists have been resourceful in devising units and methods to answer the questions. However, when the issues involve why-type questions, archaeologists have not done very well, usually choosing the same units used to answer the how-type questions.

We also examine in Chapter 8 other issues that are important in evolutionism, several of which have received fairly extensive treatment in the archaeological literature, though their exposition has not been as clear as it could be. Two such concepts that readily come to mind are *adaptation* and *selection.* The casual ways in which those terms are used in everyday speech belie both the precision in how they are used in evolutionary biology and the controversial nature of the concepts. To date, proponents of evolutionary archaeology for the most part have danced around the fact that these concepts need to be explained in language that makes sense archaeologically. Prior discussions have tended to be telegraphed statements as opposed to full-length treatments, and in some cases this brevity has made the arguments difficult to follow. Yet concepts such as adaptation, selection, drift, homology, homomorphy, and analogy are as critical to evolutionary archaeology as they are to biology. The linkage between the explanatory power of Darwinian evolutionary theory and the archaeological record is not immediately obvious; if it were, Kidder, Binford, and others probably would have picked up on it. Failure to recognize that not all archaeologists possess equal familiarity with evolutionary concepts and terminology is one reason why the resurgence in applying Darwinian evolutionary theory to archaeology that began in the late 1970s has not attracted more attention. When it has attracted attention, it is as often as not misunderstood and misapplied.

Future Directions

If evolutionism is going to play a role in archaeology, a significant portion of the discipline is going to have to become involved in the kinds of debates that up to this point have been the sole property of biologists and philosophers of biology. Archaeologists are going to have to become involved in such things as constructing units and developing methods that are useful in evolutionary studies. Perhaps of more importance, as we point out in Chapter 9, archaeologists are going to have to start applying evolutionism to the archaeological record. There is a major deficiency in recent efforts to incorporate evolutionism into archaeology, and it is one that we address continually throughout this book. The problem is that there are few instances where theory, method, and data have been combined logically in an explanatory framework. Although several important products have appeared (e.g., Dunnell and Feathers 1991; Lipo et al. 1997; Neff 1993; O'Brien et al. 1994; papers in O'Brien 1996b; Teltser 1995a; *Journal of Anthropological Archaeology*, Vol. 18(3) 1999), they have tended to focus on a narrow range of topics. This has caused critics of the approach (e.g., Boone and Smith 1998) to argue that enough time has passed for evolutionary archaeology to have yielded more products for all the intellectual investment made. Perhaps, as one of us recently pointed out (O'Brien 1996c:14), ambition *has* gotten out ahead of clearheaded applications, but the paucity of examples is also in part a function of the few archaeologists engaged in evolutionary research as well as the newness of the approach. In several respects, the situation reminds us of how biologist Richard Lewontin (1974:189) referred to the early years of population genetics:

> For many years population genetics was an immensely rich and powerful theory with almost no suitable facts on which to operate. It was like a complex and exquisite machine, designed to process a raw material that no one had succeeded in mining. Occasionally some unusually clever or lucky prospector would come upon a natural outcrop of high-grade ore, and part of the machinery would be started up to prove to its backers that it really would work. But for the most part the machine was left to engineers, forever tinkering, forever making improvements, in anticipation of the day when it would be called upon to carry out full production.

It would be gratifying to think that one day archaeologists would not need to be convinced that evolutionism offers the best means of placing the discipline on a scientific footing. We believe that this might happen, though others (e.g., Dunnell 1996) are more pessimistic. We believe it will happen in archaeology in more or less the same way it occurred in biology—through a grand synthesis of sorts between various schools of thought. There are signs that this might someday occur (e.g., Schiffer 1996), but they are still far from convincing (O'Brien et al. 1998). Regardless, for the product to be useful it must adhere faithfully to the central tenets of Darwinian evolutionary theory. It must also steer clear not only of producing untestable adaptationist arguments but also of falling into what O'Brien and Holland (1995b) refer to as "quantified essentialism," whereby variation is measured but no explanation is provided for why certain variants persisted and others did not.

To those interested in making archaeology an explanatory science, we offer this volume as a first step. And to anyone who believes that all or most of the work has already been accomplished, we offer the assurance that this is not the case. Darwinian evolutionism, as currently formulated, does not and cannot provide a cookbook-like guide to explaining the archaeological record. It does, however, offer the theory as well as some of the concepts necessary to do so. Once the theory is accepted as being applicable to the archaeological record, the really important task remains of linking the theory with archaeological data. This, to us, is where the truly exciting future of archaeology lies. Given the alarmingly rapid rate at which large segments of the archaeological record are disappearing, perhaps we had better get on with the job.

Chapter 2

Two Kinds of Science

Essentialism and Materialism

The ultimate conclusions of the population thinker and of the typologist are precisely the opposite. For the typologist the type ... is real and the variation an illusion, while for the populationist, the type (average) is an abstraction and only the variation is real. No two ways of looking at nature could be more different. (Mayr 1959a:412)

Mayr is correct: No two ways of looking at nature *could* be more different than the outlook of a typologist, to whom we will usually refer as an essentialist, and a populationist, to whom we will usually refer as a materialist. The "importance of clearly differentiating these two basic philosophies and concepts of nature [essentialism and materialism] cannot be overemphasized. Virtually every controversy in the field of evolutionary theory, and there are few fields of science with as many controversies, was ... between a typologist and a populationist" (Mayr 1959a:2). Such controversies have not been confined to biology; many similar ones occurred in archaeology after 1930, forming the heart of what Dunnell (1995; see also Lyman et al. 1997b; O'Brien and Lyman 1998) referred to as the materialist paradox—attempting to carry out materialist science under an essentialist umbrella. The failure of archaeologists to recognize the paradox led to the abandonment of the distinctive materialist view of the archaeological record held by some culture historians (Lyman and O'Brien 1997)—a view that, given enough time, might have finally led to the realization that the record could be examined in Darwinian terms. In its place archaeologists cobbled together a series of perspectives and methods based on an inappropriate model of science. This loose amalgam of tenets and procedures, which had its roots in the late 1940s (e.g., Willey and Woodbury 1942) and was nurtured during the 1950s (e.g., Phillips and Willey 1953; Willey and Phillips 1958), eventually became known as the new, or processual, archaeology (Binford and Binford 1968).

Because of the dramatically different views of reality that essentialism and materialism bring to the natural world, the distinction between them has been explored to varying degrees by biologists and philosophers of biology (e.g., Hull 1965; LaPorte 1997; Lewontin 1974b; Mayr 1959a, 1961, 1963, 1970, 1972, 1976b; Simpson 1963, 1970; Sober 1980, 1984a), and by archaeologists (e.g., Dunnell 1980, 1982, 1992a, 1992b, 1994, 1995; Lyman et al. 1997b; Lyman and O'Brien 1997, 1998; Neff and Larson 1997; O'Brien 1996a, 1996b, 1996c, 1996d; O'Brien and Holland 1990, 1992, 1995a; O'Brien and Lyman 1998, 2000a, 2000b; O'Brien et al. 1998), and by social scientists (e.g., Haslam 1998). All such treatments, to one degree or another, end up at the same point: Essentialism and materialism are not only different scientific theories but, also, as Mayr (1959a) put it, different *kinds* of scientific theories. Even if one were unsure of exactly what the terms *essentialism* and *materialism* meant, he or she might, at the intuitive level, find it inconceivable that one kind of science would work for explaining the entirety of the natural world—for example, the evolution of life as well as the composition of molecules.

Given a choice of kinds of science, which one does an evolutionist select? Unlike a number of issues with which we wrestle in this book, the verdict on this one is easy to render: An evolutionist, at least one who holds to Darwinian principles, must select materialism for the simple reason that as an overarching analytical strategy, essentialism cannot explain why organisms evolve. Why not? Because Darwin's brand of evolution involves a process that in effect keeps all organisms in the constant state of becoming something else. Do not be misled here. When we say "in the constant state of becoming something else," we are not talking about the ontogeny of an organism or the so-called life history of an artifact, though these may be quite important in the context of evolutionary inquiry. What we mean is that given sufficient generations in an ancestor–descendant lineage, the descendant will be a different *kind* of thing than the ancestor. If we adhere to this position literally, there is no point in time at which an organism, a group of contemporary organisms, or a lineage of organisms have an essential property that is so clearly defined that it compels us to categorize it a certain way—the heart of essentialism. As life evolves, it does so along a seamless temporal continuum. Life itself—from the first organisms to make an appearance in the oceans four billion years ago to all of the creatures that currently roam the earth—is also a continuous thread with no natural junctures or disruptions to the flow. This in no way implies that the flow has remained constant in volume or rate. In fact, it is most realistically characterized as a flow that varies significantly in rate and volume. One task of an evolutionist is to determine when and where the rate and/or volume of flow varied; other tasks include determining why the rate, volume, or both changed at some times and not others.

Any science that lays claim to evolution's flow as its subject matter must be based on the proposition that although the flow of both life and time can be carved up into units, those units are in no sense real (O'Brien and Lyman 1999a). Materialism does away with the notion of the existence of natural units in nature that are based on intrinsic properties. Darwinian evolutionism, the quintessential materialist strategy, is based squarely on that denial (e.g., Gould 1986; Mayr 1961, 1972); hence, any approach that searches for essences or essential tendencies in organisms is bound to run counter to that strategy. This in no way implies that types, kinds, and related units have no place in evolutionary studies, because they certainly do (O'Brien and Lyman 1998, 1999a). To think otherwise immediately banishes evolutionism to the metaphysical realm (Dunnell 1992a).

In mature disciplines, the issue of which kind of science to follow is rarely, if ever, raised. Biologists might wrangle over what a species is, but they normally do not worry about whether what they are doing falls under the essentialist or the materialist banner. The fact that they do not worry about it is not a sign that they do not care which kind of science they are doing, nor is it a sign that they do not understand the difference between the two. Rather, modern biologists, like many of their forebears, accept the fact that Darwin's theory slammed the door on essentialism, and they get on with the business of understanding how life evolves. Likewise, paleobiologists get on with the business of understanding how life evolved in the past. For their part, chemists and physicists do not seem particularly concerned with trying to figure out into which ontological category they fall, and they should not. Their realm is the essentialistic world, with its emphasis on natural kinds, and it always will be. Cosmologists might refer metaphorically to the evolution of solar systems and the like, but they understand the difference between using the term *evolution* as a synonym for change or development and using it to refer to organismic descent with modification.

Likewise, archaeologists for the most part appear unconcerned with the distinction, despite the fact that its importance was pointed out almost two decades ago (Dunnell 1982). By adopting Philip Phillips's (1955:246–247) dictum that "North American archaeology is anthropology or it is nothing," the discipline was forced to adopt the essentialist mode of science, thereby denying itself access to its one unique potential contribution to the study of the human organism—history (see Lyman and O'Brien [1997, 1998] for extended discussion). Thus, in our view, unless archaeologists are content to turn their subject matter over to humanists and social scientists, they *should* pay particular attention to the kind of science they practice. Lack of attention paid to differences between the two kinds of science has led to several problems in archaeology, not the least of which concerns the kinds of units archaeologists routinely use to measure change.

Although archaeology has a plethora of units for measuring variation, rarely is any thought given to the appropriateness of certain units for carrying out the kinds of work for which they were intended, one of which is measuring change. It is logical that the appropriateness of units should stem directly from the theory one is using (Hull 1970; Lewontin 1970, 1974b), which is part and parcel of one's ontological perspective.

What we find in archaeology is that throughout the twentieth century, both essentialism and materialism have played parts in how archaeologists have carried out their work, though materialism, when it has made an appearance, usually has been strictly a bit player (Lyman and O'Brien 1997). Despite occasional references in the literature to Darwinian evolution, coupled with the development of certain methods and units that fit comfortably within a materialist framework, Americanist archaeology has consistently, and often by design, followed an essentialist agenda (Lyman et al. 1997b), with the result that the epistemological underpinnings of the discipline in some respects more resemble those of physics than those of biology.

This seems odd to us, given that the subject matter of archaeology consists of products and by-products created by organisms and not, for example, atomic particles, but this is the problem created by failing to recognize or acknowledge the deep division between essentialism and materialism. We in fact would argue that the most significant problem facing the integration of evolutionary theory into archaeology is the failure to recognize that there is more than one kind of science (see Dunnell 1980, 1982, 1992a, 1992b, 1994, 1995; O'Brien and Holland 1990, 1992, 1995a, 1995b). In turn, there is more than one kind of explanation, one of which is lodged in the physical and hence predictive realm, the other in the historical realm. Many of the so-called new archaeologists of the 1960s and 1970s, with their adherence to the deductive–nomological model of science made fashionable by Carl Hempel (1965) and their commitment to a search for universal laws, certainly overlooked the distinction.

Having painted this bleak picture of archaeology and concomitantly a rosy picture of physics and biology, we need to point out something that at first glance will appear to contradict much of what we just said. We stated that for the most part biologists do not worry about what kind of science they practice; the overwhelming majority agree that materialism must form the basis of Darwinian evolutionism. However, this in no way implies that there is universal agreement in biology over various epistemological and methodological issues that comprise the materialist strategy. As we point out in detail later, this is far from being the case. The quote from Mayr that opened this chapter, in which he noted that most controversies in biology are "between a typologist and a populationist," refers to earlier evolutionary biology, not that being practiced today. In modern times—that is, since the evolutionary Synthesis of the late 1930s and early 1940s—the controversy is for the most part between populationists. At issue are two questions at the heart of Darwinian evolutionism: What are species and how do they originate? As we will see, a considerable amount has been written in the last four or five decades in an attempt to answer these questions, but in some respects, biology is no closer to the answers than it was in the 1940s (for recent overviews, see Hull 1997; Mayden 1997).

We emphasize that the vast majority of biologists and philosophers of biology who have contributed to the controversy over species are materialists—a fact that makes the controversy all the more interesting. All would pledge allegiance to Darwin's central tenets, yet the most basic unit in Darwinian evolutionism—the species—remains a point of considerable contention (Mayr 1996; O'Hara 1993; and references therein). Given our reading of the situation, we do not foresee any major biological agreement on the species front coming anytime soon, in the same way that we do not see eventual agreement among anthropologists on the definition of culture (Keesing 1974; O'Meara 1989). Dozens of species definitions have been proposed over the years (e.g., Mayden 1997), and more certainly will be formulated in the future. How one approaches the species concept in turn influences how one approaches the issue of speciation—the process of diversification that produces species. We believe there is a way out of this conundrum, however—one that has been proposed independently in biology and archaeology, though the solution has generally been overlooked. We discuss that particular solution in Chapter 5; here, our main objective is to identify the problem.

The major point to remember in the pages that follow is that our treatment of essentialism and materialism is set up to emphasize general differences between the two. Do not be misled into thinking that materialists always follow the canon of materialism in logical fashion when they do their work. For one thing, there is no such canon, and for another, it is extremely difficult, sometimes impossible, to abandon one's essentialist upbringing, with its emphasis on natural kinds, and to focus strictly on variation (Dunnell 1982). For another thing, humans have an incredible ability to experience what goes on around them and to retain an enormous stockpile of information that can be called on to help "explain" future situations, including observations made on data sets. This is both a blessing and a curse—a blessing because it allows us to communicate with other humans about the past and the future and a curse because what typically is referred to as experience, otherwise known as common sense, has little place in science. But how often do we view experience as a curse? Not very often; in fact, we normally view it as a scientific blessing. Archaeologists routinely use common sense as a sense-making system (Dunnell 1982; see also Kluckhohn 1939b), in the process viewing as real many of the units employed to bring order to the archaeological record (Chapter 3). Evolutionary biologists have done this as well, albeit to a lesser degree, but even so, one result is the ongoing controversy over species and speciation. If, as we point out in Chapter 5, we can keep our units straight, we can escape numerous dead ends and get about the business of explaining the historical development of variation in logical, materialist terms. As we will see, however, this often is more easily said than done.

ONE VIEW OF SCIENCE

At its base level, science is nothing more than a way of understanding and explaining how and why the world works the way it does. Science is not the only sense-making system, nor is it the one that is called upon most frequently as we attempt to understand and explain the world around us. We would give this honor to common sense, which in several respects is the antithesis of science (Dunnell 1982; Kluckhohn 1939b). Science uses "performance standards" (Dunnell 1971b, 1978b, 1978c, 1982; Schiffer 1996) to establish the correctness of a conclusion; that is, does a conclusion actually account for the phenomena for which it is supposed to account? Use of this method to establish correctness allows conclusions to be integrated into a systematic body of knowledge. Theory specifies three things: the kinds of phenomena to be examined; how phenomena are to be structured and measured for examination; and how those phenomena should interrelate, interact, and respond to one another given particular contingencies. Theory thus provides the basis for explanation. Common sense may appear to serve well in the place of theory, but its liabilities include the facts that it is subconscious, ethnocentric, and contingency bound (Kluckhohn 1939b), and it embodies no developed concept of time (O'Brien and Lyman 1999a). In short, it entails an essentialist view of the world and of all things in it.

Scientific inquiry consists of three significant parameters (Lewontin 1974a; see also Dunnell 1982): dynamic sufficiency, empirical sufficiency, and tolerance limits. A discipline is scientific if its theory is dynamically sufficient—that is, if its theory contains the proper elements in the proper structure to generate scientifically acceptable explanations. A scientifically acceptable explanation is one that works well in an empirical context; hence, an empirically sufficient theory is one that has elements or units that are directly measurable in the phenomenological world. Units or elements that are not measurable render empirical testing impossible. Finally, the fact that models of things always differ from the things themselves results in the need to establish tolerance limits—that is, how closely must our models mirror the things they were established to model? As Lewontin (1974a:8) put it,

> The problem of theory building is a constant interaction between constructing laws and finding an appropriate set of descriptive state variables [units or elements] such that laws can be constructed. We cannot go out and describe the world any old way we please and then sit back and demand that an explanatory and predictive theory be built on that description.... [T]here is a process of trial and synthesis going on ... in which both state descriptions and laws are being fitted together.

In short, then, we need the kinds of measurement units that are relevant to our theory and that allow us to monitor the empirical manifestations of the processes included in that theory. This relates to an issue familiar to all Americanist archaeologists, variously known as classification or typological-unit construction. It has been recognized for some time by evolutionary biologists that there are two basic kinds of units one might construct (e.g., Hull 1970; Mayr 1959a, 1968, 1981, 1987, 1995b) and that those units derive from two different views of reality, but explicit recognition of this dichotomy for archaeology has come about only recently (e.g., Dunnell 1982, 1995; O'Brien and Holland 1990), though a few foreshadowings appeared earlier (Dunnell 1971a, 1971b, 1978a, 1978b, 1978c; Kluckhohn 1960). We examine this issue in Chapters 3 and 5.

TWO CONTRASTING VIEWS OF REALITY

There may be but one view of science itself, but there is more than one kind of science—a multiplicity that is manifest in distinctive views of reality. Although the distinction between the two views has been addressed at various times by biologists (e.g., Mayr 1963, 1970, 1976b) and philosophers of biology (e.g., Hull 1965; Sober 1980, 1984a), it was first made, we believe, in a short article by Mayr (1959a) entitled "Darwin and the Evolutionary Theory in Biology" that appeared in a collection of essays on evolution and anthropology (Meggers 1959) marking the centennial of Darwin's (1859) *On the Origin of Species*. We are led to believe this in part by our own perusal of the literature and in part by what Mayr himself later said in a shortened version of the article that was published in a collection of his essays (Mayr 1976a): "So far as I know, the following essay ... was the first presentation of the contrast between essentialist and population thinking, the first full articulation of this revolutionary change in the philosophy of biology" (Mayr 1976b:26).[1] One might well debate whether Mayr's was indeed the "first full articulation" of the revolutionary change ushered in by Darwin, but there can be no denying the significance of what Mayr had to say in the article: Darwin changed forever the way organisms were viewed, by shifting analytical emphasis from individuals to populations and from essentialist types to materialist variation (Lewontin 1974b; Mayr 1972). But what does it mean to say that emphasis was shifted from individuals to populations and from types to variation? Don't populations comprise individuals? Are types not distinguished from each other on the basis of variation? What is so novel about Darwinian evolutionism?

Perhaps the best way of highlighting the novelty is by examining the tenets that underpin essentialism and materialism. In a general sense, the basic dichotomy between the two ontological positions is fairly easy to understand because of their polarity. For clarity, it is worth repeating that another term for essentialism is

[1]Darwin actually used the term *materialism* in his Notebook M to refer to his position (cited in Bowler 1990:85), but he did not directly contrast it with essentialism.

typological thinking, and another term for materialism is *population thinking* (Mayr 1959a; see also Dobzhansky 1951). As the names imply, essentialism/typological thinking searches for essences in things, the essences being intrinsic characteristics of natural kinds, or types. Materialism, the antithesis of essentialism, sees no intrinsic characteristics. It is not as aptly named as its counterpart, and for our purposes here it does not help that there is another -ism in anthropology with the same name (e.g., Harris 1979). Intuitively, the term *population thinking* is not much better because of the myriad ways in which the word *population* is used in the English language, though it does call attention to the fact that whatever is being examined contains multiple things.

Essentialism

Essentialism is based on the assumption that types, or kinds, have natural existences. Some kinds might be easily discoverable, while others might be more like Easter eggs—hidden in the tall grass but discoverable with the patience and persistence, not to mention the wonderment, of a child—but they are natural just the same. Accordingly, each type has a defining essence (hence the term *essentialism*) or, as we discuss below, at least a defining central tendency that is inherent in the members. At this point, we are not too concerned with the issue of kinds, or types, which we cover in more depth in Chapters 3 and 5, though it is impossible to discuss essentialism and delay completely a discussion of what kinds/types are and how they are created. Given that in archaeology the term *type* carries numerous connotations and meanings (Lyman et al. 1997b), we restrict ourselves here to *kind*. Note, however, that as used here the terms are interchangeable.

Kinds, for our immediate purposes, are units that an analyst uses to partition variation by creating piles of specimens that are alike in certain specifiable ways. Under the essentialist ontology, the analyst searches through a collection of objects to find defining characteristics, which are both inherent and recognizable, thus making them integral to the correct placement of the objects into natural kinds. Kinds can be observed at any scale—attribute of a discrete object, discrete object or attribute combination, or set of discrete objects—and include any number of individual things or events, and at some point—usually after multiple specimens have been studied—become inflexible units in that they cannot change to incorporate an individual that falls outside the boundaries of the kind. This important point was made in archaeology by Philip Phillips, James A. Ford, and James B. Griffin, three of the most prominent culture historians in Americanist archaeology:

> The characters that we have selected as determinants for the type gradually shift, the all-too familiar phenomenon of "creep," until at some point we can stretch our original type definition no further and to consider whether material "X" more closely resembles Type "B," already established at another center,

> or whether it is not sufficiently like either "A" or "B" and must be given an independent status as Type "C." These wretched hair-line decisions beset the classifier at every step. (Phillips et al. 1951:65)

Such decisions do beset the classifier at every step, and they do so for one simple reason: The boundaries of a kind emanate from the sample at hand; that is, specimens are observed, those that look alike are placed in a pile separate from other, dissimilar, specimens, and the essence of each kind is extracted from each pile. In most cases, at least a little variation can be tolerated within each kind or pile, though it is important to realize that to an essentialist the variation is, as Lewontin (1974a:5) put it, an "annoying distraction"—something to be put up with but that is viewed as having no explanatory value. To circumvent the problem of noise, one can find an average—the defining central tendency—in order to identify an archetypal specimen: the ideal, or "real," specimen against which all other specimens are compared. This is what Phillips et al. (1951) did, and despite their recognition that this was a procedure of the analyst and not a property of the piles of specimens, it got them into trouble (Lyman et al. 1997b; O'Brien and Lyman 1998). A clear statement of the procedure was made by Gordon R. Willey (1940:673), another prominent culture historian, when he indicated that the " 'type' concept is a synthetic one in that it is a symbol representative of a group of artifacts, the mean composed of the definitive attributes around which the specimens of a group cluster." Because the definitive mean was *derived from* specimens, the notion that a kind was somehow real (that is, had an essence) was reinforced. As we see later, this was the typical, though not universal, procedure for defining species in the pre–Darwinian world.

The interesting feature of essentialistic kinds is that there is no gradation between or among them—things are, for example, either green or blue, or something else. What do we do when we find something that appears to be halfway between green and blue? One option is to identify a new kind and give it a name, say, teal. A second option is to decide that the object is really more green than it is blue and thus toss it in the green kind. Although the fact is little discussed, the famous nineteenth-century plant hybridizer Gregor Mendel did this with his second- and later-generation peas, based on his understanding of trait retention and expression in relation to color and relative frequency of occurrence.[2] A third option is to decide before beginning the analysis how many natural kinds are going to be identified to house the specimens. This could be done by sorting through available specimens to get an idea of the amount of variation present. In so

[2]Sir Ronald Fisher (1936) first pointed out that Mendel's data were too good, implying that to get the color ratios he wanted, Mendel sometimes had made arbitrary decisions. Despite Orel's (1996) recent defense of Mendel, Fisher's claim has considerable merit. Regardless, Mendel was faced with the problem that confronts all essentialists, namely, what do you do with variation? Mendel obviously banished it from consideration.

doing, we might decide a priori that we need two kinds. The number of kinds used depends on two things: the problem we are trying to solve and how much noise our kinds can tolerate. For some analyses, where more noise is tolerable, we could call everything that is a shade of what we define as blue, "blue"; in other analyses, where we want finer discrimination, we could allow more kinds (this is the "lumper" or "splitter" dichotomy). The end product is the same: bounded kinds that contain varying numbers of individual specimens, each specimen within a kind being "annoyingly" but insignificantly different than the others in that kind. Importantly, the color something is becomes a defining essence of the kind of object it is. These essences are real in the sense that they are characteristics exhibited by empirical phenomena (objects). By definition, any object of a particular kind must share that essence; otherwise, it would be an individual of another kind.

Do not be misled by the fact that the analyst is the one deciding whether there should be two kinds, or five kinds, or ten kinds; the kinds still are considered real because the essences are considered real; that is, they have been derived from the examined specimens, so how could they not be real? The decision of how many kinds to use is completely separate from asking the fundamental question: Are the kinds themselves real? Simple combining and dividing has nothing to do with whether the kinds are viewed as real. The natural, and hence real, essences are still presumed to be there, regardless of how many kinds are actually identified. Kinds are thus sample-specific and accidental—a difficulty easily overlooked. Furthermore, the derivational process—a murky mental activity—reinforces the commonsensical ontology that the resultant kinds are real. You can, after all, *find* them.

Materialism

In opposition to the essentialist view that discoverable natural kinds exist, materialism does not involve inductive searches and instead creates units that are capable of performing particular jobs—not any and all jobs but rather *particular* jobs. This is not to imply that essentialist kinds can do *no* analytical work. Some of them, such as chemical elements, clearly can, and the same is true in archaeology—for example, essentialist pottery types can be used to mark the passage of time (O'Brien and Lyman 1998, 1999a). But not all essentialist kinds allow the analytical work of a materialist to be performed, because by their very nature they often preclude the study of evolutionary change—the constant process of the members of a lineage becoming something else—and force the analyst to study only similarity and difference. What we are really saying is that in some cases, essentialist kinds are useful in a materialist study, but this usefulness is largely accidental. This is because change in essentialist kinds can be only of the transformational sort. Such Midas-touch difference is specifically what Darwin disposed of in the study of organismic change. Stacking one essentialist kind on top of another might produce a temporal sequence, but whether it measures evolutionary change in the sense of an ancestor–descendant lineage is not inherent in the stack. Darwin argued that evolutionary change comprises the differential persistence of variants; essentialist kinds clearly do not allow the measurement of this sort of change, unless by accident.

If reality is a constantly changing entity, a materialist would argue that time would be best spent developing units that measure things that are always in the state of becoming something else. Thus, categories such as "royal blueness" and "baby blueness" might be analytically meaningful to a materialist in a certain situation, but outside that analytical situation—that is, when the temporal and spatial frames of reference have changed—they might no longer be useful. In our example, there is nothing "real" about "baby blueness" that makes it a defining essence. It might make analytical sense to *create* a category that we term *baby blue*, but note the difference between *creating* such an analytical category—in which case the category definition is imposed on a collection of objects—and *discovering* such a category—in which case the category definition is extracted from a sample of objects.

To categorize events or objects, the materialist selects attributes relevant to some problem, and it is possession of those attributes, and *only* those attributes, that results in the sorting of specimens into internally homogeneous, externally heterogeneous piles. Any other attributes that the objects or event might happen to exhibit can be described, but they are not integral to the decision as to whether an object ends up in pile A or pile B. They are simply hitchhikers. Importantly, specimens that share defining attributes—those that end up together in one of the analyst's piles—have been grouped together not because of some inherent, shared quality but rather because they hold in common some number of attributes *selected by the analyst* as being important analytically. Mayr (1987) refers to these attributes as properties in common, meaning that all objects placed in a pile share the defining attributes. They might also share other properties in common—or some might share those properties, whereas other specimens in the pile do not—but this is entirely irrelevant.

We might decide, based on observation, that the color of a stone tool is unrelated to function, whereas the angle of the working edge *is* related; thus, if we are interested in functional variation in stone tools, we might choose as our attributes edge angles, traces of use wear—a testable inference manifest as kinds of edge damage with particular spatial distributions—and other attributes that we propose are causally related to the property of analytical interest. We thus would not worry about the color of an artifact, though we certainly could describe the color of the specimens if we so desired. What governs our proposal is, as we said earlier, observation of how stone tools interact with their environment. Here, we need to insert a brief statement that is a prelude to extended discussion in later chapters. Our observations of dimensions of stone-tool use are observations of mechanical forces and their consequences on both the tool itself and the material being cut. The properties of a stone tool—edge angle, and so forth—and how those properties affect such things as cutting are, as we refer to them below, *immanent properties*. They are timeless and spaceless properties and contrast with *configurational properties*, which are not timeless and spaceless. This is a major point of contention between evolutionary archaeologists and behavioral archaeologists (O'Brien et al. 1998), the latter of whom search for general laws of behavior. They see no difference between mechanical properties and behavioral properties; thus, present human behaviors, if they are found to recur, can be used as proxies for past behaviors. Evolutionary archaeologists use inference of the mechanical sort, but they avoid inferences that are configurationally based.

Once a materialist has created a set of informed kinds, then empirical specimens are examined. Each particular specimen is studied to determine whether it has the attributes or properties definitive of one kind or another. This is simply a process of *identifying* which specimens have which definitive attributes and grouping those specimens that display a kind's definitive properties together. Evolutionary change can then be monitored as shifting frequencies—the differential persistence of variants—of kinds through time. As Mayr (1991b:128) notes, a "population [of specimens] changes by a slow shift of its mean value," and that mean is a statistical abstraction of the various kinds making up a sample of variants. There is another key issue involved—the identification of homologous structures—but we delay discussion of it until Chapter 6.

THE MATERIALIST PARADOX AND THE STUDY OF SPECIES

The sizable literature that exists on Darwinian evolutionism makes it clear that Darwin changed forever how organic evolution is viewed, in the process effecting a change from an essentialist to a materialist view. It is a mistake, however, to include all pre-Darwinian biologists in the essentialist camp, just as it is equally a mistake to conclude that all biologists, after reading *On the Origin of Species*, were converted to materialism (Mayr 1982a, 1991b). But there can be no doubt that Darwin's notion of descent with modification eventually forced biologists to begin viewing evolutionary change as something that is continuous as opposed to disjointed (Mayr 1982b, 1995a). Importantly, this shift in view was not immediate, and in fact, it took not quite a century for the position to be largely accepted. Also, continuity says nothing about rate of change—a subject that is still contentious in biological circles.

In broad terms, concomitant with the shift in evolutionary biology from essentialism to materialism was a change in how biologists began viewing species—that is, as populations of organisms rather than as natural kinds that have identifiable essences. After Darwin, the species itself, as opposed to the individual organisms within the species, became the primary unit of focus—an extended perspective that was entirely appropriate under the Darwinian view for two reasons. First, Darwin recognized that a species as a whole evolved, not the individual organisms within the species. Second, Darwin shifted primary analytical focus from observing differences between groups of organisms or species to determining what those differences meant in terms of evolutionary change—what he labeled as "descent with modification."

The reason for the shift was that Darwin was attempting to explain not only the diversity of life forms but also why those diverse forms all seemed to be well adapted. To do so, he proposed a unique solution: All forms shared descent from a common ancestor, but no organism was a clone of any other. The distinctly visible sets of similar organisms must have resulted from the long-term gradual action of natural selection targeting each individual, generally missing those that were better adapted, while simultaneously tending to eliminate those that were less well adapted (Mayr 1982a:116–117). Thus, in order to document the history of a species, one must document the individual variants within that species' lineage, track those variants through time, and identify the selective contexts that drove the shifts in frequencies of variants (Szalay and Bock 1991). The construction and explanation of a lineage comprises the materialist epistemology.

Constructing lineages involves the investigation of change, not simply the documentation of similarity and difference. This is the major difference between essentialism and materialism: The former treats only similarity and difference, while the latter treats change as well as similarity and difference. How could it be otherwise? Chemistry examines essential differences between various units—the difference between a carbon atom and a nitrogen atom—but it does not track changes in carbon atoms. The only way something with an essence can change is somehow to transform itself (or be transformed) into something else—that is, it must become a member of another kind by changing essences. Therefore, essentialist change is of the disjunctive kind; it can only be accomplished by a jump across bounded units. Alternatively, materialistic change is a continuous, seamless process that occurs in a boundless system—boundless in the sense that no natural kinds (units) exist.

The difference between disjunctive and seamless change is a fairly noncontentious issue in biology, though previous handling of the differences between essentialism and materialism has made it sound as if Darwin's work refuted essentialism as a legitimate position in biology because he showed that species evolve (e.g., Mayr 1959a, 1991b, 1995a). This is true in only the broadest sense; hence, we have to exercise caution in how we handle that statement. We should be cognizant of a critical point made by Elliott Sober (1980:355–356): The mere fact that species evolve does not prove they lack essences. What Darwin really showed was that evolutionary theory removed any *need* to provide species with constituent definitions (Sober 1980:360). Occasionally lost in the philosophical shuffle is the vast difference between seamless and disjunctive change. This should be the subject of primary focus, not that essential kinds cannot change.

Natural, or essentialist, kinds *can* change, but the manner in which they change is usually referred to as transmutation. Take, for example, the transmutation of elements:

> In the periodic table, we have discrete jumps—between atomic number 36 and atomic number 37 there are no intermediate atomic numbers to blur distinctions. But let us reflect for a moment on the mechanism of transmutation. Consider, as an example, the experiment which settled the question of how nitrogen can be transmuted into oxygen (Ihde 1964, p. 509):
>
> $$^{4}_{2}He + ^{14}_{7}N \rightarrow ^{17}_{8}O + ^{1}_{1}H$$
>
> In this reaction, the alpha-particle is absorbed and a proton is expelled. Let us ask of this process ... [a]t what point does the bombarded nucleus cease to be a nitrogen nucleus and when does it start being a nucleus of oxygen?
>
> There *may* be a precise and principled answer to this question which is given by the relevant physical theory. But then again there may not. I would suggest that which of these outcomes prevails really does not matter to the question of whether essentialism is a correct doctrine concerning the chemical kinds. It may well be that having a particular atomic number is a vague concept. But this is quite consistent with that (vague) property's being the essence of a chemical kind. This really does not matter, as long as the vagueness of "nitrogen" and that of "atomic number 14" coincide. (Sober 1980:357–358)

Those two kinds—nitrogen and atomic number 14—*have* to coincide because they are, by definition, autocorrelated, meaning that each defines the other *and no other*. Similarly, isotopes can transmute. Take for example the heavier carbon isotope ^{14}C and one of its lighter counterparts, ^{13}C. Both isotopes have essences, based on their respective numbers of protons and neutrons. Through the addition or subtraction of a neutron, one isotope can be transformed into another, but this is transmutation, not materialist change. The isotope ^{13}C is not always in the process of becoming ^{14}C.

The discovery that elements and isotopes transmute did not undermine the periodic table, nor did it demonstrate that elements and isotopes do not have essences. As Sober (1980:366) points out, to be nitrogen is one thing, and to be oxygen is something else entirely. As long as there are intrinsic characteristics that a set of things, and only those things, have in common, they are members of a particular kind. What is important is that they have essences, not that they can swap essences through transmutation. This point has been made with reference to evolutionary units in archaeology (e.g., O'Brien and Holland 1990, 1992), but the argument has not been well stated.

Species, unlike elements and isotopes, do evolve, though this notion became acceptable only after 1859. However, even after Darwin demonstrated the role of natural selection in effecting speciation, it was not altogether clear what species were (Bowler 1990). Some pre-Darwinian biologists handled the notion of change by avoiding it altogether through reference to a Creator who had fashioned a finite number of species when he made the earth and placed organisms on it. All that was left for the biologist to do was find and name the species. Alternatively, to those who admitted the existence of some sort of evolution, the typical view was that one species somehow was transformed into another one. In other words, one bounded species, when it had run its course, was transformed into another one.

The issue of change and what it entails is one component of the Darwinian legacy, but there is another, related one. That issue is the species concept, which is of interest here for two interrelated reasons. First, it is the ideal vehicle to use in highlighting differences between essentialism and materialism. What, exactly, is a species, according to the two ontological positions? If, as should be obvious by now, the answers to that question differ, can both be defended? But if we think about it, is this not a curious question to ask, given that modern biology holds to the materialist view? If we accept that view, it appears that we should be able to quickly discard the essentialist view of species. As we will see, however, this is not necessarily how things have turned out. This brings us to the second reason that the species concept is of interest here. Provided that we hold to the materialist position, which states that things are always in the process of becoming something else, how can we ever hope to make sense out of a historical record? Does not the materialist position actually hamper our attempts to do evolutionary studies? Accordingly, is it possible that phylogenetic analysis, whereby one works to reconstruct ancestor–descendant lineages, is really nothing more than a metaphysical exercise? In plain language, how does an archaeologist or paleobiologist remain faithful to materialism and still get any work done? The answer to this question lies in how one approaches the issue of units and in how one actually uses those units to identify and measure variation. As we will see, biologists have long wrestled with the issue of units, not altogether successfully. The following sections are intended to serve only as an introduction to the problem of species as units—a topic that we treat in considerably more detail in Chapters 4 and 5. The main point that we are trying to emphasize here is one we made earlier: It is not always easy to be a materialist, even if you are a biologist or paleobiologist.

Kinds as Units

Kinds have long played a central role in biological thought. Mayr (e.g., 1959a, 1982a) and others trace the typological metaphysic back to Plato and the notion of *eidos*, translated here as *form*—that is, as something that is real. Sober (1980) does not deny this lineage but also traces the manner in which essentialism handles variation back to Aristotle and his Natural State Model—that is, individual things have natural states that they now reside in or once resided in (see also Hull 1965). All other states in which something happens to reside are the results of outside influences acting on it. With certain exceptions, the tendency of pre–Darwinian taxonomists was to follow Platonic and Aristotelian logic and embrace types—here, species—not only as bookkeeping units but also as real, natural units that were discoverable by identifying appropriate essences. There were, alternatively, some biologists who declared species unreal—a declaration based either (1) on the absence of characteristics that could unite all the individual organisms that biologists wanted to place within a species or (2) on the fact that there was far too much variation in nature to sort through; hence, the hope of finding either an essence or a natural state, even if such existed, was illusory. Under such perspectives, species were viewed simply as convenient cubbyholes in which to place like-looking creatures.

Is what this latter group of nineteenth-century biologists was saying true? Is nature playing a trick on us by masking the true essences of species with shades of reality? If so, how do we make our way through all the variation—what we might call the mutant forms—and get at the real types? How do we, as Sober (1980:367) puts it, render the variation transparent? One way is by averaging out the variation across individuals. Not only does this suppress widely diverging variation, but it also allows one to construct an ideal archetype. This notion appealed to some nineteenth-century biologists, and they turned to mathematicians and statisticians for computational assistance. One of the more prominent was Adolphe Quetelet, a Belgian statistician who spent considerable time and effort searching for what he called the "average man," or the true representative of the social body—a representative who was unencumbered by "multifarious and idiosyncratic characteristics which make for diversity in a population" (Sober 1980:366). For Quetelet, variation was more than a distraction; it was meaningless. Variation could only be produced by accidental causes and hence was unanalyzable (Hilts 1973:219).

Quetelet focused on the arithmetic mean, which was significant, at least to him, because it was the product of constant, as opposed to unpredictable, causes in nature. Importantly, Quetelet (1835) distinguished between the arithmetic mean and the true, or "typical," mean, the latter referring to phenomena whose distributions conformed to the law of error (Hilts 1973:217). It was on this measure that he relied during his search for the "average man." In reviewing the close connection between statistics and social science during the nineteenth century, Victor Hilts (1973) summarized how, in 1850, astronomer Sir John Herschel, warmly received Quetelet's notion that the average man was more than an artifact of measurement error:

> An average may exist of the most different objects, as the heights of houses in a town, or the sizes of books in a library. It may be convenient to convey a general notion of the things averaged; but it involves no conception of a

> natural and recognizable central magnitude, all differences from which are to be regarded as deviations from a standard. The notion of a mean, on the other hand, does imply such a conception, standing distinguished from an average by this very feature, *viz.* the regular marching of the groups, increasing to a maximum and thence again diminishing. An average gives us no assurance that the future will be like the past. A mean may be reckoned on with the most implicit confidence. (Herschel 1850:23)[3]

In other words, an "average" was the result of measurement error, whereas a "mean" was real and reflective of an archetype. To use Sober's (1980:367) example, Quetelet "was impressed by the fact that the results of accurately measuring the waists of a thousand Scottish soldiers would assume the same bell-shaped distribution as the results of inaccurately measuring the girth of a single, average, soldier a thousand times." The former produced a true and useful characteristic diagnostic of the average man—the subject of interest—whereas the latter merely produced an average that was an artifact of measurement error.

Quetelet's essentialist message was clear: The average was an abstraction of a series of numbers, but the arithmetic mean was a reality. This, the (real) central tendency, was what biologists were after. Not until the work of Darwin's cousin Francis Galton later in the nineteenth century did biometricians begin to realize not only the significance of deviation from the mean but also the link between that variation and heredity. Suddenly, variation no longer was an embarrassment but a critical ingredient in the search for order in the natural world. It is in Galton's work, flawed as it was, that we see the first germs not only of population genetics but also of naturalism as well. Population genetics was one of the major players on the biology field in the first half of the twentieth century and in effect was the one that finally constructed a bridge between the naturalists and the experimentalists that resulted in the Modern Synthesis.

One issue that certainly had not been ignored prior to 1940 but assumed a central role after that date was the species concept. Despite the best efforts of the population geneticists to construct a bridge that linked the epistemological underpinnings of the Darwinian naturalists and the Mendelian geneticists, the issue of what a species was remained problematic. How biologists and philosophers of biology approached that problem gives us a glimpse of some of the difficulties raised by using a materialist perspective to study empirical phenomena.

Species: A Particular Kind of Unit

Seventy years ago, mammalogist E. Raymond Hall (1926) looked at some late Pleistocene-age marten bones found in two caves in northern California and

[3]Herschel, a committed essentialist, later referred to Darwin's descent with modification as "a theory of higgledy-piggledy" (quoted in Francis Darwin [1893:232]).

proposed that a unique subspecies of marten (*Martes caurina* [now *americanus*] *nobilis*) existed in western North America prehistorically. A decade later, he changed his mind, deciding that what he had thought were taxonomically diagnostic features of the fossils were not in fact diagnostic (Hall 1936). The matter might have ended there, but in 1970, paleobiologist Elaine Anderson (1970) studied numerous skeletons of modern North American martens (*M. americanus*) and fishers (*M. pennanti*), as well as the fossils Hall (1926) had examined, plus a number of fossils that had come to light since his examination. She concluded that there were sufficient morphometric differences between the bones and teeth of extant populations and certain of the fossils to warrant assigning a limited set of the fossil material to the taxonomically unique noble marten. She raised the original taxonomic unit comprising noble martens to the level of a species (*M. nobilis*). At least thirteen sites in North America have produced remains of noble martens (Graham and Graham 1994; Grayson 1993:184), many of which have sufficient chronological control to indicate that the noble marten existed from the late Pleistocene until about 3,000 years ago.

Twenty-one years after Anderson resurrected the noble marten as a taxonomic unit, biologists Phillip Youngman and Frederick Schueler (1991) questioned whether the fossils assigned to that unit in fact represented a distinct species. After comparing a single measurement of the mandible, three measurements of the lower first molar, two skull measurements, and four measurements of the humerus across fossils assigned to the noble marten and a set of modern comparative marten skeletons, they concluded that the fossils were in many respects so similar to modern marten specimens that there was little reason to assign the fossils to a distinct species. The only difference was one of size, and that criterion, they argued, was a poor one for distinguishing among species. Alternatively, Anderson (1994) and zooarchaeologist Don Grayson (1993:184) imply or suggest, respectively, that the noble marten still warrants recognition as a separate species because of not only its size but also the shape of its teeth. Paleobiologist Russ Graham notes that the issue has not yet been satisfactorily resolved; thus, he retains the species name "because of the morphological information it conveys" (Graham and Graham 1994:31). What is that morphological information?

Fossil bones and teeth assigned to the noble marten tend to fall between the modern pine marten and modern fisher in size. Furthermore, fossils of the major teeth of the noble marten differ from those of the two modern species not only in size but also in various shape attributes (Anderson 1994; Graham and Graham 1994; Grayson 1993; Youngman and Schueler 1991). The major difficulty in identifying unknown remains of the genus *Martes* as representing the noble marten resides in the fact that this taxon's definitive criteria have been extracted from fossils of unknown taxonomic affinity. Anderson (1970) and Youngman and Schueler (1991) measured bones and teeth of modern martens and fishers and then compared those measurements to fossil remains. If the differences between the latter and the two modern taxa were sufficient, then a new taxon was said to have been identified. The problem, then, reduces both to the nature of the comparative materials used to establish the morphometric boundaries of the two known taxa and probably to interobserver variation. Anderson and Youngman and Schueler present sample sizes, ranges, standard deviations, and means for the measurements they took on comparative specimens, but because they measured different specimens, these statistics differ.

The extant literature does not provide a consistent set of necessary and sufficient conditions that must be met by a specimen of unknown taxonomic affinity for it to be assigned to a particular taxon. At best, we could visit repositories holding the specimens, measure them ourselves, and then decide where to place them taxonomically. At worst, we could simply choose among the various options based on which investigator we think presents the better case. To those outside the field of biological taxonomy, it might appear that the taxonomic placement of the noble marten is an isolated example of picky scientists arguing over details. Such an assessment is wrong because this is not an isolated example. Taxonomic arguments, whether over martens or hominids, are commonplace in natural history. As new evidence comes in, or older evidence is reevaluated, specimens are constantly being placed in different taxonomic kinds. At issue, however, is a deeper concern, and the haranguing that goes on over the taxonomic placement of particular specimens pales in comparison. That deeper concern is over what a species is—not whether a specimen belongs in species A or subspecies B, but what those units mean epistemologically.

The question of what a species is not only has long interested biologists but also has been a point of contention in large part because of the epistemological difficulty of the issue and also because of the existence of various subfields within the broad discipline of biology. If anything, the biological literature of the last six decades has demonstrated that a single definition of a species is unworkable at best (Hull 1997; Mayden 1997; Mayr 1987, 1993). What appears to work well for an ecologist, for example, whose specimens are living organisms, does not work particularly well for a paleobiologist, whose subject matter for the most part consists of preserved hard parts of long-dead organisms. Any species definition that uses as a key criterion the ability of a group of organisms to breed and produce fertile offspring, while it may be helpful to an ecologist, would seem to be of no use to a paleobiologist because the definitive criteria do not fossilize. An ecologist can, at least conceivably, examine mating behavior and thus get some idea of species boundaries, but a paleobiologist cannot. Similarly, a biologist studying modern species can examine such things as isolating mechanisms (Dobzhansky 1937b)—genetic differences between populations that restrict or prevent gene flow—but such concepts are useless to a paleobiologist concerned with species and speciation. Paleobiologists can and do examine past physical environments in which organisms lived, and in this respect their work is similar to that of ecologists, but for the most part, the empirical world of the paleobiologist is restricted to the formal (e.g., size, shape, color, and hardness) dimension. Although paleoenvironmental differences can play important roles in decisions regarding which fossils to place in which species, of far more importance are differences in phenotypic characteristics such as the size and shape of teeth. Paleobiologists continue to label particular sets of fossils with species names despite repeated recognition over the past several decades that the conception of species as reproductively isolated populations of organisms cannot be empirically tested with the fossil record (e.g., Fox 1986; Pearson 1998a; Tattersall 1986; Trueman 1979; Weller 1961).

But still, haven't we all been taught in a basic biology course that the propensity for breeding is a basic criterion for recognition of a species? Furthermore, were we not also taught that even more important to species recognition is the fact that breeding *actually* takes place, as opposed to theoretically *could* take place, among a group of organisms? How, then, can we ignore this key criterion, even when we have no direct evidence for interbreeding? Can there actually be more than one definition of a species—one for field biologists and ecologists and one for paleobiologists? The answer is yes, and in fact, there are (and always were—see Dobzhansky 1935, 1937b; Simpson 1951) literally dozens of species definitions (for a concise review, see Mayden 1997). But is this a red herring? Regardless of the number of species definitions, how can a science, irrespective of whether it is a materialist or essentialist science, have multiple working definitions for the same thing? Physicists, after all, do not have multiple definitions of electrons. Why can biologists not settle on a mutually satisfying definition of a species?

In the simplest of terms the answer is, because we cannot see a species. We can see the organisms included in the species, but we cannot see the species itself. We cannot "see" electrons either, but we can detect their presence. In biology, the problem is compounded by the fact that not only can we not see a species, but as Mayr (1987:146; see also Mayr 1969b:27-29; Simpson 1940, 1951) points out, we also use the term *species* for two entirely different things, one being a conceptual category—in the sense of family, genus, species—and the other what Simpson calls a "taxon"—that is, the thing that is out there and comprises the individual organisms. The fact that we cannot see a species as a category but only as a taxon is not a good reason for rejecting the existence of the category. We cannot see gravity, but we still believe it exists. Furthermore, we can measure it. Biologists who do not place much stock in the species concept, referring to it as simply a bookkeeping device, do so not because they cannot see a species but rather because they do not believe such a thing really exists. They see evolution as being too seamless to have produced some bounded unit of creatures we call a species (e.g., Mayden 1997). Perhaps if we could freeze time we might be able to identify a species, but in the real world, where time is continuous, things are too fluid—too "in the process of becoming something else"—with breeding boundaries constantly changing, isolating mechanisms coming and going, and so on, for a species to be real. Biologists who hold the species concept dear would counter with the argument that if you can identify a species at any instant in time—that is, by freezing time and counting noses—then it must be real. Of course, the counter-counterargument is that time cannot be frozen, so why bother worrying about what a species is? How did a group of materialists get in such a predicament?

Prelude to the Synthesis

One might have thought that after Darwin postulated the slow, gradual evolution of species by means of natural selection, biologists would have agreed at least on what a species was, even though they might have debated the tempo and mode of evolution. However, this was decidedly not the case. By the end of the nineteenth century—keeping in mind that the discovery of Mendel's earlier experiments was still a few years off—biologists were split over the species concept and how species evolved. Species for the most part—and here, we are generalizing—were viewed as real collections of things, though there was a minority view that they were simply collections of real things. The difference between those two views is important: The former presents species with existence, while the latter views them simply as analytical devices (see Williams 1966).

With the discovery in 1900 of Mendel's work, coupled with the experimental work of the period, there developed what might be termed a school of thought—mutationism—that broadly speaking viewed variation as discontinuous and thus species as packages of traits that spontaneously arose as a direct result of mutation. Again painting with a broad brush, from around 1900 on, one can talk of two camps in biology: the naturalists and the experimentalists. The latter, as Mayr (1982a:541, 1993:131) notes, developed the (correct) notion of particulate inheritance and from this drew the (incorrect) conclusion that evolution was saltational and new species appeared abruptly. They were dominated by typological thinking and ignored the existence of populations. The naturalists, meanwhile, kept their eyes on the slow, gradual change in populations of organisms but adopted the (incorrect) view that inheritance was of the blending sort.

Because they looked for sudden jumps—saltations—the experimentalists focused squarely on the confirmation of discontinuous variation, or heterogenesis. It was this variation caused by mutation that was, in their view, the all-important element in evolution—a point summed up beautifully by William Bateson in his late nineteenth-century book *Materials for the Study of Variation*: "Variation, whatever may be its cause ... is the essential phenomenon of Evolution. Variation, in fact is Evolution. The readiest way then, of solving the problem of Evolution is to study the facts of Variation" (Bateson 1894:6). But he also noted that species "are discontinuous: May not the Variation by which Species are produced be discontinuous too?" (Bateson 1894:18). This was purely typological thinking, and it came from the notion that variation exhibits a punctuated life—continuous for a while, then discontinuous, then continuous, and so on. Darwin had originally admitted as much, though he was attacked and later softened his stance. The experimentalists kept the notion of discontinuous variation alive and used it to explain the creation of species.

The greatest saltationist of all, Hugo de Vries, noted that "the ordinary or so-called individual [continuous] variability [cannot] ... lead to a transgression of the species border even under conditions of the most stringent and continued selection" (de Vries 1901:4). Hence, speciation had to result from the rise of discontinuous variants: "The new species thus originates suddenly, it is produced by the existing one without any visible preparation and without transition" (de Vries 1901:3). This perspective forced de Vries to call every novel variant a new species—that is, a change in morphology signaled the creation of a new species. This stance, though logical, conflated the origin of species with the origin of species *characters*. This was not an untypical move on the part of the early experimentalists. Curiously, de Vries held to the notion that species pass through "mutable" and "immutable" periods, which led him to postulate that instead of species originating through a struggle for existence, they are exterminated by them.

Neither side—the naturalists, who held to slow, gradual change, and the experimentalists, who viewed evolution as a series of saltational jumps—particularly understood or listened to the other. As we noted earlier, Darwin sought to explain the origin of the diversity of life and the maintenance of adaptedness of life forms. The naturalists of the early twentieth century focused on the former, the experimentalists on the latter. The naturalists studied the horizontal, or spatial and speciational, aspects of evolution, whereas the experimentalists studied the vertical, or temporal and adaptational, aspects of evolution. Neither side thought in terms of both dimensions; thus, there "was no genuine population thinking in either camp" (Mayr 1993:131).

By at least the 1920s, the genetical work of de Vries and others was shown to be fatally flawed; the continuous variation of Darwin and the supposed discontinuous variation of the experimentalists were one and the same thing. However, the naturalists, who had maintained an active stance against de Vriesian mutationism, did not keep up with the changes in genetics and thus kept firing at a foe that was already dead. The new school of experimentalists might have killed off the older notion of discontinuous variation, but early on, the school was not particularly kind to selection and its role in producing speciation. Led by Mendelian convert Thomas Hunt Morgan (e.g., 1932) and his student Hermann J. Muller (e.g., Morgan et al. 1915), these geneticists did not deny the existence of natural selection, but basically they viewed it merely as a destroyer of variation and an extinguisher of historical lines. The Morgan-led geneticists gave short shrift to the notion that selection might be the creator of species, and in fact, they gave short shrift to a species being much more than a bookkeeping unit. In 1923, in rebuttal to the Mendelians' notion that new variants equaled new species, Morgan wrote:

> How far these new types furnish the variations that make new species may depend on what we call "species." If, as some systematists frankly state, species are arbitrary collections of individuals assembled for the purpose of classification; or if, as other systematists admit, there are all kinds of species both in nature and in books, it would be absurd for us to pretend to be able to say how such arbitrary groups have *arisen*. It is possible that some of them may not have arisen at all—that they may have only been brought together by taxonomists. (Morgan 1923:237–238)

Here, Morgan, who we would agree with Garland Allen (1980:379) was a decided materialist, was not only questioning the reality of species but saying that classifications were entirely arbitrary. A naturalist might create one system and a geneticist another, both of which would be logical and valid. But notice that Morgan also used the statement made by some systematists that there are "all kinds of species both in nature and in books" as a reason for jettisoning the notion that species are real. This double use of the term species is what Mayr (1987; see also Mayr 1969b:27–29) was referring to when he noted that the term is used to designate a taxon as well as a category. Morgan's inability to distinguish between the two uses led him to label as "absurd" any pretension of being able to say how such "arbitrary groups" might have arisen. Morgan was an empiricist of the first order, not someone who sat around speculating about evolution as a problem of history.

Of course, evolution *is* a problem of history, but what Morgan was reacting to was the fuzziness of the naturalistic approach to species definition. No amount of posturing and appeals to Darwinian wisdom could hide the fact that in the seventy or so years since Darwin published *On the Origin of Species*, the question of what a species is had not been answered satisfactorily. The naturalists had danced around the issue, while the early experimentalists had tried to attack the issue head-on by showing how mutations, acting over a short period of time, could cause speciation. For their part, the later geneticists had killed the notion that variation was disjointed but were unconvinced of the reality of species. The naturalists were slowly becoming aware of the fact that species are variable populations that are isolated reproductively from other such populations, but in the late 1920s, the landmark publications in population genetics—for example, Ronald A. Fisher's (1930) *The Genetical Theory of Natural Selection*, J. B. S. Haldane's (1932) *The Causes of Evolution*, and Sewall Wright's (1930) "The Genetical Theory of Natural Selection: A Review" and (1931) "Evolution in Mendelian Populations"—were still a few years off. These publications would demonstrate once and for all that even the slightest selective differences could bring about changes in gene frequency within a population and hence lead to evolution. Naturalists did not immediately appreciate the work of the population geneticists, but that began to change with publication of Theodosius Dobzhansky's *Genetics and the Origin of Species* in 1937. In several respects, Dobzhansky's (1937a) book marked the beginning of the Modern Synthesis (Cain 1993:9).

Given what we have said up to this point about the Synthesis, should we not expect that the unification of the experimentalists, the naturalists, and the population geneticists would have created a unified view of what a species is? We might expect that, but it did not occur. One reason it did not was that at least one group felt disfranchised when it came to the species definition. That group was the paleontologists, who could not see a way to use the emerging definition being formulated by a new group of biologists who were products of the Synthesis (for historical perspectives, see Cain 1993:10; Laporte 1991; for examples, see chapters in Sylvester-Bradley 1956). This new group was the neontologists, whose ranks were filled by old-school naturalists—ornithologists, zoologists, and the like—who at least understood the basics of what the population geneticists were saying. It was the neontologists, including Dobzhansky and Mayr, who began to dominate the post-Synthesis stage. Dobzhansky (1937a:11) summed up the position of the experimentalists when he remarked that evolution merely comprised "a change in the genetic composition of populations." He and his colleagues "were confidently asserting [by the mid-1930s] that explanations for evolutionary and systematic problems were to be found in the genetic and cytological domains" (Cain 1993:17).

Despite the enormous contributions made by Dobzhansky (see Cain [1993:9–10] and references therein), we more or less bypass discussion of the great Russian naturalist and concentrate instead on Mayr, primarily because it was he and not Dobzhansky who throughout the 1940s and 1950s was at the center of controversies over species and speciation. To give equal time to the nonneontologists, we examine the work of paleontologist George Gaylord Simpson, who, although not strictly adversarial in his relations with the neontologists, took a different view of what a species was. In point of fact, he had a difficult time reconciling the work he did with what the neontologists did—a difficulty rendered in what perhaps was his most famous quotation: Neontologists "may reveal what happens to a hundred rats in the course of ten years under fixed and simple conditions, but not what happened to a billion rats in the course of ten million years under fluctuating conditions of earth history. Obviously the latter problem is more important" (Simpson 1944:xvii)

We focus on Mayr and Simpson not because we believe that their work summarizes all aspects of the species issue but rather because they are excellent examples of two scientists who reach what seem at first glance to be fundamentally different views of reality but actually are complementary. Also, Mayr and

Simpson were for several decades two of the heavyweights in the fray over species. One might think that if the species conundrum were going to be solved, one of them would have been the one to do it. Each solved it to his own satisfaction, but the biological and philosophical literature of the last several decades suggests that that satisfaction was not particularly widespread.

Neontology and the Synthesis

One of Mayr's contributions to the fray was the popularization of the *biological-species concept*, which, he later noted, emphasized "the species as a community of populations, reproductive isolation (often based on behavioral mechanisms rather than merely on a sterility barrier), and the ecological interactions of sympatric populations that do not belong to the same species. The major intrinsic attribute characterizing a species is its set of isolating mechanisms that keeps it distinct from other species" (Mayr 1980:34). The etiology of the biological-species concept is difficult to trace, though there are hints of it in pre–Darwinian biology (Mayr 1969a:25–26), and these increase near the beginning of the twentieth century (Mayr 1959b). It was in the writings of the systematists–naturalists of the 1920s, especially ornithologists such as Bernhard Rensch, under whom Mayr studied, that the concept was developed formally. Mayr (1993:133), correctly in our view, takes credit for the "rapid adoption" of the concept after 1942.

The problem Mayr set out for himself was to bring the notion of reproductive isolation into the discussion of variation and evolutionary change. It was the origin of this isolation between populations that was critical to the problem of speciation, and it was critical for linking the Darwinian concept of slow, continuous change with both the "steady increase in the number of species which seems to have occurred in geological history" (Mayr 1959b:222) and the empirically founded conclusion that there were sharp genetic breaks between different *extant* populations or species. Extrapolating to the temporal dimension, the gradual descent with modification so often spoken of by Darwin comprised evolutionary change within a lineage—change that given sufficient time would produce a descendant population sufficiently different from its ancestral one to be assigned to a new species. This is today termed *anagenesis*—within-lineage evolution—and is distinguished from *cladogenesis*, or branching evolution, which denotes the multiplication of species within a lineage. Mayr, in an attempt once and for all to undermine both the purely genetical definition of the experimentalists, with its emphasis on changes in gene frequency, and the purely morphological definition of the naturalists, established the theoretical principles of the biological species in his 1942 book, *Systematics and the Origin of Species*.

Fifty years later, Mayr summarized his contributions to the discussion of species and speciation:

> The recognition that species are populations, not types, was particularly important in the explanation of speciation. I reported in 1942 that in birds, mammals, butterflies, and snails, that is, in all taxonomically well-studied groups, speciation invariably turned out to have been geographical. This means that a population that had been isolated by geographical or vegetational barriers had acquired genetical isolating mechanisms during this geographical isolation and that this subsequently permitted it to coexist with the parental species without interbreeding. Here my studies of island faunas were particularly important because [they] enabled me to show that geographic speciation is a continuous process: populations on the most recently colonized islands are still identical with the source population while the longer an island population had been isolated, the more different it was, until finally after a sufficient time interval, complete species status had been reached. (Mayr 1993:135)

Mayr (1993:135) went on to note that perhaps his major contribution

> was that I solved the old conundrum of how one could reconcile the sharp demarcation of species in a local fauna and flora with the Darwinian concept of gradual evolution. I demonstrated that in the local situation species are indeed sharply separated by gaps, but that if one looks at a species taxon in its total geographical representation through its entire range, one finds that most species consist of a large aggregation of local populations. Some of these, particularly those isolated at the periphery of the species range, are actually incipient species, that is, in transition from the status of local population to that of an independent new species. This refuted the old claim of Darwin's opponents that the sharp delimitation of local species, emphasized by naturalists from Linnaeus on, was incompatible with Darwin's concept of gradual evolution. The puzzle is solved by expanding the non-dimensional species of the local naturalist to the geographically variable species in a multi-dimensional approach. Furthermore this showed that it was unnecessary, indeed incorrect, to postulate speciation by saltation because geographic speciation is a gradualistic process.

Mayr may indeed have believed that he solved a major conundrum in 1942, but apparently not all biologists were so persuaded. True, extant species are separated by sharp gaps—for one thing, individuals within separate species do not interbreed across species borders—and it is also true that across a species's range there are local populations. One of Mayr's major contributions at the time was his proposition that, contra an assumption of the geneticists, it is the small, isolated "founder" populations that exhibit a greater potential to speciate than do larger populations—a point that more recent experimental evidence suggests is not true universally (e.g., Barton and Charlesworth 1984; Lande 1980; Templeton 1980, 1981). Mayr (1942, 1970) originally referred to this variously as "geographic speciation" or "allopatric speciation" (e.g., Eldredge 1971) but later termed it *peripatric speciation* (Mayr 1982b, 1982c). How could what Mayr was proposing be of any benefit to a paleontologist, whose subject matter is long dead? Mayr (e.g., 1943:133; 1949:371) answered this question repeatedly by pointing out what to him was the obvious answer:

> I always remember an experience I had 40 years ago when I lived all alone with a primitive tribe of Papuans in the mountains of New Guinea. These superb woodsmen had 136 names for the 137 species of birds I distinguished (confusing only two nondescript species of warblers). That primitive Stone Age man recognizes the same entities of nature as western university-trained scientists refutes rather decisively the claim that species are nothing but a product of human imagination.... When you study the birds in your woods and gardens, do you ever find intermediates between blue tits and great tits, or between thrushes and blackbirds, or between jackdaws and rooks? Of course you do not. Every species of bird, mammal, or other higher animal is extraordinarily well defined at a given locality.... Species are the product of evolution and not of the human mind. (Mayr 1969b:313)

In other words, the biological-species concept must be an accurate reflection of reality because two independent and differently enculturated humans classified a group of organisms the same way. Thus, species had to comprise one or more populations of interbreeding organisms that are bounded by their reproductive isolation from populations comprising other species. Mayr's message was simple: Species are real; all one has to do, be the person a Western scientist or a New Guinea woodsman, is find them. This was fine, but Mayr failed to account for the fact that both individuals were occupying the same time–space position along the evolutionary continuum at the time they were making their observations (for detailed discussion of the fallacies with such reasoning, see Hull 1992). Where did the notion of time fit into Mayr's biological-species concept? It did not fit at all. This is where paleontologist George Gaylord Simpson waded in.

Paleontology and the Synthesis

Stephen Jay Gould and Niles Eldredge (1977:15) argue, as do Mayr (1980:37) and others (e.g., Cain 1992, 1993; Laporte 1991, 1994), that it was through the efforts of Simpson that paleontology became an integral part of the Modern Synthesis, and we would agree—up to a point. His first book, *Tempo and Mode in Evolution*, published in 1944, was written to demonstrate that what paleontologists were observing in the fossil record—that is, the macroevolutionary events visible there—was consistent with the microevolutionary findings of the geneticists; no special processes or causes had to be invoked to explain the fossil record. Simpson knew well the split between naturalists and geneticists, and he was keenly aware of the rift between biologists in general and paleontologists. As Laporte (1991:81) put it, Simpson knew that "[b]ones-in-stones were a world apart from flies-in-bottles." By the time *Tempo and Mode in Evolution* was published, Simpson had been arguing for nearly twenty years that the temporal dimension of biological phenomena was central to understanding those phenomena in an evolutionary sense (Cain 1992). He already had begun to escape the typological thinking of his mentors, especially vertebrate paleontologist Richard Swann Lull, his adviser at Yale, by adopting statistical techniques for assessing the phylogenetic significance of variation within sets of associated fossils (e.g., Simpson 1937a, 1937b, 1937c), and he had explored the problem of types or taxonomic units such as species (Simpson 1940). These early forays into epistemology and ontology were incorporated, in perfected form (Laporte 1983, 1994), into *Tempo and Mode in Evolution*.

In the introduction to that book, Simpson admitted that his attempted synthesis of genetics and paleontology

> may be particularly surprising and possibly hazardous. Not long ago, paleontologists felt that a geneticist was a person who shut himself in a room, pulled down the shades, watched small flies disporting themselves in milk bottles, and thought that he was studying nature.... On the other hand, the geneticists said that paleontology had no further contributions to make to biology, that its only point had been the completed demonstration of the truth of evolution, and that it was a subject too purely descriptive to merit the name "science." The paleontologist, they believed, is like a man who undertakes to study the principles of the internal combustion engine by standing on a street corner and watching the motor cars whiz by.... It is not surprising that workers in the two fields viewed each other with distrust and sometimes with the scorn of ignorance. (Simpson 1944:xv–xvi)

Simpson did what no other paleontologist before him had done: He started from the principles of neontological Darwinism, then moved to the principles of genetics and its modern emphasis on small, continuous change as opposed to the earlier emphasis on disjunctive change, and finally asked if the major features of the fossil record were consistent with the neoevolutionary version of Darwinism (Gould 1980b:158). In retrospect, Simpson's volume was a masterpiece of argument for the marriage between naturalism and genetics and is still viewed as a classic guide to the conduct of paleobiological research (e.g., Eldredge 1985; Laporte 1983). But did his work have the immense impact that some (e.g., Gould 1980a; Laporte 1983; Stanley 1979) claim? We think not. Keeping in mind (1) that paleontologists have always been in the minority in biology in terms of numbers; (2) the disciplinary politics of the early 1940s, which relegated paleontology to an earth science rather than a life science (Laporte 1991:98); (3) the general disinterest in time on the part of most biologists; and (4) a still-strong interest in genetics (Stanley 1981:107–108), how could the contribution of paleobiologists to the Synthesis have been anything but minimal? Space played a central role in the neontologists' views of species and speciation (e.g., Mayr 1942, 1949, 1959b), but time was given virtually no part to play. The weight of contrary views resulted in Simpson publishing a heavily revised version of *Tempo and Mode in Evolution* in 1953 under the title *The Major Features of Evolution*. Absent from this revised edition were many of his innovative insights of the 1930s and 1940s—insights that have gained increasing attention in recent years (Eldredge 1985; Laporte 1983; Stanley 1979, 1981). So what of the species concept for the early Simpson?

Mayr (1982b:1120) complained that "Simpson was not alone in his neglect of the problem of speciation. Most paleontologists, during the 100 years after the publication of the *Origin*, entirely ignored the problem of the origin of organismic diversity." Such a response was predictable, given that Mayr's fieldwork involved the study of living organisms. It was Mayr who "developed the concepts of allopatric speciation and the founder effect," and it was Mayr who consistently emphasized "the fine-grained spatial relationships of populations within narrowly resolved time" (Laporte 1994:157). For their part, paleontologists might have read some of what the population geneticists had to say, but in our opinion, few of them—Simpson was an exception—understood the genetical literature at the level that, say, the neontologists did. Paleontologists also tended to ignore what the neontologists had to say. Simpson (1951:287) recognized this when he lamented that "on the whole, paleontologists have been rather slow to grasp or accept the population concept of taxonomic groups"—hence, our earlier comment that Simpson's work had only limited impact. But historians suggest this was only one of several problems (Eldredge 1985; Laporte 1983; Stanley 1981).

Joel Cracraft (1979:23), for example, echoing Mayr, claimed, "Paleontologists abstracted and used those aspects of genetics and systematics that best fitted their preconceptions of the evolutionary process and, in fact, did not incorporate into their writings the most important theoretical construct the 'New Synthesis' had to offer—namely, the notion that species originate via allopatrically isolated populations after the instigation of some form of geographic isolation." This, of course, ignores the fact that the paleontologists were working with fossils, so their materials "were intrinsically coarse-grained, both in space and, especially, in time. [Thus] it would have been very unlikely for Simpson-as-paleontologist to have formulated, say, the concept of allopatric speciation" (Laporte 1994:157–158). In other words, paleontologists could not easily adopt the biological-species concept (e.g., Sylvester-Bradley 1956) because, as Simpson made abundantly clear,

> a species as a subdivision of a temporal, or vertical, succession is a quite different thing from a species as a spatial, or horizontal, unit and it cannot be defined in the same way. The difference is so great and, to a thoughtful paleozoologist, so obvious that it is proper to doubt whether such subdivisions should be called species and whether vertical classification should not proceed on an entirely different plan from the basically and historically horizontal Linnaean system. (Simpson 1943:171)

This passage is key to understanding Simpson's view of the world. He did not deny that species were real; rather, he did not see the value in spending much time worrying about the problem. To him, Mayr's (1949:371) "non-dimensional" biological species—what Simpson termed a "spatial, or horizontal, unit"—was one thing; Simpson's own paleontological species—a "subdivision of a temporal, or vertical, succession"—was another. From the context of the term "non-dimensional," we suspect Mayr (1949) was indicating that an interbreeding population of organisms comprising a biological species has a location in time and space rather than a distribution and thus can be reproductively isolated from geographically or temporally distinct populations. Mayr (1949:372) had little difficulty integrating his and Simpson's views, noting that the "fact that species are the products of evolution and continue to evolve seems only rarely to become a source of practical difficulties for paleontologists. In the majority of cases the working paleontologist is dealing with a non-dimensional situation." In other words, a set of several fossil samples sufficiently temporally removed from one another would comprise a set of species by their apparent morphological discontinuity.

Simpson (1951:286) later labeled as "arbitrary" his procedure for breaking the vertical succession into species (see also Simpson 1943:172) and characterized it as comprising situations "when organisms are placed in separate groups *although the information about them indicates essential continuity*" (emphasis added). Criteria used in drawing arbitrary boundaries included stratigraphic and morphological discontinuities (Simpson 1943, 1951). Resulting groups were not "artificial or unreal: they are natural groups approximating populations that once existed in nature" (Simpson 1943:176). But how can "natural" groups be obtained if the whole process of slicing a vertical succession is arbitrary? Simpson would have answered that question by pointing out that regardless of how one breaks the continuum up, the resulting chunks will contain populations of organisms. Maybe they will not contain *all* the organisms that made up a population—in fact, they rarely will—but they will contain enough organisms so that a population is approximated. To Simpson, this population was a species. When one went high enough in a column of sediment that the organisms looked different enough from the ones below in terms of morphological features, truncate the first species and start a new one. In some cases, stratigraphic disruptions might provide a convenient indication of where to make the species break. Better yet, morphological change might correlate positively with stratigraphic boundaries, thus providing a double reason for stopping one species and beginning another.

Cracraft summed up the situation between the paleontologists and neontologists, and although his summary is over two decades old, without changing the tenor of his remarks, he could be writing today. We quote him at some length because of the clarity of his assessment.

> An examination of the theoretical postulates within paleontology and neontology over the past 50 years suggests that such a synthesis was philosophically and biologically difficult, if not impossible. The confusion and argumentation, I believe, support the view that the synthesis was more propaganda than a

> marriage pregnant with harmony. The primary reason for the lack of unification in paleontological and neontological evolutionary theory is the philosophical and biological debate over the reality of species as discrete evolutionary units. Paleontologists have maintained historically (1) that the reality of neontological species is illusionary; (2) that when projected back into evolutionary time, distinctions between "descendent species" (whether modern or not) and their "ancestral species" cannot be made; and therefore, (3) that species are only arbitrarily designated segments of an evolutionary continuum.... Paleontologists almost universally have adopted a nominalist philosophy. To accept gradualism *and* the reality of discrete species would be positively schizophrenic.
>
> Most neontologists, on the other hand, have viewed species as "real" biological entities, certainly discrete in space but ambiguously discrete in time. With the time dimension ostensibly impenetrable to analysis, neontologists have said little about the unity of species in time, although they have been influenced by their paleontological colleagues. (Cracraft 1979:23; emphasis in original)

The "paleontological colleagues" to whom Cracraft was referring included Eldredge and Gould, who in a series of papers throughout the 1970s (e.g., Eldredge 1971, 1972; Eldredge and Gould 1972, 1974; Gould and Eldredge 1977; see also Eldredge and Gould 1997; Gould and Eldredge 1988a, 1993) proposed to replace the phyletic gradualism of the paleontologists with a punctuated-equilibrium model that centered on the process of allopatric speciation. We take up the topic of punctuated equilibrium in more detail in Chapter 4; it suffices here to note that Eldredge and Gould posited that the fossil record was not as "gappy" as it appeared. Rather, paleontologists were viewing long periods of stasis followed by rapid bursts of speciation. Critics assailed Eldredge and Gould for invoking shades of Richard Goldschmidt's "hopeful monsters"—the strange mutations that represented the "decisive step in evolution" (Goldschmidt 1940:183). Gould (1977:30) himself said as much—"Macroevolution proceeds by the rare success of these hopeful monsters, not by continuous small changes within populations"—but his unfortunate reference was more metaphorical than it was a throwback to Goldschmidt's notion that it was a single, large-scale mutation that gave rise to a new species. Clearly, Gould and Eldredge have consistently adhered to the species-as-populations concept, though they have been inconsistent in their proposal of specific mechanisms responsible for stasis and rapid evolution (see Coyne and Charlesworth 1997).

Returning to Cracraft's point, it is indeed difficult to see how paleontologists and neontologists could ever hope to effect a synthesis as long as there were deep epistemological differences over the species concept. It is evident, however, that by the early 1940s, Simpson was intent on doing just that, and his book *Tempo and Mode in Evolution* (Simpson 1944) was a step toward that goal (for a cogent review of the book and its history, see Laporte 1983). Simpson argued successfully that the macroevolutionary phenomena familiar to neontologists and paleontologists were not at odds with the findings of geneticists. In effect, his argument preserved once and for all the role of natural selection in effecting evolutionary change. Simpson might have been interested in demonstrating the fit between genetics and paleontology, but this, he later claimed (Simpson 1984), was not his major goal. In fact, he bristled at the thought that anyone who read the book carefully could miss his major point, which was "to explore and in a way to exploit the fact that paleontology is the only four-dimensional biological science: time, 'tempo,' is inherent in it. Thus the aim of this book, which I think it accomplished, was to bring this dimension squarely, methodologically, into the study of evolutionary theory" (Simpson 1984:xxii).

Over the next two decades Simpson published several more books that dealt with macroevolutionary theory, including *Major Features of Evolution* in 1953, and *Principles of Animal Taxonomy* in 1961. In these volumes, as he had in *Tempo and Mode in Evolution*, Simpson continually pointed out the need for careful consideration of time relative to the species concept, in the process moving farther away from Mayr's biological-species concept, which he accepted but for which he could find no paleontological application (Simpson 1951). Being a paleontologist, he had no choice but *not* to adopt Mayr's definition. What good was a definition that was based on criteria one could not see in the empirical record? As Weller (1949:681) noted, genetic-based definitions such as the biological-species concept were "beyond the reach of paleontology."

Moreover, Simpson saw no reason to jettison the notion of slow, gradual change in evolution—a gradual change that eventually produced new species. Granted, the tempo of evolution could shift—he acknowledged as much in the title of his 1944 book, later even noting that Eldredge and Gould's punctuated equilibrium was based on his earlier notions (Simpson 1984:xxv)—but over vast geological time, it was the slow, gradual tempo that had characterized evolution. Mayr (1993:136) makes a similar claim with respect to his own earlier ideas (Mayr 1954), which later became known as peripatric speciation (Mayr 1982b, 1982c). Finally, despite the focus of the biological-species concept on interbreeding and the impermeable boundaries of the included gene pool, formal attributes (e.g., morphology) tended to provide the definitive criteria of most recognized and named species anyway, whether fossil or extant (Simpson 1943:164–165, 1951; see also Sylvester-Bradley 1956; Wright 1950)—something even Mayr (1943: 134–135) admitted.

Speciation

Perhaps the most significant aspect of Simpson's work was his clear separation of speciation from what he termed *phyletic evolution*. Speciation, when it occurred, was a process of splitting—"a process of differentiation within populations and of the rise of genetic isolation between populations formally part of the same species. Subsequent divergence is not, strictly speaking, speciation but an aspect of phyletic evolution" (Simpson 1953:380). Simpson characterized speciation as either the result of genetic isolation between two populations, which is more or less in line with the position of the neontologists, or the result of the "origin of new characters and their [subsequent] distribution or differentiation among and within populations" (Simpson 1953:379), which is decidedly a non-neontological position. Importantly, Simpson differentiated between phyletic evolution and speciation, noting that "when an ancestral species splits into two or more descendent species by segregation of existing variation, phyletic evolution is not an essential part of the picture; and when a single population undergoes extensive, cumulative change, there is no splitting" (Simpson 1953:384). Cracraft (1979:20) interpreted Simpson's statement to mean that the "process of isolation does not involve significant morphological change—that only comes later with phyletic evolution within isolates." We would agree, but consistent with Simpson's emphasis on both tempo and mode, we would extend the interpretation to include the following: Phyletic evolution—the slow accumulation of change within a lineage—has little or nothing to do with the multiplication of species. This point will be explored in subsequent chapters.

In 1961, Simpson published a definition of a species concept that he felt was suitable for the delimitation of fossil species. He termed it the *evolutionary species* and defined it as "a lineage (an ancestral–descendent sequence of populations) evolving separately from others and with its own unitary role and tendencies" (Simpson 1961:133; see also Simpson 1951). Mayr (1993:133) later attacked this definition, pointing out that the "replacement of the clear-cut criterion of the biological species (reproductive isolation) by such undefined vague terms as 'evolutionary tendencies' and 'evolving separately' does not permit discrimination between good species and isolates ... [n]or does it even permit the delimitation of an 'evolutionary species' within a phyletic lineage." Maybe this is true, but Simpson was only reacting to his own frustration at not being able to see a biological species in the fossil record. His use of the term *lineage* underscores the fact that, to Simpson, evolution was the production of separate historical lineages, each comprising ancestors and descendants and each with its own "tendencies." It was the job of the paleontologist to track the waxing and waning of these lineages through the fossil record—literally, to track morphological variation through time and across space.

To Simpson, in remarkable anticipation of what later became known as cladistics, the lineage was central to evolutionary study (e.g., Simpson 1951). The tempo of change within a lineage might speed up at times and cause the fossil record to appear as if species were fixed when the rate was slow, but that was not a problem in the study of lineages as historical phenomena. Many apparent gaps created by rapid evolution in the continuum would be filled in as the fossil record became better known. For the moment, such increases in the rate of change provided convenient locations to draw boundaries between species, provided sufficient difference in morphology was apparent (e.g., Simpson 1943:170). Mayr (e.g., 1969a:35), not surprisingly, did not speak of rates—that required detailed consideration of time—but rather explained the invisibility of transitional specimens—subspecies—in terms of his model of geographic speciation: "Since much speciation occurs in peripheral isolates, the discovery of strata with [morphologically] intermediate populations (incipient species) is highly improbable and will occur only rarely."

To Simpson, species were little more than bookkeeping devices used by the analyst to make some sense out of an enormous pile of variation and to facilitate communication among investigators. To create species, one simply stacked specimens in a time column—the correct temporal order being derived generally via superpositioning—and then decided where to make the breaks between species. In other words, when the analyst decided that there was too much noise, or that the species was beginning to extend over a too-long temporal span, then perhaps it was time to terminate the species and start another one. Such a procedure was perfectly acceptable when one adhered to the following tenet: "An evolutionary species is defined as a separate lineage.... If you start at any point in the sequence and follow the line backward through time, there is no place where the definition ceases to apply" (Simpson 1961:165). These were the words of a person who obviously did not place too much stock in the temporal discreteness of species. As if that statement were not enough, Simpson also said, "If the fossil record were complete, you could start with man and run back to a protist still in the species *Homo sapiens*. Such classification is manifestly useless and somehow wrong in principle. Certainly the lineage must be chopped into segments for purposes of classification, and this must be done arbitrarily ... because there is no nonarbitrary way to subdivide a continuous line" (Simpson 1961:165). Keep this statement firmly in mind when later we discuss the concept of archaeological types. We find that Simpson had an ontological and epistemological partner in James A. Ford. And like Simpson, Ford was attacked for his arbitrariness in how he sliced time into analytical chunks.

Did Simpson, as is sometimes claimed, absolutely deny the reality of species? No, he firmly believed, as did Mayr (e.g., 1949), that species existed "nondimensionally"—that is, as populations of interbreeding organisms in one place and time (Simpson 1951:288–289). What he did reject was their *temporal* discreteness. Mayr (1949:372) acknowledged this as well and correctly argued that an expanded—that is, from nondimensional to dimensional—conception of species did not invalidate his biological-species concept. Furthermore, he, like Simpson, conceived of a sample of fossils in a stratum as being just as nondimensional—"sympatric-synchronous" in his words (Mayr 1949:371)—as a population of interbreeding organisms comprising a species and living today in Iowa. But

Mayr's focus on nondimensional species in his fieldwork caused him to ignore the role of time in his conception of species and to focus on their discontinuities in space (e.g., Mayr 1987). To Simpson, variation through time was continuous, and there was no theoretical warrant for subdividing that continuum into real units. This did not mean that species did not exist; again, it meant that since they could not be seen, why worry about it? Simpson was interested in creating units that were useful paleontologically—that is, units that allowed the construction and tracking of evolutionary lineages. The parallels between the conceptual differences of Simpson and Mayr, and the conceptual differences of culture historians and anthropologists are striking.

Explaining Lineages in *Tempo and Mode in Evolution*

Figure 2.1 illustrates Simpson's three major modes of evolution—speciation, phyletic evolution, and quantum evolution. Notice the continuous flow of time in each diagram, including that showing the quantum evolution mode. Simpson (1944:199) defined speciation as "the local differentiation of two or more groups within a more widespread population. On the smallest scale, this process involves group differences so minor and so fluctuating that they are given *no taxonomic designation and have no clear evolutionary significance*, although the event may prove that they are the beginning of changes that do become permanent and important" (emphasis added). The analyst makes the decision as to what is and is not "evolutionarily significant." Simpson recognized two patterns of speciation—one in which a single widespread population diversifies within various adaptive subzones (Figure 2.2a) and the other in which already diverse daughter populations relocate into subzones at various times (Figure 2.2b). In more modern language, we would refer to Simpson's speciation as *cladogenesis*.

Phyletic evolution, or in modern terms *anagenesis*, involved "the sustained, directional ... shift of the average characters of populations. It is not primarily the splitting up of a population, but the change of the population as a whole. Obviously it can give rise to new species just as well as the different mode here called speciation, but it is not the typical mode of speciation, and it is less clearly seen at that level than at higher levels" (Simpson 1944:202). Simpson proposed three patterns of phyletic evolution—a first in which a well-adapted population changed little over time (Figure 2.3a); a second in which a population became narrower (more specialized) in terms of its adaptation to a zone, eventually becoming extinct (Figure 2.3b); and a third in which a population slowly changed in response to a shifting adaptive zone (Figure 2.3c). Simpson (1944:203) was of the opinion that "[a]side from isolated discoveries that contribute less directly to the study of evolution, nine-tenths of the pertinent data of paleontology fall into patterns of the phyletic mode." In other words, 90 percent of what a paleontologist studies is slow, steady change in the morphological structure of organisms.

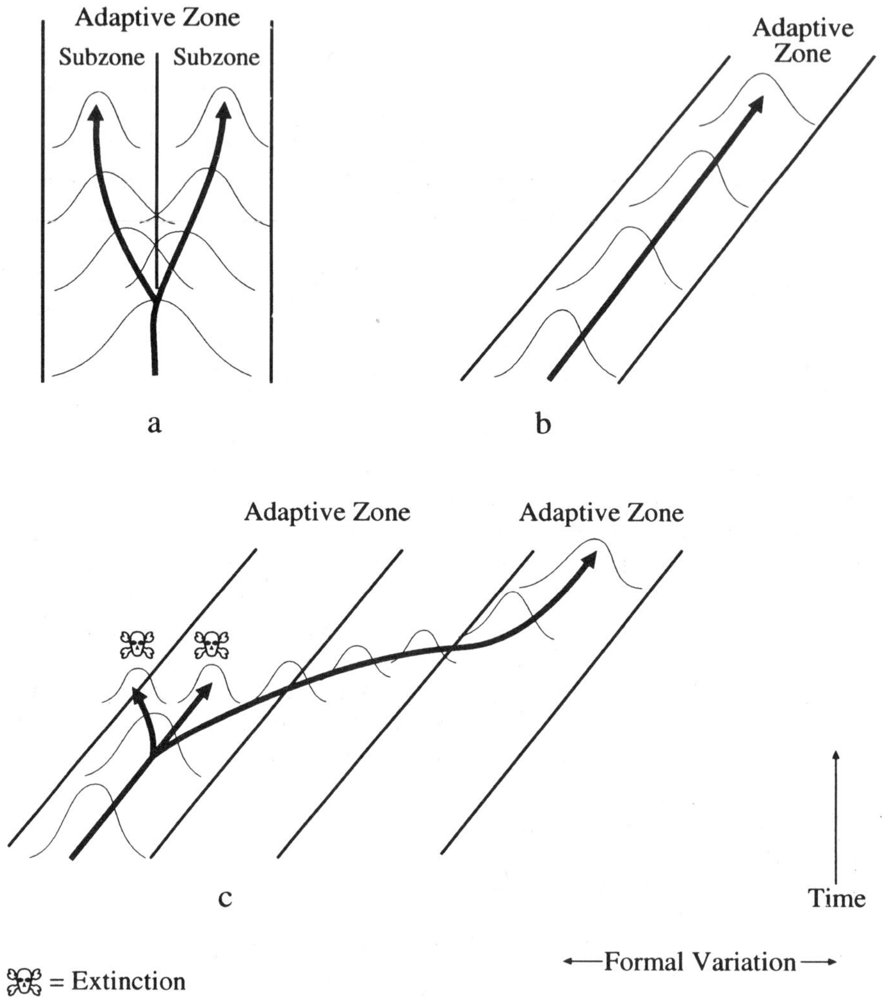

Figure 2.1. George Gaylord Simpson's three major modes of evolution: (a) speciation, (b) phyletic evolution, and (c) quantum evolution. Note that time is continuous in all three. In speciation, two or more groups are differentiated within a larger population. In phyletic evolution, or anagenesis, instead of a split there is a sustained, directional shift of the average characters of populations. In quantum evolution, there is a relatively rapid shift of a population in disequilibrium to an equilibrium distinctly unlike the ancestral condition. Simpson suspected that up to 90 percent of evolution was of the phyletic sort (after Simpson 1944).

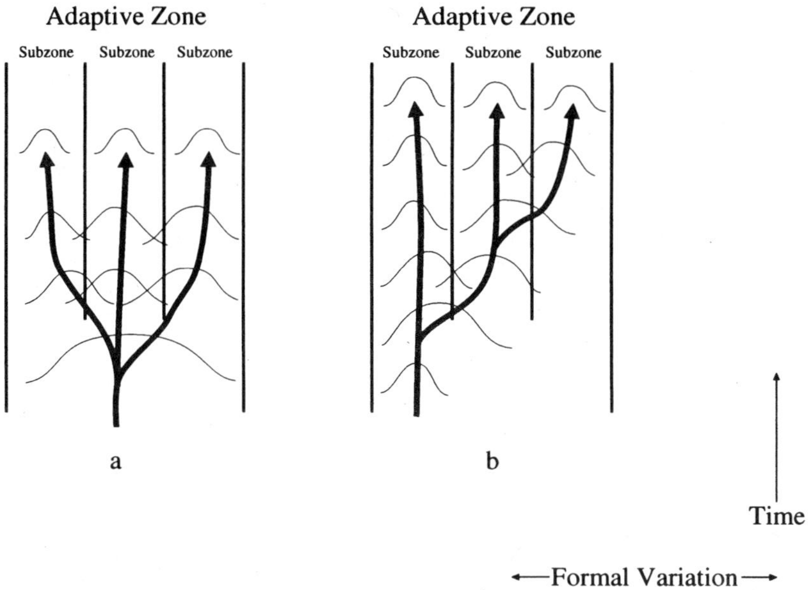

FIGURE 2.2. Simpson's two patterns of speciation: a, a single, widespread population diversifies within various adaptive subzones; b, already diverse daughter populations relocate into subzones at various times (after Simpson 1944).

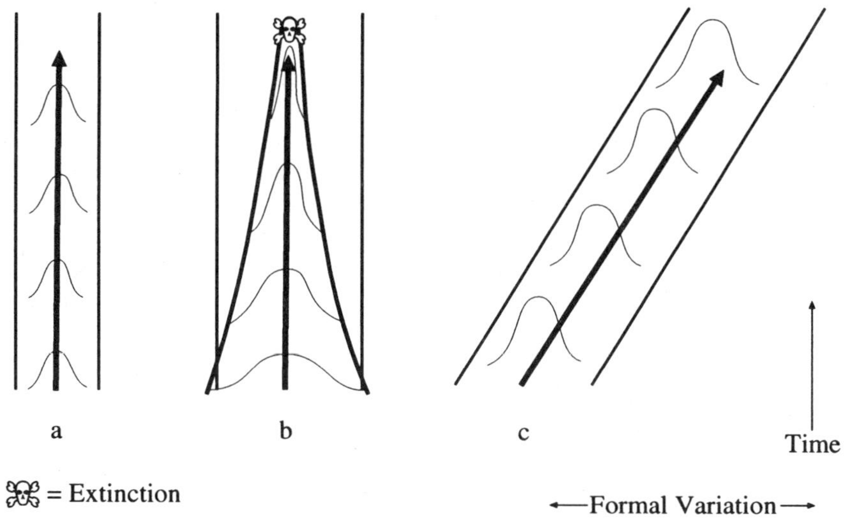

FIGURE 2.3. Simpson's three patterns of phyletic evolution: a, a well-adapted population changes little over time; b, a population becomes narrower (more specialized) in terms of its adaptation to a zone, finally becoming extinct; c, a population slowly changes in response to a shifting adaptive zone (after Simpson 1944).

Simpson acknowledged that (1) after a certain amount of time, so much change might have taken place that organisms at that point were so unlike their ancestors that they no longer could have interbred with them (not that the ancestors were around), and (2) at that point, the organisms were a separate species, but this was irrelevant to what he saw as the business of paleontologists. For example, he drew an analogy between the paleontologist's temporal transition of organisms—a chronocline—and the neontologist's spatial character gradient—a chorocline (Simpson 1943:173–176).

As important as phyletic gradualism was to Simpson, he devoted more coverage to what he, with apologies to physicists, termed *quantum evolution*, which he defined as "the relatively rapid shift of a biotic population in disequilibrium to an equilibrium distinctly unlike an ancestral condition. Such a sequence can occur on a relatively small scale in any sort of population and in any part of the complex evolutionary process" (Simpson 1944:206). Using the adaptive landscape metaphor of Sewall Wright (1932), Simpson (1944:207) identified three phases of quantum evolution: "(1) an inadaptive phase, in which the group in question loses the equilibrium of its ancestors or collaterals, (2) a preadaptive phase, in which there is great selection pressure and the group moves toward a new equilibrium, and (3) an adaptive phase, in which the new equilibrium is reached" (Simpson 1944:207). Quantum evolution clearly was a "larger scale" phenomenon than speciation and phyletic evolution and was "applicable in situations in which subthreshold actions produce no reactions but superthreshold actions produce reactions of definite (not necessarily equal) magnitude (this magnitude being strictly the quantum involved)" (Simpson 1944:206).

Despite identifying quantum evolution as a unique mode, Simpson (1944: 206) noted that it "may be involved in either speciation or phyletic evolution, and it has been mentioned that certain patterns within those modes intergrade with quantum evolution." But clearly he believed that quantum evolution was "the dominant and most essential process in the origin of taxonomic units of relatively high rank, such as families, orders, and classes. It is believed to include circumstances that explain the mystery that hovers over the origins of such major groups" (Simpson 1944:206). Simpson, in the sentence preceding the one just cited, pointed out that quantum evolution "can give rise to taxonomic groups of any size, and the sequences involved can be (subjectively) divided into morphological units of any desired scope, from subspecies up" (Simpson 1944:206). So in one sentence Simpson was saying that taxonomic units of any size, from subspecies up, were "subjective"—that is, artificial—units. Then, in the next sentence, he speculated that quantum evolution was "the dominant and most essential process in the origin of taxonomic units of relatively high rank, such as families, orders, and classes" (Simpson 1944:206). Note that he did not say something to the effect that quantum evolution was "the dominant and most essential process responsible for the variation that *taxonomists distribute hierarchically into tax-*

onomic units"; rather, he said that it was "the dominant and most essential process in the *origin of taxonomic units* of relatively high rank" (emphasis added). Here, Simpson was caught in one corner of the materialist paradox: Are taxonomic units real or not?

Was he speaking metaphorically of the reality of higher taxonomic units? We suspect he was, but he failed to make this clear. Look at how he explained the origin of a "major" taxonomic group:

> The origin of such a group involves the rise of a distinctly new adaptive type or, on the adaptive grid, the shift from one main zone to another. In these terms, quantum evolution is interzonal, in contrast with zonal phyletic and subzonal speciational evolution. The essential difference, however, is not so much the transfer from one zone to another, which may also occur in phyletic evolution, at least as regards movement between approximated zones, but a form of discontinuity. In phyletic evolution equilibrium of the organism–environment system is continuous, or nearly so, although the point of equilibrium may and usually does shift. In quantum evolution equilibrium is lost, and a new equilibrium is reached. There is an interval between the two equilibria, the biological analogue of a quantum, in which the system is unstable and cannot long persist without either falling back to its previous state (rarely or never accomplished in fact), becoming extinct (the usual outcome), or shifting the whole distance to the new equilibrium (quantum evolution, strictly speaking). (Simpson 1944:207)

In speaking of adaptive "types" Simpson appears to have been referring to essentialist units; his "transitional" type was, for example, midway, plus or minus, between two other zones. We say "appears" because he failed to make clear that his adaptive "types" were no more "real" than species were. If time was continuous, as was variation in tooth height, then for analytical purposes the adaptive landscape could be carved up into units, but those units were not "real." Similarly, his evolutionary "states" no doubt were not real in any essentialist sense; rather, they were merely relatively stable phenomena bounded by instances of relatively rapid evolutionary change. Simpson never actually claimed his states were real, but references to such things as "falling back to its previous state" are suggestive of such a stance. If time and variation are continuous, how could variable organisms ever revert to a previous state? Conceptually, this could occur only if variation is bundled into units (types), averages are taken, and names are given to the types. Then, if at three points in time a bundle of variation that is being tracked looks first like A, then like B, and finally like A again, then clearly the bundle has fallen back to its previous state. In fairness to Simpson, he pointed out that this rarely occurred, with the more likely outcome extinction, but the mere admittance of such a possibility demonstrates that he at least visited another corner of the materialist paradox. We can begin to discern why he seems to have occasionally fallen prey to the paradox by exploring how he discussed quantum evolution.

Simpson (1944:207) noted that with respect to quantum evolution, "the most disputable point is whether the transition is instantaneous, in closer analogy with the quantum of physics, or whether quantum evolution occurs at rapid but finite rates." He quickly noted, however, that "[s]trong reasons have been adduced for believing that the latter is the case." In accordance with this perspective, he chose as an example "a somewhat marginal case in which quantum evolution and phyletic evolution intergrade" (Simpson 1944:209)—the case of North American equid hypsodonty (the development of high-crown teeth), which he believed was caused by the emergence during the Miocene of a population of horses more adapted for grazing than for browsing; that is, a small population of leaf-eating browsers increased in overall size and developed higher-crown teeth than their ancestors had—exactly, as Eldredge (1995a:22) points out, "the requisite dental machinery for eating the newly evolved tough siliceous grasses" of the western United States. Simpson argued that genetic drift—a concept introduced by Wright (1931, 1932)—gradually through many generations shifted the population away from browsing and toward grazing, thus making the increase in tooth height adaptively valuable. In Simpson's scheme, this slow, steady increase was typical of phyletic evolution. Then, selection took over and rapidly altered the teeth of the horses. At that point, there was no alternative for the population with the higher teeth; the game had become in effect an "all-or-none, sink-or-swim evolution—and it had to move rapidly if it were to work at all" (Eldredge 1995a:23). These were the words Simpson (1944:209) used to explain the scenario:

> An effect of this change was that the browsing peak approached the as yet unoccupied grazing peak. There was considerable variation at all times, evidently with mutations both toward and away from hypsodonty. Selection pressure against hypsodonty above the optimum existed, but was relatively weak, because this condition was unnecessary rather than decidedly disadvantageous, such as hypsodonty much below the optimum. Some of the more hypsodont variants reached a point on the adaptive landscape at the base, or on the lowest slopes, of the grazing peak. It became possible for them to supplement their food supply by eating some grass, a relatively harsh food and highly abrasive to the teeth, although if they had eaten only grass, their teeth would not have outlasted the normal reproductive period, and they would have been opposed by very strong selective pressure.
>
> This point was a threshold. It initiated a sort of trigger effect that set off an evolutionary quantum reaction, although this differed from a major quantum effect and partook also of the aspects of phyletic evolution, because it occurred within a continuously large population and involved only a relatively simple change in a few organs rather than a transformation of the whole organism.

Simpson illustrated his quantum evolution mode as shown in Figure 2.4. Note that in the inadaptive phase, a portion of the equid population is just begin-

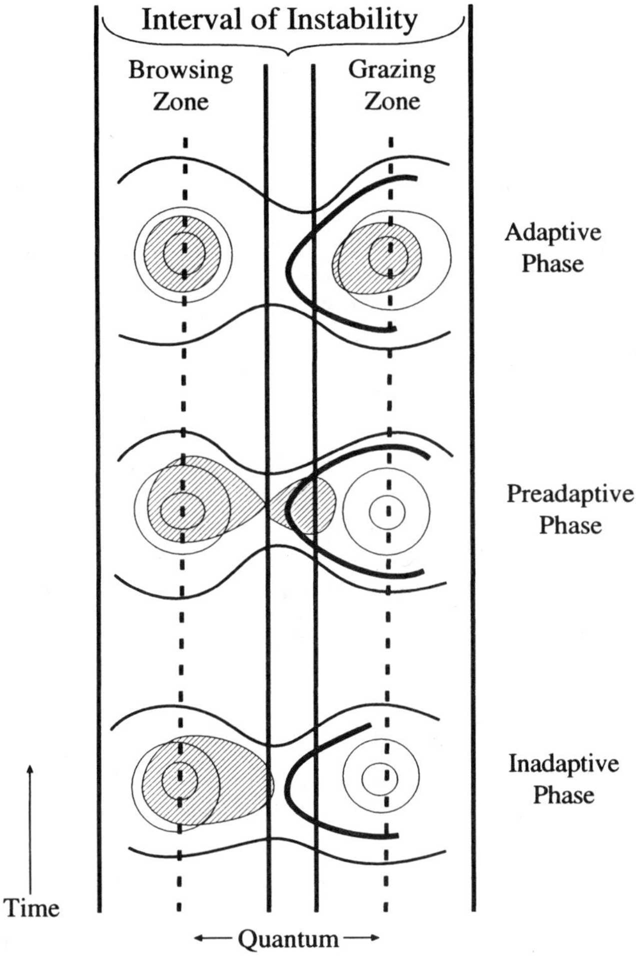

FIGURE 2.4. Simpson's three phases of quantum evolution: an inadaptive phase, in which a group loses the equilibrium of its ancestors or collaterals; a preadaptive phase, in which there is considerable selection pressure and the group moves toward a new equilibrium; and an adaptive phase, in which the new equilibrium is reached. The example Simpson used was the proposed development of a population of Tertiary-age horses in North America that was more adapted for grazing than it was for browsing. Fairly quickly, hence the term "quantum," that population split off from the ancestral population and became fixed in the grazing zone (after Simpson 1944).

ning to enter the "interval of instability," and in the preadaptive phase, it is entering the "grazing zone." By the adaptive phase, the split between populations is complete, with the descendant population firmly ensconced in the grazing zone and the ancestral population remaining in the "browsing zone." Had speciation occurred, with the descendant population being one species and the ancestral population another? Maybe, but again, to Simpson it really did not matter. What was important was that over time selection had moved some of the descendants of the original population to a new adaptive peak—a term used by Wright (1931) to refer to an economic state that a population resides in, similar to what modern biologists would refer to as a *niche*. Although Simpson did not deny that speciation had occurred, he did not consider it essential to the story line.

As a bit of an aside, and to bring Simpson's views up to date, we note that the idea of an adaptive landscape borrowed from Wright plays a major role in the modern notion of punctuated equilibrium (for a synopsis, see Lewin 1986). This is not all there is to the latter notion, but Simpson (1944) anticipated some of it well with his notion of quantum evolution. For example, the rapidity of quantum evolution results in a poor fossil record of its occurrence; the same holds for punctuated equilibrium (Gould and Eldredge 1977). Furthermore, Simpson believed the rapid invasion of new adaptive zones, such as is made possible after a major extinction event, can result in an "explosive evolution" during which cladogenesis occurs (Figure 2.5). Many of Simpson's (1944) ideas were criticized by reviewers, and because the neo–Darwinian synthesis "hardened" during the 1940s and 1950s (Gould 1983b), the features of quantum evolution were hardly mentioned again until resurrected in the 1970s (see Chapter 4).

Paleontology and the Materialist Paradox

Although Simpson saw himself as a neo-Darwinian, how could he in fact be one when he, according to Mayr (1982b:1120), was not interested in Darwin's major topic—the origin of species? In our view, Simpson *was* interested in such origins, as his discussion of quantum evolution, for example, makes clear. His focus on time, however, prompted him to deny the vertical discreteness of species, making him somewhat of a heretic. But was Simpson an essentialist in materialist clothing, or was he a materialist who became ensnared in the materialist paradox—that is, a materialist who could not always escape essentialist thinking? If we had to classify Simpson, we would come down decidedly on the side of materialist, because it is entirely misleading to think of what Simpson was doing in purely essentialist terms.

For one thing, Simpson, more so than any of his contemporaries, never lost sight of the importance of time in the evolutionary equation. Second, he constantly emphasized the continuity of the fossil record, despite the appearance of gaps in it. Some of those gaps, he guessed, were the results of sampling error or poor

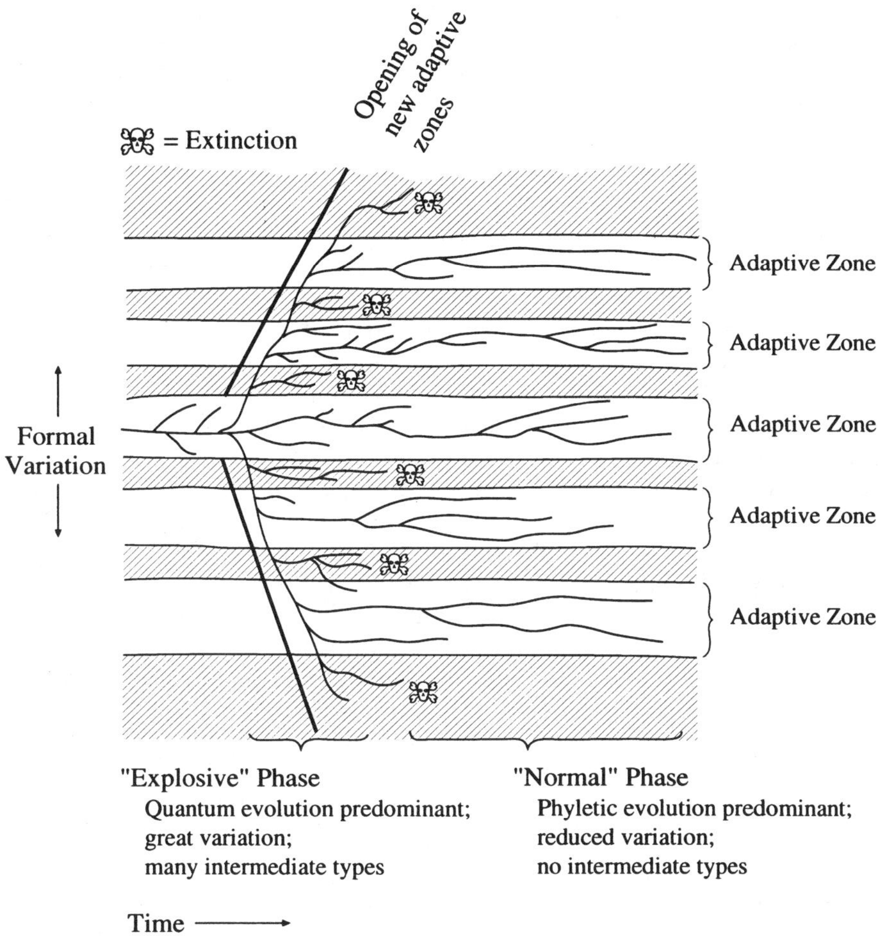

FIGURE 2.5. Simpson's depiction of the two phases of development of a population that rapidly invades a new adaptive zone. At first, there is an explosion of variation, with many intermediate types, then a period of phyletic evolution, with no intermediate types. Although he did not use the term, Simpson was identifying what became known as *cladogenesis*, or branching evolution. Also, though he did not pursue the issue, his "normal" phase obviously represents the occupation of adaptive zones by species formed during the "explosive" phase (after Simpson 1944).

preservation, but he also believed that some of the apparent "gaps" could be explained in terms of quantum evolution—more or less the same point made by Eldredge and Gould in introducing their notion of punctuated equilibrium. He also emphasized the presence of "transitional" forms in the record, as he did "transitional" stages that punctuated (our term, not his) long periods of stasis. As we see later, there is nothing inherently anti-Darwinian about transitional specimens in an empirical sense; it is when they become real in an ideational sense—something that is impossible under a materialist perspective—that we have a logical error. It is a logical error because, under materialism, there is no such thing as transitional specimens, as materialism holds that there are no natural kinds. With a transitional specimen, we have to ask the question, "Transitional between what and what?" If there are no "whats," then there can be no transitional specimens. Or as Bob Leonard once pointed out to us, under a materialist perspective, we could treat *all* specimens as transitional. He is correct: Either treat none of them as transitional or treat all of them as transitional. There can be no alternative ideationally.

At the risk of sounding redundant, we need to repeat a point we just made regarding the utility of the concept of "transitional" specimens. Leonard's comments apply only to ideational units and not to empirical units. Thus, his comments in no way suggest that the term *transitional* has no meaning in an evolutionary sense, because it does. Its utility is as a shorthand device to call attention to specimens that fall morphologically between two essentialist categories and that give us some clue that there is perhaps heritable continuity between the two categories. Given that almost the entire archaeological record and much of the biological record is known through reference to essentialist categories—types and species, for example—it seems ill-advised at this point not to make use of the concept of transitional specimens as an aid for constructing what we might think are hereditarily based lineages. Thus, in later chapters we use the term *transitional*, but we employ it solely as an exploratory device.

Why Simpson failed to avoid the materialist paradox is easy to discern. He was defining his evolutionary species empirically, using statistics to find significant differences between stratigraphically bounded samples of fossils. Criteria used to distinguish what were called species within a lineage, while admittedly an accident of sampling, delineated a sample of "sympatric–synchronous" (Mayr 1949) organisms that "constituted the species at a given time" (Simpson 1937a: 21). In other words, species were real in this nondimensional sense. Here, we think, is where many misunderstood Simpson's overall materialist epistemology. The misunderstanding was reinforced by Simpson himself, when he argued that quantum evolution produced taxonomic units of various ranks, when he spoke of adaptive zones and adaptive types, and when he discussed transitional types. These terms were all used metaphorically by Simpson, and while he failed to make this usage explicit, it is clear from his belief that a particular species existed before, during, and after speciation (Simpson 1951:294)—a cladistic viewpoint—that he was decidedly not an essentialist.

Simpson regularly grappled, at least conceptually, with speciation. This is abundantly clear in Figure 2.6 (Simpson 1951:295), an illustration that, to Simpson (1951:296), showed how easy it was to recognize speciation once the fossil record was sufficiently well known. The fossil record was, however, too often incomplete

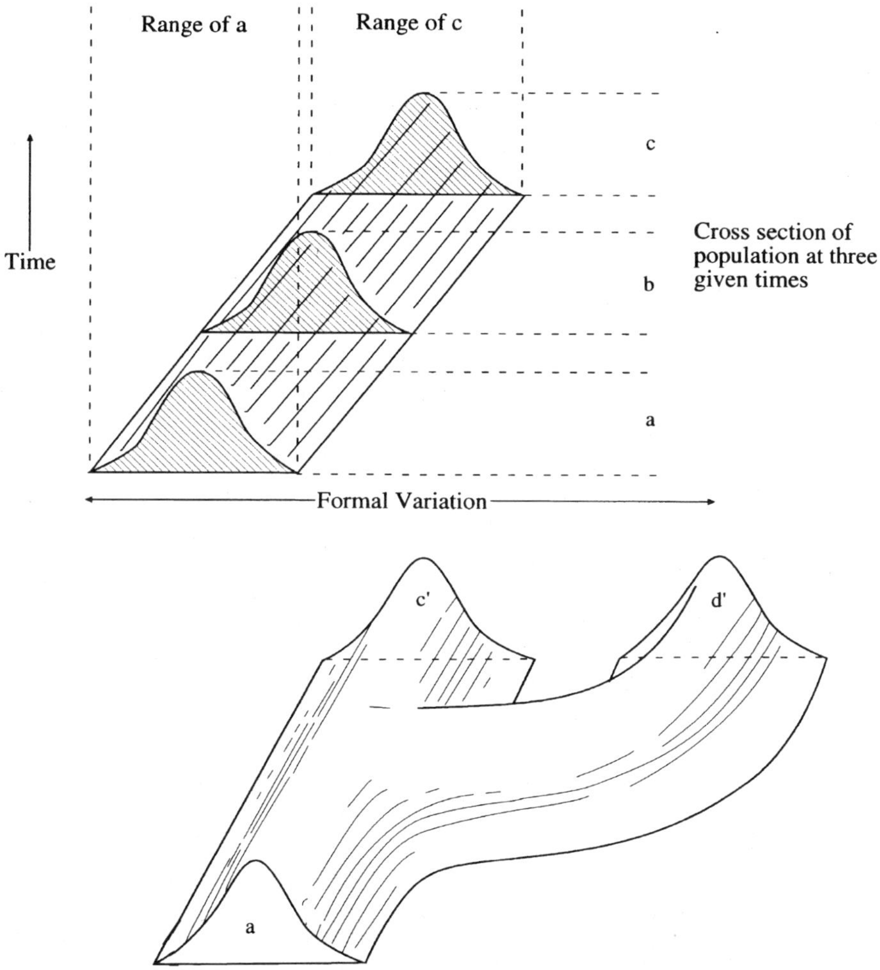

FIGURE 2.6. Simpson's views on speciation and species recognition. If the fossil record is fairly complete, Simpson argued, there is no problem in recognizing species, though in the example shown at the top, the overlap in formal variation exhibited by species b relative to both species a and c is problematic, resulting in species b being labeled as a "transitional" species. In the example shown at the bottom, an incomplete fossil record, or even a more complete one for that matter, would not allow us to see the true relation among the taxa (after Simpson 1951).

and/or of insufficient temporal resolution to allow such distinctions. Look closely at the upper diagram in Figure 2.6. Samples a and c are sufficiently distinct formally to be given separate species names. Add sample b (comprising a "transitional" type), and the distinction between a and c blurs. Now look at the lower diagram in Figure 2.6. If samples are found at a, c′, and d′, three species would probably be named. The branching, or speciation, event is where things get interesting. For one thing, this is another way to illustrate quantum evolution (see Figure 2.1), though Simpson (1951) did not label it as such. When a sample is found at the point of branching, whether the frequency distribution is unimodal or bimodal in terms of the attribute(s) measured is critical. Early in the branching episode, it will be unimodal; later, it will be bimodal. The problems thus reduce to having samples of sufficient size to detect the bimodality, and such samples must also be available all along the continuum so that speciation—the divergence, or branching—can be detected. This is a tall order given the usually coarse temporal resolution of the stratigraphic record studied by paleontologists and the difficulty in determining if a bimodal distribution is real in the sense of denoting cladogenic, or branching, speciation, or if it is a result of sampling error.

In the absence of ideal samples, Simpson and his contemporaries could seldom deal with speciation, and his species took on a disjointed appearance along a temporal continuum. This gave scenarios of evolutionary lineages a discontinuous but nonbranching appearance. The temporal duration of stratigraphically bounded samples of fossils was critical to interpretation, but in being ignored, it forced analysis of sets of fossils spanning various temporal durations. This contributed to the punctuated appearance of the lineage (Lyman and O'Brien n.d.a, n.d.b)—punctuations that appeared at the boundaries between strata. The reliance by Simpson and other paleontologists on geological phenomena to denote "sympatric–synchronous" sets of fossils forced them to adopt the evolutionarily untenable position that "paleontological taxonomy should be consistent with stratigraphic conclusions" (Bell 1950:493). In other words, species appeared to transform from one to another at boundaries between strata. This same problem has long plagued archaeology (Lyman and O'Brien 1999a, n.d.b; O'Brien and Lyman 1998, 1999a).

Eldredge (1979:8) squarely identified the problem when he noted that in "one guise or another, the transformational approach to evolution ignores species and hence is under no obligation to integrate speciation theory into the paradigm." Simpson would have responded—accurately, we believe—by stating that his view of evolution was hardly transformational. He also would have agreed that speciation *cannot* be ignored and would have countered Eldredge's charge with, "If I can't often see it, why bother with it?" What do you do with something about which there is considerable agreement that it is real, yet it cannot be seen? Surely, such things as geographic barriers and isolating mechanisms exist and, when acting in concert, can cause speciation and the production of new species. Simply because we cannot see something or measure it easily, are we to throw up our hands and either pretend it does not exist or ignore it? No, we just have to admit that there is no empirically obvious solution to the problem and either muddle along the best we can with what we have or attempt to resolve the issue. Simpson chose the former; paleontologists have yet to adopt a resolution suggested over thirty-five years ago by an archaeologist—a resolution we explore more fully later.

Chapter 3

The Materialist Paradox in Archaeology

> Exigencies of language require us to think and talk about pottery types as though they had some sort of independent existence. "This sherd *is* Baytown Plain." Upon sufficient repetition of this statement, the concept Baytown Plain takes on a massive solidity. The time comes when we are ready to fight for dear old Baytown. What we have to try to remember is that the statement really means something like this: "This sherd sufficiently resembles material which *for the time being* we have elected to call Baytown Plain." Frequent repetition of this and similar exorcisms we have found to be extremely salutary during the classificatory activities. (Phillips et al. 1951:66; emphasis in original)

Two statements by eminent Americanist archaeologists of the twentieth century lead us to suspect that almost from the beginning, the discipline has wanted to be, without using the term, a *materialist* science. Notice what Gordon R. Willey, writing in the early 1950s, had to say about the basic aims of archaeology: "The objectives of archeology ... are approached by the study and manipulation of three basic factors: form, space, and time" (Willey 1953:361). A few years later, Albert C. Spaulding (1960:439) repeated this dictum, using slightly different wording: "[A]rchaeology can be defined minimally as the study of the interrelationship of form, temporal locus, and spatial locus exhibited by artifacts. In other words, archaeologists are always concerned with these interrelationships, whatever broader interests they may have, and these interrelationships are the special business of archaeology." The objectives as stated by Willey and Spaulding are the same as those that paleobiologists have—the study of change in form through time and across space. Archaeologists use artifacts instead of bones, teeth, and shells, but the goal is the same. As we saw in Chapter 2, there are fundamental epistemological disagreements in biology over such things as what a species is and how and when speciation occurs, but clearly post-Synthesis paleobiologists and neontologists had at least one foot in the materialist camp. Conversely, archaeologists, despite having occasional materialist urges, have grounded their "special business" squarely in essentialism.

The terms *essentialism* and *materialism* were not yet common in the scientific vocabulary of the 1940s, but if they had been, George Gaylord Simpson probably would have been among the first to claim allegiance to the materialist position. Had the differences between the two positions been delineated, it would have been interesting to see where American archaeologists placed themselves. Our guess is that many would have called themselves materialists for some of the same reasons that Simpson would have: They had an interest in changes in form through time and across space. Paleobiologists were interested in functional change through time—changes in phenotypic characteristics for which functions had been worked out. Archaeologists, too, were interested in functional change, but it was an interest that was conflated with their interest in form. This conflation of form and function was a trap that paleobiologists saw early on, and for the most part, they avoided it. They also knew well the difference between analogous and homologous structures (e.g., Boyden 1943, 1947; Scharrer 1946), and they devised procedures for identifying each—something with which few archaeologists were prepared to deal, though several of them at least were aware of the problem (e.g., Kroeber 1931a).

In retrospect, one might argue that paleobiologists working in the 1940s after the Modern Synthesis had a much easier time of it than did their archaeological colleagues. For one thing, paleobiologists were dealing with a kind of evolution that, although it produced change, usually did so gradually, or so it was initially thought. To be sure, the evolutionary tempo could speed up, but the overall tempo was nothing compared to that with which archaeologists were dealing. Looking into the past, archaeologists saw whole cultures evolving, and they saw them evolving rapidly (e.g., Gladwin 1936)—at least the archaeological record made it appear as if rapid evolution had occurred. This changed the rules of the game considerably. For example, the careful, deliberate manner in which paleobiologists approached the question of homologous versus analogous features might have worked in biology, but how could it work in archaeology, where the evolutionary tempo was speeded up by several orders of magnitude? Homologous and analogous features—the former the result of relatedness and the latter of convergence—were present in the record, but how did one go about sorting out one from the other? And did cultural evolution not produce its own unique form of analogous and homologous features—ones that were distinctly different than those produced by biological evolution? Given this kind of reasoning, plus the overwhelming difference in rate of change in the archaeological record as compared to the fossil record, it is no wonder that archaeologists did little more than use evolutionary metaphors. Why take the time to study epistemological issues or to learn complicated analytical procedures when they did not appear to be applicable to cultural phenomena? This was a serious mistake on the part of archaeologists, but it is an understandable one.

The inattention that archaeologists paid to what was going on in biology during the 1930s and 1940s led them to rely on the same evolutionism—cultural evolutionism—that had been around in anthropology since the last half of the nineteenth century. This was not an unnatural occurrence given that Americanist archaeologists were trained in departments of anthropology and thus had considerable exposure to the writings of Herbert Spencer, Lewis Henry Morgan, and Edward B. Tylor—either exposure in a positive way or exposure through vilification—but it eventually plunged archaeology headlong into the materialist paradox. It was constant epistemological and methodological confusion over evolution that was behind the unique development of Americanist archaeology during the twentieth century. We cannot say that this confusion *caused* Americanist archaeology to unfold as it did, because at any point various paths were open, but it indeed limited the number of potential paths that the discipline could follow. The path that was selected eventually led to a dead end, at which point archaeologists cast about for a way out. As we document in Chapter 4, in 1962, a new door—processualism—appeared and offered what archaeologists saw as an escape route. Because of processualism's appeal, it would be several decades before some archaeologists began to suspect that they were yet again at a dead end—one that was much narrower than the one into which culture history had led them. As it was in earlier decades, evolutionism was a step along this new path, though it was of an entirely different kind than that offered by Darwinism (Lewontin 1974b).

Our objective here is not to present a detailed summary of evolutionism in Americanist archaeology (for extended discussion, see Lyman et al. 1997b; Lyman and O'Brien 1997) but rather to focus on a few select aspects of the history and development of archaeology in the twentieth century, as archaeologists developed methods of accessing the past. In the first part of this chapter, we pay particular attention to how archaeologists have approached the issue of time. Highlighting the fits and starts that the discipline has gone through in carving up time is an excellent vehicle for viewing the major problems it encountered when trying to do materialistic science using an essentialist outlook. Biologists might have an epistemological problem with the species concept, but even so great a problem pales in comparison to the predicament Americanist archaeology got itself into in its quest for a scientific study of formal variation and how variants waxed and waned over time. This predicament came into sharp focus during the culture-historical period in Americanist archaeology.

Since our primary emphasis is on time, we necessarily bypass several important approaches that figured prominently at one point or another as archaeologists tried to make sense out of the material record. Many of the approaches used to address questions about culture change were formulated early in the twentieth century, and by the end of the second decade had begun to coalesce into a set of methods that, after they were formalized and axiomatized in later years, constituted the first formal paradigm for Americanist archaeology—a paradigm commonly termed *culture history* (Strong 1952; Trigger 1989; Willey and Sabloff 1993). That paradigm fell from favor in the early 1960s, yet many of its central tenets remain in newer approaches and continue to be fundamental within the discipline.

Culture historians on several occasions came close to developing an evolutionary archaeology in a Darwinian sense—meaning that their thinking approximated a materialist metaphysic. References to biological evolution, or at least to evolutionary metaphors, as explanation for how and why cultures change were fairly common in anthropology and archaeology during the early twentieth century (e.g., Kidder 1915; Kroeber 1931a). Furthermore, some of the founding members of the culture-history paradigm in Americanist archaeology—a few of whom originally were trained as biologists—attempted to incorporate various notions of biological evolution into their archaeological research (e.g., Colton 1939; Gladwin and Gladwin 1934). Finally, as we noted in Chapter 1, at least one founding member, A. V. Kidder (1932:8), explicitly noted that the "ultimate goal" of culture history was to solve "the problems of cultural evolution."

Twentieth-century Americanist archaeology grew up around an ad hoc consensus concerning evolution, but in the end, it produced empirical generalizations that, because they were formulated in the absence of theory, were incapable of serving as explanations. The essentialist stance adopted by archaeologists produced a commonsensical understanding of the archaeological record in which interpretations were expressed in simple English words that in some cases had agreed-upon meanings and in other cases did not. There was no attempt, such as occurred in biology (and paleontology through the efforts of Simpson) in the 1930s and 1940s, to develop a uniquely archaeological theory of change. Although the more programmatic culture-history literature extolled the virtues of an evolutionary approach, archaeologists did not really use evolutionism in any sort of Darwinian sense, instead relying on cultural evolutionism to understand change over time.

By the late 1940s, culture history was in its heyday, but already more than a few chinks were starting to appear in its armor. The pattern-oriented reconstructions of static cultures produced by culture historians (e.g., Willey and Woodbury

1942) were coming under increasing fire from archaeologists and anthropologists, who saw the ultimate goal of archaeology as being not the study of time but rather, as Robert Braidwood (1959:79) put it, the study of the "Indian behind the artifact." We believe this had been the ultimate goal of culture history all along but that it had been put on hold while the temporal dimension was made analytically visible (O'Brien and Lyman 1999a). Once time was sufficiently under control in particular areas, efforts then turned to interpreting chronological sequences. Such efforts were doomed because the artifact units in use were inappropriate and attempts were made to interpret sequences of cultures rather than to explain why the archaeological record looked the way it did. Interpretation and explanation are both sense-making systems, but the former plays almost no role in science, despite the fact that the two are often conflated (e.g., Watson 1986; Wylie 1995; see O'Brien 1996a).

Failure by culture historians to deal with the construction of cultural units in rigorous fashion, in conjunction with a strong conviction toward a non-Darwinian view of cultural evolution, presented an opportunity ripe for exploitation. The processual archaeology of the 1960s represented a shift away from explaining homologous similarities in the archaeological record to focusing on and explaining analogous similarities between living cultures and those of the past (Binford 1962, 1965). There was a shift from ideographic history as explanation to a search for nomothetic laws as the source of explanation, though the focus was still on cultural units rather than on the archaeological record. As a result, some of the commonsensical, ethnologically based speculations about units and processes used by culture historians to rationalize their largely empirical, trial-and-error procedures assumed a dominant role in the paradigm that succeeded culture history. Ideas about evolution that had been summarily recycled for decades by culture historians now made sense. But the basic tenets of processualism were weakened for the same reason that those underpinning culture history collapsed: They were not grounded in a theory that actually explained variation in terms external to the variation itself. Unlike culture history, however, processualists often ignored time in their efforts to make the past accessible. But this is getting ahead of the story. To put later events in perspective, we need to go back and examine the roots of the problem.

ON THE EVE OF CULTURE HISTORY

We place the birth of culture history in the middle of the second decade of the twentieth century—an arbitrary decision based more on an interrelated series of events than on the appearance of any single product. If we try to pinpoint the event any more closely, we end up obscuring the fact that what happened around 1915 in the American Southwest was simply the end result of a rather long gestation period, something on the order of at least three, and maybe four, decades. To understand why culture history appeared when it did and had the look it did at the time, we need to understand something about the intellectual climate from which the paradigm emerged.

Human Antiquity in North America

Several contentious issues confronted prehistorians at the end of the nineteenth century, but none was more rancorous than the dispute regarding the initial date of human occupation of North America. By the beginning of the twentieth century, Bureau of American Ethnology (BAE) prehistorians and their allies, led by William Henry Holmes, had effectively undermined the position that humans had inhabited the continent during the last glacial period (Meltzer 1983, 1985, 1991). Holmes in particular demonstrated that there was no good reason to suspect that certain stone objects from the Northeast, which admittedly were crude in appearance, were temporal equivalents of Paleolithic[1] artifacts from Europe (Holmes 1892, 1897). Holmes's discounting of purported glacial-age tools, coupled with the fact that no indisputable associations between humans and extinct animals were known in North America, led to the complete rejection of great time depth to human tenure on the continent. Holmes and other opponents were wrong, but it would be several decades before convincing proof would be forthcoming (Figgins 1927).

As a result of the failure of glacial-age proponents to mount an effective case, the majority of prehistorians at the turn of the century were working under the assumption that the North American archaeological record was shallow relative to time. Prehistorians were unsure how short the chronology was, and for several decades, they did not worry too much about it. The large-scale survey of mounds in the eastern United States conducted by Cyrus Thomas for the BAE (Thomas 1891, 1894) had demonstrated that the earthworks were made by ancestors of ethnographically documented Native Americans and not by an extinct race of mound builders. This connection furthered the notion of a short chronology and led to the use of the ethnographic present as an analog of the past—a patently essentialist method that was later referred to as the direct historical approach (Steward 1942) and became the cornerstone of BAE archaeology (O'Brien and Lyman 1997). Prehistorians realized that culture change had occurred in North America, but the lack of an American Paleolithic, and hence the lack of the same

[1]Sir John Lubbock (Lord Avebury) in 1865 had distinguished between the Paleolithic, or Old Stone Age, and the Neolithic, or New Stone Age. The former included chipped-stone artifacts associated with extinct mammals, whereas the latter included more finely crafted—"polished," or ground—stone tools associated with modern mammals (Daniel 1981; Van Riper 1993). In Europe, the temporal difference was based on marked stratigraphic differences (Daniel 1975; Van Riper 1993) as well as associated fauna and differences in tool form.

periods of culture that had been discovered in Europe, meant that a uniquely Americanist set of archaeological methods and explanatory models would have to be developed if the study of culture change was to become a viable subject of inquiry. As Franz Boas (1902:1) noted at the beginning of the twentieth century, "In the study of American archaeology we are compelled to apply methods somewhat different from those used in the archaeology of the Old World."

Artifact Categorization

One might well ask why prehistorians, despite their traditional belief in the short chronology, did not begin to realize that the variation evident in artifact classes such as pottery and flaked-stone objects might have been the result of something other than tribal differences and the like. Although some prehistorians were willing to assign more time depth to the archaeological record than most of their colleagues, even they placed more emphasis on ethnic differences as having produced the majority of the variation. By the beginning of the twentieth century, the variation evident in the archaeological record was ever-increasing—a result of the decades of work that the federal government had sponsored in various areas of the United States, particularly the Southwest. As more artifacts and more variant forms of artifacts became known, it was increasingly difficult to keep track of where particular kinds of things came from and for individual investigators to communicate with one another about the particular forms in their respective collections. As Dunnell (1986) points out, the development of archaeological categorization in the second half of the nineteenth century marked the emergence of archaeology as a distinct area of inquiry. Categorization theoretically serves two functions: to structure observations so that they can be explained, and to provide a set of terminological conventions that allows communication; in Americanist archaeology, it more often has served the latter than the former.

Early classification schemes were based on readily perceived differences and similarities among specimens. Artifact form, because it can be observed directly, received the lion's share of attention, though in many instances, form and function were conflated. Differences between piles of artifacts, each of which contained similar specimens, often were explained in terms of ethnological differences. Any variation in artifact form within each pile of specimens had no perceived explanatory value and was viewed simply as noise that resulted either from different levels of skill in manufacturing or from raw-material quality. As U.S. National Museum anthropologist Charles Rau observed early on, "Classification of the arrowheads with regard to their chronological development is not attempted, and hardly deemed necessary. North American Indians of the same tribe (as, for instance, the Pai-Utes of Southern Utah) arm their arrows with stone points of different forms, the shape of the arrowhead being a matter of individual taste or convenience" (1876:159). Other prehistorians disparaged ethnic, or spatially based, classifications because particular forms were found in widely separated areas (e.g., Fowke 1896). The widespread occurrence of certain forms might have been viewed differently had their been some warrant for the categorization, but there was none, and prehistorians turned instead to common sense in search of attributes that could be used as a basis for sorting.

In the late nineteenth century, classification systems for pottery lagged behind those developed for stone artifacts because most classifications used shape, and ceramic specimens more often than not were represented by sherds rather than by intact vessels. One bright spot was William Henry Holmes, who worried about how to explain variation in vessel form. He "was imbued with anthropological evolutionism, particularly the cultural evolutionism of Lewis Henry Morgan [and], like many of his colleagues, employed the Morgan stages of humanity (i.e., savagery, barbarism, etc.) as if they were matters of fact" (Meltzer and Dunnell 1992:xxviii). Regardless of his affinity for Morgan's scheme, Holmes had little interest in time, and his knowledge of the archaeological record prompted him to doubt the possibility of documenting evolutionary progress prehistorically. He "frequently found it difficult to see the evident change [in artifact forms] as progressive. Consequently, he was forced to seek other (e.g., racial, environmental, and diffusionist) explanations for the differences" (Meltzer and Dunnell 1992: xxxvii).

The Age–Area Hypothesis

By the second decade of the twentieth century, a few prehistorians began asking if perhaps some of the spatial distribution of artifact forms could be linked to time, even if it was not deep time. Ethnologists had long attempted to derive methods for measuring short-term change from culture traits, one result of which was formulation of the age–area concept, which made three interrelated assumptions: Culture traits will disperse in all directions from the point of origin—like the "ripples" (Murdock 1948:294) emanating from a raindrop that hits a puddle; all culture traits will disperse at the same rate—again, like ripples; and the larger the geographic area over which a culture trait is found, the older the trait. Thus, mapping the geographic distribution of culture traits evidenced by archaeological remains should indicate something of their history. The age–area concept was supported in some quarters of anthropology and archaeology (e.g., Kroeber 1931b) through reference to an identical concept in biology (e.g., Willis 1922), though it was by no means completely accepted even within that discipline (Gleason 1923).

The Culture-Area Concept

Particular artifact forms might be found in widely separated areas of North America, but by the early years of the twentieth century, archaeologists came to appreciate that many other forms were more localized. Difference in geographic location of artifacts was obvious, but of what significance was the difference? Their answer was that the variation must be ethnic; furthermore, it might be tied directly to environmental differences. What came to be known as culture areas—bounded geographical units created on the basis of artifact content and viewed as real expressions of ethnic affiliation (e.g., Wissler 1914, 1916a, 1917a, 1923, 1924)—had their roots in the culture-classification work of Otis T. Mason (1896, 1905), in Cyrus Thomas's (1894) regional groupings of mound forms, and in Holmes's seminal papers (e.g., Holmes 1886a, 1886b, 1886c, 1903) on pottery classification.

Criticisms of the age–area concept by Julian Steward (1929) and others were severe. They pointed out that the concept failed to recognize independent invention, or what might be thought of as convergence—a process that everyone agreed took place, even if its relative importance might be disputed. Rather, the age–area concept identified similarities in different cultures as homologous structures, that is, as the results of trade, diffusion, and the like. These criticisms did not, however, dissuade Clark Wissler (1919) and others from developing the notion of culture areas, no doubt because of the loose parallels perceived between cultural evolution and biological evolution. A. L. Kroeber (1931a) later argued that culture traits were analogous to species of organisms and that cultures, which comprise suites of traits, were analogous to faunas and floras, which comprise suites of animal and plant species, respectively. For Kroeber (1931a:149), although a "culture complex is 'polyphyletic' [and] a genus is, almost by definition, monophyletic ... the analogy does at least refer to the fact that culture elements [traits] like species represent the smallest units of material which the historical anthropologist and biologist respectively have to deal with." Boas (1932:609) echoed Kroeber's statement a year later. Therefore, data on the geographic distribution of culture traits, as with data on plant and animal taxa, allow "inferences as to the origin and areal history of the group" because "the Age and Area principle seems the same in biology and cultural anthropology" (Kroeber 1931a:150). Kroeber's comment that species represent the smallest units of material with which biologists deal obviously was incorrect. What about genes or the organelles in cells, both of which were well-known biological phenomena when Kroeber made that comment? These are smaller units than species. And what about the individual organisms that not only make up a species but also are the units on which selection operates?

With respect to writing culture history, Kroeber recognized the obfuscating factor of independent invention—analogous similarity—and noted that anthropologists had too often not sufficiently tested their diffusionary explanations: "The fundamentally different evidential value of homologous and analogous similarities for determination of historical relationship, that is, genuine systematic or genetic relationship, has long been an axiom in biological science. The distinction has been much less clearly made in anthropology, and rarely explicitly, but holds with equal force" (Kroeber 1931a:151). Unfortunately, Kroeber, for want of a theory, never developed these insights into an analytical procedure for anthropology or archaeology. We describe in Chapter 6 what he did do.

Superposition and Stratigraphy

The reasoning that eventually led to the formal development of the culture-area concept and the age–area hypothesis was all well and good, but to some archaeologists, it sounded too speculative, and by the second decade of the twentieth century they were beginning to ask how the age of artifacts could be measured directly, without reference to such things as the then-incipient notion of culture areas. Taking their cue from geologists and paleontologists, these skeptics realized that the principle of superposition and stratigraphic observations provided such a means. How they came to this conclusion and the ramifications it had on the growth of culture history are of critical importance in understanding the nature of the materialist paradox. For that reason, we need to examine not only the stratigraphic method itself but also the assumptions that went along with its use. The twentieth-century lodging of essentialism in Americanist archaeology was a direct outgrowth of the way in which archaeologists approached the issue of superposition and stratigraphic excavation. For a few years, it appeared that archaeology might actually create a materialist science, but the effort failed and essentialism soon assumed its previous place of primacy (Lyman et al. 1997b). What in 1916 was a confirmational strategy for suspected chronological orderings quickly became a creational strategy for such orderings (Lyman and O'Brien 1999a, n.d.b). At that point, Americanist archaeology was firmly back in the essentialist camp. To set the stage for what happened in the fateful second decade of the twentieth century, we backtrack a bit and begin the story of superposition and stratigraphy in the late nineteenth century.

The earliest of what we would refer to as stratigraphic excavations—the collection of artifacts from vertically discrete units of sediment that are being excavated—in North America (Lyman and O'Brien 1999a) were undertaken not by prehistorians but by naturalists—for example, physician Jeffries Wyman, who examined shell middens in New England (Wyman 1868) and Florida (Wyman 1875), and William Healey Dall, a conchologist who excavated shell mounds in the Aleutian Islands (Dall 1877). Later, in 1901–1902, prehistorian Charles Peabody excavated Edwards Mound in Coahoma County, Mississippi (Peabody 1904), noting that stratigraphic observation indicated there were two periods of mound construction and use. A year later, Peabody and Warren K. Moorehead (1904) excavated Jacobs Cavern in McDonald County, Missouri, using what was at the time a sophisticated excavation strategy consisting of square units and vertical levels (O'Brien 1996a). Nels Nelson (1906) excavated stratigraphically a decade earlier than generally recognized.

Other prehistorians knew about these stratigraphic procedures, but few used them in their work. Why? Just after the turn of the century Boas (1902:1) wrote that "it seems probable that the remains found in most of the archaeological sites of America were left by a people similar in culture to the present Indians." This critically important statement indicates why Americanist archaeologists did what they did between about 1900 and 1915. The usual perception of historians of archaeology (e.g., Browman and Givens 1996; Strong 1952; Willey 1968; Willey and Sabloff 1993) is that there was a "stratigraphic revolution" around 1915, at which point archaeologists, who suddenly had become interested in keeping track of time, developed (or at least adopted) stratigraphic excavation techniques to do so. This is incorrect: Naturalists and prehistorians were excavating stratigraphically in the late nineteenth century, and they knew that the vertical proveniences of things such as artifacts and burials were related to time. They simply assumed, however, as did Boas, that the time span was so short that it did not matter. Variation in material culture did not seem to correlate with time.

Our view of history stems in part from differences of opinion over the term *stratigraphic excavation*. Browman and Givens (1996:81), for example, distinguish between "post facto stratigraphic observation" and "actual stratigraphic excavation," equating the former with the identification of "archaeological strata ... in the walls of trenches excavated as single [vertical] units" and the latter with the identification of "archaeological strata [that are] microstrata of geological units" and the employment of those strata as "data recovery units" (Browman and Givens 1996:80). A straightforward way to make strata visible is to expose them through excavation, but simply because one does *not* peel back one stratum at a time does not indicate an absence of stratigraphic excavation (Lyman and O'Brien 1999a; O'Brien and Lyman 1999a). The term *excavation* is irrelevant to the issue of stratigraphy. Paleontologists make stratigraphic observations all the time without excavating. They find an exposure, clean it up a bit, and collect the fossils sticking out of the face. Archaeologists often do the same thing, so attaching the term *excavation* to "stratigraphic" obscures the real issue: Were archaeologists using discrete and distinct depositional units as artifact-recovery units prior to the so-called "stratigraphic revolution?" The answer is unquestionably "yes" (Lyman and O'Brien 1999a).

But there is still something missing from this discussion: the measurement of time that results from analyzing a collection of superposed artifacts. After adding this criterion to any of the previous definitions of stratigraphic excavation, the relevant issue becomes not whether archaeologists excavated in or collected artifacts from discrete vertical units prior to about 1912, but whether they sought to measure time via superposition and stratigraphic interpretation. Working in 1914 Fred Sterns (1915:121) observed that "the proof of [cultural] sequences must be grounded on stratigraphic evidence, and stratified sites have been very rare [in North America]. Hence such a site has a high scarcity value and warrants special study even though it be otherwise of minor importance." One such site, a shell mound in Maine, was excavated by Frederic Loomis and D. B. Young (1912) in 1909: "[T]he heap was [plotted] in sections five feet wide, and as each section was worked, every find (of a tooth, tool, bit of pottery, etc.) was recorded, both as to its horizontal position and vertical depth.... [Vertical s]ections of the heap were plotted from time to time" (Loomis and Young 1912:19). They noted that in some strata, "the shells were very much broken up, apparently due to tramping and building fires on them. Where the shells were but little broken, and free from ashes, they would seem to indicate rapid accumulation, and offered but little in the line of finds.... Where the layers were made of ashes and finely broken shells, the period of accumulation was longer and agreeing with that, the numbers of articles found in these layers was also greater" (Loomis and Young 1912:19–20).

Clearly, Loomis and Young collected artifacts and faunal remains from particular depositional units, and they did it three years prior to Manuel Gamio's (1913) much ballyhooed work in the Valley of Mexico—just as Mark Harrington's (1909a, 1909b) work preceded Loomis and Young's by eight or nine years. Loomis and Young knew superposition measured time, but they were not asking questions that demanded chronological control and thus did not use stratigraphic excavation to address such questions, because their mind-set paralleled that of Sterns (1915:125), who observed that "an important fact arguing against any great difference in time between the upper and lower ash-beds is that the pottery and the flint and bone implements found in these two sets of fireplaces show absolutely no difference in type." This was simply a rewording of Boas's (1902) earlier statement that as a result of the shallow time depth then ascribed to the American archaeological record, there were no significant differences between historically documented cultures and prehistoric ones. Simply put, this mind-set was why no one consistently or rigorously asked chronological questions, despite the fact that many archaeologists collected artifacts from distinct depositional units. This would soon change, and the change can be traced directly back to the work of Friedrich Max Uhle in California.

Measuring Time with Form?

Uhle used stratigraphic principles to a minor extent during his extensive fieldwork in Peru (Uhle 1903; see Rowe 1954:20; Willey and Sabloff 1993:79). Rowe (1954:20) indicated that Uhle used stratified collections—that is, superposition—to *confirm* the chronology of pottery styles he had developed based on other criteria such as historic documentation. If what Rowe says is true, this makes Uhle one of the first archaeologists in the Americas to have done so. That strategy, as we point out later, did not become popular until around 1915. However, it is Uhle's work in the San Francisco Bay area, published in "The Emeryville Shellmound" (Uhle 1907), that is best known to archaeologists, because it was there that he excavated parts of the mound in stratigraphic fashion. Aided by paleontologist John C. Merriam (Woodbury 1960a), Uhle geared his excavations to "obtain a view of all the strata contained in the mound" (Uhle 1907:8). He believed the shell mound represented several centuries' worth of occupation and that "cultural differences should be indicated in the successive strata" (Uhle 1907:36). Although he excavated "stratum by stratum," Uhle noted that "the dividing lines [between] the various strata (I to VII) were chosen arbitrarily from the several visible lines of structure" (Uhle 1907:8). Those strata then were tied to lower strata (VIII to X) by excavating a separate trench that contained the missing strata. Uhle (1907:15) indicated that the "lines of stratification mark clearly the gradual development of the strata of the mound." He was able to note the stratigraphic provenience of burials, to examine variation in the artifact assemblages associated with each stratum, and to estimate the age of the deposits.

Uhle's work, which in terms of methodological sophistication was well ahead of its time, was summarily dismissed by Kroeber (1909:15), who stated that "an independent examination of the material on which [Uhle's opinion of distinct progression and development of civilization having taken place during the growth of the deposit] is reared, tends to [negate] rather than to confirm it." Kroeber (1909:15) observed that although "finely worked objects" occurred in shallow stratigraphic contexts and "rough stone fragments predominate in [the] low layers," most artifact types were found throughout the stratigraphic sequence. This indicated to Kroeber that while there was "some gradual elaboration and refinement of technical processes ... it was a change of degree only, and one in no way to be compared even for a moment with a transition as fundamental as that from palaeolithic to neolithic" (Kroeber 1909:16; see also Rowe 1975:158). In other words, human occupation of the West Coast of the United States had not paralleled developments in Europe; hence, the essentialist stages Paleolithic and Neolithic, which were the cornerstones of European archaeology, obviously were not applicable. Furthermore, since there was no epochal change on the order of that seen in Europe, time was inconsequential in the equation.

Over a decade later, Kroeber (1923b:140) lamented that archaeological evidence of cultural change and time depth in California was hindered by "the absence of pottery from the greater portion of the state"; the "little evidence of stratification" in the state's sites; and inattention on the part of archaeologists to "depth, position, or collocation" of artifacts. He noted that he saw no evidence of major cultural change in Nels Nelson's materials from the Ellis Landing shell midden on the edge of San Francisco Bay (Nelson 1910), and he reiterated his earlier assessment of Uhle's work at the Emeryville shell midden: The few artifacts available showed no "marked cultural changes"; change was in "quality and finish ... rather than [the introduction of] new types"; and "the principal types ... occur in all strata" (Kroeber 1923b:141). For Kroeber (1909:15), these observations indicated that the "same modes of life ... were followed in the periods represented by the earliest and the latest strata." These were the words of a pure essentialist interested in large-scale change. Or, at the least, Kroeber would have argued, let me see the introduction of a few new types of artifacts somewhere along the line.

In part because archaeologists have only skimmed what Kroeber wrote, there is a common misconception that he was uninterested in chronology. In truth, he knew full well the value of data relative to the stratigraphic context of artifacts, regarding such information as an "urgent need" if one was to acquire "information as to cultural and chronological relations" (Kroeber 1909:39, 41). Thus, he did not discount the value of Uhle's stratigraphic method; rather, he discounted the significance of the chronological indications of cultural change Uhle documented. For one thing, the few thousand years granted to human antiquity on the coast of California—and everywhere else in the Americas—rendered "any evidence of radical change" or "notable development" unlikely (Kroeber 1909:16). But most of the problem resided in Kroeber's conceptions of culture history and how it was being measured. Being Boas's student, it is not surprising that Kroeber elaborated on his mentor's (Boas 1902) earlier comment regarding the culture history of the Americas:

> The civilization revealed by [archaeology] is in its essentials the same as that found in the same region by the more recent explorer and settler.... [N]either archaeology nor ethnology has yet been able to discover either the presence or absence of any important cultural features in one period that are not respectively present or absent in the other.... [A]rchaeology at no point gives any evidence of significant changes in culture.... [D]ifferences between the past and present are only differences in detail, involving nothing more than a *passing change of fashion* in manufacture or in manipulation of the same process. (Kroeber 1909:3–5; emphasis added)

Rowe (1962:399–400) argued that in 1909, Kroeber was visualizing "cultural change in terms of major shifts in technology and subsistence, [and] any changes of less moment [such as those Uhle documented] were insignificant." Uhle seems to have favored a gradualistic, nonessentialist form of cultural evolution, whereas Kroeber favored an essentialist form such as that represented by Europe's Paleolithic–Neolithic–Bronze Age–Iron Age sequence. This essentialist viewpoint of change was typical at the turn of the century. We saw it with prehistorians attached to the BAE, and it was displayed even more prominently in Kroeber. Change through time, if it was worth measuring, was marked by major differences in materials, functional differences in particular. With respect to his data from Emeryville (Table 3.1), Uhle (1907:39) remarked, "It is evident that the character of the objects in the upper strata is entirely different from that of the implements which are found in the lower [strata]." But these categories of artifacts are generally descriptive or technological—"flaked stone" or "rubbed

Table 3.1. Max Uhle's Data for Frequencies of Artifact Types by Stratum in Emeryville Shellmound, California

Stratum	Rubbed-stone implements[a]	Obsidian implements	Flaked-stone implements other than of obsidian	Knife-like implements	Rough awl-like implements
I	2 [5][b]	2 [2]	–	[6]	– [8]
II	21 [10]	11 [5]	6 [10]	[13]	– [16]
III	3 [7]	4 [4]	4 [7]	[13]	– [11]
IV	4	2	4	[5]	– [6]
V	4 [4]	1 [1]	5 [2]	[4]	– [5]
VI	– [1]	– [1]	3	[1]	– [2]
VII	– [1]	2 [–]	6	[1]	– [1]
VIII	1 [7]	1 [4]	24	1 [9]	– [3]
IX	– [3]	1 [2]	62	4 [4]	5 [5]
X	– [2]	– [1]	17	– [2]	4 [3]

[a]Numbers in brackets "denote the number of objects which might have been expected as the proportional content of one of the middle strata" (Uhle 1907:39).
[b]Mortars and pestles excluded.

stone"—or functional—"knife-like" and "awl-like." It is no wonder that Kroeber perceived no significant cultural change here such as would realign the discipline within a decade. Uhle's categories were decidedly different from those that Kroeber himself would in a few short years use to display the passage of time and culture change.

In taking stock of the discipline on the eve of the so-called stratigraphic revolution, Roland B. Dixon (1913:551) pointed out various apparent instances of cultural change, but he also indicated that many of these were little more than "increasing perfection of the products of a uniform culture"—a reiteration of Boas's and Kroeber's position. When archaeological cultures appeared to be significantly different, Dixon (1913:553) assumed they represented "successive cultures" rather than samples of a continuum of change. He suggested that the successions reflected "numerous and far-reaching ethnic movements, resulting in a stratification of cultures, such that later have dispossessed and overlain earlier" (Dixon 1913:559). His view was thus of the essentialist sort: Internal cultural evolution was precluded, and all difference observed in the archaeological record was explained in commonsense, ethnological terms such as ethnic difference.

Commentators on Dixon's paper presaged events to come, noting such things as the following: "There is always danger of mistaking analogy for genealogy [and] misconstruing the phenomena of parallelism and of convergence" (MacCurdy, in Dixon 1913:569) and "Chronology is at the root of the matter, being the nerve electrifying the dead body of history. It should be incumbent upon the

American archaeologist to establish a chronological basis of the pre-columbian cultures" (Laufer, in Dixon 1913:577). The message was clear: Establish temporal control so that one could begin to sort out analogous from homologous similarities. That challenge would be met in admirable fashion shortly after Dixon's remarks were published, but meeting it had nothing to do with archaeologists beginning to excavate stratigraphically as is often stated (e.g., Browman and Givens 1996). Rather, it had to do with a change in how formal variation in artifacts was studied.

CULTURE HISTORY AND MEASURING TIME WITH ARTIFACTS

The stage was set in 1915 for a remarkable turnabout in how Americanists used artifacts to measure time—artifacts that were either excavated from sites or collected from the surfaces of sites. The shift comprised a change in how artifact categories were created. Prior to 1915, archaeologists were searching for differences in sets of culture traits of the sort that would suggest major qualitative differences in cultures—sets of variation that could then be construed as having occupied different temporal positions aligned with a progressive evolutionary model of cultural development such as that proposed by Lewis Henry Morgan (1877) in his classic treatise, *Ancient Society*. Differences between cultures would be on the order of those being reported in Europe—differences based on the presence or absence of culture traits manifest as technological or functional categories of artifacts. Anything of less magnitude than full-scale replacement of such traits was dismissed as being insignificant. This view of cultural development changed abruptly between 1914 and 1916, when archaeologists shifted their scale of observation from the presence/absence of cultural traits to the frequencies of specific trait variants. It was at that point that materialism made its initial entry into Americanist archaeology.

Time as a Continuous Dimension

The short span encompassing the years 1912 to 1915 was a fateful one for Americanist archaeology. During that period, three field projects were undertaken in the Southwest that at the time of initiation would have appeared to have little in common. Certainly they were not originally viewed as integrated components of a master design for bringing chronological control to the Southwest, but in the end, the three projects, taken in the aggregate, demonstrated how to solve chronological problems using a materialist approach. The first project began in 1912, when Clark Wissler, curator of anthropology at the American Museum of Natural History, sent Nels Nelson (1913, 1914, 1915, 1916) to the Rio Grande Valley of New Mexico specifically "to take up the historical problem in the Southwest to determine if possible the relations between the prehistoric and historic peoples" (Wissler 1915:395). More or less at the same time, prompted by his dissertation research (Kidder 1915), A. V. Kidder (1916, 1917) began to address similar questions of chronology at Pecos Pueblo, just northeast of the Galisteo Basin where Nelson was working. The third project was initiated by Kroeber (1916a, 1916b), who visited Zuñi Pueblo, west of where Nelson and Kidder were working, to carry out ethnographic field work for the American Museum. Through an inspired bit of reasoning, and almost totally unrelated to his ethnographic research, Kroeber conducted some innovative archaeological research that was to have a dramatic effect on how the archaeological record was viewed. Leslie Spier (1917a,1917b)—also at Wissler's direction—continued Kroeber's work near Zuñi and provided an empirical test of Kroeber's findings. The innovative work of these four archaeologists comprises the roots of the materialist study of time and culture change.

Nels C. Nelson at San Cristobal

In the report of his later excavations in the Galisteo Basin, Nelson (1916:162) stated that by the beginning of the 1914 season, he suspected he knew the chronological order of five kinds of pottery that were commonly found on sites throughout the basin. Apparently, Nelson's initial efforts to locate undisturbed stratigraphic deposits were thwarted (Nelson 1916), but late in the first season he found some at Pueblo San Cristobal. Nelson, having worked with Old World archaeologists Otto Obermaier and Henri Breuil in Spain in 1913 and seen "levels marked off on the walls" of the excavations (Nelson, in Woodbury 1960b:98), duplicated some of the European techniques when he excavated San Cristobal. He dug the site in arbitrary 1-foot-thick levels rather than in natural stratigraphic units, segregating and counting sherds from each level. In "Chronology of the Tano Ruins, New Mexico" (Nelson 1916), Nelson presented the absolute abundances of each of five basic types of pottery from each of the ten 1-foot-thick levels, and in an accompanying table he adjusted observed sherd frequencies to account for different excavation volumes—both rather innovative procedures for the time, but ones that Uhle (1907) had used at Emeryville Shellmound (Table 3.1) and Nelson (1906) had mimicked when he dug part of that same shellmound in 1906.

Nelson (1916:167) interpreted the absolute abundances of some of his pottery types from San Cristobal as approximating "normal frequency curves" (Figure 3.1). Such a pattern was to be "expected," Nelson (1916:167) suggested, because a "pottery style" had come "slowly into vogue, attained a maximum and [then went through] a gradual decline ... to extinction." Nelson rationalized such frequency distributions as the waxing and waning of an item's popularity. This either was an extraordinary insight for the time or, if it was a commonly accepted notion, no one prior to Nelson had bothered to put it down in print in such clear language. With explicit reference to Nelson's work, Wissler (1916b:194) indicated that "we do not

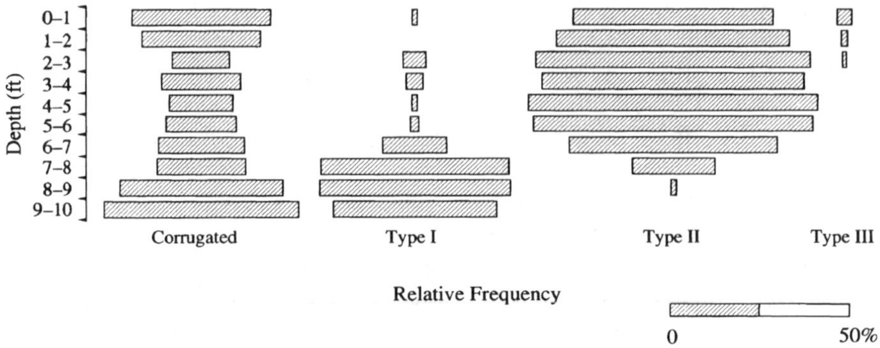

FIGURE 3.1. Nels Nelson's pottery data from Pueblo San Cristobal, New Mexico, showing the waxing and waning popularity of types. Note that Nelson believed, correctly, that the frequency of corrugated ware was not a good indicator of age. Note also the essentially unimodal frequency distribution of his types I–III (after Nelson 1916).

know the particular causes that give us this [unimodal frequency] form of curve." There was no explanatory theory to account for such frequency distributions, and instead, the observed frequency distributions were accounted for by common sense. This notion that the frequency of a *variant* of a culture trait was a measure of its popularity—what we have elsewhere termed the *popularity principle* (Lyman et al. 1997b:43)—was to become a central tenet of culture history. If it had remained a chronological principle and not picked up a sizable load of conceptual baggage along the way, culture history might have sidestepped the materialist paradox. But that was not to be the case.

Given that collecting artifacts from distinct vertical proveniences was not new, what, if anything, was so novel about Nelson's approach to chronological questions? Nelson (1916:162) stated explicitly that when he went back to San Cristobal in 1914, he sought to test a suspected local sequence of four pottery types: "By the opening of the [1914] season, it was reasonably certain, both from internal evidence and from various general considerations, what was the chronological order of the four apparent pottery types, but tangible proof was still wanting." He knew or suspected the relative chronological positions of pottery types, but only by excavating at San Cristobal was he able to establish their relative chronological positions; all previous superpositional indications of chronology were "incomplete and fragmentary, each showing merely the time relations of two successive pottery types at some place or other in the total series of four or five types" (Nelson 1916:163). The important innovation found in Nelson's work was his demonstration that pottery types altered in absolute frequency through time in a pattern that he characterized as "very nearly normal frequency curves" (Nelson 1916:167). He was able to measure culture change using not the then-typical qualitative differences in artifact assemblages such as the presence or absence of pottery—a culture trait—as had been done by Harrington (1909b) and Wissler (1909), but rather by documenting in revolutionary fashion the *changing frequencies* of pottery types or variants of the culture trait of pottery (Lyman and O'Brien 1999a; O'Brien and Lyman 1999a). Stratigraphic provenience confirmed that those frequencies in fact measured the passage of time.

Wissler (1916b:195–196) suggested that the frequency changes in artifact types documented by Nelson comprised changes in "specific styles in ceramic art [and represented] stylistic pulsations." Use of the word *style* may have been accidental, but the word came to denote precisely such kinds, or types, of artifacts—those that could be used to measure time. Initially, they waxed in abundance after first appearance, reached a peak abundance, and then waned to disappearance. Later, they would be equivalent to index fossils that denoted relatively brief time periods. The temporal connotation of artifact *styles* did not change until the 1960s and 1970s, when archaeologists began to try to render meaning other than time from the categories of artifacts that had been called styles. We return to this topic in Chapter 4, but before leaving Nelson, we need to examine other statements he made.

Nelson had found superposed remains before the critical 1914 field season in the Southwest, but he noted that in such cases "there is often no appreciable [chronological] differentiation of remains" (Nelson 1916:163). When he found evidence of chronological differentiation, it was between types at the ends of a continuum of several pottery types; thus, he lamented that such instances were "merely clean-cut superpositions showing nothing but time relations" (Nelson 1916:163). However, when two types in the continuum were found stratigraphically mixed together, "one gradually replacing the other[, this] was the evidence wanted, because it accounted for the otherwise unknown time that separated the merely superposed occurrences of types and from the point of view of the merely physical relationships of contiguity, connected them" (Nelson 1916:163).

This statement is important because it reveals that Nelson was thinking about culture change in terms that simply did not mesh with the thinking of many of his colleagues. Whereas they were thinking in essentialist terms of culture traits, the meaning of which was derived from ethnologically informed common sense, Nelson was thinking in materialist terms of analytical units used to measure time *and* that had historical and, by implication of overlap, *heritable* continuity. As a result, he replaced, at least in his own mind, the then-prevalent notion that culture change could be modeled as a flight of stairs, each step representing a static evolutionary stage and each riser representing a rather abrupt transformation from one stage to the next, with a model that viewed culture change as a gradually ascending ramp (e.g., Nelson 1919a, 1919b, 1919c, 1932)—albeit a ramp that moved through progressively more advanced stages (Figure 3.2). Plotting frequencies of

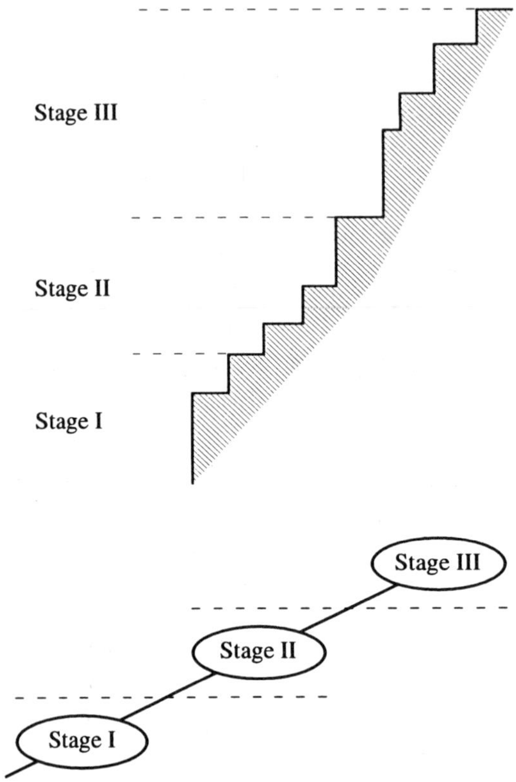

Figure 3.2. Two ways of looking at time: top, as a series of steps, with the risers representing sudden, punctuated change; bottom, as a continuous ramp. Stages are recognized in each based on the sudden (top) or gradual (bottom) appearance of new features.

pottery styles against time, which was rendered as geologically vertical space, would illustrate the gradual cultural evolution Nelson sought and eventually allow one to document the relative ages of the cultural stages (e.g., Nelson 1937). This was not only a revolution in analytical method, it was also a revolution in metaphysic. It was decidedly *not* a revolution in excavation strategy, despite received wisdom to the contrary.

A. V. Kidder and Pecos Pueblo

In 1915, A. V. Kidder (1915:461) noted that an announcement of Nelson's work indicated the latter's excavation had revealed a deposit "so stratified that the relative ages of [four distinct types of pottery] could be ascertained," but he also lamented that "[w]hat the wares were has not yet been announced." Based on his dissertation research, Kidder (1915:452) suspected what the sequence of pottery wares, or types, was, but "as yet no stratified finds have given us absolutely conclusive proof of this." His early suspicions regarding temporal sequence were founded on (1) the association of certain pottery types with "nearly obliterated ruins of obviously greater age than any others in the region" (Kidder 1915:452–453); (2) the geographic distribution of certain pottery styles interpreted in light of the age–area concept (Kidder 1931:4); and (3) suspicions regarding the evolutionary progression of pottery styles and technologies (Kidder 1915:453–456). He concluded his discussion by noting the necessity of finding stratified sites in order to draw "reliable development or historical conclusions" (Kidder 1915:461). Superposed collections would serve to *confirm* a hypothesized local sequence rather than to create one.

A year later, Kidder (1916:120) reported that Pecos Pueblo, New Mexico, was chosen for study because historical documents indicated that it had been occupied from 1540 until 1840, and surface finds included "practically all types of prehistoric wares known to occur in the upper Rio Grande district." Occupation was thus believed to have been continuous and from "very early times." No other site then known in New Mexico and "available for excavation" seemed to have that attribute:

> [I] hoped that remains [at Pecos] would there be found so stratified as to make clear the development of the various Pueblo arts and to enable students to place in their proper chronological order numerous New Mexican ruins whose culture has long been known but whose relation to one another has been entirely problematical. This hope was strengthened by the fact that Mr. N. C. Nelson ... had recently discovered very important stratified remains at San Cristobal a few miles to the west. (Kidder 1916:120)

Similar deposits at Pecos would allow comparative analyses and the extension of Nelson's chronology, which ended at 1680, when San Cristobal was abandoned, into the middle of the nineteenth century (Kidder 1916:120).

Early in his excavations, Kidder (1916:122) recognized that pottery types in the lower levels of his trenches were "markedly different from [those] at the top and that there were several distinct types between," and his summary chart (Kidder 1924) (Figure 3.3) arranged them in terms of their vertical distribution (see also Kidder 1916, 1931, 1936b; Kidder and Kidder 1917). He indicated that although he had earlier proposed a chronology "based on technological and artistic grounds" (e.g., Kidder 1915, 1917), Nelson "was attacking the same problem on the much sounder basis of stratigraphy, a method which had not been used, save sporadically, in southwestern archaeology.... Its importance cannot be overlooked" (Kidder 1936b:xix). Kidder mimicked at Pecos at least Nelson's use of vertical provenience of artifacts.

Kidder, unlike his predecessors Nelson and Gamio, both of whom used arbitrary levels, excavated portions of Pecos after the first field season in natural stratigraphic layers. Kidder was a student of Egyptologist George Reisner, who in turn had been a student of W. M. Flinders Petrie (Browman and Givens 1996;

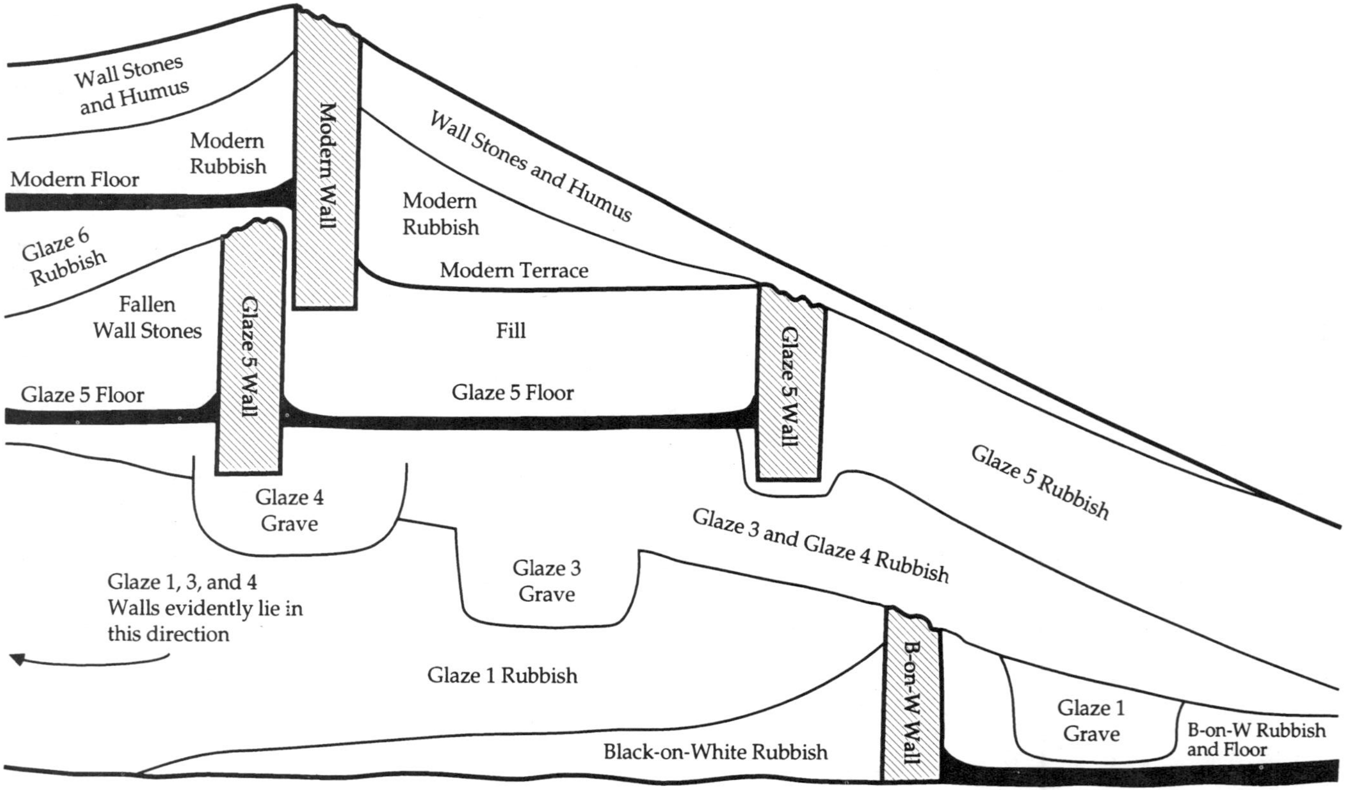

Figure 3.3. A. V. Kidder's cross section through one of the midden deposits at Pecos Pueblo, New Mexico, showing the stratigraphic positions of pottery types and architectural features (after Kidder 1924).

Stiebing 1993:82), the person responsible for popularizing stratigraphic excavation in Egypt. Kidder's biographers (e.g., Givens 1992:50; Willey 1988:307; Woodbury 1973:43) and historians of archaeology (e.g., Willey and Sabloff 1993:103–105) have suggested or implied that Nelson influenced Kidder relative to paying attention to superpositional relations of artifacts, though we think it is more accurate to suggest that it was Kidder who perfected the technique and modified it to focus on natural stratigraphic units (see Wauchope 1965:151).

Although Kidder is remembered for peeling back individual strata and collecting artifacts from each unit, he made two other seldom noted but nonetheless important contributions to the birth of culture history. Recall that Nelson (1916: 167) described frequency distributions as "very nearly normal frequency curves" and noted that such distributions were to "be expected." Kidder mimicked Nelson's analytical technique but also modified it. Kidder not only listed the absolute frequencies of pottery types against their vertical provenience in tabular form—the mimicking part—but he also graphed the changes in *relative* frequencies of his pottery types against his excavation levels (Kidder and Kidder 1917)—the innovative part. By using relative frequencies, Kidder circumvented the problem of variation in volume excavated per vertical unit in a manner different than Nelson.

We have redrawn one of the Kidders' graphs in Figure 3.4. When inspecting these graphs, Kidder and Kidder (1917:341) noted that many—but not all—types displayed "approximately normal frequency curves." They echoed Nelson and Wissler, and interpreted such curves as "indicating that each [type] had a natural rise, vogue, and decline" (Kidder and Kidder 1917:349). Although the graphs implied a materialist metaphysic, the included interpretive statements were not theoretically informed explanations; they were merely commonsensical interpretations of observations that were expressed in words everyone could comprehend. In this case, the words constituted the *popularity principle*. Thus, one of Kidder's less-often noted contributions to culture history involved the perfection of an analytical technique grounded in materialism—a technique that later came to be known variously as "ceramic stratigraphy" (Willey 1938, 1939) or "percentage stratigraphy" (Ford 1962; Willey 1939:142; see Lyman et al. 1998b; O'Brien and Lyman 1998, 1999a).

Kidder's other seldom-noted contribution was of a decidedly different sort: creating a chronological ordering using changes in pottery designs. Kidder (1931:7) later claimed to have "attempted a seriation, on comparative grounds, of the material available." For his dissertation research, Kidder (1915, 1917) performed what Rowe (1961:327) later characterized as "ordering by continuity of features and variation in themes," and it was essentially what John Evans (1850, 1875), William Matthews Flinders Petrie (e.g., 1899, 1901), and A. L.-F. Pitt-Rivers (1875a, 1875b), had earlier done in England. This was quite different from what Kroeber was doing at the same time.

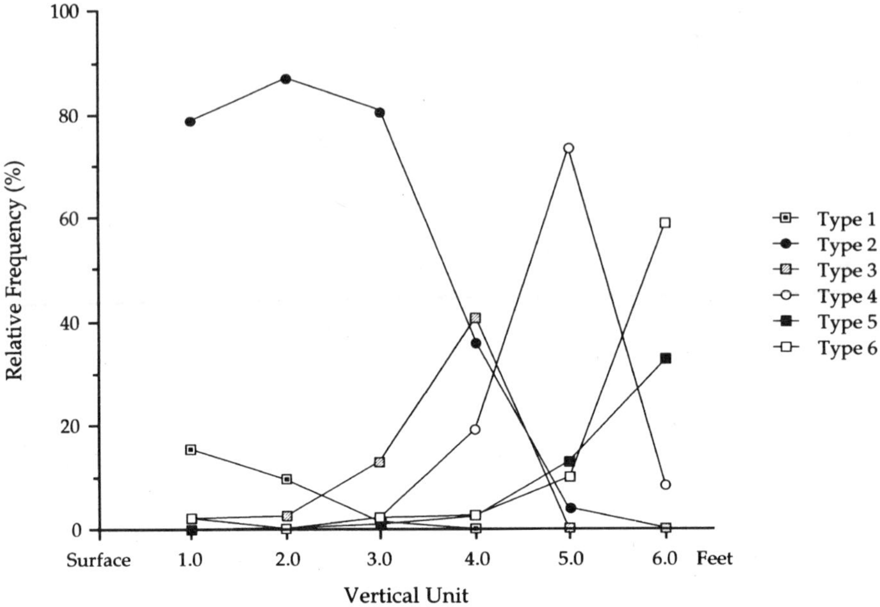

Figure 3.4. A broken-stick graph of data from Pecos Pueblo, New Mexico, showing the fluctuating frequencies of types over vertical space (time) (after Kidder and Kidder 1917).

Kidder's (1915) original pottery sequence was based on his suspicions regarding the evolution of various design and technological features, and he demonstrated how such an evolutionary and thus temporal sequence could be worked out (Kidder 1917). Importantly, he cautioned that the "only safe method for the working out of developments in decorative art is to build up one's sequences from chronologically sequent material, and so let one's theories form themselves from the sequences" (Kidder 1917:369). The theory is more appropriately termed a *hypothesis* regarding the ordering in a chronological sequence. The hypothetical chronological sequence (Figure 3.5) was tested, in 1916–1917 at Pecos, through the study of superposed collections. The term *phyletic seriation* is a reasonable one for labeling what Kidder did when generating the hypothetical sequence (Lyman et al. 1997b; O'Brien and Lyman 1999a). The term *phyletic*, in the same sense in which Simpson (1944) used it in paleontology (Chapter 2), denotes a suspected line of heritable descent—a point we take up in considerably more detail in Chapter 6. Although Kidder no doubt had a rule in mind for creating the ordering—probably something on the order of Morgan's (1877) cultural evolution—he also knew that the chronological implications of such a seriation had to be tested and that any notion of progress would simultaneously be tested. His

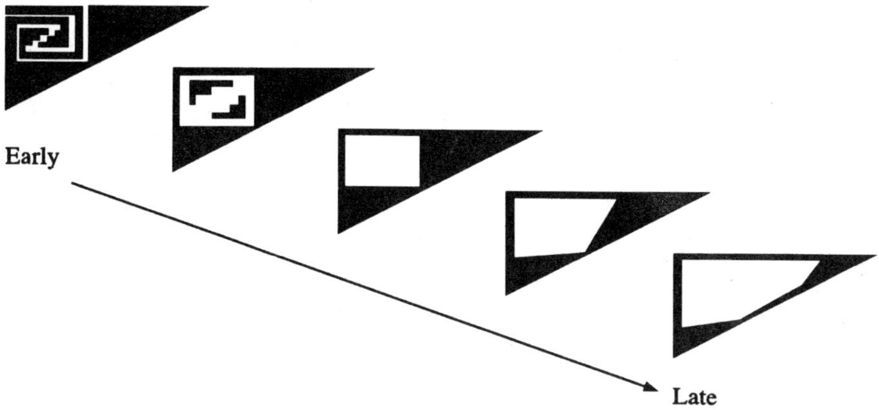

FIGURE 3.5. A. V. Kidder's illustration of the evolution of a ceramic design on pottery from Pecos Pueblo, New Mexico. Kidder proposed that through time, the pottery design became less intricate, changing from a stepped motif, to a pair of stepped motifs, and finally to a series of ever-larger white spaces (after Kidder 1917).

seriations thus were very much phyletic in an ancestor–descendant sense. Importantly, he, like Nelson, used variants of the culture trait of pottery to measure time and heritable descent.

A. L. Kroeber and Zuñi Pueblo

Rarely is the origin of a major innovation in science clear. Not only are these matters usually clouded by competing versions of similar age, but initial presentations tend to be post hoc, functional accounts that lack discussion of the conditions and reasoning that led to the innovation. Not so with frequency seriation, the chronological ordering of artifact assemblages based on changes in the relative abundance of artifact types. A. L. Kroeber (1916b), in his remarkable paper "Zuñi Potsherds," gives a discursive account of precisely how he came up with the idea of frequency seriation, making it quite clear that earlier explorations of similar ideas (e.g., Evans 1850, 1875; Petrie 1899; Pitt-Rivers 1875a, 1875b; Uhle 1903) played absolutely no role in the development of the method, as is often claimed or implied (e.g., Browman and Givens 1996; Praetzellis 1993; Willey and Sabloff 1993).

During the summer of 1915, Kroeber volunteered to help American Museum personnel with their research in the Southwest (Wissler 1915:397). While walking across the countryside around Zuñi Pueblo, Kroeber began collecting sherds from the surfaces of more than a dozen prehistoric sites. He noticed that some collections tended to be dominated by "red, black, and patterned potsherds," whereas other collections were dominated by white sherds (Kroeber 1916b:8). He concluded that there "could be no doubt that here, within a half hour's radius of the largest inhabited pueblo [Zuñi], were prehistoric remains of two types and two periods, as distinct as oil and water. The condition of the sites indicated the black and red ware ruins as the more recent" (Kroeber 1916b:9). Based on historical evidence and on the condition of the sites, Kroeber (1916b:9–10), like Kidder, concluded that concerning "the type and period of white ware and the type and period of black and of red ware, the latter is the more recent [belonging] in part to the time of early American history; the former is wholly prehistoric." Thus Kroeber had two lines of evidence to indicate that he was dealing with temporal differences in the variants of the culture trait of pottery he found: (1) historical data, which indicated that "[s]everal of the ruins were inhabited in Spanish times" (Kroeber 1916a:43), whereas other ruins were said by his informants to have been inhabited "long ago" (Kroeber 1916b:9), and (2) "Type A [historic] ruins normally include standing walls, and loose rock abounds. All Type B [prehistoric] sites are low or flat, without walls or rock, and ... it seems more likely that this condition is due to the decay of age, or to the carrying away of the broken rock to serve as material in the nearby constructions of later ages" (Kroeber 1916a:43).

Kroeber (1916b:8) "attempted to pick up all sherds visible in certain spots [of each site], rather than range over the whole site and stoop only for the attractive ones." He did not excavate: "I have not turned a spadeful of earth in the Zuñi country. But the outlines of a thousand years' civilizational changes which the surface reveals are so clear, that there is no question of the wealth of knowledge that the ground holds for the critical but not over timid excavator" (Kroeber 1916b:14). But he was cautious. Kroeber (1916b:20, 21) believed that for his proposed chronological classification, the "final proof is in the spade," and he lamented that "in the present chaos of knowledge who can say which of these differences [in frequencies of sherd types] are due to age and which to locality and environment?" Thus, like his contemporaries, Kroeber assumed that superposed collections were the as-yet-unavailable confirmational proof of the suspected chronology.

Kroeber arranged his surface collections to derive not just a two-period sequence but a five-period (actually six, if one includes modern Zuñi pottery) cultural sequence—what he referred to as "shorter epochs" (Kroeber 1916a:44). These are shown in Table 3.2. The innovative aspect of Kroeber's work has often been overlooked. In modern terms, Kroeber's "frequency seriation" (Dunnell 1970) began with corrugated ware as the oldest and most frequent type. Sites—actually, the collections of pottery from individual sites—were arranged so that the relative abundance of that type decreased, with two exceptions, monotonically (Table 3.2). The basis for this arrangement was Kroeber's (1916b:15) impression that corrugated ware, given its rare association with modern pottery types and its regular association with decayed ruins, decreased in frequency as time passed; this allowed him to arrange "the sites in order accordingly."

Table 3.2. A. L. Kroeber's Frequency Seriation of Pottery Sherds from Sites around Zuñi Pueblo, New Mexico

<table>
<tr><th>Period</th><th>Site</th><th>Corrugated</th><th>Three color</th><th>Black on red</th><th>Any red</th><th>Black</th></tr>
<tr><td>Present</td><td>Zuñi</td><td>0</td><td>12</td><td>1</td><td></td><td></td></tr>
<tr><td>Late A</td><td>Towwayallanna</td><td>1</td><td>8</td><td>3</td><td rowspan="8">22</td><td rowspan="8">53</td></tr>
<tr><td></td><td>Kolliwa</td><td>—</td><td>7</td><td>2</td></tr>
<tr><td></td><td>Shunntekya</td><td>2</td><td>7</td><td>2</td></tr>
<tr><td></td><td>Wimmayawa</td><td>2</td><td>4</td><td>1</td></tr>
<tr><td></td><td>Mattsakya</td><td>3</td><td>4</td><td>3</td></tr>
<tr><td></td><td>Kyakkima</td><td>4</td><td>3</td><td>2</td></tr>
<tr><td>Early A</td><td>Pinnawa</td><td>10</td><td>1</td><td>8</td></tr>
<tr><td></td><td>Site W</td><td>24</td><td>—</td><td>1</td></tr>
<tr><td>Late B</td><td>Hattsinawa</td><td>27</td><td>—</td><td>5</td><td>10</td><td>19</td></tr>
<tr><td></td><td>Kyakkima West</td><td>12</td><td>—</td><td>4</td><td>8</td><td>—</td></tr>
<tr><td>Middle B</td><td>Shoptluwwayala</td><td>40</td><td>—</td><td>2</td><td>3</td><td>7</td></tr>
<tr><td></td><td>Hawwikku B</td><td>49</td><td>—</td><td>6</td><td>12</td><td>9</td></tr>
<tr><td>Early B</td><td>Te'allatashshhanna</td><td>66</td><td>—</td><td>—</td><td>—</td><td>5</td></tr>
<tr><td></td><td>Site X</td><td>71</td><td>—</td><td>—</td><td>3</td><td>1</td></tr>
<tr><td></td><td>Telnatluwwayala</td><td>72</td><td>—</td><td>—</td><td>2</td><td>—</td></tr>
<tr><td>Uncertain</td><td>He'itli'annanna</td><td>—</td><td>—</td><td>—</td><td>—</td><td>3</td></tr>
<tr><td></td><td>Site Y</td><td>—</td><td>—</td><td>—</td><td>—</td><td>—</td></tr>
</table>

Source: Kroeber 1916b.

The relative abundance of Kroeber's "Three Color" type increased monotonically once it appeared in the sequence and was most abundant in the modern Zuñi assemblage. Frequencies of his "Black on Red" type merely tagged along and fluctuated in abundance, but his "Any Red" and "Black Ware" types tended to decrease monotonically. Kroeber (1916b:16–18) initially identified ten pottery types but lumped five of them into two types (and ignored two others) for purposes of his seriation. Ultimately, he used the relative frequencies of only three types to seriate his Period A sites and presented only the summed site frequencies for the other two types (lumped from the original five). Kroeber (1916b:20) justified this lumping by noting that variations in the frequencies appeared to be a result of sampling error. Such lumping indicates that Kroeber conceived of his types as something created by the analyst as opposed to something to be discovered. They were units that—if properly constructed—allowed measurement of differences in time. But there was no theoretical warrant for such an inference. Rather, the basis of the arrangement was the observed association of different pottery types with ruins inferred, but not proven, to be of different ages.

Kroeber's (1916a:44) chronologically sensitive pottery types indicated that his short "epochs ... shade[d] into one another," and there was "no gap or marked break between periods A and B," his two main periods. On the one hand, Puebloan culture change was a flowing stream. The two major periods might have originally been "as distinct as oil and water" (Kroeber 1916b:9)—an essentialist notion—but they had originally been distinguished on the basis of criteria—primarily the degree of deterioration of associated ruins—other than those used in the seriation. Kroeber (1916b:15) believed that the two major periods "can normally be distinguished without the least uncertainty, and the separateness of the two is fundamental, [but] nevertheless they do not represent two different migrations, nationalities, or waves of culture, but rather a steady and continuous development on the soil." This was a clear expression—the clearest up to that point in Americanist archaeology—of the materialist position. The warrant for Kroeber's frequency seriation was the popularity principle discussed by Nelson (1916) and Wissler (1916b) at the same time that Kroeber was at Zuñi.

Leslie Spier and the Seriation of Zuñi Potsherds

Kroeber's desired proof that the ordering of Zuñi potsherds was indeed chronological was provided by Leslie Spier, who joined Kroeber's proportional frequencies with Nelson's superposed levels (Spier 1917a). Based on Nelson's, Kroeber's, Kidder's, and some of his own stratigraphic work, Spier (1917b) knew the basic ceramic sequence: "It seems reasonable to believe that we are dealing with no other phenomenon than the several phases of a single pottery art" (Spier 1917a:281). Citing his own superpositional data and Kroeber's (implied popularity) principle of ordering corrugated ware from having a high relative abundance early and being absent late in the temporal sequence, Spier (1917a:281) used "fluctuations in this type for a first grouping, a preliminary seriation of the data from superficial [i.e., surface] samples," arguing that "it might prove fertile then to arrange these data according to their percentages of corrugated ware in sequence from lowest to highest" (Spier 1917a:281–282). He never questioned the basis of this (popularity) "principle for the seriation of the data," indicating only that the principle was "to be subjected to the method of proof of concurrent variations" (Spier 1917a:281). The latter statement, however, referred to the correctness of the resulting arrangement rather than to the inference that the arrangement represented the passage of time.

After noting that the seriation based on the frequency of corrugated ware resulted in the recognition that painted wares apparently preceded glazed wares in time—all inferential and implicitly founded on the popularity principle—Spier elaborated on his method of proof: "The test of such a seriation as an historical series will lie in the observed [frequency] seriation of the accompanying wares; for, when a group of three or more distinct, but mutually dependent, values are ranked according to some postulated sequence for one, and the other values are found to present serially concurrent variations, it may be concluded that the result is not fortuitous" (Spier 1917a:282). After arranging his collections on the basis of consistent frequency change in one type, Spier then determined if the other types included in his seriation consistently increased or decreased in the arrangement of assemblages. He used regression analysis to test the fit in his seriation and suggested that his seriation was valid "but not certain"; that is, the arrangement was good, but whether it represented the passage of time was unclear. He tested the latter proposition using superposition.

Spier found some of his seriations to be corroborated by Kidder's (e.g., 1916) superposed sequences and others by his own excavations. He also noted that "parallel development" might invalidate his sequence (Spier 1917a:305). In his summary, Spier (1917a:326) suggested that there was "no reason to doubt that samples of potsherds collected from successive levels of the ash heaps present us with valid chronological indices," and he argued that he had demonstrated "that it is possible to collect surface samples approximating in [chronological representativeness] those from refuse heaps." Of course, he knew which way time was going in his seriation based on superposition and on the general nature of the frequency distribution of particular kinds of pottery that resulted from the work of Nelson, Kidder, and others. His test of concomitant variation in types used in the seriation, along with regression analysis and analysis of residuals, were innovative. Truly, Spier's (1917a:328) work was, as he said, "an exposition of archaeological method." In his review of Spier's work, Kidder (1919:301) stated that

> Mr. Spier's presentation and statistical handling of his material are essentially sound. His ranking of sites on the basis of their percentages of wares (which [Nelson's] stratigraphical work had shown to be chronologically significant) is a new and valuable contribution to method.... Good method can only lead to good results; Mr. Spier's conclusions are thoroughly satisfactory. His work will stand as the basis for all future archaeological study of the Little Colorado drainage.... [It is] fundamental.

Time as a Discontinuous Dimension

Within the span of two years, 1916 and 1917, Nelson and Kroeber, followed quickly by Spier and Kidder, showed beyond any doubt that culture change could be analytically detected by examination of shifts in frequencies of particular kinds of units—styles of artifacts—and they confirmed that such shifts in fact were temporal by using superposition. One would suspect, then, that they might have pursued this truly innovative way of approaching the archaeological record—one founded in materialism—in their later work, but they did not. Kidder, for example, in his later reports on Pecos (Kidder 1924, 1931, 1932, 1936a, 1936b), did not replicate his initial effort at assessing pottery chronology and development based on design-type frequencies. Willey and Sabloff (1993) and Givens (1992) imply that Kidder (1931) used percentage stratigraphy in later work, but in fact, the application they refer to was done by Charles Amsden (1931). In most of his later work, Kidder used particular pottery types as index fossils (O'Brien and Lyman 1999a), thus abandoning percentage stratigraphy as a way of measuring time.

In a retrospective look at the discipline, made in the 1930s, Frank H. H. Roberts (1935:3–4) correctly pointed out where part of the problem resided:

> Recently [Kroeber's, Nelson's, Spier's, and Kidder's method of examining the percentage representation of pottery types] has fallen into discard. Just why this should be the case is not apparent. It is true that under certain conditions it is not an infallible source of evidence, particularly in chronological studies based solely on surface material. Nevertheless it is helpful in outlining the main characteristics of a district and in indicating where intensive work should be undertaken. In a study of the ceramics of a single site it has more than enough merit to warrant its retention in archaeological procedure. It graphically demonstrates the true nature of the pottery complex. One explanation for the failure to make use of the system is, perhaps, that the workers have become so absorbed in a detailed study of pottery *per se* that they have forgotten the important factor of giving percentages. It is only from such data that the real significance of each group [read *type*] in the series [of pottery types] can be judged.

Roberts was perhaps a bit too negative, as a few individuals working in the Southwest were studying variation in the relative abundances of pottery types and were using percentage stratigraphy (e.g., Martin 1936) or seriation (e.g., Martin 1938, 1939) in their efforts to measure time. But overall, Roberts was correct: Archaeologists were paying more attention to the pottery itself and less attention to its role in chronology construction. This resulted from the fact that they did not have to rely on abundances of types anymore to track time. They now had superposition and stratigraphy to do that for them. Remarks made at the time reveal this clearly.

In his seminal synthesis of Southwestern archaeology, Kidder (1924:45) stated that the "ideal form of chronological evidence is provided by stratigraphy, i.e., when remains of one type are found lying below those of another.... To take full advantage of stratigraphic evidence ... the investigator must select for study those phenomena which most accurately reflect changes in culture or, what amounts to the same thing, chronological periods. Pottery has so far provided the most useful material for such studies [as it] is a highly sensitive register of cultural change." Nowhere in this volume does Kidder mention phyletic seriation, percentage stratigraphy, or frequency seriation. But his message is clear on two counts. First, artifact types, if properly constructed by the analyst, can be used as index fossils, and second, these monitor not only temporal change but cultural change—from one essentialist unit to the next—as well. All one had to do was stack the units up in the proper order temporally, and stratigraphic excavation was the technique of choice. As Neil Judd wrote in the late 1920s, "Chronology is the key that will unlock many secrets of American prehistory and stratigraphy is the stuff of which chronology is made" (Judd 1929:408). Nelson (1937) made virtually identical remarks a decade later. No one mentioned that it was percentage stratigraphy, seriation, and, particularly, certain ways of classifying artifacts that had originally made culture change visible and measurable.

What a few years earlier had been a strategy for testing and confirming hypothesized chronological sequences was now a creational strategy. Thus, by the early 1920s, Wissler's (1917a) impression that the presence of strata—that is, empirically discrete and distinct depositional units—denoted a discontinuous occupation by multiple successive cultures seems to have been generally accepted within the discipline. Gerard Fowke (1922:37) pointed out that the "intermittent character of occupancy is ... shown by the distinct segregation of numerous successive layers of kitchen refuse," and Kroeber (1925:927) argued that "the correct [excavation] procedure [is] to follow lines of deposition in instituting comparisons." The impression that strata/vertical units could be used to denote discrete "occupations" reinforced otherwise nebulous notions of a sequence of "cultures"—phenomena that were clearly visible ethnographically, and therefore real—in the archaeological record. Confusion over the nature of time—that is, viewing it simultaneously as both a continuous flow and as a punctuated continuum—was ever-increasing. Kidder's remarks on the pottery from Pecos Pueblo represent a case of how archaeologists grappled with the conception of time as a continuum but monitored its flow with discontinuous units:

> The division of the Glaze ware of Pecos into six chronologically sequent types is a very convenient and, superficially, satisfactory arrangement. For some time I was very proud of it, so much so, in fact, that I came to think and write about the types as if they were definite and describable entities. They are, of course, nothing of the sort, being merely useful cross-sections of a constantly changing cultural trait. Most types, in reality, grew one from the other by stages well-nigh imperceptible. My groupings therefore amount to a selection of six recognizable nodes of individuality; and a forcing into association with the most strongly marked or "peak" material of many actually older and younger transitional specimens.... This pottery did not stand still; through some three centuries it underwent a slow, usually subtle, but never ceasing metamorphosis. (Kidder 1936b:xx)

Kidder's comment reveals the fundamental paradox in the early foundations of the culture-history paradigm: The conceived materialistic "slow, usually subtle, but never ceasing metamorphosis" of artifact forms through time was monitored using discontinuous, typological, essentialist, "*recognizable* nodes of individuality." If Kidder had made clear that he recognized there were no breaks to the temporal continuum but that he was *artificially* creating units—types—to allow him to measure that continuum, then he could have kept both feet in the materialist camp. But he did not make clear that distinction—undoubtedly, it was unclear in his own mind—and as a result became mired in the same paradox that ensnared his colleagues. Similarly, Nelson (1932:105) claimed that the history of a culture was conceived as a flowing stream, and the flow could be monitored by observing "a few cross-sections of the flow taken at strategic points." That was a materialist statement. But when archaeologists began viewing those cross sections as discrete types, occupations, cultures, and the like, and began correlating the occupations with distinct depositional units—strata—they fell prey to the materialist paradox.

A fine example of this problem—one built into the culture-history paradigm from the time when superposition became a creational method rather than a confirmational one—is found in efforts by Gordon Willey and Richard Woodbury to construct a chronology of pottery types for northwestern Florida:

> Any pottery type is based on a number of stylistic features found in combination, but changes occur over time, and transitions are often so gradual as to prevent sharp distinctions. However, the periods into which the pottery types have been grouped are each based on one or more "key" or "marker" types, which have been found to be sufficiently restricted in range and distinctive in appearance to allow their occurrence to be quite precisely determined. It is also hoped that the "periods" will prove to represent distinct cultures when the bare skeleton of ceramic chronology has been given flesh and body in the form of a full and "functional" culture description. (Willey and Woodbury 1942:236)

The role of "marker" types is largely that of an index fossil; they are used to denote a particular time period—the briefer the better—within a particular area (O'Brien and Lyman 1999a). More importantly, the statement by Willey and Woodbury exemplifies the conflation of a conceptual model of gradual, continuous cultural change—a flowing stream—with the ethnologically informed perception of cultures as discrete units. Index fossils, or marker types, could be used to measure the passage of time—a materialist approach—but surely they could also be correlated with an ethnographic unit such as a "culture"—a purely essentialist approach. As Kroeber (1925:931) remarked, artifact types "must, of course, be interpreted as periods." Even though such a correlation was not testable, it served as an ad hoc, commonsensical warrant for breaking up the cultural continuum into what were otherwise clearly *arbitrary* chunks. These chunks were referred to as cultures after 1916, just as they had been before the turn of the century (Lyman et al. 1998b), and they became the cornerstones of regional reconstructions (e.g., Ford 1936a; Kidder 1924). Unclear at the time was that these units were products of two diametrically opposed ontologies. Failure to appreciate this fact created an immediate problem for Americanist archaeologists and eventually led to the fall from disciplinary favor of culture history as a paradigm.

Although Willey and Woodbury's effort to construct a chronology is an excellent example of confusing gradual, continuous cultural change with the ethnologically informed perception of cultures as discrete units, perhaps the classic example resulted from the collaborative efforts of Philip Phillips, James A. Ford, and James B. Griffin (1951) in their study of the ceramic-period record of the Mississippi Valley from roughly the Missouri–Arkansas border on the north to the Arkansas–Louisiana border on the south. The study was a landmark both substantively and methodologically (Dunnell 1985a; O'Brien and Dunnell 1998; O'Brien and Lyman 1998), and in many respects it captured the conflict between essentialism and materialism. The greatest value of the monograph, *Archaeological Survey in the Lower Mississippi Alluvial Valley, 1940–1947* (Phillips et al. 1951), is found in the dialogue among its three authors, whose differences of opinion—founded in conflicting ontologies—are highlighted in various sections.

Our immediate interest in the project is what it tells us not only about how time was being approached but also about the issue of artifact types, which formed the basis of various culture-historical methods of carving up time. Phillips et al.'s discussion of typology serves as an excellent introduction to the detailed treatment the topic receives in Chapter 5. All three authors worked on the classification of the pottery, but the final decisions on what to include and how to phrase the discussion resided with Griffin (Dunnell 1985a; O'Brien and Lyman 1998). The authors conceived of a pottery type as demanding "hair-splitting," arbitrary decisions, because actual specimens graded more or less continuously in their formal attributes from one to another—similar to the problem of identifying species (Figure 2.6). A type to Phillips et al. was a stylistic norm that shifted so gradually in terms of formal attributes across space and through time that the shifts were undetectable, making the distinctions of types arbitrary (Phillips et al. 1951:64) (Figure 3.6). In that manner, their types were ideational units, the time–space boundaries of which were supposed to be limited in order to allow them to write culture history—specifically, to measure time–space differences. But the theoretical content of their types was compromised by their commonsensical

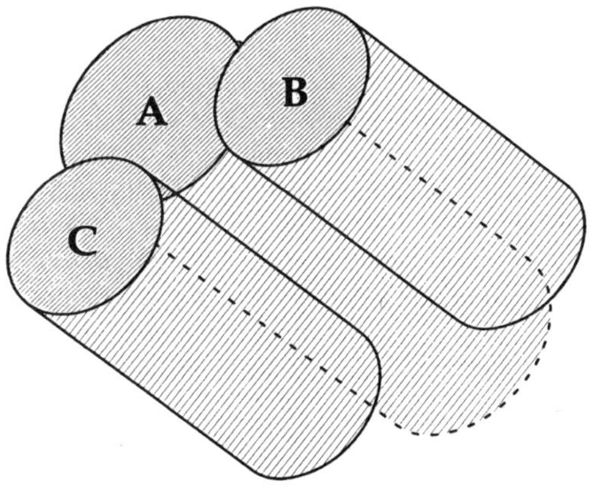

FIGURE 3.6. Model of the way in which Philip Phillips, James A. Ford, and James B. Griffin conceived of pottery types. Each letter denotes a type, the circle around it denotes the range of formal variation within the type, and the length of the cylinder represents the temporal range of a type (after Phillips et al. 1951).

rationale; their types would "finally achieve cultural meaning" when, through trial-and-error extensional redefinition, they showed approximate "correspondence to ethnographic distributions in time and space" (Phillips et al. 1951:64).

Artifact types were supposed to measure time; to determine whether they actually did so, one checked their distribution relative to one another against superposition. The ethnographic implications of the types created by Phillips et al., however, were not testable because there were no formal attributes that clearly or theoretically gave types those kinds of properties, nor were there any theoretically informed tests to ensure that the constructed types had them. Like Alex Krieger (1944) and others with similar views, Phillips et al. reverted to commonsensical rationalization. For example, change in "ceramic features" was conceived as gradual because types were supposed to reflect prehistoric norms, or what they termed "the consensus of community opinion" (Phillips et al. 1951:62) of what a proper-looking pot was. They reasoned that "the characters we select as criteria for type definition ... are bound to correspond to characters that might have served to distinguish one sort of pottery from another in the minds of the people who made and used it" (Phillips et al. 1951:63). Maybe so, but Phillips et al. never presented a warrant for determining what went on in the heads of the vessel manufacturers. Of course, even they probably would have admitted that, when it came down to it, such information was impossible to generate from a pile of sherds, but as we see in Chapter 5, others would not.

Phillips et al. subjectively chose geographic "popularity" centers, out from which they expected types to gradually drift or "creep" into other types—a reflection not only of the extensional character of their type definitions but also of the notion that culture is transmitted and shared:

> Each community that had reached a certain level of sophistication in pottery-making will be found to have been maintaining side by side several different vessel styles.... Between these centers, styles vary and trend toward those of other centers in rough proportion to the distances involved, subject of course to ethnic distributions and geographic factors.
>
> Thus we have in mind the concept of a continuously evolving regional pottery tradition, showing a more or less parallel development in and around a number of centers, each of which employs a number of distinct but related styles, each style in turn being in process of change both areally and temporally. With this remarkably unstable material, we set out to fashion a key to the prehistory of the region. Faced with this three-dimensional flow, which seldom if ever exhibits "natural" segregation, and being obliged to reduce it to some sort of manageable form, we arbitrarily cut it into units, Such *created units of the ceramic continuum* are called *pottery types*. (Phillips et al. 1951:62–63; emphasis in the original)

Thus, their types were arbitrary constructs founded on their notions of cultural change; types—rationalized as stylistic norms—were slices through the braided stream of cultures' shared ideas.

Phillips et al. constantly warned archaeologists of the dangers of blindly accepting and using their ceramic typology without revision. This warning indicates that their types were not classes in the true sense of the word (Chapter 5) but rather groups of objects, the definitions of which were extensionally derived from the particular piles of sherds. The authors also provided one of the most insightful comments on ceramic typologies ever written—the quote used to open this chapter—in that it identified the same paradox identified by J. O. Brew (1946) a few years earlier: the difficulty of conceiving of change in materialist terms but structuring observations in essentialist terms. In reviewing the development of Americanist archaeology (Lyman et al. 1997a, 1997b; O'Brien and Dunnell 1998; O'Brien and Lyman 1998, 1999a), one begins to realize that Phillips et al.'s warning to keep the sherds separate conceptually from the types was not repeated often enough. Otherwise, archaeologists might have taken notice of the materialist paradox.

The extent of the paradox was so great that even Phillips, Ford, and Griffin were pulled into it, to such an extent that Ford was made to publish some of his work elsewhere (Ford 1952). Whereas Ford (e.g., 1935a, 1949) viewed culture change as gradual and steady—a "placid stream of pottery continuity" (Phillips et al. 1951:427; see also Ford 1951:91–100)—Phillips and Griffin conceived of the flow as jerky or punctuated. This latter view emanated in part from the inherent discontinuity of stratigraphic units (Phillips et al. 1951:428); thus, the breaks in the flow that Phillips and Griffin perceived tended to correspond to stratigraphic boundaries (O'Brien and Dunnell 1998; O'Brien and Lyman 1998). But this is exactly as Griffin and Phillips had expected the archaeological record to be arranged—nice, orderly units that had highly visible discontinuities between them (see especially Phillips 1970). For Ford, however, pottery

> was developing in a continuum throughout its entire history in the Mississippi Valley, that whether new types evolve by modification of older ones or come in as new ideas from the outside, they take their place in an uninterrupted cultural flow. The logical consequence of such a view is that, in most cases a "mixed" pottery complex represents a single brief span of time on the continuum, an "instant" for all practical purposes, when both elements of the mixture were being made and used side by side. The importance of this postulation for the seriation method can hardly be exaggerated. Ford does not deny that mixed complexes sometimes do result from reoccupation of sites. Such collections he frankly banishes from his graphs and says so....
>
> Griffin and Phillips, on the other hand, while not rejecting the general theory of continuity, are inclined to feel ... that there are more instances of mixture through reoccupation of sites than Ford has recognized. In particular ... they have tended to see indications of at least one significant break in the otherwise placid stream of pottery continuity at the point where the tempering material shifts from clay to shell, in other words between the Baytown and Mississippi periods. They feel that, by including mixed collections on the graphs, Ford has effected a spurious transition that seems to prove his continuity hypothesis, but in reality leaves the question open. (Phillips et al. 1951:427)

Several decades of excavation in the meander-belt portion of the Mississippi Valley would demonstrate that although there was a change from clay-tempered pottery to shell-tempered pottery, in many localities that change was not as abrupt as Griffin and Phillips had forecast.

Phillips and Griffin were willing to admit that seriation and stratigraphy were "independent methods of analysis" (Phillips et al. 1951:239), and they correctly noted that the stratigraphic exposure of an excavation unit was accurate for that exposure only and should not be extrapolated over an entire site or over the surrounding region (Phillips et al. 1951:291). But Phillips and Griffin's essentialist notions of "abrupt cultural change" demanded in the end that they view many sites as stratigraphically "mixed" when they showed no such abrupt change (Phillips et al. 1951:291–292). Similarly, Ford's surface collections were viewed by Phillips and Griffin as mixed; thus, when seriated they consisted of false "transitional" collections "that are actually the result of reoccupation" (Phillips et al. 1951:292). Here, we are using the term *seriation* loosely (Lyman et al. 1998b; O'Brien and Lyman 1998, 1999a). What Phillips et al. meant by seriation was noting the percentages of pottery types in a surface assemblage and then using the percentages to place a site in a master chronological sequence that had been constructed through stratigraphic means. How Ford's colleagues *knew* the surface collections and the collections from their stratigraphic tests were mixed highlights their essentialist stance. They could not empirically test such an inference except to cite those cases of clear stratigraphic separation of types as accurate reflections of reality and to discount as mixed those cases where the types were associated stratigraphically.

An excellent example of Phillips and Griffin's perspective is given in the discussion of the pottery from the Rose Mound site in Cross County, Arkansas, where they excavated a single 2-meter-square unit in arbitrary 10-centimeter levels. Twenty-two levels were excavated, and the authors noted that "the stratification of this [unit] has a particularly important bearing on the interpretation of the pottery stratigraphy" with regard to the shift from sand-tempered to shell-tempered pottery, or what was then termed Baytown to Mississippian pottery (Phillips et al. 1951:286–287). A graph of the type percentages constructed as a bar chart (for which Ford was famous [Lyman et al. 1998b; O'Brien and Lyman 1998]) was superimposed on a composite stratigraphic profile that showed the location of the boundary between the two lowest stratigraphic units (Figure 3.7). Phillips et al. (1951:288–289) concluded (at least Phillips and Griffin did) that

> the effect of this slope [of strata boundaries] on the pottery stratigraphy is to introduce a spurious transition of a new and insidious kind.... The resulting pottery graph could not fail, therefore, to show a smooth transition in [pottery types]. To judge correctly the stratigraphic relationship between [the two stratigraphic zones, the five arbitrary levels containing the boundary between the strata] would have to be eliminated from consideration. *The resulting*

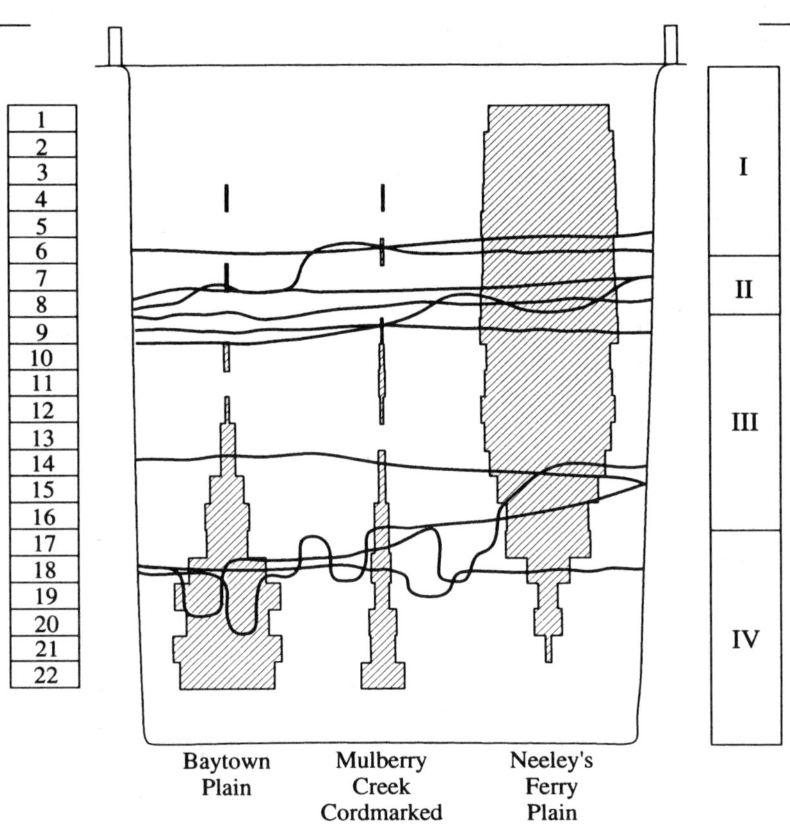

Figure 3.7. Composite profile diagram of Cut A, Rose Mound, Cross County, Arkansas, with relative abundances of three pottery types by excavation level superimposed. The column of numbers in boxes on the left refer to 10-centimeter excavation levels. Roman numerals on the right denote idealized strata obtained by averaging the four walls of the 2-meter-square excavation unit. Heavy undulating lines in the center are composites of strata boundaries using all four profiles (after Phillips et al. 1951).

> *conclusion is that there was an abrupt shift from a pure Baytown to an equally pure Mississippi pottery complex.* (emphasis added)

In reality, there was no such abrupt shift. Rather, through time, shell gradually replaced clay as the leading temper in the study area. The replacement was never complete, with the use of both tempers continuing up into the historical period. If Phillips and Griffin were aware of this, they were silent about it, perhaps because of the minority of clay-tempered pottery in later assemblages. Even they would have admitted that the abrupt replacement of clay with shell did not necessarily mean that everyone made the switch simultaneously. Diffusion, one of the stalwart mechanisms of culture change, could always be called on to account for the spread of a trait out from the initial center of activity. Ironically, it was the diffusionist views of Ford—who held out against Phillips and Griffin in terms of the clay–shell issue, preferring to view the replacement as an in situ development—that so irritated Phillips and Griffin (O'Brien and Lyman 1998) that they would not let him publish them in the survey monograph. Ford published them a year later as *Measurements of Some Prehistoric Design Developments in the Southeastern States* (Ford 1952), in which he attempted to align various chronologies from the Southeast.

Ford believed that the similarities of those various chronologies in terms of both sequence and pottery type denoted the various channels of the braided stream of the cultural continuum. He thus sought not only to "align ceramic chronologies in adjacent geographical areas" (Ford 1952:318) but also to "trace evolving strains [idea streams] through time and across space and measure them qualitatively and quantitatively" (Ford 1952:319). Ford (1952:322) explicitly noted that he wanted to "follow some of the more prominent streams of ideas," and he cited Kroeber's (1948:260) illustration of cultural evolution as "a tree in which the branches grew back together as readily and as frequently as they separated" (Ford 1952:322). We can label this *reticulate evolution*, differentiating it from *phylogenetic evolution*, the model that applies to organic evolution (Figure 3.8). Kroeber (1931a), Boas (1932), and Steward (1941) were concerned that this difference between cultural and biological evolution rendered the latter an inappropriate model for the former (Lyman and O'Brien 1997); this same difference has prompted several modern paleobiologists (e.g., Gould 1996, 1997a) to come to the same conclusion. Ford did not see that as fatal and in fact argued that the reticulations—the merging events—were important phenomena requiring explanation.

In his design-element monograph, Ford used pottery types to align excavated levels from various sites in each of the eight regions of the Southeast he examined (Figure 3.9). He then correlated the various chronologies by "focusing attention on the patterning formed by type maximums and by placing less weight on the apparent starting and stopping points of type occurrences" (Ford 1952:331). In remarkable anticipation of modern paleobiology's concern with accurately measuring the duration of a species' existence (e.g., Foote 1997a; Solow and Smith 1997), Ford essentially ignored the end points of each type's temporal duration because of potential problems such as mechanical mixing—a problem that he also saw as affecting his ability to produce nice, smooth curves that illustrated the inception, maximum appearance, and disappearance of a type. To counteract outside effects, he smoothed his type-frequency curves after ordering the collections in each regional chronology, acknowledging that the final arrangement probably was not correct in detail but that there probably were no major errors to

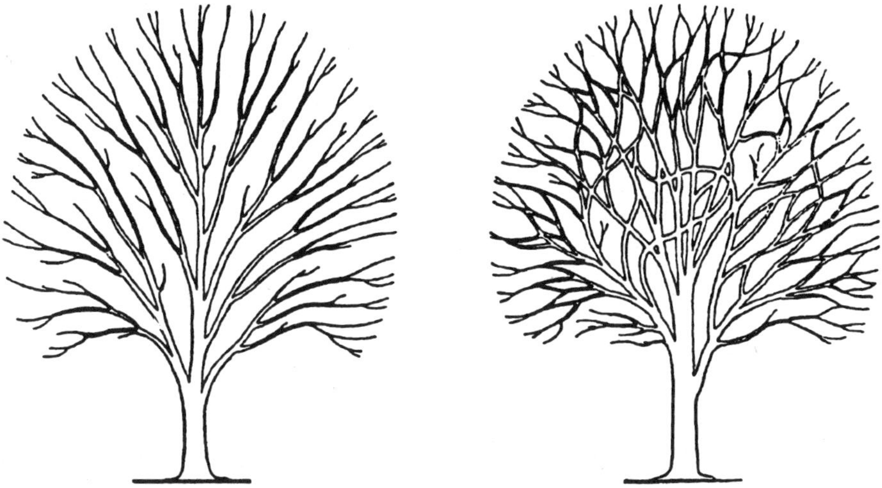

FIGURE 3.8. A. L. Kroeber's depiction of organic evolution (left) and cultural evolution (right). Note the resemblance of the reticulate tree of cultural evolution to Ford's notion of culture as a (braided) stream (after Kroeber 1948).

the general order, and begging the reader for acceptance of the arrangement for the sake of discussion.

In aligning the various regional chronologies, Ford (1952:329) was attempting to track the movement of what he termed "cultural influences." This simply followed an axiom of culture history that had been around since before the turn of the century and which was formally stated by Willey (1953:363): "Typological similarity is [an] indicator of cultural relatedness." Once areal chronologies were aligned, Ford (1952:343) examined the histories of what he termed "eight ceramic decorative traditions," each representing a "selected stream of ideas across space [that had developed] through time." He talked about one design type leaving "no direct descendants" (p. 347), of one design type evolving from another (p. 350), and of particular design types and elements displaying "close kinship" to one another (p. 355). In other words, formally similar types were phyletically related, with their similarity being of the homologous sort. There was no theoretical basis for such interpretations, only Ford's commonsensical notion of the braided stream of cultural evolution.

Phillips and Griffin were not the only archaeologists who were unimpressed with Ford's work. In one of the great statements from the culture-historical period, Albert Spaulding (1953a:591) compared Ford's smoothing of the frequency curves of types to a "bloody amputation." Spaulding (1953a:589) viewed Ford's measurements as nothing more than "counting and ranking" of pottery types—a

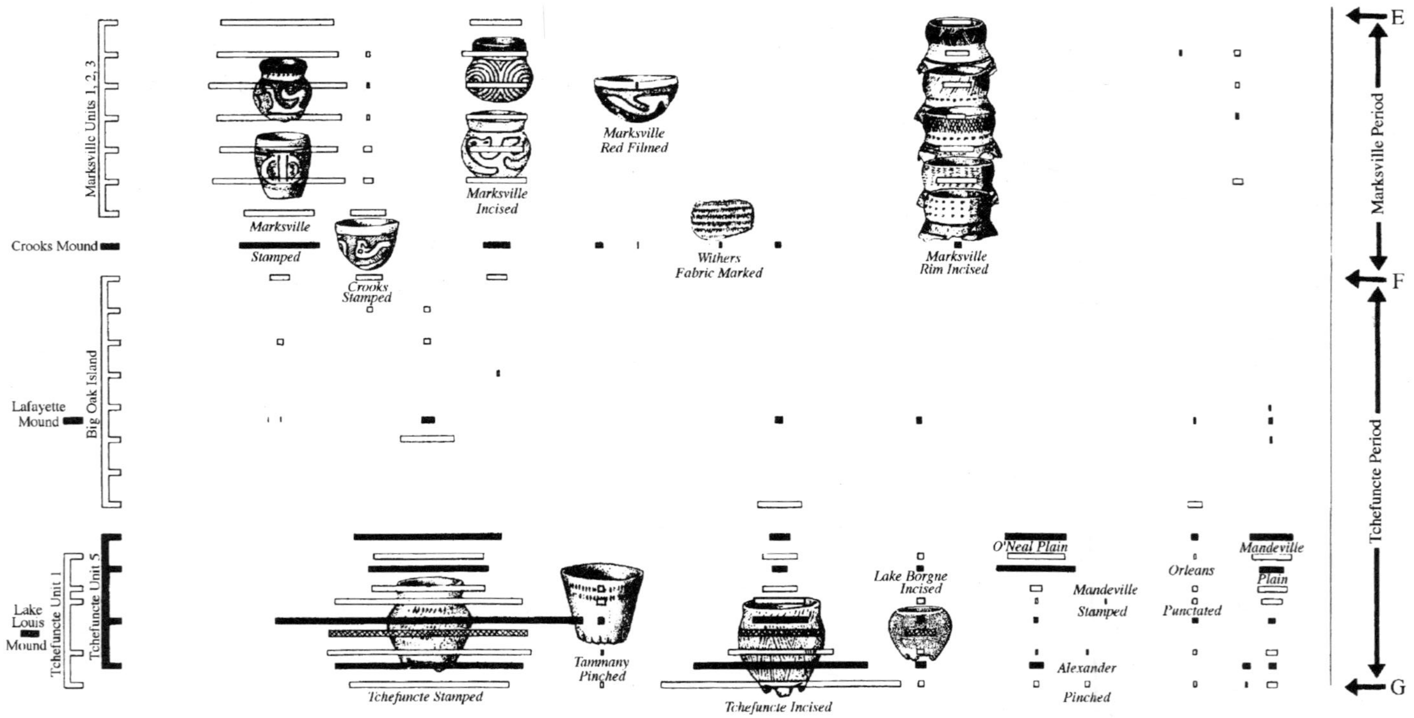

FIGURE 3.9. James A. Ford's graph of ceramic chronology for the Mouth of the Red River area based on excavations in Mississippi and Louisiana. Pottery types are represented by the vertical columns; bars represent the percentage of each type in a particular excavation unit at various sites (left-hand column). Arbitrary periods are shown in the right-hand column. Only the earlier 40 percent or so of the ceramic sequence is shown here (after Ford 1952).

procedure that in his opinion was neither measurement nor scientific. Ford, according to Spaulding (1953a:590–591), could "only assert that his final ordering is the best possible under the circumstances and only reproduce the graphs to substantiate the assertion. The absence of any mathematical expression of degree of fit leaves the skeptical reader with no recourse other than reproducing the component histograms and trying new arrangements himself, a task which is made difficult by the absence of the original counts on which the percentages were calculated."

Spaulding (1954a:113) later pointed out that in Ford's study, "time relationships were in large part inferred by means of relative frequencies of ceramic types[, but type] counts and locus, not time, were the empirical data of the study. [Thus a] general murkiness of exposition, of which the infelicitous use of measurement is an example, is a persistent stumbling block in understanding the arguments advanced." Spaulding was saying, in so many words, that one cannot hope to measure geographic distance with units such as years, that one cannot measure time with units such as kilometers, and most importantly, that one cannot measure temporal change in *cultural ideas* with *arbitrary* units called types. To Spaulding (1953b, 1954b), statistical analysis was the only archaeologically useful means of creating artifact types, because it allowed the analyst to get close to what prehistoric artisans had in mind when they created objects. Thus, in his review of Ford's design-element study, Spaulding (1954a) was questioning the basis for Ford's inference that his arrangements of type frequencies in time and space—when those types were arbitrarily defined—actually measured the temporal flow of streams of cultural ideas. Ford had no theoretical basis for such an inference and could only suggest that Spaulding's empirical challenge was founded on "a lack of familiarity with certain characteristics of the mechanics of cultural development and diffusion which were used as basic assumptions" (Ford 1954b:110). Ford never made clear what those mechanics were, but they must have been the popularity principle and, by implication, heritable continuity: People had ideas, those ideas were transmitted and became popular, then they died out—were transmitted less often—and were replaced by other ideas, some of which diffused from other localities.

CULTURE HISTORY AND THE CLASSIFICATION OF ARTIFACT AGGREGATES

Early twentieth-century Americanist archaeologists worried not only about how to measure time and how to classify artifacts in such a manner as to allow them to measure time, but also about how to build units larger than types of discrete objects—units that would allow comparison of the cultural manifestations they studied. They worried about how to classify *cultures*, as those were represented by the sets, or aggregates, of artifacts that were regularly found associated with each other. Ultimately, archaeologists wanted to determine how such aggregates of forms were related in time and space and how they were related to each other developmentally. A number of schemes for classifying artifact aggregates were proposed over the years; our focus here is limited to three of the better known ones. The three systems share a few commonalities, one of which is that they all are taxonomic structures, but they differ radically from one another. One omitted time from initial consideration. The other two attempted to discover phylogenetic history, but one was based on a biological analog and the other on an ethnological analog.

The Midwestern Taxonomic Method

By the 1930s, Americanist archaeologists had come to something of an impasse over the means and terms used to describe and discuss assemblages of artifacts. The term *culture* was ubiquitous in the role of a grouping unit, but it varied tremendously in scope and meaning from one application to the next. The Midwestern Taxonomic Method (MTM), which from the start was intimately associated with William McKern (1934, 1937, 1939, 1940, 1942, 1943, 1944), was created in the 1930s to rectify these problems by providing a broadly useful means of categorizing assemblages. In outline, the method is quite simple. The building blocks of the method are *components*, which are assemblages of associated artifacts believed to represent specific occupations of given locales by specific groups of people. Thus, a component is not equivalent to a site unless a place has experienced only a single occupation (McKern 1939, 1944). In practice, components comprise a list of traits and are defined on the basis of some of those traits. Similarity in trait lists is used to assign components to groups, five levels of which are recognized. From least to most inclusive, these are *focus*, *aspect*, *phase*, *pattern*, and *base*. Components belonging to a single focus share most traits, and the traits (e.g., particular designs on ceramics) are specific. Foci that share many, usually more general, traits (e.g., technique of ceramic decoration) than those used for establishing foci are grouped as an aspect, and so on until one reaches the base, where only a few, often inferential, traits (e.g., agriculture) are held in common. Three kinds of traits are distinguished: *Linked traits* are common to more than one unit; *diagnostic traits* are limited to a single unit; and *determinants* are traits that occur in all members of a unit but in no other unit (Figure 3.10).

McKern was explicit that the units had no intrinsic temporal or spatial meaning; such meaning had to be acquired independently of the method, which was founded purely on the formal similarity of the units (Spaulding 1949, 1957). In modern terms, the method is an application of numerical phenetics, or numerical taxonomy, without recourse to any actual measurement of similarity beyond simple inspection. Indeed, McKern's contemporaries were quick to point out that the failure to measure similarity quantitatively was a major flaw in the method (Kroeber 1940, 1942), but it also was a major plus in that components were easily

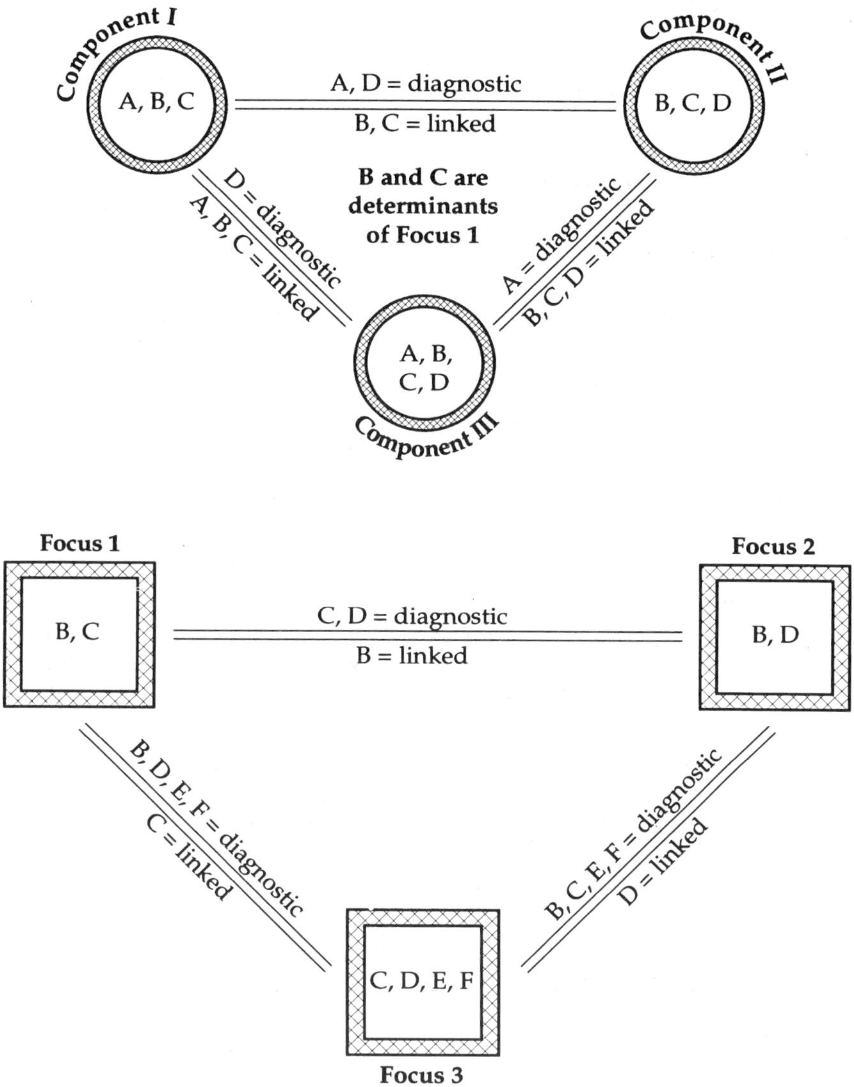

FIGURE 3.10. The analytical relations among traits (capital letters), components (circles), and foci (squares) in the Midwestern Taxonomic Method. Because components I, II, and III share traits B and C, they are linked traits in this comparison; other traits are diagnostic. Traits B and C are determinants of focus 1; components not containing traits B and C are not members of focus 1 (after Lyman et al. 1997b).

classifiable. Inspection of components assigned to a focus or higher-level unit resulted in the identification of traits common to all *and* limited to members of the unit. Such identifications provided criteria for the addition of future members: "Classification is nothing more than the process of *recognizing* classes, each class identified by a complex of characteristics" (McKern 1939:304; emphasis in the original). Once the set of determinants has been determined, "any manifestation found to show this complex of [determinant] traits may be classified by definition" (McKern 1940:19). This meant that the particular traits that met the conditions—they were present in all inspected units but in no others—depended on the components polled. Thus, the list changed as knowledge of the archaeological record grew and more components were examined. All foci, aspects, and so forth, were nothing more than historical accidents comprising assemblages known at the time the determinants were derived. Interestingly, such extensional derivation of determinants is reasonable only if one presumes cultures are static and that they change abruptly from one state to another, directly contradicting the view of culture change as more or less gradual and continuous.

A *component* was defined as "the manifestation of any given focus at a specific site" (McKern 1939:308). Units of this sort were often referred to by McKern as "manifestations" of one of his larger units. A component was a "real" unit made up of a set of associated artifacts. McKern (1940:18) offered an ethnological rationale for the trait lists derived from components: "A complex of traits determined at a site [read *a component*] may represent the customs and ways of living (culture) of a local group representing a single variety, or band, of Indians, sharing an identical lot of cultural habits." This commonsense warrant is also an inference, but it is one that is impossible to test in a nontautological manner without the help of a theory. One important issue remains, however. Trait lists are constructed from a set of associated artifacts, but how does one determine if two or more artifacts are associated? The answer is, of course, by using some sort of depositional (empirical) unit—such as a stratum—that has visible boundaries in order to conclude that a set of artifacts contains items that not only are associated archaeologically but were associated systemically (Schiffer 1972) as well.

Several archaeologists observed, not surprisingly, that the cultural continuum was obscured by the MTM. Jesse Jennings (1947:192–93), in what we would label as a materialist statement, complained:

> When a continuous flow of cultural traits can be seen to persist from period to period, the new periods being different in the possession of new traits, but retaining many old ones, the injection of a phase, focus, or pattern, before admitting a chronological and/or genetic relationship seems to serve no purpose beyond adding to the suspense.... [F]orcing a clear, sharp stratigraphic sequence of cultures whose distant cultural equivalents are known, and whose chronological position is amply demonstrated [, into] the [Midwestern Taxonomic Method's] classificatory pigeon-holes is wasted effort.

Comparing trait lists, demonstrating similarities and differences between such lists, showing their relative chronological positions, and thereby establishing the relationships of the lists, "paid the historical debt [and is] what archaeology is expected to do," according to Jennings (1947:193). Spaulding (e.g., 1957) thought this was nonsense, as the MTM was explicitly concerned solely with the formal dimension; the temporal and spatial dimensions were to be added later, and then the phylogenetic relations of the units would be determined. Most archaeologists jettisoned the MTM, preferring a system that looked at time first. If it did other things as well, fine, but primary emphasis had to be on time. But do not be misled into thinking that the reason archaeologists discarded the MTM was because it did not view time as a continuum; it was jettisoned because it did not include time at all.

Although its terminology passed from the scene in the 1960s, units such as "Archaic," "Woodland," and"Mississippian"—units still very much alive—either originated as higher-level units in the MTM or were passed on to us through it. Many phases currently in use in Americanist archaeology, particularly in the Southeast, are the same constructs first proposed as foci in the MTM (O'Brien 1995, 1996a). Thus, many of the basic units of prehistory are historical accidents that exist solely because particular assemblages were known when the MTM was applied in a given area (Fox 1992, 1998; O'Brien and Fox 1994a, 1994b). The impact of the MTM on contemporary ideas of prehistory remains enormous, if largely unremarked (e.g., Chang 1967; Trigger 1968). One of its greatest impacts was on the archaeology of the American Southwest, though the basis of the taxonomy used there was very different than that which underlay the MTM.

Southwestern Taxonomy

The Globe Conference of 1931, convened by Harold S. Gladwin and held at Gila Pueblo, Arizona, was organized to deal with Hohokam materials from southern Arizona in a manner not unlike that in which Kidder's Pecos Conference was handling much of the rest of the Southwest. A system of pottery nomenclature and classification loosely based on a biological model had been adopted a year earlier (Gladwin and Gladwin 1930), and from this emerged a "comprehensive scheme by which relationships and relative chronology could be expressed" (Gladwin and Gladwin 1934:8–9). Harold Gladwin and his wife Winifred believed that all prehistoric groups in the Southwest had passed through the same cultural stages but at different times in different places—classic Morgan–Tylor unilinear cultural evolution. The Gladwins' system of nomenclature was taxonomic in structure and consisted of four basic units. *Roots* represented groups of people; *stems* represented distinct geographic areas occupied by specific groups of people (root); *branches* corresponded to "culture areas" (at a much lower and finer level than Kroeber's [1939] culture areas) and represented "the various [cultural] developmental series [of phases] which have sprung from roots" (McGregor 1941:6); and *phases* were time- and space-bound cultural variants within branches that contained characteristic types of artifacts and architecture (Gladwin and Gladwin 1934:9–10).

Shortly after the original formulation appeared in print, Gladwin (1936:258) felt that because the use of "a generic and a specific name for pottery types implied a biological analogy [it] was a mistake. The idea is being carried too far along biological or zoological lines, and men do not realize the profound differences which exist between zoological species and the things which have been made by men and women. Zoological species do not cross and intergrade[, and biological evolution] is so slow as to be hardly distinguishable." He believed cultural evolution was much more rapid and involved "merging and cross-influences.... We are really dealing with [cultural] varieties rather than species, and, in consequence, there are bound to be a great many intermediate and transitional types" (Gladwin 1936:258). Gladwin clearly was correct about differences in evolutionary rates between biological and cultural evolution; where he made his mistake was in equating what archaeologists called a type—its name—with a biological genus and species.

By the 1930s, literally hundreds of pottery types had been created, and standard practice was to use a binomial system in proposing a type name—St. John's Polychrome, for example. In turn, some archaeologists viewed type names as biological analogs, though there was no logical reason to do so. Some undoubtedly used the analog metaphorically, but others placed considerable stock in such nomenclature, arguing that it was possible to arrange pottery types phylogenetically. Thus, not only were type *names* used analogically, but also the types themselves.

Gladwin's scheme was modified by Harold S. Colton and Lyndon L. Hargrave (1937; see also Colton 1939, 1942) by "fusing it with McKern's" method (Kluckhohn 1939a:160) for the purpose of understanding not only chronological issues but cultural-developmental issues as well. This was "a step in the right direction," proclaimed Clyde Kluckhohn (1939a:160), who had been worried about determining which traits were diagnostic of (essentialist) cultural units; he even used the term "type fossils" (in quotes). On the one hand, given Kidder's (e.g., 1915:453; Kidder and Kidder 1917:348) earlier discussions of pottery types as having not only ancestral–descendant relationships but also the capacity to become extinct, it is perhaps not surprising that Gladwin at least initially took the biology-like approach that he did. On the other hand, Colton was a biologist by training (Brew 1946; Gallagher 1978), and Hargrave had been trained as an archaeologist but had a deep avocational interest in biology, particularly ornithology, and had spent most of his career working for and with Colton (Carothers 1978). Both Kidder's and Gladwin's writings, and Colton and Hargrave's intellectual expertise and interests, probably contributed greatly to the latter two adopting Gladwin's ideas even as Gladwin was abandoning them.

Colton (1939, 1942) in particular followed Gladwin's lead. Colton's (1939:3) goal was "to outline the cultural units of Northern Arizona and show ... their relationships." He believed that use of archaeologically accessible material culture traits should permit one to establish how "closely related" prehistoric cultural units were (Colton 1939:5). Marked differences in material culture denoted distantly related (if at all) entities that "might loosely [be called] tribes" (Colton 1939:5), and similar material cultures denoted closely related tribes. Following McKern (1934, 1937, 1939), "the characteristics that distinguish these 'tribes' are called determinants" (Colton 1939:5). Following Gladwin and Gladwin (1934), a phase was "a fundamental unit in the culture of a human population of an area" made up of "contemporaneous prehistoric sites [the] determinants [of which show] great resemblances to one another"; "genetically related [phases] are grouped together into a branch[, which we] can think of ... as representing an Indian tribe" (Colton 1939:6).

For Colton, following Gladwin, related branches made up a stem, and related stems constituted a root. Thus, unlike the MTM, which expressly disavowed such phylogenetic implications, Gladwin—at least initially—generally, and Colton particularly, offered a theoretical warrant of sorts for their groups. As it turns out, Colton seems merely to have borrowed the units and grouping procedure of the MTM, applied some of Gladwin's terms to some of the units, and ultimately applied to the units a set of evolutionary terms that were generally understandable.

Reviewers focused on Colton's phylogenetic interpretations rather than on his methods. On the one hand, Erik Reed (1940:190) thought Colton's scheme was a reasonable one: "The genetic and temporal approach seems more desirable in a region such as this where chronology is relatively well-known." On the other hand, Julian Steward (1941:367) remarked,

> It is apparent from the cultural relationships shown in this scheme that strict adherence to a method drawn from biology inevitably fails to take into account the distinctively cultural and unbiological fact of blends and crosses between essentially unlike types.... It is true that cultural streams often tend to be distinct, but they are never entirely unmixed and often approach a complete blend. A taxonomic scheme cannot indicate this fact without becoming mainly a list of exceptions. It must pigeon-hole.... The method employed inevitably distorts true cultural relationships.

In a later statement, Colton (1942:34) explained that the scheme he had borrowed from Gladwin and used for his 1939 monograph was "a classification of prehistoric human groups" and was not a classification of cultures. This was an interesting distinction in that Colton had his eye on local groups, not on loose amalgams of groups shoved under the anthropological umbrella of a "culture." Thus he hinted at a theory of transmission—person to person within a group—but did not clarify the fact that ideas, rather than genes, were transmitted. Although

Colton may have implicitly recognized the significance of what he was proposing—and certainly he possessed the requisite background in biology to have done so—his lack of exposition of transmission mechanisms incorporated within an explicit and well-developed theory of cultural development failed to convince his contemporaries that what he was doing was somehow warranted. This was in many respects an unfair criticism, but certainly an understandable one in light of archaeology's general distrust of adding nonsomatic features to the evolutionary mix.

In the Southwestern systems, the phase was the basic unit of interest. For Gladwin (e.g., 1937) and others, a "phase was assumed to have a certain temporal uniformity [and] the search went on to discover what the period of regularity was" (Olson 1962:459). In the 1930s, "a dominant pottery type was the prime indicator for phases" (Olson 1962:460); such types were, following Kidder, defined based on specimens at hand and subsequently refined in the light of new evidence until they became type, or index, fossils that occupied very limited time–space positions (O'Brien and Lyman 1999a). Phases and larger units were identified on the basis of the presence or absence of those index fossils.

Stratigraphic excavation allowed a phase's characteristics to be extracted from the vertical—and thus time-sensitive—spatial units of archaeological sites. Either natural stratigraphic units or arbitrary levels provided sets of artifacts that were by definition temporally associated. From those assemblages could be extracted a list of traits—manifest as artifact types—that had demonstrably (after some trial-and-error refinement) limited temporal distributions. In turn, comparing these temporally diagnostic traits across the geographic space occupied by multiple sites allowed one to select those traits that were linked—held in common by multiple components—and those that could serve as determinants—distinguished by vertically superposed components. Those index fossils came to designate cultures. Such a progression toward index fossils was not unique to the Southwest; Ford was using index, or marker, types in the Southeast as early as the mid-1930s (Ford 1935c, 1936a).

Index fossils had long been useful in paleobiology for precisely the same reason they were and are useful in archaeology: They allowed one to correlate sets of material or strata across broad geographic ranges (Hancock 1977; Mallory 1970; Rudwick 1996). Generally, one used stratigraphic discontinuities to specify a set of fossils that, through extensional definition, provided the necessary and sufficient conditions for unit membership, being sure to include within a fossil species only a range of variation in form similar to that evidenced by extant species (O'Brien and Lyman 1999a). The necessary assumption for correlation is that "similar organisms in different [strata] imply some kind of equivalence in their enclosing matrices" (Eldredge and Gould 1977:25), such as the age of the depositional event comprised by the correlated strata. "Empirical paleontology" suggested some degree of stasis in species units over time and space, allowing biostratigraphic correlation (Eldredge and Gould 1977:29). The empirical archaeological record was interpreted similarly and resulted in the development of two units—components and phases—that conceptually warranted the basis of such correlation across that record. Such units were central to the third scheme for classifying aggregates of artifacts.

The Willey–Phillips System

Gordon Willey and Philip Phillips waded into the fray in the early 1950s with a new proposal for carving up the time–space landscape. Their first statement, in a paper entitled "Method and Theory in American Archaeology: An Operational Basis for Culture-Historical Integration" (Phillips and Willey 1953), was an attempt to formalize the classification of aggregates. That paper was combined with a later one (Willey and Phillips 1955) to form the basis of a book, *Method and Theory in American Archaeology* (Willey and Phillips 1958) that summarized much of the conceptualization of Americanist archaeology and prehistory in the 1950s. Their 1953 article appears to have grown out of Willey's (1953) perception of a growing disparity between practice and the original tenets of the MTM, which over the years had evolved into a haphazard system for measuring time and space. The MTM was founded strictly on formal similarity, with no pretext that similarity denoted historical relatedness. This tenet was anomalous, because as we noted earlier, it had long been axiomatic in Americanist archaeology that typological similarity was an indicator of cultural relatedness. Thus, despite the protestations of McKern, "a common or similar history" for units judged to be similar was automatically implied (Willey 1953:363–364), and archaeologists used it that way. How could the intended and practiced uses of the MTM be reconciled?

To begin, Willey drew a contrast between what he termed two archaeological theories, and it is worth quoting his words at some length:

> [T]heories of culture change and continuity are fundamental to archaeological studies.... [T]he treatment of archaeological assemblages in any historicogenetic system [e.g., Colton 1939; Gladwin and Gladwin 1934] has a basis in theories of continuity and change. Even if space and time factors are not formally observed, principles of continuity and change are expressed in the degrees of trait likeness or unlikeness which are the mechanics for establishing the genetic lines binding the assemblages together....
>
> The processes by which, or through which, cultural continuity and change are maintained or accomplished have not received study and reflective thought commensurate with the way these concepts have been invoked by American archaeologists. "Evolution" and "diffusion" have been tag names employed, but these are broad categories rather than specific explanations, and there have been few clear theoretical formulations along these lines.... [Concerning the Gladwin and Colton schemes] there is a rather simplistic

> evolutionary or genetic analogy at work here. To be sure, there is some universal basis for expressing the development of human culture in this fashion.... [But in some cases] it appears that evolutionary theory has been very naively applied....
>
> Diffusionist theory in American archaeology has probably received more analysis, or analytical speculation, than has evolutionist theory. It is at the core of most archaeological interpretation. Trade, migration, gradual borrowing, and idea or stimulus diffusion have all been advanced in specific instances.... As with evolutionary hypotheses, theories of diffusion may be legitimately brought forward to explain various patternings in space–time distributions. (Willey 1953:368–369)

Phillips and Willey (1953:617) suggested that "an archaeological culture is an arbitrary division of the space–time–cultural continuum" and thus followed some of their predecessors in conceiving of culture history as a braided stream—a materialist conception. They began their discussion by noting that any archaeological unit is "the resultant combination of three unlike basic properties: space, time, and form" (Phillips and Willey 1953:617). This was followed by explicit definitions of three spatial units of different scales—*locality* (often a site), *region*, and *area*—the first two of which they aligned, admittedly tentatively, with a community (or local group) and a tribe (or society), respectively. This reflected their essentialist viewpoint: Their units had some reality in a social or anthropological sense that could perhaps be discovered. Their formal units began, not surprisingly, with *components* and *phases*. The latter was chosen over McKern's *focus* because Phillips and Willey preferred the "stronger temporal implication" of *phase*—precisely the point Gladwin (1936) had made earlier.

A component was approximately equivalent to the sociological units of band, neighborhood, and village, and a phase was approximately equivalent to a society. This alignment of archaeological units with sociocultural ones also revealed that Phillips and Willey were thinking in essentialist terms: Their component and phase units were justified because they had some objective and empirical reality. There was no theoretical basis for such interpretations of components and phases, though the interpretations rendered in ethnographic terms provided a commonsensical warrant for the archaeological units in lieu of a theoretical one. Recall that it was Phillips (1955:246–247) who two short years later coined that oft-repeated phrase "New World archaeology is anthropology or it is nothing." Such a position had already cast a cloud over culture history—a then-unappreciated sign that a storm was brewing.

Following McKern (1939), Phillips and Willey (1953:619) defined a component as "the manifestation of a given [phase] at a specific site." A phase was conceived of as being temporally, spatially, and formally "sufficiently characteristic to distinguish it from all other units similarly conceived" (Phillips and Willey 1953:620). Components could be "combined into phases because analysis reveals cultural uniformity amounting to practical identity" (Phillips and Willey 1953:630). Unlike in the MTM, a Phillips–Willey phase could be defined on the basis of a single component. The necessary and sufficient conditions for membership in that phase could then be modified on the basis of newly discovered components. For example, when several local sequences—where "a local sequence is a series of components found in vertical stratigraphic relationship in a single site" (Phillips and Willey 1953:623)—are correlated, the correlation is generally "accompanied by a progressive generalization of the definitions [of phases] until their original usefulness is impaired.... [I]n the process the original formulations [of local sequences] are retailored to fit the wider spatial and (perhaps) deeper temporal requirements" (Phillips and Willey 1953:623).

In short, phases were units that were largely dependent on the first component or set of components specified. The problem was that these units, once put into the literature, never went away. As time passed, and as other investigators made reference to them, they became more real. For example, look at a fairly recent correlation between archaeological phases as mapped by Phillips (1970) in the central Mississippi River valley and ethnic provinces identified by Hernán de Soto when he traversed the region in the early 1540s: "The De Soto expedition crossed the Mississippi near Commerce Landing in the province of Quizquiz (east Walls phase), crossed to Aquixco (west Walls phase), went inland to Casqui (Parkin phase), back east to Pacaha (Nodena phase), visited the Ste. François Mountains and the Greenbrier phase, retraced their steps to Casqui, and went south to Quigate (Kent and/or Old Town phases) and then farther west to the Ozarks (Quapaw phase) before turning south toward Little Rock" (Morse and Morse 1983:313). Such a correlation is totally unwarranted, especially in light of the fact that the phase boundaries are based on nothing more than a few variations in such things as pottery-type percentage and the amount of shell temper found in the paste of ceramic vessels (O'Brien 1995, 1996a; O'Brien and Wood 1998).

Despite Phillips and Willey's attempt to provide a solution to the time–space conundrum, the usual problem remained: How were sets of things to be identified so that their similarities could be measured? Phillips and Willey's first requirement was to identify the boundaries of the sets. What helped here was the suggestion that "a site or a level within a site" could be a single component and thus represent a phase (Phillips and Willey 1953:620). Hence, spatial boundaries, whether naturally or arbitrarily defined, dictated what went into a component and thus into a phase; all artifacts within the boundaries were included within the sets of things to be compared. Otherwise, Phillips and Willey were as obtuse as their predecessors: Phases are classes of components, and components are empirical manifestations of phases. Thus, one must have the phases in order to recognize or identify components, but one needs components to write the phase definitions. The result is that a phase is defined on the basis of one component. This is acceptable given Phillips and Willey's warrants for their units: Phases are somehow real, being approximately equivalent to real societies, as required by their notion "that you cannot hope to shed light on processes by means of abstractions that have no theoretically possible counterparts in cultural and social 'reality' " (Phillips and Willey 1953:629). Their "theory," unfortunately, is not theory in the usual sense but rather a hypothesis concerning a possible relation between an archaeological unit and an ethnological unit. And, as with the components and foci of the MTM, the only way Phillips and Willey's components could provide definitive criteria of their phases was if the continuum of culture change over time were somehow discontinuous, else even a component of short duration might be a mixture of an ancestral and a descendant phase.

What, precisely, then, was a component? That a component was supposed to be equivalent to a band, neighborhood, or village provides a clue. It was, as in the MTM, the manifestation of a single occupation; thus, it was a set of temporally, spatially, and behaviorally associated artifacts. As Willey (1953:363–364) noted,

> The concept of the assemblage is implicitly grounded in the historical validity of the artifact–feature complex as a *unit*. Such a unity, by the very nature of its internal associations, bespeaks spatial–temporal correlates.... [T]here is [therefore] covert historical theory in the assemblage concept.... The unity of the assemblage, if historical unity can be assumed, must lead to the conclusion that we are dealing with the remains of an integrated cultural complex in the case of the *component*.

The locations of the temporal and spatial boundaries of an assemblage were vague because that is all they could be in the absence of theory. Stratigraphic boundaries were thus typically called on in lieu of a theoretically informed specification of boundaries. In paleontology, "ideally, the boundaries of taxa are wholly arbitrary [given the evolutionary continuum]; in the real world, we rely on convenient [stratigraphic] breaks in the fossil record" (Eldredge and Gould 1977:29). The reasoning of most archaeologists is similar—stratigraphic boundaries are empirical and do not require theoretical specification. As Rouse (1953:59) noted, stratigraphic excavation provides collections of artifacts that "are segregated by layers of refuse and subdivisions thereof, in order that the succession of *occupations* may be determined" (emphasis added). This echoed Wissler's (1917a), Fowke's (1922), and Kroeber's (1925) earlier remarks that a single continuous occupation will occur in an unstratified deposit, whereas multiple occupations will each occur in a separate stratum. The creation of artificial boundaries, as when one excavates in arbitrary levels, was to be avoided in order to "prevent contamination of one classified culture type by another" (Phillips 1942:200).

To Phillips and Willey, the apparent stability of phases was not totally an artifact of the geological boundaries of their member components. The "material traits [of a phase] can, under certain circumstances, be remarkably stable" (Phillips and Willey 1953:622). One could construct local sequences, regional sequences, and period and area chronologies that increase the spatial scale and "still maintain contact with the primary stratigraphic data" (Phillips and Willey 1953:624) but simultaneously reflect Kidder-like peaks of occurrence. This is the essentialist metaphysic expressed as a post hoc warrant for the construction of phases and similar units. To extend the metaphor, the materialist's braided stream of culture no longer flows gradually and continuously; its rate of flow is jerky—generally slow but punctuated by temporally brief, fast stretches (see Spaulding 1957). This notion that there are periods of stasis in the flowing braided stream of cultural history is one significant legacy of the culture-history paradigm and one to which we will return.

In their later formulation, Willey and Phillips (1958:24) indicated that as "typological and stratigraphic analyses become more refined, it often becomes desirable to subdivide phases into smaller (primarily temporal) units, and it seems best to regard these as *subphases*." Subphases can be distinguished by "differences [in] a few specific items of [phase] content or where such differences are expressible only in variations in frequency" (Willey and Phillips 1958:24). As new information came in, phases could be chopped up into smaller units and redefined in light of the new information. In effect, Willey and Phillips were trying to bring the level of their focus down to smaller chunks of the time–space continuum that more accurately reflected small groups of people, a gambit that Phillips (1970) continued with his type–variety system of pottery nomenclature for the lower Mississippi River valley.

But the central problems remained. How was one to reconcile the intended use of the MTM and how it was actually being used, and how was one to employ the explanatory "theories" of evolution and diffusion? How was one to determine evolutionary relations between phase units and then explain them? Phillips and Willey (1953) suggested that the use of *traditions* and *horizons*, or *horizon styles*, would reflect the braided stream of the evolution of cultures and allow one to correlate phases. As Rouse (1954:222) noted, a metaphor could be drawn between the use of horizons and traditions as integrative devices for archaeological materials distributed across an area and a rectangular piece of cloth, the side edges representing the geographical limits of the area and the top and bottom edges representing the temporal limits:

> The warp threads of the cloth consist of a series of regional traditions running from the bottom towards the top of the cloth, while the weft is composed of a number of horizon styles which extend from one side of the cloth towards the other. The cloth is decorated with a series of irregularly arranged rectangles, each representing a single culture [read *phase*], and these are so colored that they appear to form a series of horizontal bands. (Rouse 1954:222)

But Rouse (1955) saw as too simplistic Willey's dictum that typological similarity denoted cultural relatedness. Rouse wanted to determine the phylogenetic relations among phases rather than merely track the distributions of a few artifact classes, as the horizon and tradition units did. To illustrate this, he identified three techniques for correlating phases. First, one might use an MTM-like procedure to group phases that shared traits. Why the traits were shared was a separate issue. Second, one could note similarities in the time–space distributions of phases. Identical or adjacent distributions of two or more phases "establishes contemporaneity and contiguity, or lack thereof, and nothing else" (Rouse 1955:717). To argue that contemporaneous phases were phylogenetically related "because they share a given horizon style ... is on the genetic rather than the distributional level of interpretation, for it requires an assumption that the style has diffused from one phase to the others with little or no time lag" (Rouse 1955:718). Third, one might trace the "genetic" relations among phases by establishing that the phases had been in "contact" temporally and spatially by using horizons and traditions that comprised homologous types (Rouse 1955:719). One needed to distinguish between analogous and homologous similarity, the linchpin to this kind of comparison, to ensure the relations were "genetic." We discuss later how homologous similarity was distinguished archaeologically. Here, the critical point concerns Willey and Phillips's thoughts on Rouse's comments.

Willey and Phillips (1958:31) argued that Rouse's "genetic" relations could only "be revealed and expressed by means of integrative concepts that are culturally determined." The phrase "culturally determined" was critical. Horizons and horizon styles, by definition, reflected cultural transmission or diffusion over space. A tradition was "a (primarily) temporal continuity represented by persistent configurations in single technologies or other systems of related forms" (Willey and Phillips 1958:38) that operated at Rouse's (1955) "genetic level of interpretation" and reflected transmission or heredity across time. Willey and Phillips's conception of cultural development was well captured by the braided-stream metaphor. Each trickle was a tradition that to varying degrees met and mixed with other trickles as denoted by horizons and horizon styles. The problem, as Rouse had pointed out, was that such a conception assumed that the typological similarities denoted by horizons and traditions were of the homologous sort. Willey and Phillips (1958) did not comment directly on this critical issue. In fact, they did something quite different.

For Willey and Phillips, horizons and traditions provided the empirical warrants for discussing the historical development of cultures. They were "integrative" units that denoted "some form of historical contact" *rather than* "implications of phylogeny" (Willey and Phillips 1958:30). In Willey and Phillips's (1958:30–31) view, the writing of culture history demanded "culturally determined" integrative concepts such as horizons and traditions, *not* phylogenetic ones. Willey's (1953:368) suggestion that "principles of continuity and change are expressed in the degrees of trait likeness and unlikeness which are the mechanisms of establishing the *genetic* lines binding the assemblages together" (emphasis added) thus was purely metaphorical. But cultural or historical "relatedness," when couched in a temporal framework aimed at studying the developmental lineages of cultures such as that envisioned by Willey and Phillips, cannot fail to be phylogenetic in the sense of Darwinian evolution. The generally disciplinewide abhorrence of anything Darwinian and/or genetic resulted in this contradiction internal to the Willey–Phillips scheme going unrecognized. Thus, it is not difficult to figure out which of Willey's (1953) two explanatory "theories"—evolution and diffusion—was favored by the majority of culture historians. The processual archaeology that emerged in the 1960s tried to bring evolution back into the interpretive tool kit, but it was not the phylogenetic, Darwinian sort that culture history had discarded.

AT THE END OF THE 1950s

In their later formulation, Willey and Phillips (1958:16–17) modified their original wording and indicated that they now preferred "the concept of an archaeological unit [read *culture*] as a provisionally defined segment of the total continuum, whose ultimate validation will depend on the degree to which its internal spatial and temporal dimensions can be shown to coincide with significant variations in the nature and rate of cultural change in that continuum." Archaeologists were to make every effort to "understand precisely what quantities of space and time are involved in the formulation" of such culture units (Willey and Phillips 1958:17). This concern attended the notion of the *tempo* of culture change—is it slow or fast, is it gradual and continuous or punctuated and discontinuous, is it some combination of both, and if so, when is it one and when is it the other? Given that archaeologists had only recently been provided with a nearly universally applicable absolute chronological method—radiocarbon dating—it is not surprising that questions concerning the tempo of culture change could only now be seriously asked and answered. Dendrochronology had provided some earlier insight into such matters, but it was limited temporally and geographically. Radiocarbon dating had no such limitations, and answers to questions of duration and tempo forced much rethinking of the archaeological record. It also hastened the demise of culture history because to most archaeologists radiocarbon made chronology building a thing of the past (O'Brien and Lyman 1999a). Now, whole sets of artifacts could be dated absolutely. The method advocated by Willey and Phillips, which was built on over forty years of tradition, ensured that archaeologists viewed those sets of artifacts as cultures. Americanist archaeology was firmly in the grip of essentialism. What archaeologists forgot, but paleobiologists did not, was that, as Gould (1986:60) put it, history matters. And in archaeology, as in paleobiology, it matters a great deal.

Chapter 4

The Place of History in Modern Paleobiology and Archaeology

> Since the 1940's [and the publication of George Gaylord Simpson's (1944) *Tempo and Mode in Evolution*], paleontology has undertaken several long excursions in other directions—particularly into the functional morphology of individual taxa and the reconstruction of fossil communities. These studies, at their best, have been elegant and persuasive; they have certainly been of great value. But they have not produced new evolutionary theory; they have not asserted the theoretical independence of paleobiology; they have, in essence, shown that fossil organisms and assemblages work much as modern ones; they have, in this sense, continued the tradition of paleobiology as a derivative science, indebted for all its insights to the evolutionary biology of modern organisms and ready to provide nothing but simple confirmation in return. But why be a paleontologist if we are condemned only to verify imperfectly what students of living organisms can propose directly? (Gould and Eldredge 1977: 148–149)

> Ignoring evolution simply means that explanations will be *incomplete*.... Evolution matters because history matters.... Reconstructing genealogical relationships is the goal of a *historical science*. (Sober 1993:7, 15; emphasis in the original)

During the last few decades, several paleobiologists have explicitly noted parallels between their field of inquiry and archaeology. C. W. Harper (1976:269), for example, suggested that testing "fundamental biological generalizations by studying the fossil record [is] analogous to testing sociological generalizations by examining the data of archaeology." Steven Stanley (1979:8) found this analogy "unfair ... because sociologists can fully observe in their lifetimes many of the processes they seek to analyze." Stanley was referring to the difference between what have come to be known among biologists as microevolutionary and macroevolutionary processes, which we consider in more detail later in this chapter. Stanley's assessment of Harper's analogy as unfair rests on the assumption that cultural (or social) evolution is driven only by microevolutionary processes, whatever those might be for cultural lineages. Such processes generally are characterized in biology as genetic and visible within a petrie dish or population of fruit flies studied during the lifetime of the researcher. We perceive this to be part of the problem in archaeology—using cultural processes visibly operating today among various ethnic groups as the sole source of explanations for the past, though this is only part of a larger problem originating in archaeology's use of anthropology in general.

HISTORY: THE HEART OF THE MATTER

Paleobiologist Niles Eldredge (1989:184) noted that paleobiology and archaeology are both "charged with documenting the actual results of history—be that history the fruits of the evolutionary process as preserved in the sequence of faunas and floras in the fossil record, or the history of human sociocultural evolution as preserved in the archaeological record." We find this nonprocess-restricted equation of the two disciplines to be useful, though we do not agree with a literal reading of it with respect to the matter of "sociocultural evolution." Furthermore, documenting (pre)history is only part of the job paleobiologists undertake, just as it is only part of the job for archaeologists. The other significant part of their respective jobs is to explain those histories—why they unfolded as they did.

History is critical to any evolutionary study. The plethora of statements to this effect in biology (Nitecki and Nitecki 1992), especially with respect to identifying character states as "adaptations," continues to grow (for overviews, see Brandon 1990; Burian 1992). There also is some recent explicit recognition of the importance of history to anthropological research generally (e.g., Mace and Pagel 1994). Robert Boyd and Peter Richerson (1992:179–180) are clear about this when they state that "Darwinian theory is both scientific and historical. The history of any evolving lineage or culture is a sequence of unique, contingent events." They grapple with the question of "What makes change historical?" and with the notion of making historical explanation scientific, and we agree with their conclusion that in "the biological and social domains, 'science' without 'history' leaves many interesting phenomena unexplained, while 'history' without 'science' cannot produce an explanatory account of the past, only a listing of disconnected facts" (Boyd and Richerson 1992:201).

Archaeology's singular claim to unique status within the human sciences is its access to portions of past phenotypes. Ethnographers, sociologists, psychologists, historians, and others who study humans are limited to living organisms or written records. Only archaeologists have access to the entire time span of culture, however it is defined. Clark Wissler, for one, *could* have argued that the archaeological record was unnecessary to his use of the culture-area concept and the age–area hypothesis in writing history (see the review in Kroeber 1931b), but he quickly employed that record to confirm directly his ideas about culture history (e.g., Wissler 1919). The significance of this action is found in an instructive parallel in paleobiology. Modern biologists who undertake cladistic analyses might protest that the fossil record is unnecessary to determine the phylogenetic history of organisms, but this position is losing ground as paleobiologists more frequently use the fossil record to help test cladistic hypotheses (e.g., Benton and Hitchin 1997; Clyde and Fisher 1997; Donoghue et al. 1989; Fisher 1994; Forey 1992; Huelsenbeck and Rannala 1997; Norell and Novacek 1992; Novacek 1992; Patterson 1981; Smith 1994; Wagner 1995).

Historical questions are the most obvious ones archaeologists can ask. This, of course, is hardly a strong warrant for asking them (Eldredge 1989). Wissler's efforts, and those of some cladists, indicate we may not need the prehistoric record to produce answers to some historical questions. However, we believe archaeologists should ask historical questions not only because they have access to data that provide a direct test of historical hypotheses (whether founded in cladistical analyses [e.g., Mace and Pagel 1994], Wissler's age–area hypothesis, or some other model) but also because answers to historical questions are critical to gaining a complete understanding of cultural manifestations occupying particular positions in time and space.

Thus, we take as archaeology's most important goals the writing and explaining of the history of human phenotypes (Lyman and O'Brien 1998; O'Brien et al. 1998). From an evolutionary perspective, to "*explain* means to identify a mechanism that causes evolution and to demonstrate the consequences of its operation" (Bell 1997:1; emphasis in the original). From a Darwinian perspective, two of the more important mechanisms are selection and drift (transmission), and the causes precede the consequence of the working of the mechanisms. Selection and transmission are historical mechanisms; they operate every moment, at some times more strongly or more rapidly than at others, creating the varying tempo of evolutionary change over time (e.g., Gould et al. 1987). But what is history other than the passage of time? Robert O'Hara (1988:144) provides a useful discussion:

> [G]enerally speaking a *chronicle* is a description of a series of events, arranged in chronological order but *not* accompanied by any causal statements, explanations, or interpretations. A chronicle says simply that *A* happened, and then *B* happened, and then *C* happened. A *history*, in contrast to a chronicle, contains statements about causal connections, explanations, or interpretations. It does not say simply that *A* happened before *B* and that *B* happened before *C*, but rather that *B* happened *because* of *A*, and *C* happened *because* of *B*.... *Phylogeny* is the *evolutionary chronicle*: the branched sequence of character change in organisms through time.... [H]istory, as distinct from chronicle, contains a class of statements called *narrative sentences*, and narrative sentences, which are essential to historical writing, will never appear in [chronicles]. A narrative sentence describes an *event*, taking place at a particular time, with reference to *another event* taking place at a *later* time.... Just as narrative sentences distinguish history from chronicle, *evolutionary narrative sentences* distinguish evolutionary history from evolutionary chronicle (emphasis in the original).

Critical points of O'Hara's discussion are that (1) false or inaccurate chronicles cannot result in accurate history, and (2) narrative sentences provide the explanations of why chronicles are the way they are. Culture historians such as A. V. Kidder, Irving Rouse, and James Ford, who were vilified by the processual archaeologists of the 1960s and 1970s, recognized these distinctions decades ago (Lyman and O'Brien 1997; Lyman et al. 1997b). However, they could not escape the same problem that plagues evolutionary biology today—one identified by O'Hara (1988) when he distinguished between the explanation of *states* and the explanation of *events* of change. The former comprises essentialist, or typological, thinking; the latter, materialist, or population, thinking (Hull 1965; LaPorte 1997; Mayr 1959a; Sober 1980) and distinguishes Darwinian evolution as not only a different theory of change but also a different *kind* of theory. It demands a different ontology. Culture historians failed to recognize this and attempted to explain the difference in culture *states*—culture types—in anthropological terms, not realizing that they could have been explaining events of change in Darwin's materialist terms.

Showing that a particular phenotypic trait has a positive fitness value is critical to evolutionary studies (see Novacek [1996] for a recent review). In archaeology this requires the mechanical properties of artifacts to be measured (O'Brien and Holland 1995a; O'Brien et al. 1994) in a manner similar to that in which one determines that a panting dog is regulating its body temperature (e.g., Mayr 1983). Does a particular kind of pottery work better within the particular time–space position it occupies than some other kind of pottery? If so, why? How does that particular *state* of pottery work in that particular selective context? Archaeologists with Darwinian leanings have spent considerable time answering these questions (e.g., Braun 1987; Dunnell and Feathers 1991; O'Brien et al. 1994), but they comprise only one set of questions that must be answered; the second set is historical and particularistic. What was the selective environment in which a phenotypic trait was found, and what were the selective environments that led to its appearance? What was the history that led to that trait's establishment (Why does a dog pant instead of perspire)? These are questions about the history of change in phenotypic traits, and they give archaeology its evolutionary bent. Answering the questions regarding pottery state requires the use of immanent properties and processes, or an essentialist ontology; answering the questions regarding pottery *change* requires the use of configurational properties and processes, or a materialist ontology (Simpson 1963, 1970; Szalay and Bock 1991).

A potential objection to such a position is found in Robert Bettinger and Peter Richerson's (1996:226) statement that "given time's ravages, few archaeologists will ever be privileged to participate in constructing a 'how actually' explanation." We think they are attempting to make two points here. First, they are stating that the historical chronicles and narratives of evolutionary archaeology are merely plausible stories. We agree, with the qualification that the stories constructed under evolutionary archaeology are theoretically informed and thus are testable rather than inductively generated inferences (contrary to Broughton and O'Connell 1999). Second, Bettinger and Richerson are arguing that the real story will never be known. Their distinguishing between "how possibly" and "how actually" explanations suggests they find little satisfaction with the former, characterized by O'Hara (1988:149) as statements regarding "how a change *may* have taken place," and would much prefer the latter, or how a change "*did* take place" (O'Hara 1988:150; emphasis in the original). Bettinger and Richerson's point is that the latter is impossible to attain. We doubt paleobiologists would disagree, but neither would they throw up their hands and change professions.

Bettinger and Richerson (1996) reference Robert Brandon (1990) when making the distinction between "how possibly" and "how actually" explanations. Brandon's (1990:176–184) point was that "how possibly" explanations are quite valuable to Darwinism and in many cases can be distinguished only epistemologically from explanations of the "how actually" sort. When a "how possibly" explanation accounts for numerous observations and provides an empirically and logically—both founded in theory—coherent explanation, it attains the status of a "how actually" explanation yet remains testable in light of new evidence. Brandon (1990:183) acknowledges that we may never know when we have truly answered "how actually" questions, though he also states that "no one can fairly describe [such a 'how possibly' explanation] as merely an imaginative bit of story telling." Or as Daniel Dennett (1995:319) puts it, the power of Darwin's "theory of natural selection is not the power to prove exactly how (pre)history was, but only the power to prove how it could have been, given what we know about how things are."

The Spectre of Lamarck

Evolutionary archaeology has had its fair share of critics from within the discipline (e.g., Boone and Smith 1998; Spencer 1997, 1998), a few of whom have perceived at least limited value to such an approach (e.g., Rosenberg 1994, 1998; Schiffer 1996), but a more formidable foe—at least in terms of his knowledge of evolutionism in general—is Stephen Jay Gould, who in several papers has argued that Darwinian evolutionism is not applicable to cultural phenomena. We introduce Gould's arguments here because they contain several elements—direction and rate of change—that set the stage for what we discuss in the remainder of the chapter. We later return to some of the objections raised by Gould and see what archaeologists have to say on the (in)applicability of Darwinian evolutionism to the study of humans and the products they made, used, and threw away.

In one of his earlier dismissals, Gould (1987a:18) stated: "I am convinced that comparisons between biological evolution and human cultural or technological change have done vastly more harm than good—and examples abound of this most common of intellectual traps.... Biological evolution is powered by natural selection, cultural evolution by a different set of principles that I understand but dimly." He then spelled out what he found to be the roots of the problem:

> First, cultural evolution can be faster by orders of magnitude than biological change at its maximal Darwinian rate—and questions of timing are of the essence in evolutionary arguments. Second, cultural evolution is direct and Lamarckian in form: the achievements of one generation are passed by education and publication directly to descendants, thus producing the great potential speed of cultural change. Biological evolution is indirect and Darwinian, as favorable traits do not descend to the next generation unless, by good fortune, they arise as products of genetic change. Third, the basic topologies of biological and cultural change are completely different. Biological evolution is a system of constant divergence without any subsequent joining of branches. Lineages once distinct, are separate forever. In human history, transmission across lineages is, perhaps, the major source of cultural change. (Gould 1987a:18).

For Gould, the "fundamental" differences between cultural and biological evolution render any effort to draw analogies between them "the nonsense of false analogy" and comprises a "false application" of Darwinian rules and principles (Gould 1987a:18). He used similar wording a decade later to make nearly the same points.

> [M]ost importantly, human cultural change operates in a manner that precludes a controlling role for natural selection.... [F]irst, biological evolution proceeds by continuous division of species into independent lineages that must remain forever separated on the branching tree of life. Human cultural change works by the opposite process of borrowing and amalgamation... If

> we seek any biological analogue for cultural change, I suspect that infection will work much better than evolution.
>
> Secondly, human cultural change runs by the powerful mechanism of Lamarckian inheritance of acquired characters. Anything useful (or, alas, destructive) that our generation invents can be passed directly to our offspring by direct education. Change in this rapid Lamarckian mode easily overwhelms the much slower process of Darwinian natural selection, which requires a Mendelian form of inheritance based on small-scale, undirected variation that can then be sifted and sorted through a struggle for existence. Genetic variation is Mendelian, so Darwinism rules biological evolution. But cultural variation is largely Lamarckian, and natural selection cannot determine the recent history of our technological societies. (Gould 1997a:22)

Gould's comments are curious. He does not, for example, define "Darwinism," though his remarks imply that it comprises the mechanisms of natural selection and inheritance, and little else. Such an inference is corroborated by a paper he published earlier in which he noted that the term *Darwinism* connoted different things to different people, but that the "essence of Darwinism lies in a claim that natural selection is the primary directing force of evolution" and that evolution is gradual and selection creates adaptations (Gould 1982a:381). But if, as Richard Lewontin (1983:67) put it, Darwin's theory was "a radical epistemological break with past theories" of history because it focused on variation, how can Gould suggest what he does?

Gould (1987a, 1997a) identifies three differences between the overall processes of biological and cultural evolution, and because of these differences, he concludes that natural selection does not or cannot operate on cultural phenomena. Each of the three differences he chooses were pointed out long ago by anthropologists (see the review in Lyman and O'Brien 1997). That cultural evolution is reticulate whereas biological evolution is only branching (Figure 3.8) is precisely the reasoning used by A. L. Kroeber, Julian Steward, and other anthropologists to suggest that biological evolution could not be applied to cultural phenomena. Here, the problem is equating a human society with a reproductively isolated biological species. The significance of the difference loses force when it is realized that Gould—like Kroeber and Steward—does not mean to imply that evolution cannot occur at the subspecific level via recombination. As a process, hybridization occurs frequently among biological units, including species. It was suggested several decades ago in biology that hybridization was an important source of new variants (e.g., Lewontin and Birch 1966)—a view that is now generally accepted (Arnold 1997; Endler 1986a, 1998). Paleobiologist Niles Eldredge (1999) has made a rather compelling case for considering a *population* a critically important unit of evolution. This is easy to grasp conceptually when it is realized that speciation—the branching or diversification of a species lineage into several species—involves a minimum of two populations of a species that, through one process or another, become reproductively isolated from one another. Those populations may start to diverge (genetically, morphologically, adaptationally) if gene flow between them is slowed or stopped, but they may rejoin if the barrier to gene flow is removed, thereby halting the speciation process and also effectively comprising an instance of interpopulation hybridization.

Perhaps the easiest way to make the argument we present in the preceding paragraph is to consider Figure 4.1, based on a figure published by Robert O'Hara (1993:240) and comprising what he terms an "evolutionary chronicle" showing a "fabric of relationships." Each circle comprises an evolutionary unit, lines connecting circles denote reproductive interaction, each row of circles represents a generation, and time passes from bottom to top. Note that each evolutionary unit in a descendant generation receives input—inherits from—two ancestral units. In the final generation in the graph, Generation 22, there are three distinct sets of more inclusive evolutionary units labeled A, B, and C. Units A and B diverged—stopped interacting reproductively—in Generation 13, and Units B and C diverged after Generation 9. Units A and B did not interact reproductively between Generations 3 and 13, but did so in Generation 13. Otherwise, there are greater or lesser amounts of hybridization between the units denoted by the circles. The most important thing to think about when examining Figure 4.1 is simply this: Think of the circles as individual genotypes, then as individual organisms, then as populations of organisms. The graph works equally well at any scale. Similarly, think of each circle as an attribute of a discrete object, as an artifact type, and as a particular kind of aggregate of artifact types, such as a tool kit. Finally, think of the circles as individual humans that transmit ideas rather than genes; then think of them as bands of humans or societies, then as cultures. That "cultures" hybridize can no longer be considered a valid argument against the applicability of Darwinian theory to cultural units of whatever scale. O'Hara's graph is an excellent model of both biological and cultural evolution and of how the units of both kinds of evolution may or may not interact reproductively.

This brings us to Gould's (1997a) comment that cultural evolution is much faster than biological evolution. We pointed out in Chapter 1 that precisely this same point was made by Harold Gladwin (1936:158) when he remarked that "[biological] evolution is so slow as to be hardly distinguishable. The evolution of culture ... was stepped up to an almost incredible speed." But Gladwin does not make explicit why this might be important; Gould does. Cultural evolution is so fast that cultures can change before natural selection can act. In response, we simply make a point we have made before (Lyman and O'Brien 1998). Why was DDT outlawed? Why must our automobiles now meet certain emission standards? Because people decided that was the way things should be, you might respond, but *why* did we decide these things? In simple terms, because people believe that the continued addition of DDT and automobile-generated pollutants to our world *would ultimately result in our demise as a species*. We anticipated the result that

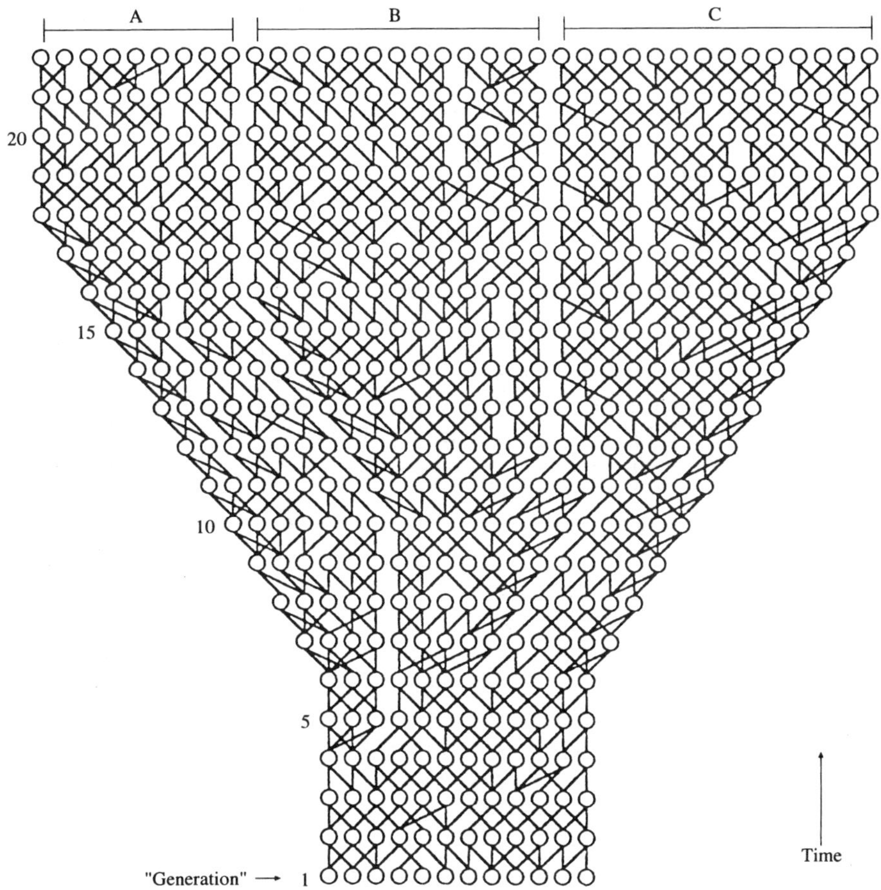

FIGURE 4.1. A model of an evolutionary history displaying relationships among evolutionary units. Each circle comprises an evolutionary unit, lines connecting circles denote transmission between them, and each row comprises a generation. In Generation 22, three distinct sets of more inclusive evolutionary units—A, B, C—are indicated. Notice that those three inclusive units comprise lineages that, during an earlier time, hybridized. The evolutionary units denoted by the circles can be conceived of at any scale—gene or attribute, population or type, species or culture (after O'Hara 1993).

natural selection might produce and, because we did not like that possible result, we changed our behavior, our culture. We might have anticipated the wrong result in both cases—that DDT and pollutants might eventually kill us—but their absence will increase our chances of survival. Only time, and natural selection, will tell if we made the correct choice. And more importantly, if we made the correct choice, we did so because of our understanding of natural selection. Natural selection is, then, the final, or "ultimate" (Mayr 1961), arbiter. For this reason, we find it difficult to agree with Gould's argument that natural selection plays no role in cultural evolution, though many biologists and anthropologists would agree with him on this matter.

One of Gould's reasons for denying a Darwinian model of culture change has also been mentioned by some modern anthropologists (e.g., Rosenberg 1994, 1998; Spencer 1997, 1998). Gould (1996:222) indicates that the

> uniquely and distinctively Lamarckian style of human cultural inheritance [what he refers to as "education"] gives our technological history a directional and cumulative character that no natural Darwinian evolution can possess.... [C]ultural change is potentially progressive or self-complexifying because Lamarckian inheritance accumulates favorable innovations by direct transmission, and amalgamation of traditions allows any culture to choose and join the most useful inventions of several separate societies.

This mode of inheritance—named after Jean-Baptiste de Lamarck, whose notion of an evolution fueled by directed variation was overturned by Darwin—works with "explosive rapidity" and "powers human cultural change by a mechanism unknown in the slower world of Darwinian evolution" (Gould 1996:220, 221). The potential rapidity of cultural change relative to the rate of biological change was identified by archaeologist Harold Gladwin (1936:158), who noted that it was the reticulate nature of cultural evolution—cultures hybridize—that allowed an increased rate of change: "Zoological species do not cross and intergrade; evolution is so slow as to be hardly distinguishable. The evolution of culture ... was stepped up to almost incredible speed, and on every side we find evidence of merging and cross-influences."

Gould's (1997a) argument is that natural selection will typically have insufficient time to operate on available variation because that variation can change as a result of intragenerational transmission of new variants—the "explosively rapid mechanism of cultural heredity"—occurring prior to the pressures of selection, resulting in change in the pool of extant variants. In simple terms, transmission is faster than selection. We see two interrelated problems with this view. First, it emanates from the narrow (but unabashedly biological) perspective of intergenerational differences in the *reproductive* success of organisms and ignores what has been termed the "replicative success" of artifacts (Leonard and Jones 1987). If artifacts have different fitness values—that is, if one variant of artifact works better than another in a particular recurring situation—then (1) it is likely the more fit variant will be replicated (reproduced) more often than the less fit one, and (2) because tools are part of the extrasomatic means by which humans adapt, then the possessors of less fit tools may well reproduce, over the long term, less than those who possess better tools.

The second problem is that selection does not have an intrinsic rate. Selection varies in strength in the sense that it can alter the tempo of evolution, in one case resulting in, say, anagenetic evolution from an ancestral to a descendant species in several hundred generations, and in another case resulting in such a change over several thousand generations. The key here is that only by using some scale of temporal units against which selection can be aligned can selection be said to have a rate. Our choice of generations as the temporal unit is purposeful because it is just this unit that Gould has implicitly used. Only this unit allows the argument that intergenerational selection does not work as fast as intragenerational transmission. Although it could hardly be otherwise, do not be misled. This does not mean that selection works only at the intergenerational scale. Shifting the scale of the temporal unit from a human generation to tool use-lives, virtually all of which are shorter than a generation of hominids, eliminates Gould's second problem.

The viral model proposed by Ben Cullen (1993, 1996) for the transmission of cultural phenomena is an insightful bit of reasoning along the lines of Gould's (1997a) suggestion of an infection analog, but as presently construed by anthropologists, it rests heavily on human choice, or what is referred to in biology as directed evolution. Other arguments that echo Gould's (1996, 1997a) have become common in the 1990s as archaeologists continue to design and discuss ways to make sense out of the archaeological record. Some of the rhetoric has deep roots in the processual archaeology of the 1960s, but the catalyst for much of it resides in the paleobiological literature of the 1970s and 1980s. One benchmark of the neo–Darwinian Synthesis that took place was the founding in 1946 of the Society of the Study of Evolution and the initial publication of its journal *Evolution* (Antonovics 1987). As Mayr (1947:i) noted in the foreword to the first issue, the Synthesis had resulted in the "coming together of evolutionists from all fields [and] initiated a new era of evolution research." Forty years later, biologists wondered aloud if the synthesis had been beneficial (e.g., Antonovics 1987; Endler 1986b; Endler and McLellan 1988). Paleobiologists also wondered (e.g., Gould 1983b), perhaps because they were among the first researchers with interests in evolutionary biology to detect, in the 1970s, potential weaknesses in the synthesis. We begin with these events before turning to what happened in archaeology at about the same time.

PALEOBIOLOGY AS THE SCIENCE OF ORGANIC EVOLUTION

George Gaylord Simpson was one of the architects of the Synthesis, which, as we saw in Chapter 2, comprised the integration of population genetics and neontological biology. Simpson's role was to carry the integration into the realm of paleontology, which to that point had been largely the tail on the biological kite (to borrow Steward's [1944] metaphor of archaeology as a tail on the kite of anthropology). In *Tempo and Mode in Evolution*, Simpson (1944) illustrated how neo–Darwinism could be used to explain the fossil record in terms of biological evolution. He argued that although naturalists and experimental geneticists might see evolution at work in a petrie dish or among a population of organisms in the wild, they could not see larger-scale evolutionary processes such as variations in tempo and mode, nor could they document and explain long-term trends in the paleontological record. His notion of quantum evolution was a nod to mode, and for tempo he outlined three rates of evolutionary change that a paleobiologist might find reflected in the fossil record.

Simpson (1944:3) posed the question, "How fast, as a matter of fact, do animals evolve in nature?" He then stated a set of "basic theorems" about rates of evolution:

> (1) The rate of evolution of one character may be a function of another character and not genetically separable even though the rates are not equal. (2) The rate of evolution of any character or combination of characters may change markedly at any time in phyletic evolution, even though the direction of evolution remains the same. (3) The rates of evolution of two or more characters within a single phylum [or clade] may change independently. (4) Two phyla of common ancestry may become differentiated by differences in rates of evolution of different characters, without any marked qualitative differences or differences in direction of evolution. (Simpson 1944:12)

Simpson's explicit acknowledgment of variation in the rate of evolutionary change seems to have been less controversial at the time than his notion of quantum evolution. Eldredge (1992) and Gould (1980a, 1995b) document why Simpson later backed off from this early insightful view to adopt more standard view (e.g., Simpson 1953). Although Simpson's notions were, in somewhat different form, resurrected in the 1970s (Laporte 1983), the Synthesis of the 1940s "hardened" (Gould 1983b) around several central tenets at the expense of others: Evolution is mainly gradual and continuous, and change is largely adaptive, or driven by selection (Antonovics 1987; Gould 1983b), though Richard Burian (1988:252) notes that critics of the Synthesis have "considerable difficulty in getting a precise fix on the testable content of the synthetic theory," as evidenced by "contradictory criticisms of [it]."

The shift in paleontology from what a modern archaeologist might categorize as a form of antiquarianism to a science of biological evolution began in 1944 with Simpson's *Tempo and Mode in Evolution* and culminated in 1976 with the publication of the first issue of the journal *Paleobiology*. After ten years of publication, the then-editors remarked that the journal had been founded "for the publication of research papers and essays in the area of biological paleontology. The purpose of the journal was, and remains, the uniting of paleontology with modern biology" (Sepkoski and Crane 1985:i). A diverse range of topics was included under the journal's title, but the driving focus comprised an "attempt at a biological interpretation of the fossil record" (Sepkoski and Crane 1985:i). Scanning the table of contents for the tenth-anniversary issue that these comments introduced confirms precisely this fact. Prior to initiation of the journal, the Paleontological Society had sponsored the *Journal of Paleontology*, characterized in the tenth-anniversary issue of *Paleobiology* as "heavily given over to description of taxa" (Eldredge and Novacek 1985:65). *Paleobiology* was devoted to almost everything else a paleontologist might choose to pursue, but with a decided emphasis on *explaining* the history of organisms from a Darwinian perspective. There were other things happening in the discipline as well. Probably the most contentious issue in the 1970s and 1980s was the suggestion that biological evolution was not gradual and continuous; rather, its flow was discontinuous, with long periods of stasis punctuated by events of rapid change and, particularly, diversification.

Punctuated Equilibrium

Punctuated equilibrium was introduced in the early 1970s (Eldredge 1971; Eldredge and Gould 1972; Gould and Eldredge 1977; Stanley 1975, 1976, 1979). Some of its elements had seen earlier exposure (for a historical review, see Harper 1975), especially in Simpson's (1944) "quantum evolution" (see also Kurtén 1960, 1965), though architects of punctuated equilibrium, while admitting general similarities (Eldredge 1992; Gould 1995b), found them fairly insignificant (Eldredge 1995a; Gould 1982b; see also Laporte 1983). In particular, Simpson's quantum evolution concerned taxa higher in the Linnaean system than species; punctuated equilibrium explicitly includes species as well as genera, families, and so on (Eldredge 1995a). As with many suggestions regarding new theories or major modifications to old ones, punctuated equilibrium was not immediately adopted or embraced by everyone who was studying biological evolution. Some found great value in the concept (e.g., Cracraft 1982; Johnson 1975, 1982; Stanley 1982; Vrba 1980, 1983, 1985), even though they argued among themselves about details—for example, Cracraft (1985) versus Vrba (1984), Gould and Eldredge (1977) versus Stanley (1975)—and others did not accept it (e.g., Gingerich 1984b; Hoffman 1982; Levinton and Simon 1980; Ridley 1980; Schopf 1981, 1982; Stebbins and Ayala 1981).

Arguments over the pros and cons of punctuated equilibrium were carried out in *Science*—see Harper (1975, 1976) versus Stanley (1976), Gingerich (1983, 1984c) versus Gould (1984a), Schopf and Hoffman (1983) versus Gould (1983a)—*Nature* (Boucot 1982; Charlesworth and Lande 1982; Cohen and Schwartz 1983; Ginzburg and Rost 1982; Kat and Davis 1983; Lindsay 1982; Williamson 1981, 1982, 1983), and other journals (e.g., Charlesworth 1982; Fryer et al. 1983, 1985; Gould 1982c; Williamson 1985a, 1985b). The early controversy resulted in various conferences, one in October 1980 (Levinton and Futuyma 1982; Schopf 1980) and another in December 1981 (Novacek 1982), on rates and kinds of evolution that were directed toward hashing out some of the disagreements, but success was limited (Futuyama et al. 1981; Lewin 1980). The architects found themselves having to explain the nuances of the concept to audiences that did not seem to understand what they had proposed (e.g., Eldredge 1974; Eldredge and Gould 1974; Gould 1982b). Even the value of the concept as a stimulus for research, regardless of its correctness, was debated (Gingerich 1984a; Lister 1984; Rhodes 1983, 1984). After all this discussion, one might think that Roger Lewin's (1986) article "Punctuated Equilibrium Is Now Old Hat" was an accurate signal regarding who had won the battle and when. This was not the case (Levinton 1986); the authors and supporters of the concept had to continue to defend it through the 1980s (e.g., Eldredge 1989; Gould 1984b, 1989b; Gould and Eldredge 1986, 1988a, 1988b; Stanley 1982, 1989), though they were not shy about suggesting it had established itself as a necessary part of biological evolutionary theory, particularly for paleobiological research (e.g., Eldredge and Gould 1988).

Twenty-five years after it was proposed, some biologists still argue that punctuated equilibrium seems to vary from one conceptual manifestation to the next (e.g., Coyne and Charlesworth 1997), with the architects responding that this is untrue (e.g., Eldredge and Gould 1997; Gould 1989b). Other biologists (e.g., Huxley 1982; Penny 1983; Rhodes 1983; Thompson 1983) suggest that an abrupt or rapid rate of evolution was never precluded by Darwin; this point, too, is disputed by the architects (Gould and Eldredge 1983; Stanley 1982). Whatever the case, in the past several years many paleobiologists have stopped arguing about the possibility of noncontinuously gradual evolutionary rates and have focused on two things: producing detailed measurements of the rates of change and durations of periods of stasis, and deciphering the formational history of the fossil record—taphonomy—and how that history may or may not influence our interpretations of the rates and modes of evolution reflected in the record. We take these facts as a sign that, on the one hand, conceptually at least, the notion of relatively abrupt evolutionary change has been generally accepted. On the other hand, whether change is gradual or abrupt in any particular case can only be determined when appropriate empirical data are brought to light—a fact the architects of the notion readily acknowledge (Gould and Eldredge 1993). But as we will see, the tempo of change is only half of the equation; the other half is the mode of change.

Data Quality

Some early commentators suggested that the data on which the notion of punctuated equilibrium was originally based fail to support it (Brown 1987), though the architects disagree (e.g., Eldredge 1999). The quality of the data brought to bear on questions of evolutionary tempo and mode—the point that ultimately led to the postulation of punctuated equilibrium (Eldredge and Gould

1972)—must be such as to provide unambiguous tests of evolutionary tempos and modes. As is well documented by Eldredge and Gould (1972, 1977), Darwin worried that the fossil record was incomplete and filled with gaps, as had Charles Lyell before him and Simpson during the twentieth century. Thus, paleontologists would have a difficult time analytically detecting the continuous, cumulative, gradual evolution of a lineage—termed "phyletic gradualism" by punctuated equilibrium's architects. Eldredge and Gould (1972; Gould and Eldredge 1977) argued that the "theory" of phyletic gradualism had so strongly come to dominate evolutionary thinking of the 1960s and 1970s—Gould's (1983b) "hardening" of the neo–Darwinian Synthesis of the 1940s—that data contradicting it were ignored. In particular, data suggesting morphological (and by implication, genetic) stasis over long time spans and instances of environmental change were discounted, but so, too, were data suggesting the abrupt replacement of what were argued to be ancestral species by what were believed to be daughter species. In an extended response to their early critics, Gould and Eldredge (1977) meticulously demonstrated the pernicious effects the blinders of phyletic gradualism were having on the discipline.

The architects stated that one needs evidence of "bed-by-bed gradualism within taxa [or species]" to demonstrate an instance of phyletic gradualism (Gould and Eldredge 1977:122). One might suspect that a continuous stratigraphic record is required. There are other requirements, such as controlling spatial variation, but a potential weakness resides in the requirement of stratigraphic continuity. Because of the problem as to whether a particular fossil sequence represents gradual change or stasis punctuated by abrupt change, one must have a detailed measurement of elapsed time. Time, for Gould and Eldredge, is rendered as a fossiliferous stratigraphic column comprising depositional units, or "beds." In their evaluation of various suggested instances of phyletic gradualism, they equated each depositional unit with the same absolute unit of time when they regressed absolute vertical position in a column of sediment (stratigraphic depth) against changes in the morphological features of fossils (Gould and Eldredge 1977) and when they subjectively used vertical position as a proxy measure of absolute time. They presumed a constant rate of sediment deposition over an entire stratigraphic column, in one case choosing the median of three estimates of depositional rates (Gould and Eldredge 1977:133) to characterize an entire column; thus, they conflated relative (superposed position) and absolute (calendric) time, or ordinal and interval scales of measurement. Any apparent gradualistic trend, instance of stasis, or punctuation event could, therefore, be a function of this conflation—a point made by various critics of punctuated equilibrium (e.g., Dingus and Sadler 1982; Ginzburg and Rost 1982; Ridley 1980).

Given the poor resolution of the absolute time scale in many geological columns, one might allow Gould and Eldredge's conflation of two rather different scales of measurement to stand, but they did not acknowledge the possible effects of varying rates of sediment deposition within a column on their estimates of rates. Instead, they effectively discounted its significance when they argued that "allopatric speciation" will appear, in the fossil record, "in a geological instant" (Gould and Eldredge 1977:121); that is, change rendered as the appearance of new taxa will be geologically abrupt, even though it may occur gradually in "ecological time" (Gould and Eldredge 1977:121). Thus, they turned to evidence of stasis—"Stasis is data" (Gould and Eldredge 1977:129) became their mantra—as the source of corroboration of punctuated equilibrium: "[W]e regard stasis as more amenable to study. Morphological breaks can usually be attributed equally well to punctuated equilibria or to an imperfect record" (Gould and Eldredge 1977:148). Stasis can be detected regardless of the absolute age of the samples compared, as long as it is clear that the samples are of different ages (e.g., Kelley 1983). But if the *rate of change* rather than stasis is the subject of interest, then absolute time must be tightly controlled. It also must be tightly controlled if one seeks to determine the duration of periods of stasis. The critical aspect of measuring stasis and punctuation events, then, involves the recognition of species.

Recognizing Species

The procedure for identifying multiple taxa, particularly species, relates directly to how one conceives of evolutionary tempos. Gould and Eldredge (1977: 115–116) stated that "tempos can be observed and measured: modes must be inferred, usually from empirical distributions of tempos." This suggests that we must measure the rate of change in order to determine the mode of change. However, because events of change are "geologically instantaneous," the focus should be on stasis of forms (Gould 1982b:82–85). What is observed to be gradual change in "ecological time" will appear as a "geological microsecond" (Gould and Eldredge 1977:121), or punctuated change. Any given instance of geologically abrupt change may represent an instance of punctuation, or it may represent an incomplete fossil record: "It takes unusual stratigraphic resolution to prove that a morphological break records a true evolutionary punctuation, but stasis can be studied in conventional geological sections with their missing data of more than 90 percent—for [a] 10 percent [sample over some temporal interval], widely dispersed over the interval, is a good sample" (Gould 1982b:86). Thus, stratigraphic continuity is *not* required to detect morphological stasis; "stasis only requires adequate sampling, not continuity" (Gould 1985:7).

What, then, are punctuation and stasis? Operationally, punctuation—"geologically instantaneous" change—is indicated when it comprises "1 percent or less of [a species'] later existence in stasis" (Gould 1982b:84, 1982c:137). Stasis is "mild, directionless fluctuation" in morphology (Gould and Eldredge 1977:127). More specifically, "the norm for a species during the heyday of its existence as a large population is morphological stasis, minor non-directional fluctuation in form, or minor directional change bearing no relationship to pathways of alteration in subsequent daughter species" (Gould and Eldredge 1977:117). The phrase preceding the first comma can only be operationalized—a "norm" identified—if one knows the frequency distribution for the variants comprising a species. The phrases after the first comma can only be operationalized after daughter species are identified. In other words, one has to define a species in order to determine if it is morphologically stable. Thus, the problem of recognizing instances of stasis reduces to a classification problem (Chapter 5). Other researchers (e.g., Brown 1987; Levinton 1983; Schopf 1981, 1982; Turner 1986) have come to exactly the same conclusion arguing from different bases.

Eldredge (1995b:40) documents how punctuated equilibrium resolves the differences between Simpson's quantum evolution account of discontinuity between taxa at levels higher than species evidenced in the fossil record and Dobzhansky's (1937a) and Mayr's (1942) "finer scale" perception of discontinuity between species. Simpson perceived evolutionarily successive species as comprising temporal continua, as did Dobzhansky and Mayr. But the latter two saw discontinuity at the species level along the spatial dimension, whereas Simpson saw discontinuity at the genus or higher taxonomic level along the temporal dimension: "Punctuated equilibia, reduced to its barest essentials, says that the empirical patterns of discontinuity of the sort that Simpson saw as typical of the origin of 'higher taxa' actually pertain as well to minimally diagnosable clusters of phenotypes: 'species' " (Eldredge 1995b:41). Thus, punctuated equilibrium adopts Simpson's notion of discontinuity, but unlike his quantum evolution, it also acknowledges discontinuity at the species level. This characterizes Dobzhansky's and Mayr's position, but unlike their views, punctuated equilibrium explicitly acknowledges stasis within species.

So, what is a species? In their seminal paper, Eldredge and Gould (1972: 92–93) made two basic points. First, in their view, the concept of a biological species and its entailments are critical: A species is a reproductively isolated set of organisms that tends to speciate (diversify) allopatrically or peripatrically. Second, this conception of species results in the definition of entities that are limited to a particular time and place. Therefore, this conception of species "has no objective application to evolving [phyletic] continua, only an arbitrary one based on subjective criteria for division. [The concept cannot] be transplanted bodily into our temporal taxonomy. [But the concept] abounds with implications for the operation of evolutionary processes" (Eldredge and Gould 1972:93). So far, so good. But in their first two papers on punctuated equilibrium (Eldredge and Gould 1972; Gould and Eldredge 1977) Gould and Eldredge did not tell us how to recognize a species in the fossil record. Given only those two papers, we might guess that they followed paleontological tradition and operationalized the notion "species" as comprising a set of fossils sufficiently similar in morphology potentially to represent a biological species, yet displaying no more morphological variation than the organisms comprising a living species (e.g., Raup and Stanley 1978; Simpson 1943). The result would be a "chronospecies," or an "intergrading [and] successional species" carved out of an evolutionary continuum in a "necessarily subjective and arbitrary" manner (Stanley 1979:13). A chronospecies "is a segment of a lineage judged to encompass little enough evolution that the individuals within it can be assigned a single species name. In practice, a typical chronospecies does not exhibit a great deal more total variability, from end to end, than is found among the living populations of a single species" (Stanley 1981:14).

Some find the subjectivity and arbitrariness of designating chronospecies objectionable. Joel Cracraft (1981a:458) was explicit in his objection: "Subdividing a phyletic lineage into species ... is arbitrary and does not delimit evolutionary taxonomic units having objective ontological status. Few, if any, interesting biological problems can be investigated with such an approach to species." One might measure individual specimens spanning some temporal duration without identifying them as members of particular species, but such a procedure is flawed, according to the architects of punctuated equilibrium, because it is followed by another step if one subscribes to phyletic gradualism: "New subspecies, species, and even genera are simply subsets (arbitrarily delineated) of an evolving continuum. Under this view, the difficulty of defining taxa is a consequence of the [conceived] evolutionary process itself: ideally, the boundaries of taxa are wholly arbitrary; in the real world, we rely on convenient breaks in the fossil record" (Eldredge and Gould 1977:29). Eldredge (1979:16) suggests this method of species identification is flawed because one graphs morphological changes against stratigraphic (temporal) position and then "encircles segments of lineages most clearly linked continuously, and names taxa on the basis of these clusters. [This demands that one assume that m]orphology evolves inexorably, and we chop it up to name taxa. [This, in turn, demands that we assume e]volution is phyletic" (Eldredge 1979:16).

Simpson (1943:174) termed the change through time a "chronocline," where a cline represents a character gradient. A chunk of a chronocline comprises a chronospecies. The difficulty with identifying a chronospecies resides, then, in first identifying a chronocline, or temporal gradient in a character or attribute. As pointed out by Kevin Padian (1989:75), some characters "change more or less uniformly through time, but others change not at all, and still others vacillate with no clear trend. This is ... one reason to be suspicious of the evolutionary utility of clines: no criterion for identifying a cline seems to be in force. A cline is simply a gradient in character state along a continuum, and it may be broken, temporarily reversed, or stepped. Furthermore, there is no criterion for a cline's magnitude and no control on its probability." This cogent criticism can, as we will see, be reworded a bit and applied with equal force to another method used to identify fossil species.

An example of identifying species within a phyletic continuum is shown in

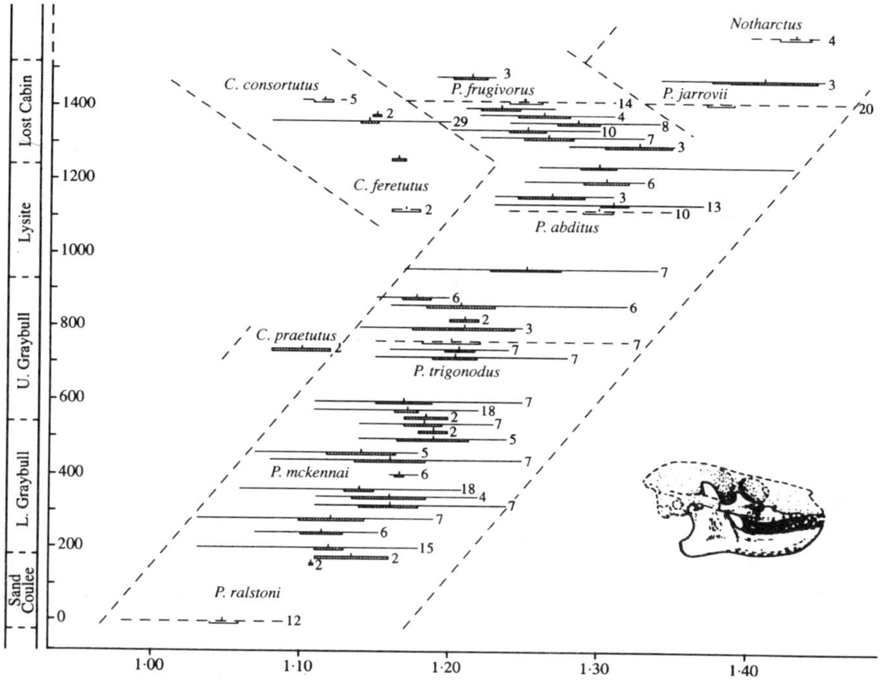

FIGURE 4.2. Stratigraphic distribution of primate *Pelycodus* and the related genus *Copelemur* in early Eocene sediments of the Big Horn Basin, Wyoming. Horizontal axis is $\log_{10}$ (length × width) in millimeters of the lower first molar. Names along the vertical axis are formations, and numbers are stratigraphic position (in feet) above the base of the Willwood Formation. Horizontal line is the sample range; vertical slash is the mean; solid bar is the standard error of the mean; small number is the sample size. Dashed lines show the inferred phylogenetic relations of species of North American *Pelycodus* (after Gingerich and Simons 1977).

Figure 4.2, originally published as an example of phyletic gradualism (Gingerich and Simons 1977; see also Gingerich 1979a). Referring to similar figures published by the same author (e.g., Gingerich 1976a, 1976b), Gould and Eldredge (1977:131) noted that the "dotted guidelines that Gingerich draws about his data are gradualistic interpretations, not literal renderings of the evidence." To explain what the critics are saying, let us look at the relation between a chronospecies defined on the basis of paleontological evidence and a biological species defined on the basis of reproductive isolation. In the latter, species are groups of related, potentially interbreeding individuals. Each individual is slightly different from every other one, and because the individuals making up a species change over time (thereby forming lineages), the average morphology of the members may either remain stable over time, as in Lineage 1 in Figure 4.3 or change over time, as in

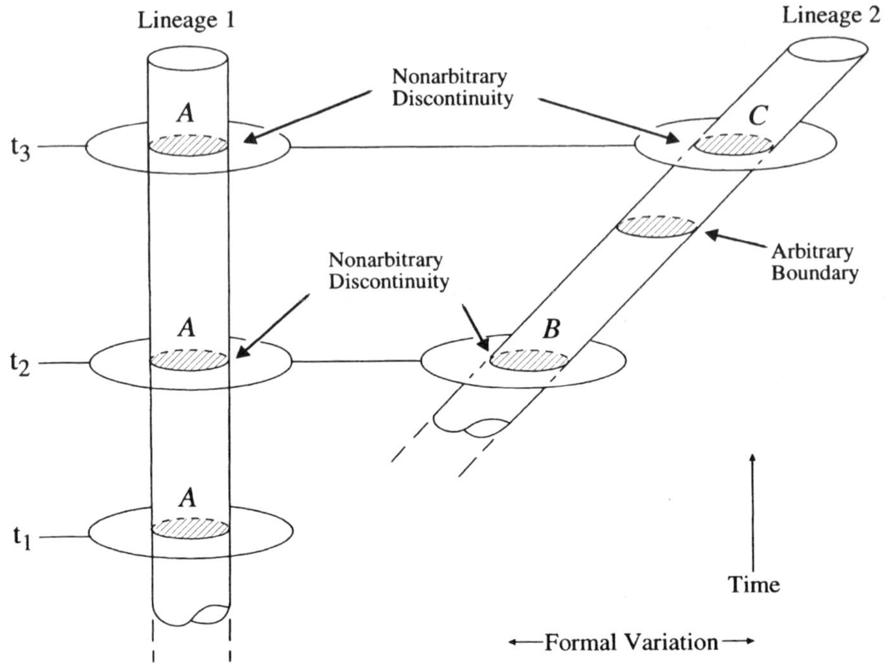

FIGURE 4.3. Diagram showing the relation among three species (A–C) in two lineages (1, 2) sampled at three different times (t_1, t_2, t_3). Species in the fossil record have both a biological dimension, where nonarbitrary morphological discontinuities are used to infer reproductive isolation at any given time, and a time dimension, where successive morphologically changing units of an evolving lineage must be separated at an arbitrary boundary (after Gingerich 1976a).

Lineage 2 in Figure 4.3. The peripatric-speciation model holds that reproductive isolation created by geographic isolation of a parent and daughter population will eventually result in the biological reproductive isolation of the two populations, their morphological divergence, and the creation of two lineages; that is, if enough time has passed since the two populations became isolated from one another, individuals in the two will comprise two separate lineages that cannot interbreed successfully, and they will be readily distinguishable based on their morphological differences.

Thus, at time t_2 in Figure 4.3, there are nonarbitrary reproductive discontinuities between species A and B; likewise, at time t_3, there are similar discontinuities between species A and C. There also are nonarbitrary morphological discontinuities between the species in each pair that can be measured by a paleobiologist. The distinction between the two successive chronospecies B and C in Lineage 2 of Figure 4.3 is arbitrary in cases where the fossil record is sufficiently complete to show the transition from one species to another. Gingerich (1979b:48) put it well: "This need for an arbitrary boundary does not mean that differences between two successive species do not exist. Rather, the problem is analogous to one of keeping time. The fact that one o'clock grades continuously into two o'clock does not mean that the two hours are the same; we make a necessarily arbitrary boundary between them and recognize that they are different." Of course, as we noted earlier, the geological record's stratigraphic boundaries are often used to denote boundaries between chronospecies, though they need not be. If continuity in gradually shifting morphology is detected without stratigraphic boundaries, a distinction made between species B and C by the analyst is completely arbitrary.

It is the arbitrariness of boundaries such as that between species B and C that Cracraft, Eldredge, Gould, and others find offensive. But it is precisely such arbitrary boundaries that are demanded by phyletic gradualism. The problem reduces to this: Phyletic gradualism conceives of evolution as continuous, as modeled in the phylogeny on the left of Figure 4.4, in which the rate of ticking of time's clock does not vary as one reads up the graph vertically (morphological variation is graphed horizontally in uniform units). Units of the temporal–morphological continuum must be arbitrary chunks—chronospecies—of the continuum. Punctuated equilibrium conceives of species units as real because they are

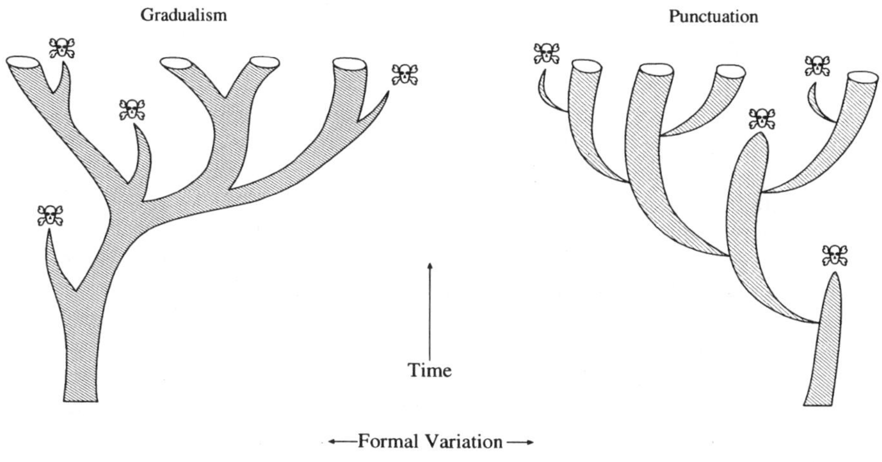

FIGURE 4.4. Models of phyletic gradualism (a) and punctuated equilibrium (b). Time passes in uniform increments along the vertical axis, and morphology varies in uniform increments along the horizontal axis. Chronospecies must be arbitrary chunks of the evolutionary continuum of phyletic gradualism; species are the real vertical branches of the discontinuous evolution of punctuated equilibrium (after Lewin 1982).

static entities over time that have initiations, durations, and terminations, as modeled in the phylogeny on the right of Figure 4.4. Notice that in the punctuation diagram some branches emerge from preceding stems quickly, with minimal morphological variation, and then rapidly become morphologically stable over relatively long spans of time. It is precisely those vertically long and horizontally restricted branches that comprise *real* rather than arbitrary chronospecies. They are produced, as Brian Charlesworth (1982:133) notes, by "rapid evolution in small [isolated] populations involving both the acquisition of the inability to interbreed with the rest of the species (*speciation*) and morphological change," followed by long periods of morphological stasis.

Recognizing the difference between phyletic gradualism and punctuated equilibrium demands, it is thought, a difference in method of species identification because of the fundamental difference in how evolutionary change is viewed by each. The method of species identification employed by gradualists is decidedly *not* the method used in punctuated equilibrium. As Eldredge (1982:339) notes, the issue centers around ontology: "[S]pecies are individuals in an ontological sense," according to punctuated equilibrium. Punctuated equilibrium assumes that species are real, distinct entities, not arbitrary parts of an evolutionary continuum. Gould and Eldredge (1977:117) were explicit about this early on, treating "species as 'real' units in nature, not as arbitrary elements in a continuum ranging from individual to kingdom." So, how are they detected in the fossil record? Eldredge (1979:16) argued

> that it is possible to recognize and diagnose species at any one point in time, to distinguish such taxa from others living sympatrically (including synchronically, of course) as well as allopatrically and/or allochronically. It is possible to specify in what respects those species differ among each other and, in most cases, those species-specific *differentia* are found in other samples, be they older, younger, or elsewhere. In other words, species are real entities with both geographic and stratigraphic distributions (few species are known from but a single bedding plane). Temporally, there is no significant change in these species-specific *differentia*—after all, it is this very continuity, which allows us to recognize species in more than one place at more than one time. Other morphological features within the stratigraphic distribution of a species—features which were not cited as *differentia*—often *do* show sequential change within the history of that species. [Punctuated equilibrium] conclude[s] that species are distinguished both spatially and temporally by virtue of a number of specifiable attributes, which tend *not* to change inexorably as time goes by (italics in the original).

Remarks by Gould provide insight to the procedure and hint at potential problems with it. He stated that "species are recognized morphologically by a total gestalt, not a list of differentia.... [N]o expert sees only the list, even if he defines taxa by it.... Taxa ... are recognized by their total appearance; they are only defined by an abbreviated list of differentia" (Gould 1976:177). Apparently, by "recognition" Gould meant a procedure resulting in an inductively derived summary of attributes, or character states, used to create a unit, in this case a species; by "definition" he meant specification of the necessary and sufficient conditions for membership within the species unit. Keeping the things separate from the differentia is critical, for as Elliott Sober (1993:145) notes, "A kind of thing is different from things of that kind"; different kinds will result in the placement of different combinations of things in discrete piles. Differentia denote the kind of thing and allow one to *identify* a newly discovered specimen of unknown identity as belonging to one species unit or another, that is, as a thing of a particular kind. So, how does one come up with differentia? That the procedure comprises inductive pattern recognition is clear in other statements made by Eldredge and Gould (1977:29):

> Implicit in [plotting] the stratigraphic range of any "species" ... is the existence of one or more morphological features serving simultaneously to (1) cluster a series of population samples over a certain segment of time and space, and (2) form a basis for distinguishing such a cluster from other similar ones. This leads to an important conclusion: by the mere recognition of *any* nontrivial stratigraphic range of *any* morphologically defined taxon at or near specific rank, we are necessarily implying a stability or stasis in species-specific *differentia*. All characters that are modified in the process of speciation and allow us to recognize "new" species must remain in a more or less recognizable state throughout the biochron of the species, or we would not be able to recognize it through time (emphasis in the original).

The internal stability of species as units and the discontinuities between them presumed by punctuated equilibrium allow their recognition within the fossil record. For example, as Gould's student Patricia Kelley (1983:595) put it, morphological "stasis or nondirectional fluctuation characterizes established species." And Elisabeth Vrba (1980:63) noted "the essential proposition" of punctuated equilibrium is stasis within a species. Furthermore,

> for species to be real the property of distinctness at any one time is not enough. They must be objectively distinguishable segments of lineages in time as well. That is, they must have origins, histories and terminations.... The notion of equilibrium within discrete species suggests that a species should possess at least some characters which separate it from other species in time and space; and which remain constant in time.... The history or duration is marked by stability in species-specific characters. (Vrba 1980:65)

The two preceding quotes indicate that were species not stable morphologically through time and across space, they could not be recognized in the fossil record. As Thomas Schopf (1982:1145) pointed out, the data requirement that fossil samples come from large geographic areas (Eldredge and Gould 1977;

Gould and Eldredge 1977) ensures that "any observed change in morphology is not attributed 'merely' to migration into an area." Punctuated equilibrium assumes a species is *mostly* stable over time; that is, there is no gradual cumulative or directional change in morphology over time such as phyletic gradualism suggests, though some noncumulative or nondirectional fluctuation in form is allowed (e.g., Kelley 1983). Stability as well as fluctuation in form are empirically determined, given how formal properties are measured; that is, definitions of taxa are extensionally derived, no doubt because of the early and continuing utility of such units for stratigraphic correlation (Eldredge and Gould 1977). As paleobiologist Paul Pearson (1992:125) notes, "Stratigraphic utility remains ... the driving force for taxonomy." But Pearson (1998b:317) also points out that

> the "species" of paleontology are recognized in the fossil record by the typological method. In effect they correspond to a set of morphologically similar specimens that are comparable to the holotype of the species (i.e., typological species are morphospecies). The appearances and disappearances of such morphospecies in the fossil record will only correspond to genuine speciation events and extinction events if, on the one hand, speciations are sudden and punctuational and, on the other hand, stasis predominates. If gradual evolution occurs, a lineage that tracks morphospace without branching or becoming extinct may nevertheless be perceived by taxonomists as a series of overlapping morphospecies that intergrade with one another, possible inducing a step-series type pattern.

Pearson's remarks beg the question of what is meant by Vrba's "objectively distinguished segments of lineages." What about theoretical problems? For example, how can basic stability of species morphology be theoretically warranted? Gould (1983c:362) believes past hesitancy to develop such warrants "arises from an exaggerated fear of 'essentialism.'" This fear, in his view, resulted in the architects of the Synthesis

> rejecting the idea that species have'essences'—defined in a useful modern way as genetic and developmental coherences that resist selective pressures of the moment, and impose a higher level, macroevolutionary, constraint upon change within local populations. The radical anti-essentialism of modern [post–Synthesis] evolutionary biology leads us to ignore or deemphasize the interplay of internal design and external selection.
>
> Developmental programs and genetic systems are "essential" properties "distributed" to individuals by virtue of their membership in a species. (Gould 1983c:362–363)

As one might suspect, architects of the Synthesis think Gould is a bit off the mark; Mayr (e.g., 1963, 1970, 1982b), for one, has long spoken of the "cohesion of the genotype" resulting from various internal—to use Gould's word—constraints. Regardless, it is beyond our scope to explore this issue in greater depth here. What is important in the present context is how species are recognized or identified in the fossil record. *If* species are morphologically stable, as presumed by the punctuated-equilibrium model, *then* species-specific differentia should be readily detected (if not patently obvious in a sample of sufficient size) in the fossil record. But because evolutionary change in morphology can fluctuate in direction, or be slow or fast along a particular direction of change, species must be identified extensionally; that is, the "differentia" of a species must be extracted from the particular initial set of fossils under study. Once the differentia are in hand, then other fossils of unknown taxa can simply be identified as members of one or more species or, if necessary, one or more new sets of differentia can be extracted to create one or more new species.

Our point is simple. Given the inductive procedure of species recognition, differentia are an accident of sampling, and a phenomenon similar to an archaeologist's "typological creep" (e.g., Phillips et al. 1951) could become a problem, as has been recognized by some paleontologists (e.g., Maglio 1971). Precisely this difficulty has been detected—fluctuation over analytical time in the definitive criteria and some fluctuation over analytical time in the specimens included within a species—for fossil species. Such fluctuation may cease when a sufficiently large sample is available (Hughes and Labandeira 1995) and the definitive criteria stabilize, but this leaves unaddressed the requisite number of fossil specimens (see Schopf 1982). More important, it does not indicate which of the virtually infinite number of characters displayed by fossils we should choose as differentia. If the criticism of using the concept of chronocline when identifying chronospecies is faulty, then it seems the concept of using differentia for identifying morphologically static species is valid. In the former, one chooses characters that change continuously, gradually, and unidirectionally; in the latter, one chooses characters that do not change over long time spans (Brown 1987). The remarks of Eldredge and Gould, quoted earlier, suggest that the differentia should be characters that do not change states over the temporal duration of the species. In our view, then, the procedure for recognizing fossil species advocated by the architects of punctuated equilibrium represents a set of analytical blinders. What is presumed—species stability—is found, *given how species are recognized analytically*. Gould and Eldredge are correct that a similar problem plagues how species are recognized under the phyletic-gradualism model, though they sweep the problem under the carpet when they indicate that a change in ontology and epistemology is all that is necessary to operationalize the punctuated-equilibrium model (e.g., Eldredge 1982). We agree with Richard Fortey's (1985:17) statement, made in his review of the two models, that "theory influences the way we make our observations."

Before moving on, it is important to note two things. First, Fortey (1985:17) observed that both models "should be permitted to compete against the data on equal terms" (see also Fortey 1988; Vaupel Klein 1994). The means we advocate of allowing such competition avoids both the blinders of the inductive pattern-recognition procedure of punctuated equilibrium and the arbitrariness of the chronospecies units produced by the phyletic-gradualism model, and it also allows the measurement of rates of change (when plotted against interval-scale time) as well as changes in diversity and alterations of the mode of change. Second, in anticipation of our later discussion, we note that the architects and users of both models of evolution—phyletic gradualism and punctuated equilibrium—have not considered seriously the procedures of systematics. The several preceding paragraphs should make it abundantly clear how truly critical the issue of systematics is to issues of evolutionary tempo and mode. But before we consider systematics, we need to consider other aspects of tempo and mode.

Modes and Rates of Change

In their first major defense and restatement of punctuated equilibrium, Gould and Eldredge (1977:116) indicated that in their model "speciation [rapid cladogenesis] is orders of magnitude more important than phyletic [gradualism] as a mode of evolutionary change." Gould (1982b:84) later reiterated that punctuated equilibrium was "a specific claim about speciation and its deployment in geological time; it should not be used as a synonym for any theory of rapid evolutionary change at any scale." Simpson's (1944) quantum evolution was, according to Gould (1982b:84–85), precisely the latter—"rapid anagenetic change [that, in contrast to cladogenesis, did not comprise] speciation and its consequences." In stark contrast to quantum evolution, punctuated equilibrium "holds that accumulated speciation is the root of most major evolutionary change" (Gould 1982b:85). A few years later, Gould and Eldredge (1986:144) stated that punctuated equilibrium "has always been rooted in the [nontraditional] claim that Mayrian peripatry [allopatric or peripatric speciation], in its most conventional form, translates correctly into geological time as (in most cases) a bedding plane, not as phyletic gradualism."

The reason that the cladogenic aspect of the punctuated-equilibrium model had to be reiterated was to ensure that it was not confused with saltationism, or the belief that evolution depends on the appearance of macromutations that exhibit significant disjunctions with their parents. Rapid anagenetic change—saltation—is the antithesis of cladogenesis, or the diversification of one lineage or species into several—what Mayr (e.g., 1982b) referred to as the multiplication of species. These two forms of geologically rapid change are, supposedly, readily distinguished in the fossil record:

> Examples of stasis alone and simple abrupt replacement, although conforming to expectations of punctuated equilibrium, are not direct evidence for our mechanism: for stasis might just be a lull in anagenetic gradualism, and replacement might represent rapid transformation without branching, or migration of a distant (phyletic or geographic) relative rather than evolution *in*

> *situ*. A good test of punctuated equilibrium requires (in addition to the obvious need for documented rapidity in an interval known to be sufficiently short) both a phyletic hypothesis to assert sister-group relationship of the taxa involved, and survival of putative ancestors to affirm an event of true branching [cladogenesis] rather than rapid phyletic transformation. (Gould and Eldredge 1993:225)

Gould and Eldredge (1986, 1993; Eldredge and Gould 1997) contend that the basic punctuated-equilibrium model has not changed in the twenty-five years since its introduction (see Chapter 7); it concentrates "morphological change in geologically instantaneous episodes of cladogenesis, thereby compelling an explanation of [long-term] trends as the differential success of species" (Gould and Eldredge 1986:146). In other words, the creation of new forms, specifically species and higher taxa, by way of diversification is the critical aspect of punctuated equilibrium (hence the term *macroevolution*; e.g., Stanley 1979), not the creation of new forms of organisms within a phyletic lineage of particular species (known as microevolution). Thus, "[h]igher-level [e.g., species-level] selection is an addition to, not a refutation of, the conventional Darwinian view of selection upon organisms" (Gould and Eldredge 1986:146–147). Through a series of papers, Gould and colleagues (e.g., Gould et al. 1977) demonstrated that diversification, or cladogenesis, did seem at least in part to drive evolutionary change. They did this by showing that clades, or monophyletic groups, were initially and relatively abruptly taxonomically rich, but that over time the richness of species decreased, producing what are called "bottom-heavy clades" (Gould et al. 1987). We return to the relevance of this observation for archaeology in Chapter 7, but here it is worthwhile to summarize some of the critiques of the cladogenic aspect of the punctuated-equilibrium model.

In what Gould (1989b:129) characterized as "the most cogent critique of punctuated equilibrium yet published," J. R. G. Turner (1986:185) suggested that punctuated equilibrium consists of three aspects: (1) periods of morphological stasis alternating with brief periods of "extremely rapid change"; (2) instances of "large morphological changes [that] are (almost) always cladogenic ... and therefore very little significant evolutionary change occurs by the neo–Darwinian [1940s Synthesis] processes of gradual or rapid change *within* [a lineage]"; and (3) long-term trends and patterns of change that are driven by a form of selection operating between species rather than between individuals within a species. The first aspect is, Turner (1986:185) argued, readily subsumable under the Synthesis of the 1940s, as is the third aspect, though it comprises a significant "extension" of the Synthesis. It is the second aspect of punctuated equilibrium that Turner (1986: 194) perceived as weak, remarking that "there is something decidedly odd about the hypothesis that punctuational [saltational] change cannot occur without leading to cladogenesis." Turner (1986:194) noted that "punctuational change is 'associated' with cladogenesis only in the tautological sense that until there are two or more branches of a tree, we cannot see them becoming different from one another." As Schopf (1982:1146) pointed out, the problem for evolutionary studies is that "the assumption of the nature of the entities called species becomes an assumption in interpreting observed morphology." We return to this point later, noting here that there is even more to it than that.

Turner (1986:200) posited that punctuated equilibrium focuses on morphospecies—a term we can use to distinguish the species recognized by punctuated equilibrium (Schopf 1982:1146) from those recognized by phyletic gradualism and termed chronospecies—rather than on biological species (reproductively isolated populations). The concept of morphospecies ignores the existence of sibling species—species that are reproductively isolated in nature but which are morphologically indistinguishable—and also ignores morphological variation resulting from ecophenotypic responses of individuals within a population that might be interpreted by a paleontologist as representing distinct biological species (e.g., Palmer 1985). The notion of reproductive isolation, following from the notion of peripatric speciation that informed the construction of the punctuated-equilibrium model, forced its architects to conclude that cladogenesis and concomitant morphological change occur simultaneously as a result of isolation. In other words, "morphological change and cladogenesis are in effect correlated 1:1" (Turner 1986:203). But, Turner noted, the existence of sibling species indicates that morphological change is unnecessary to speciation. This, plus the fact that phyletic change has been documented, renders "the full punctuated equilibrium package ... a decidedly questionable assumption" (Turner 1986:200–201). The first and third aspects of punctuated equilibrium are reasonable and provide significant expansions to the neo–Darwinian Synthesis; the second aspect is questionable, in Turner's view, though one may adopt the other two aspects without adopting the second.

Given the mantra "stasis is data," it is not surprising that Gould (1989b:130) finds Turner's argument regarding sibling species misplaced and suggests that if "periods of stasis include multiple species, then [punctuated equilibrium's] argument is only strengthened—for the constraints upon change then affect several related forms." Furthermore, according to Gould (1989b:130), punctuated equilibrium calls not only on morphology to distinguish species but also on "geographical distribution and, especially, sympatric occurrence of species as a sign of biological distinctness." These criteria "provide the major, and conceptually simple, test for cladogenesis" (Gould 1989b:130), or Turner's (1986) second aspect of punctuated equilibrium. The contemporaneity of ancestral and descendant species is the major empirical indication of cladogenesis. Without this, "we do not know whether the punctuation has been caused by cladogenesis as punctuated equilibrium proposes, or by anagenesis at punctuational tempos" (Gould 1989b:130).

Gould is being rather clever here, in essence stating that you can either believe in cladogenesis—branching evolution—in which case you believe in punctuated equilibrium, or you believe in "anagenesis at punctuated tempos," in which case you are a saltationist. Actually, as Gould well knows, anagenesis can be slow or rapid, so saltationism plays no role necessarily in the cladogenesis–anagenesis issue. Of more importance, however, is Gould's failure to point out that the cladogenesis–anagenesis issue is little more than a red herring in examining tempo and mode in evolution. Chapter 7 is devoted to this issue, but we need to point out here that cladogenesis fits comfortably under both views of evolutionary tempo. It certainly plays a singular role in punctuated equilibrium, but it fits just as comfortably under phyletic gradualism. So does anagenesis, which is anathema to punctuated equilibrium. Gould does not ever admit what the real issue is: Does punctuated equilibrium, with its modes of cladogenesis and peripatric speciation, explain the paleobiological record better than phyletic gradualism, with its modes of cladogenesis and anagenesis, *either* of which could be at work at a given moment? That is the only question that matters.

Jeffrey Levinton (1983) focused on different aspects of punctuated equilibrium than did Turner (1986), suggesting the controversy regarding punctuated equilibrium, which is "set squarely within traditional evolutionary arguments ... is neither revolutionary nor very different from the sort of debate going on in evolutionary biology for the last 50 years" (Levinton 1983:112). This is because Mayr's (e.g., 1942, 1963) theory of geographic, or peripatric, speciation "*is* the theory [of punctuated equilibrium], for all practical purposes" (Levinton 1983: 113; emphasis in the original). This in turn is because the two principle assertions of punctuated equilibrium, according to Levinton (1983:113), are that the history of a species is dominated by stasis, and change is concentrated at times of speciation. Stasis occurs as a result of mechanisms of genetic homeostasis, and abrupt cladogenesis is the result of genetic revolutions within peripherally isolated small populations that, because of their brief temporal duration and narrow geographic occurrence, are unlikely to be found in the fossil record. Remains of organisms that are transitional between species will be found rarely if at all.

Recall that Turner (1986) found the one-to-one correlation between punctuational change and cladogenesis tautological. Levinton (1983) underscored the supposed tautology by noting that there are two sorts of stasis being discussed—character stasis and species stasis—and two sorts of change—character change and species change. Given that prehistoric species are identified as a suite of characters evidenced by fossils, and different species are defined by different character states, then only a change in characters can denote a change in species. Therefore, "morphological evolution can readily be related to speciation since one is the source of the other's inference" (Levinton 1983:115). Gould (1985:5) countered this argument by noting that the punctuations "are an ecologist's gradualism. Our theory is rooted in this very recognition—in pointing out that our predecessors had scaled improperly in assuming that ecological gradualism extrapolated to phyletic continuity, and in arguing instead that the proper geological extension of peripatric speciation is punctuated equilibrium." Shifting the ontological scale from organisms to species—to what Gould (1985) referred to as the "second tier"—forces one also to shift the epistemological (methodological) scale to the study of the "irreducible properties of populations" (Gould 1985:6), or what are formally termed "emergent properties" (Gould 1985:3)—attributes of an aggregate of individuals that emerge as properties of the aggregate and that are not found as properties of the individuals making up the aggregate (for brief discussions of the history of the term in ecology, see Edson et al. 1981; Salt 1979). These can include things such as "population structure, size, [and] migration dependent upon population structure" (Gould 1983c:359). As we will see, such a shift in scale is critically important to the implications of punctuated equilibrium and how it expands our conceptions and perceptions of evolutionary processes.

Given that punctuated equilibrium comprises a shift in scale (e.g., Vaupel Klein 1994), its most vocal advocate does not deny that phyletic gradualism occurs in nature (Gould 1985, 1989b). To produce an "adequate test" of the tempo and mode of the evolution of a particular lineage or clade therefore requires "good geographic coverage, long sequence of closely spaced samples, unambiguous definition of taxa, and adequate biometrical testing on sufficiently large samples" (Gould and Eldredge 1977:125). We would add that a measure of time, preferably of interval scale, also is required. Gould (1997c) recently halted a potential misinterpretation of some of his own data as confirming phyletic gradualism when such a temporal measurement scale was available. The fine-scale temporal resolution associated with this case of anagenetic change in species (Goodfriend and Gould 1996) is *not* contradictory to punctuated equilibrium, but rather "represents a geological eye-blink in the fullness of evolutionary time" (Gould 1997c:18). This case comprises "a strong confirmation of punctuated equilibrium" because the fossils involved were found "on what would become a single bedding plane in the geological record" and thus "occurred in a geological moment and represented a punctuation, not a gradual sequence of fossils" (Gould 1997c:18, 64). In short, at the temporal—geological—scale within which species would typically be recognized in the fossil record, change would appear abrupt; only the remarkable absolute temporal resolution of the case at hand revealed that change was in fact rather gradual. The fine-scale temporal resolution provided by this sample comprises, then, in Gould's (1997c:12) words, "the paradox of the visibly irrelevant." As had been foretold by Fortey (1985), falsification of punctuated equilibrium would be difficult.

Implications of Punctuated Equilibrium

Gould (1987b:3, 157) once wrote that the

> theory of punctuated equilibrium ... is not, as so often misunderstood, a radical claim for truly sudden change, but a recognition that ordinary pro-

> cesses of speciation, properly conceived as glacially slow by the standard of our own life-span, do not translate into geological time as long sequences of insensibly graded intermediates (the traditional, or gradualistic, view), but as geologically "sudden" origins at single bedding planes.... [Thus,] punctuated equilibrium, as its essential statement, accepts the literal record of geologically abrupt appearance and subsequent stasis as a reality for most species, not an expression of true gradualism filtered through an imperfect fossil record.

That the fossil record comprises an incomplete record of evolutionary change was, of course, Darwin's (1859) view, as it has been the view of twentieth-century paleontologists (e.g., Durham, 1967; Newell 1959; see also Benton and Storrs 1994). In taking the fossil record at face value, punctuated equilibrium has important implications for conceptions of Darwinian evolution, but this means only that the "modern synthesis is incomplete, not incorrect" (Gould 1982a:382).

Similar to the notion found in the neo–Darwinian Synthesis that the appearance of genetic novelties—whether caused by mutation or recombination—is undirected with respect to the selective environments in which those novelties appear, punctuated equilibrium holds that novel species—or morphological variants—appear during relatively abrupt events of cladogenesis. Such appearances provide the fodder for selection and evolutionary change; speciation thus occupies "a central role in constructing evolutionary novelty.... Speciation supplies the raw material of change" (Gould 1985:6). Given that the punctuated-equilibrium model is parallel to the Synthesis in many respects yet differs in scale, one of its often-mentioned implications is that evolutionary processes and effects are hierarchical (e.g., Arnold and Fristrup 1982; Eldredge 1990, 1996; Gould 1980a, 1982a, 1985, 1995a; Vrba and Eldredge 1984; Vrba and Gould 1986). Thus, (1) the ontological shift in scale from conceiving of evolution as working on organisms comprising populations, to working on populations comprising species, to working on species comprising genera and higher taxonomic levels, and (2) the concomitant epistemological shift from monitoring attributes of individual organisms to monitoring emergent properties of aggregates of organisms such as populations and species, has broadened the scope of evolutionary theory. The term *species selection* is one result (e.g., Barnosky 1987; Cracraft 1985; Gould and Eldredge 1988b; Vrba 1984), but perhaps more importantly, the model has produced significant insights into the process of evolution itself.

Selection, Sorting, Spandrels, and Exaptation

Gould (e.g., 1980a, 1982a, 1985, 1990; see also Eldredge 1979) identifies two aspects of the neo–Darwinian Synthesis that have tended to blind paleobiologists to the interpretive significance of apparent gaps in the fossil record—perceived as the absence of transitional, or missing-link, specimens and the abrupt appearance of new species—and to the interpretive significance of apparent long-term morphological stasis of species once they appear. By adopting the view that evolution was (1) continuous and gradual and (2) anagenetic, the gaps must represent a lack of preservation, and change had to be solely adaptive. Natural selection's role was one of not only removing the less fit but also of creating forms that were remarkably well fit. To ensure that selection enjoyed this role as a creative mechanism, the introduction of new variants had to be undirected—to appear without respect to their utility in the selective environment in which they originated—and new variants had to be like but just a bit different from their predecessors. This reinforces the notion that the pace of evolution is gradual and comprises a long-term, cumulative building of fit organisms. According to Gould (1980a, 1982a, 1983b, 1985), the Synthesis "hardened" around these central tenets through the 1950s and 1960s (but see Mayr 1991b).

The shift in ontological and epistemological scales comprising punctuated equilibrium first expanded the notion of rates from always and only gradual to varying from stasis to relatively rapid. That the rate of evolution might vary had long been recognized, at least since Simpson's *Tempo and Mode in Evolution*, but that a zero rate—that is, stasis—might occur was seldom if ever acknowledged (Gould 1982a, 1983b). The shift in scale also showed that adaptationist explanations emanating from the crystallized Synthesis were perhaps too readily drawn (Gould and Lewontin 1979; but see Mayr 1983). Contrary to the Synthesis, punctuated equilibrium prompted the thought that not every character, or attribute, of an organism had an adaptive function; some might merely be "spandrels" (Gould 1997b).

Spandrels—an architectural term used metaphorically (Gould and Lewontin 1979) to refer to biological features—comprise a "class of forms and space that arise as necessary byproducts of another decision in design, and not as adaptations for direct utility themselves" (Gould 1997b:10750). In other words, a spandrel is an "automatic byproduct of other architectural decisions and therefore nonadaptive in origin. Spandrels are not adaptations, despite their availability for later fruitful use and whatever their coordinated utility in the original structure" (Gould 1997b:10752). Spandrels can comprise a particular form of feature, its position, its constitution, and/or its frequency in a structure (Gould 1997b:10752). Thus, a spandrel might simply "spread through clades by nonadaptive hitchhiking" (Gould 1985:7). In Sober's (1984a:98–100) words, selection *for* a trait that enhances fitness is the cause that may result in the effect of selection *of* a trait that is a mere spandrel (see also Sober 1993:82). Gould and Vrba (1982:6) coined the term *exaptation* to help account for the evolutionary history of spandrels and defined it as comprising "characters, evolved for other usages (or for no function at all), [that were] later 'coopted' for their current role [or function]." Thus, modern evolutionary thinking now holds that character states categorized as adaptive *must* have been shaped by selection (Baum and Larson 1991; Brandon 1990; Burian 1992; Gould 1986; Leroi et al. 1994; O'Hara 1988; Sober 1984a; Taylor 1987), else they are exaptations—character states that may not have been shaped by natural selection but now influence fitness (Gould 1991a; Gould and Vrba 1982). This possibility underscores the critical aspect of history in evolutionary studies and the importance of distinguishing between analogous and homologous similarity among variant characters (see also Gould 1986; O'Brien and Holland 1992).

Recognizing spandrels and exaptations as categories of morphological features displayed by organisms provides an explanation for the observations of geneticists that some combinations of genes appear to be functionally equivalent and others appear to have little functional significance at all (Kimura 1979, 1983, 1991; King and Jukes 1969). Thus, some significant portion of evolutionary change does not appear to comprise adaptive change. Some argue that the " 'selectionist' [Gould's neo–Darwinist-hardened Synthesis] and 'neutralist' views of molecular evolution are competing hypotheses within the framework of the synthetic theory of evolution" (Stebbins and Ayala 1981:967), but the dichotomy seems contrived. The neutral theory accounts for such things as genetic drift (Kimura 1983); the hardened Synthesis does not, according to Gould's (e.g., 1982a) rendition of it. As the Synthesis matured, Mayr (1963:586) remarked that the "proponents of the synthetic theory maintain that all evolution is due to accumulation of small genetic changes, guided by natural selection, and that transspecific evolution is nothing but an extrapolation and magnification of the events that take place within populations and species." Gould (1980a:122) points out how detailed genetic analyses reveal "that alleles often drift to high frequency or fixation, and that many common variants are therefore neutral or just slightly deleterious." As a result of such observations, "the specter of stochasticity has intruded upon explanations of evolutionary *change*. This represents a fundamental challenge to Darwinism, which holds, as its very basis, that random factors enter only in the production of raw material, and that the deterministic process of selection produces change and direction" (Gould 1980a:122). Simulations of speciation and clade richness suggest an important role for random processes (e.g., Gould et al. 1977, 1987), a topic to which we return in Chapter 7.

By holding that "evolutionary trends cannot represent a simple extrapolation of allelic substitution within a population," punctuated equilibrium maintains that "evolution is concentrated in events of [successful] speciation" (Gould 1980a: 125). Evolutionary trends are the result of speciation because "the sorting of species (differential success) ... determines the direction of change (trends)" (Gould 1985:6). So what is sorting? "Sorting is a simple description of differential representation; it contains, in itself, no statement about causes" (Vrba and Gould 1986:217). One mechanism of sorting is natural selection; another is drift. Either mechanism can produce a differential representation of variants of organisms through time and the appearance of direction. One job of paleobiologists, then, is to determine which mechanism was responsible for any particular anatomical feature or character state observed in the fossil record. Importantly, one must also recognize that selection at one scale or level can produce sorting at a lower or higher scale or level; sorting "can occur as an effect of causes at other levels" (Vrba and Gould 1986:219). The selection of particular alleles can effect sorting of variant organisms carrying those alleles, and vice versa. Hence, there may in fact be selection for species-level traits—species selection (e.g., Cracraft 1985; Vrba 1984)—or the sorting of species may result from selection operating at a higher or lower scale.

Recognizing that some variant characters are neutral with respect to adaptedness leads to the suggestion that "co-optation of neutral and non-adaptive features [may characterize] the origin of major … structural innovations, [thereby providing] an understanding of trends in unselected characters" (Gould 1990:9). Carrying along through time a set of adaptively neutral features is advantageous because it results in an organism being preadapted to unanticipated contingencies (see also Gans 1979, 1988, 1993). Thus, one must distinguish between exaptive features and features that were historically built by selection for the function they now serve. The latter features are *adaptations*. Mayr (1991a:3) indicates the "process of selection is, in principle, an optimization process, but owing to its probabilistic nature, to its constraints, and to the frequency of stochastic processes, [and, we add, historical contingencies] it cannot achieve optimality." Thus, Gould (1982a: 383) suggested that the "constraints of inherited form and developmental pathways may so channel any change that [the] channel itself [may represent] the primary determinant of evolutionary direction" (see also Gould 1989a; Mayr 1988:108). Hence, "no elephant will ever fly" (Gould 1980c:40).

The adaptationist program of the neo–Darwinian Synthesis (Gould and Lewontin 1979) has tended to downplay two important aspects of evolutionary change suggested by punctuated equilibrium. First,

> the possible routes of selection are channeled by inherited morphology, building material, and the amount and nature of variation itself. Though selection moves organisms down channels, the channels themselves impose primary constraints on the direction of change. [Second,] selection on one part of a structure may impose a set of correlated and nonadaptive modifications on other parts of an integrated body. Many features, even fundamental ones, may be nonadaptive either as developmental correlates of primary adaptations or as "unanticipated" structural consequences of primary adaptations themselves. (Gould 1980c:44)

In short, evolution tinkers with available variation to build a workable solution (Jacob 1977). What it has to work with is constrained by various factors; as well, those constraining factors will channel the direction of evolution: "[D]evelopmental and architectural side-consequences and forced correlations can be as numerous and as prominent as the selected basis of a trend itself" (Gould 1990:8).

Species Duration and Time Averaging

Given that "stasis is data," it is not surprising that under the onslaught of punctuated equilibrium, paleobiologists turned attention to estimating the temporal duration of the existence of particular species. One must know when on an interval scale of time a species first appeared and when it disappeared. Gaps in the temporal record create difficulty in estimating durations of species and also in rates of change. Attempts to deal with a gappy fossil record began quickly, on the heels of the introduction of the punctuated-equilibrium model (e.g., Dingus and Sadler 1982; Sadler 1981; Sepkoski 1975), and today such studies (e.g., Foote 1997a; Foote and Raup 1996; Solow and Smith 1997) have attained remarkable sophistication in modeling the effects of stratigraphic incompleteness on estimates of taxa duration. The fundamental epistemological problem of identifying taxa in the fossil record, although recognized early in the history of punctuated equilibrium, appears to be receiving much less attention (but see Hughes and Labandeira [1995] for an important exception).

Schopf (1982) outlined the influences of classification—species identification—on estimates of species duration. If the *differentia* of taxa change over analytical time, then the duration of those taxa no doubt will change over analytical time as well, because the particular fossils included in a given taxon will change; thus, the strata containing the oldest or youngest fossils (or both) of a taxon will change. Schopf (1982:1148) found that as taxa become better "characterized," their temporal ranges become shorter. If the fossils of a particular clade are morphologically complex, the taxa within that clade will tend to have relatively short durations because the fossils can be sorted into a greater number of taxa. Finally, Schopf noted that in the absence of fine stratigraphic resolution, a fossil taxon may be assigned to a particular stratum spanning a relatively large chunk of the temporal continuum, which effectively sets a minimum duration for that taxon.

The focus on stratigraphic completeness and detecting temporal durations of taxa provided taphonomic researchers with an explicit link to evolutionary theory (Behrensmeyer and Kidwell 1985): How might the formation of the fossil record, particularly differential preservation, bias estimates of the appearance, extinction, and duration of a species? This link contributed to the recognition of a related problem: If one were interested in rates of change, samples of fossils representing relatively synchronic chunks of the temporal continuum were required to ensure that temporal variation was not being masked by what came to be known as "time-averaging" (e.g., Behrensmeyer and Schindel 1983; Schindel 1980, 1982). This problem exists when a sample comprises a relatively large chunk of the time span under study. How might such samples influence inferences regarding paleoecology (e.g., Peterson 1977)? More important from the perspective of writing evolutionary histories, how might these sorts of samples obscure or enhance the analytical visibility of morphological stasis or change? This epistemological problem, too, seems to have received only minor attention as the basic ontology of punctuated equilibrium was adopted by paleobiologists. We are left with the view that species are real in the sense that they are morphologically static entities over relatively long temporal spans.

Some questions raised during the debates over punctuated equilibrium have not been answered. How much geographic space must be sampled to ensure against the emigration and immigration of a species creating the appearance of that species's termination or initiation, respectively, in a locality (Schopf 1982)? The lack of an answer would seem to cast some doubt on estimates of species duration. But despite such problems, much of the rancor of the late 1980s abated by the early 1990s as researchers got back to the business of trying to make sense out of the fossil record (Novacek 1996). One can accept the notion of punctuated equilibrium, favor the notion of phyletic gradualism, prefer some updated version of Simpson's *Tempo and Mode in Evolution*, or grapple with all of them, given the historically contingent nature of evolution (Chapter 8). Whichever stance one takes, the point thus far is that the emergence in the 1970s of paleobiology as the explicit use of biological evolutionary theory to explain the fossil record resulted in efforts to modify that theory to account for the fossil record in its own terms. This in turn produced some significant insights relative to variables that had to be incorporated into the theory. As we have indicated in several places, the central problem remaining involves the measurement of variation and the use of species as the unit of measurement. What was going on in archaeology during this critical period in paleobiology? Specifically, how was variation viewed, and what kinds of units were being used to measure it? Was history even of analytical interest?

PROCESSUAL ARCHAEOLOGY

In the introduction to *Method and Theory in American Archaeology*, Gordon Willey and Philip Phillips (1958:5) indicated that "Culture-historical integration is ... comparable to ethnography with the time dimension added." They went on, however, to lament that "[s]o little work has been done in American archaeology on the explanatory level that it is difficult to find a name for it." They concluded by applying the term "processual interpretation" (in quotes), which they defined as "any explanatory principle that might be invoked.... [P]rocessual interpretation is the study of the nature of what is vaguely referred to as the culture-historical process. Practically speaking, it implies an attempt to discover regularities in the relationships given by the methods of historical integration" (Willey and Phillips 1958:5–6). The integrative methods of culture historians involved constructing sequences of phases, identifying horizons and traditions, and using the axiom that typological similarity denoted historical relatedness. The archaeological units used had sociocultural connotations and thus allowed the analyst to invoke sociocultural processes observed in modern cultures by anthropologists and ethnographers as the mechanisms of change (e.g., Meggers 1955). Archaeological units having sociocultural connotations were necessary because they were "intelligible in *both* the cultural and social aspects of the [human] behavior that is our subject matter" (Willey and Phillips 1958:49; emphasis in the original).

The shift away from an earlier focus on establishing sequences of artifact types to explaining in anthropological terms sequences of cultures represented by artifacts prompted one practitioner to proclaim that there was emerging in the 1950s a "New American archaeology" that was "tending to be more concerned with culture process and less concerned with the descriptive content of prehistoric cultures" (Caldwell 1959:304). The culture processes were the ones witnessed by anthropologists—diffusion, migration, and the like. The focus on homologous similarity to tell time was the obvious one to use when attempting to detect such processes in the archaeological record, but the regular failure of culture historians to analytically distinguish between homologous and analogous similarity, as noted by Rouse (1955), led to problems. Beginning in 1962 and extending over the next several years, Lewis Binford (e.g., 1962, 1964, 1965, 1967, 1968a, 1968c) outlined a research program intended to overcome these perceived problems in Americanist archaeology. To warrant disciplinary adoption of his suggested solution, Binford chose to emphasize what he perceived as the fallacies of culture history.

Alternatives to Normative Theory and the Aquatic View of Culture

Culture historians had, according to Binford (1965:203), adopted normative theory and perceived as their field of study "the ideational basis for varying ways of human life—culture.... The archaeologist's task then lies in abstracting from cultural products [artifacts] the normative concepts [shared ideas] extant in the minds of men now dead." Anticipating writers of the 1980s (e.g., Boyd and Richerson 1985), Binford (1965:204) argued that normative theory construed culture as a transmission mechanism—a means of inheriting ideas: "Learning is the recognized basis of cultural transmission between generations and diffusion the basis of transmission between social units not linked by regular breeding behavior." Culture construed as ideas and as a mechanism of transmission meant that "any attempt to break up this cultural 'whole' is considered arbitrary and thought of as a methodological expedient" (Binford 1965:204). In other words, archaeological units such as cultural foci and phases were, like the species of the paleobiological gradualists, arbitrary constructs of the analyst. Mechanisms of change such as innovation, drift, and diffusion resulted in constant, gradual change, and differences and similarities between cultural foci and phases were thought to signify "cultural relationships," which in turn "resolve into one general interpretive model" (Binford 1965:204).

The interpretive model, to Binford, comprised an aquatic view of culture. Reversing the order of his sentences,

> Culture is viewed as a vast flowing stream with minor variations in ideational norms concerning appropriate ways of making pots, getting married, treating

> one's mother-in-law, building houses, temples (or not building them, as the case may be), and even dying. These ideational variations are periodically "crystallized" at different points in time and space, resulting in distinctive and sometimes striking cultural climaxes which allow us to break up the continuum of culture into cultural phases.... Discontinuities in rates of change or in formal continuity through time are viewed as the result of historical events which tend to change the configuration of social units through such mechanisms as extensions of trade, migration, and the diffusions of "core" ideas. (Binford 1965:204)

According to Binford, the aquatic view, which culture historians were forced to adopt given their subscription to normative theory, left them "competent to pursue the investigation of culture history, a situation which may partially account for [their] failure to develop the explanatory level of archaeological theory" (Binford 1965:204–205). How, for example, could one possibly explain arbitrary chunks of a historical continuum in anthropological and ethnological terms (e.g., Spaulding 1957)? First, the chunks had to be real—in an essentialist sense—if they were to be so explained, just as the species of punctuated equilibrium had to be real. But there was more to it than that.

Binford's solution to the problems he perceived with the culture history of the 1950s involved discarding the notion of a cultural lineage as a flowing stream of ideas that changed through time and varied over space, and adopting White's (1959b) notion that culture is humankind's extrasomatic—nonbiological—means of adaptation. This required a new systematics that was based on a different concept of culture, just as the shift from phyletic gradualism to punctuated equilibrium demanded a new systematics of species. Cultures were viewed as systems of interacting parts: Culture is "multivariate, and its operation is to be understood in terms of many causally relevant variables which may function independently or in varying combinations. It is our task to isolate these causative factors and to seek regular, statable, and predictable relationships between them" (Binford 1965:205). But archaeological "cultures" were still recognized in the same sense that culture historians had recognized them (Meltzer 1979, 1981b), a topic we take up in a later chapter. This failure to alter epistemology would result in various problems that were identified earlier (e.g., Plog 1974) but which were never dealt with satisfactorily (Meltzer 1981b). The net result of Binford's suggestions was that processual archaeology became what we categorize here as a two-step analytical protocol, the first step being a prerequisite to the second.

The First Step: Cultural Reconstruction

Binford's (1962) seminal paper "Archaeology as Anthropology" set the tone for processual archaeology. The two disciplines named in the title had common goals—"to *explicate* and *explain* the total range of physical and cultural similarities and differences characteristic of the entire spatial–temporal span of man's existence" (Binford 1962:217). Archaeology's contribution was clear: "We as archaeologists have available a wide range of variability and a large sample of cultural systems. Ethnographers are restricted to the small and formally limited extant cultural systems. Archaeologists should be among the best qualified to study and directly test hypotheses concerning the process of evolutionary change" (Binford 1962:224). In remarkable parallel to Gould and Eldredge's (1977) comment with which we began this chapter, Binford (1968b:268) later stated that "archaeological research might be expected to yield explanations for some observations of cultural phenomena made exclusively through archaeological data." But he also noted, "If archaeologists and ethnologists are to work with common problems, their observations must be geared toward gathering data on the same variables, despite the obvious differences in their fields of observation. [Thus,] they must work in terms of comparable sociocultural units" (Binford 1968b:271). Up to that point, according to Binford (1962:217, 224), archaeology had operated in a "theoretical vacuum" and not only had failed "to contribute to the furtherance of the aims of anthropology but [also had] retard[ed] the accomplishment of these aims." He suggested the solution was to view a culture as a *system* of articulated parts—a change in one part would prompt a processual change in others—that served as the means by which the human organism adapted. The study of culture processes, then, comprised the study of cause-and-effect relations among cultural variables (Binford 1968a, 1968c). That archaeologists were to contribute to anthropology meant that they were "faced with the methodological task of isolating extinct socio-cultural systems as the most appropriate unit for the study of the evolutionary processes which result in cultural similarities and differences" (Binford 1964:440).

The success of processual archaeology hinged on this first step of reconstructing the dynamics of a prehistoric culture at its particular position in the time–space continuum. One result was that archaeology's focus shifted from the use of style (homologous similarity) of artifacts to measure time and cultural relatedness to function (analogous similarity) of artifacts in order to monitor the different forms cultures might take as adaptations. Binford (1962:220) characterized *style* as denoting group identity or ethnicity. This followed the culture-historical notion that a different style denoted a different time period and thus a different culture and ethnic group. Artifact function was treated rather differently. It was characterized variously as *technomic*—the function of an object in technological situations, *sociotechnic*—the function of an object in social situations, and *ideotechnic*—the function of an object in ideological situations (Binford 1962), or as primary (technomic) and secondary (sociotechnic and ideotechnic) functional variation (Binford 1965:206). Focusing on the functional properties and meaning of artifacts was perfectly in line with the definition of culture as humankind's extrasomatic means of adaptation.

The critical step of reconstruction had seen minimal development early in the 1960s, though Robert Ascher's (1961a, 1961b) and Raymond Thompson's (1958a, 1958b) efforts are noteworthy. Similarly, little effort had been made to develop analytical procedures for the determination of the functions of artifacts, other than to outline simple sorts of formal analogies such as inferring that all bilaterally symmetrical, bifacially flaked, and basically triangular lithic objects a few centimeters long were projectile points. Thus, Binford (1962:219) indicated that he considered "the study and establishment of correlations between types of social structure classified on the basis of behavioral attributes and structural types of material elements as one of the major areas of anthropological research yet to be developed." In short, material correlates of human behaviors and of sociocultural phenomena documented by ethnographers had not yet been identified by culture historians, except in the commonsensical fashion of a rather lackadaisical use of ethnographic analogy.

Attention turned rather quickly on the heels of Binford's arguments to finding material analogs of human behaviors and cultural phenomena. Once correlations between artifacts and behaviors were established, one seemingly had rather direct access to reading the dynamics of a working cultural system from the statics of the archaeological record. This made cultural reconstruction attractive to some culture historians (e.g., Martin 1971), as it was precisely what they had been searching for (Lyman et al. 1997b). Thus, Robert Ehrich's (1965:6) illustrative suggestion (and it was only illustrative, as he thought such assumptions were unwarranted [Ehrich 1963:20]) in the early 1960s that "we need only have postulated that either the potters were women living in a patrilocal society, or men living in a matrilocal one, to have a situation in which diffusion took place through the spread of pottery-makers rather than of the pots" was to shortly thereafter become a viable axiom (e.g., Deetz 1965; Hill 1966, 1970; Longacre 1964, 1966, 1970).

This initial step in the analytical protocol of processual archaeology was possible because, in Binford's (1962:219) view, the "formal structure of artifact assemblages together with the between element contextual relationships should and do present a systematic and understandable picture of the *total-extinct* cultural system" (emphasis in the original). In other words, the meaning of which would be disputed two decades later (e.g., Binford 1981), the "loss, breakage, and abandonment of implements and facilities at different locations, where [human] groups of variable structure performed different tasks, leaves a 'fossil' record of the actual operation of an extinct society" (Binford 1964:425). Archaeologists merely lacked the appropriate methods for rendering meaning in terms of cultural processes from the archaeological record: "The practical limitations on our knowledge of the past are not inherent in the nature of the archaeological record; the limitations lie in our methodological naiveté, in our lack of development of principles determining the relevance of archaeological remains to propositions regarding processes and events of the past" (Binford 1968a:23). Development of such methods comprised the second step in the analytical protocol of processual archaeology.

The Second Step: Deduction and Hypothesis Testing

Those subscribing to the second new archaeology—that of the 1950s (e.g., Caldwell 1959; Meggers 1955; Phillips 1955; and Willey and Phillips 1958), the first having comprised the so-called stratigraphic revolution (e.g., Wissler 1917b)—according to Binford (1968a), had begun to despair of ever being able to study cultural processes. Binford's version of a third "new" archaeology offered a solution to perceived problems. He, too, was interested in culture processes, which he defined as "the dynamic relationships (causes and effects) operative among sociocultural systems, [the] processes responsible for changes observed in the organization and/or content of the systems, or ... the integration of new formal components into the system" (Binford 1968a:14). Given this definition, the solution was obvious. To processualists, "explanation ... within a scientific frame of reference is simply the *demonstration* of constant articulation of [functional] variables within a system and the measurement of concomitant variability among the variables within the [cultural] system. Processual change in one variable can then be shown to relate to changes in other variables, the latter changing in turn relative to changes in the structure of the system as a whole" (Binford 1962:217; emphasis in the original).

Potential catalysts for processual change were to be sought "in systemic terms for classes of historical events such as migrations, establishment of 'contact' between areas previously isolated, etc." (Binford 1962:218). The historical *processes* of cultural change were thus exactly those of the culture historians (e.g., Willey and Phillips 1958). But whereas the culture historians had achieved some success in identifying instances of such processes at work in the past, the problem for processualists was one of answering the "why" question. *Why* did classic lowland Maya culture collapse, for example? Perhaps an immigration was the particularistic historical mechanism, but *why* the immigration and *why* the collapse (Binford 1968c)? To answer these questions required, it was thought, the establishment of "a set of general laws [that] connects the 'causes' with their 'effects' in such a way that if we know that the earlier events have taken place, we would be able to predict the event we wish to explain" (Binford 1968c:268).

The first step in reconstruction was necessary if the static archaeological record was to be explained in dynamic anthropological and processual terms. This did not signify, however, the continued sophomoric use of ethnographic analogy. Rather, something else was called for. Processual archaeologists favored

> a firmer basis for determining the behavioral correlates of material culture, [and noted that] both refinements in [archaeological] data collection and

> increased ethnographic knowledge cannot by themselves increase our knowledge of the past.... Fitting archaeological remains into ethnographically known patterns of life adds nothing to our knowledge of the past. In fact, such a procedure denies to archeology the possibility of dealing with forms of cultural adaptation outside the range of variation known ethnographically. In view of the high probability that cultural forms existed in the past for which we have no ethnographic examples, reconstruction of the lifeways of such sociocultural systems demands the rigorous testing of deductively drawn hypotheses against independent sets of data. (Binford 1968a:13)

> [T]he "interpretation" of the archeological record by the citation of analogies between archeologically observed phenomena and phenomena from a known behavioral context simply allows one to offer his *postulate* that the behavioral context was the same in both cases. In order to increase the probability that the postulate is accurate, a number of testable hypotheses must be formulated and tested. (Binford 1968b:269; emphasis in the original)

> [E]thnographic data can play two basic roles in archeological investigation: first, they serve as resources for testing hypotheses which seek to relate material and behavioral cultural phenomena; second, they may often (but need not always) serve as the basis for models of particular social relations which are postulated to have been the context for an observed archeological structure. (Binford 1968b:270)

Two things are contained in these statements. First, reconstruction hinges on established correlations between artifacts and behaviors; ethnographic analogy, in the sense that it had been used by previous archaeologists, particularly culture historians, was unacceptable to processualists. Rather, one had to build up a case that multiple interrelated variables were always and everywhere correlated (e.g., Binford 1967). Not everyone noted this critical distinction (e.g., Anderson 1969; Klejn 1973), and some who did were not convinced it provided a workable program (e.g., Munson 1969; Sabloff et al. 1973; Wobst 1978). Second, to establish the correlation required testing: "Scientific verification for archeologists is the same as for other scientists: it involves testing hypotheses systematically" (Binford 1968b:269). This part of the second step—hypothesis testing—was believed to allow processualists to conclude that the " 'non-material' aspects of culture are accessible in direct measure with the testability of propositions being advanced about them" (Binford 1968a:22).

Clearly, the distinction between configurational and immanent properties then being discussed by George Gaylord Simpson (1963, 1970) was unknown to the processualists; even had it been known, we suspect the processualists would have ignored it as unimportant given the fervor with which they embraced the newly emerging procedure of archaeological inference (e.g., Watson 1966, 1976). That the distinction was critically important is clear when one considers procedural studies of use-wear on stone tools (e.g., Frison 1968; Wilmsen 1968a,

1968b). Culture historians had paid minimal attention to such attributes, but processualists initiated studies of the mechanical properties of stone and how it responded to use as a tool material (e.g., Hayden 1979). Similarities in the overall shape of stone tools found in the archaeological and ethnographic records were no longer sufficient for rendering analogically based assessments of function. Focusing on immanent properties revealed various problems of equifinality (e.g., Knudson 1979; Miller 1982), but whether edge damage on stone tools was the result of use or some other factor was testable (Dunnell 1978b).

Something rather similar was emerging in paleontology at this same time as paleobiologists began to explore what they considered to be the functional attributes of fossils (e.g., Rudwick 1964). This was a step that they hoped would establish a new "science of form ... within evolutionary theory [that] studies adaptation" (Gould 1970:77). Functions were attributed to the parts of fossil organisms "by a principle of analogy" (Kitts 1974:461). Importantly, paleobiologists recognized and kept distinct the roles of homologs and analogs: "It is by analogy, and not homology, that we argue from modern to fossil functions" (Gould 1970:80). But they retained the notion of homolog to help write phylogenetic history. Even more important, paleobiologists also recognized the critical distinction between immanent and configurational properties. As Martin Rudwick (1964:33) remarked, "the range of our functional inferences about fossils is limited not by the range of adaptations that happen to be possessed by organisms at present alive, but by the range of our understanding of the problems of engineering." Processual archaeologists ignored this critical distinction and also largely ignored the notion of homologous similarity embodied in culture history's conception of style, ultimately discarding the latter for a new conception (e.g., Wobst 1977), a topic we address in Chapter 6.

Ultimately, processual archaeologists took as their twin rallying cries the testing of hypotheses and deductive reasoning, and the view of culture as a system of interrelated parts that served as an extrasomatic means of adaptation. Subsequent to Binford's (e.g., 1968a, 1968b) early passing references to deductive reasoning, the first formal statement of the processual protocol is found in a brief treatment by John Fritz and Fred Plog (1970) in *American Antiquity*. They indicated that explanation can be found in the "classification" of artifacts as "axes," "arrowheads," and the like, and that such classifications are founded in "a set of ideas or beliefs which function as laws" and thus "conform to the Hempel–Oppenheim model" of science (Fritz and Plog 1970:408). Relations between human behavior or culture and "characteristics of the archaeological record ... might be said to constitute archaeological theory" (Fritz and Plog 1970:408). Furthermore, empirical testing and confirmation of hypotheses were seen as providing these relations with "the status of laws so that their explicit use is possible" in explanatory efforts (Fritz and Plog 1970:409).

Numerous publications in the 1970s focused on the protocol of processual archaeology (e.g., Flannery 1967; Hill 1972; Klejn 1977; LeBlanc 1973; Morgan 1973, 1974; Read and LeBlanc 1978; Redman 1973; Renfrew 1973; Renfrew et al. 1982; Smith 1977; Tuggle et al. 1972; Watson et al. 1971, 1974). Was inductive reasoning totally unacceptable and only deductive reasoning acceptable (Hill 1972; Smith 1977)? Would conceiving of cultures as systems of interacting and interrelated parts forming an adaptive whole provide the kind of explanations processualists sought (Lowe and Barth 1980; Plog 1975; Salmon 1978, 1980)? Even philosophers of science commented on the issues (e.g., Levin 1973, 1976; Nickles 1977; Salmon 1975, 1981, 1982; Salmon and Salmon 1979; Watson 1976). General models of cultures as evolving adaptive systems appeared (e.g., Flannery 1972, 1973), and collections of case studies (e.g., Binford and Binford 1968) as well as individual studies (e.g., Hill 1970; Longacre 1970) were published and reviewed, sometimes with favor (e.g., Butler 1973; Martin 1971), sometimes with disfavor (e.g., Courbin 1982; Stanislawski 1973), and sometimes with a hopeful but skeptical "let's wait and see" attitude (e.g., Leone 1971; Muller 1973).

The general direction processual archaeology took comprised (1) ethnoarchaeological research, because mere ethnographic data typically had little to say about the kind of archaeological record created by the behaviors documented by ethnographers, and (2) experimental archaeology, to allow critical variables to be manipulated and obfuscating variables to be controlled (e.g., R. A. Gould 1978a, 1978b, 1980a, 1980b; Kramer 1979). Both resulted in the construction of hypothetical correlations between particular human behaviors and particular artifact forms, combinations, and distributions. Then, additional ethnoarchaeolcgical and/or experimental research was undertaken to test these hypotheses. Confirmation of the hypotheses indicated the presence of law-like generalizations—if not in fact laws, or atemporal and aspatial correlations between artifacts and behaviors—that allowed one to "explain" the archaeological record in human behavioral and sociocultural terms. Some archaeologists liked what they saw; others perceived limited value to such a research protocol (e.g., Dumond 1977; Johnson 1972b; Kushner 1970; Leone 1971; Muller 1973); still others did not like it all (e.g., Bayard 1969). Discussion concerning what, precisely, could be inferred from the record appeared regularly (e.g., Allen and Richardson 1971).

The critiques and debates had several notable outcomes. One involved the proper role of analogical reasoning in archaeology (e.g., R. A. Gould and Watson 1982; Murray and Walker 1988; Stahl 1993; Wylie 1982, 1985). Another was the creation of what might be construed as a particular kind of processual archaeology known as behavioral archaeology. The major spokesman for this form of archaeology—Michael Schiffer (1972, 1976, 1987)—was concerned that although the archaeological record might reflect human behaviors, as Binford had suggested, the reflection was often distorted. Debate ensued between the two over the precise nature of their particular visions of archaeology (Binford 1981; Schiffer 1985).

Behavioral archaeology had as a goal, among other things, the first step in the procedural protocol of processual archaeology: to reconstruct human behaviors that had created the archaeological record. Perhaps not surprisingly, one of the first major statements of the principles of behavioral archaeology (Schiffer 1976) was criticized as comprising "methods of archaeological inference rather than a kind of archaeology. [It] provide[s] no sense of theoretical direction [and thus] had a tendency to turn into mechanical archaeology" (Goodyear 1977:670). Albert Goodyear (1977:671) pointed out that behavioral archaeology seemed intent on producing "a sort of archaeological periodic table of human behavior"—in our terms, no distinction between immanent and configurational properties and processes was acknowledged—and ended his comments by observing that "the question remains of how far we can go considering method divorced from theory."

At about this same time, Binford (1977:9) concluded a brief introductory chapter to a volume entitled *For Theory Building in Archaeology* with the observation that "in the absence of progress toward usable theory, there is no new archaeology, only an antitraditional archaeology at best." Binford (1977:1) began his essay with the observation that "a self-conscious use of deductive methods is a prerequisite for scientific achievement." He then went on to point out that "impatient and enthusiastic excursions into the application of 'scientific' methods and rhetoric in the absence of any substitutive or original theory" suggested "an apparent faith [on the part of archaeologists] that the application of logical methods will result in the generation of theory" (Binford 1977:6). Thus, the modern "challenge is in theory building" (Binford 1977:6), which was to be a two-level process. What Binford termed "middle-range theory" comprised the determination of the particular kinds of dynamic conditions that produce particular kinds of static archaeologically observable effects. What he termed "general theory" comprised building an "understanding of the *processes* responsible for change and diversification in the organizational properties of living systems" (Binford 1977:7; emphasis in the original).

Binford (1977:5) was not particularly clear on how general theory was to be built, remarking only that "one does not build theory by accumulating universal facts or empirical generalizations, no matter how complex they may be." His remark that "the problem that any scientist must understand is how one moves from ideas [theory] to facts or observations" (Binford 1977:2) was a clue that the two must go hand in hand. One needs an observational language (formal theory regarding systematics) and an explanatory language (explanatory theory), both written in the same terms. Precisely this point was being made in biology and the philosophy of biology at nearly the same time:

> The two processes of constructing classifications and of discovering scientific laws and formulating scientific theories must be carried on together. Neither can outstrip the other very far without engendering mutually injurious effects. The idea that an extensive and elaborate classification can be considered in

> isolation from all scientific theories and then transformed only later into a theoretically significant classification is purely illusory. (Hull 1970:32).

> [T]he problem of theory building is a constant interaction between constructing laws and finding an appropriate set of descriptive state variables such that laws can be constructed. We cannot go out and describe the world any old way we please and then sit back and demand that an explanatory and predictive theory be built on that description.... [T]here is a process of trial and synthesis going on ... in which both state descriptions are being fitted together. (Lewontin 1974a:8)

Through the late 1970s and the 1980s, most processual archaeologists tended to ignore the call for the construction of general theory and instead focused their attention on determining how they should be using the deductive method to archaeology's greatest advantage. One reason for this was that many of them believed explanatory theory already was available.

The Unnecessary Third Step: Explanatory Theory

The explanatory theory that underpinned culture history (e.g., Willey and Phillips 1955) also underpinned the new archaeology of the 1960s; that theory resided in anthropology, not in archaeology. Recall Binford's (1977:7) interest in "the organizational properties of *living* systems" (emphasis added). The lack of a uniquely archaeological theory of culture development, which was the unattained desire of the culture historians, plus a failure to explicitly incorporate homologous similarity in their musings, despite statements to the contrary (e.g., Spencer 1992), denied processual archaeologists access to testable explanations of culture *history*. This was not appreciated at the time because archaeology was viewed as an anthropological research endeavor, the subject of which was *cultural* evolution. Thus, through the 1970s and into the 1980s, archaeologists tried to figure out how to align archaeology with Whitean cultural evolution (e.g., Binford 1969, 1972; Braun and Plog 1982; Carneiro 1970, 1972, 1973). Leslie White had stated that regularities of cultural dynamics occurred throughout human tenure on the earth. An archaeologist found them by following Binford's two-step procedure and thus could contribute to anthropology generally by discovering laws concerning how cultures work and change. But processual archaeologists did not attempt to build an explanatory theory for the archaeological record; rather, they used existing empirical generalizations derived from ethnology and attempted to mimic a particular version of science. Efforts were needed in figuring out how to make archaeology a *scientific* research endeavor in reconstructing the dynamics of prehistoric cultures. As we argue later, the version of science adopted by processual archaeologists reinforced the essentialist ontology that had placed their intellectual predecessors in jeopardy.

One highly influential book written during this period was *Explanation in Archaeology: An Explicitly Scientific Approach* (Watson et al. 1971), which proved to be so popular that it was revised and published over a decade later as *Archaeological Explanation: The Scientific Method in Archaeology* (Watson et al. 1984). Patty Jo Watson and her coauthors, Charles Redman and Steven LeBlanc, argued that archaeology could be scientific and that the standard an archaeologist used to measure whether he or she had reached that goal was explanation. To get there would require rigorous adherence to the scientific method: "Archaeologists should begin with clearly stated problems and then formulate testable hypothetical solutions. The degree of confirmation of conclusions should be exhibited by describing fully the field and laboratory data and the reasoning used to support these conclusions. This is what we mean by an explicitly scientific archaeological method" (Watson et al. 1984:129).

It is difficult to disagree with the statement that archaeologists should state problems clearly and describe data as completely as possible, just as it is difficult to argue against testing the implications of hypotheses. But Watson, Redman, and LeBlanc had other views on what science is and how it works:

> [S]cience *is* based on the working assumption or belief by scientists that past and present regularities *are* pertinent to future events and that under similar circumstances similar phenomena will behave in the future as they have in the past and do in the present. This practical assumption of the regularity or conformity of nature is the necessary foundation for all scientific work. Scientific descriptions, explanations, and predictions all utilize lawlike generalizations hypothesized on the presumption that natural phenomena are orderly. (Watson et al. 1984:5–6; emphasis in the original)

> The ultimate goal of any science is construction of an axiomatized theory such that observed regularities can be derived from a few basic laws as premises. Such theories are used to explain past events and to predict future ones. Good theories lead to prediction of previously unsuspected regularities. Logical and mathematical axiomatic systems are essential as models of scientific theories, but no empirical science has yet been completely axiomatized. As Hempel indicates, it may ultimately turn out for any science, or for all sciences, that the goal is actually unattainable. (Watson et al. 1984:14)

Central to Watson, Redman, and LeBlanc's arguments was philosopher Carl Hempel's account of how one arrived at explanation—what Hempel termed the *deductive–nomological* approach. Hempel's (1965) basic tenet was simple: Whatever is to be explained is derived logically from one or more universal statements, or laws, keeping in mind that certain boundary conditions might apply—hence Watson et al.'s (1984:5) above-cited definition of science as the "belief by scientists that past and present regularities *are* pertinent to future events and that under similar circumstances similar phenomena will behave in the future as they have in the past and do in the present." This is true of immanent properties—the central features of essentialist science—but not of configurational properties, the very being of which are space and time specific. Thus, the adoption of Hempel's version of scientific protocol reinforced the essentialist ontology. The use of analogical reasoning to reconstruct past cultures was readily subsumed under such an ontology (e.g., Binford and Sabloff 1982), but the general failure of processualists to recognize the important distinction between immanent and configurational properties and processes (e.g., Sabloff et al. 1987)—the former appropriate under the deductive–nomological model, the latter inappropriate—resulted in the past being no different configurationally than the present. The cultural anthropologist's sociocultural units, such as tribes, chiefdoms, and the like that are visible today could be found in the past (e.g., Rosenberg 1994; Spencer 1997).

By the middle of the 1970s, it was becoming clear that the deductive–nomological approach was dying a natural death (see review in Gibbon 1989). There were attempts to keep it alive, for example by linking it to Ernst Nagel's (1961) concept of *bridging laws*, but these also died out except among archaeologists, who began making bridges between the archaeological present and the archaeological past through such things as ethnographic analogy and ethnoarchaeology (e.g., Binford and Sabloff 1982; Fritz 1972). Archaeologists were searching for generalizations in the present and using them as proxies for the past. How else were they going to find the laws that Hempel said were there—the very laws that, once discovered, led to the formulation of "axiomatized theory" and thus ultimately to explanation? This was the foundation of Watson, Redman, and LeBlanc's "past and present regularities," but what escaped notice was that such laws concern only immanent properties and processes, not the configurational properties and processes sought by the processualists. As we noted earlier in this chapter, paleobiologists avoided this pitfall, and we believe they did so because they approached the fossil record with a well-developed theory firmly in hand, one written progressively more explicitly and completely in paleontological terms, as indicated by the extended discussion of punctuated equilibrium and its intellectual relatives. Processual archaeologists had no such theory; the alleged one they had—Whitean cultural evolution—was not truly a theory. It couched the mechanism of change—human urges and intentions—within the phenomena to be explained, and no attempt was made to write the theory in archaeological terms. These flaws went unrecognized.

Processualists believed they now had access to the past through the present. They could find patterning in their archaeological data sets and interpret the patterning in terms of what some referred to as behavioral correlates (e.g., Schiffer 1976). Or conversely, they could use present human behavior as a guide to what to look for in the prehistoric archaeological record. If one found enough correlates between the past and the present, then surely laws could be constructed to account for the similarity in pattern. Any slight deviations could be explained away in terms of slightly different "boundary conditions," to use Hempel's term, that had impinged on the creators of the past and present signatures. The end result of this exercise was scientific explanation, defined as interpretation by way of subsumption of a particular archaeological case under a law formulated in an ethnological setting. Thus, Watson (1986:452) equated archaeological interpretation with "describing and explaining the real past." And in a retrospective look at the processual archaeology of the 1970s, Charles Redman (1991:301, 303) remarked,

> A scientist assumes that the world is knowable, and that it operates in an orderly manner, which can be understood by reference to widely accepted rules or laws. Whether human phenomena are amenable to general laws, similar to those that explain the physical properties of the world, is yet to be demonstrated, but the pursuit of general principles, statistical as well as universal, remains the cornerstone of a social–scientific approach. For archaeologists this has meant, among other things, a general acceptance of uniformitarian principles....
>
> There *are* patterns in the past, and these patterns *are* reminiscent of things we can understand in the present—*uniformitarianism does work*! (emphasis in the original)

A recent review of processual archaeology finds it a quite acceptable program, if somewhat diverse in its practice and products (Spencer 1997). Underpinning the entire program is implicit acceptance of Whitean cultural evolution—written in anthropological terms—as *the* explanatory theory. What many choose to call tribal societies, for example, will eventually, and apparently inevitably (if given enough time), evolve into what are termed chiefdoms (Spencer 1997). Such a classification of cultures and/or societies is fraught with difficulty (Leonard and Jones 1987), but this has not prompted much thoughtful reflection. Similarly, the incorporation into processual archaeology of what has come to be known variously as behavioral ecology or evolutionary ecology (for an overview, see Bettinger 1991; for general statements, Smith and Winterhalder 1992), while providing some important insights into culture as humankind's extrasomatic means of adaptation, has tended to focus narrowly on how cultures work rather than on the uniquely historical aspects of how cultures develop and evolve through time (Lyman and O'Brien 1998). When the latter is considered, it generally is within some framework traceable to Leslie White or Julian Steward. Given such a research protocol, we return to Gould and Eldredge's lament that opened this chapter and wonder what we are going to learn from the archaeological record that we cannot learn from studying extant peoples and cultures.

In our view, the most significant factor that prompts us to reiterate the lament is found in the title of Binford's (1962) seminal article, "Archaeology as Anthropology." Twenty years after this paper was published, a new journal, the *Journal of Anthropological Archaeology*, was initiated. The first sentence Robert Whallon (1982:1) penned in his "Editorial Introduction" was "Anthropological archaeology aims primarily to explain the organization, operation, and evolution of human cultural systems.'' The lead article in the first issue was, appropriately, authored by Binford (1982). A book with a similar title—*Anthropological Archaeology*—was published two years later (Gibbon 1984). Earlier, we noted some of the parallels that can be found between the emergence of paleobiology, or "biological paleontology," and anthropological archaeology, but the two are quite different in the realm of theory. The former employs a theory—Darwinism—founded in history and a materialist ontology and written in paleontological terms; the latter employs an empirical generalization that is rooted in essentialism, written in anthropological and ethnological terms, and based on modern phenomena we call cultural stages, which are stacked up one on the other on the basis of a hypothesis of increasing complexity or the like (Dunnell 1989). This leads to various conceptual gymnastics (e.g., Carneiro 1972), the conflation of sequences and lineages (e.g., Schaeffer et al. 1972), and, given the Whitean view of cultural evolution and definition of culture as the extrasomatic means of adaptation, the Panglossian functionalism noted by Yoffee (1979) and others (O'Brien and Holland 1992).

Suggestions that processual archaeology might be too dependent on anthropology were few and far between in the heady days of the 1970s. James Deetz (1970:122–123), for one, suggested that "perhaps it's time we stopped trying to find post-nuptial residence, descent, marriage patterns in our data because these in fact are classificatory rubrics which are about third or fourth order abstractions themselves which the ethnographers are responsible for." A few years later, George Gumerman and David Phillips (1978:185) lamented, "Computer packages and philosophers of science have been 'used' in the most mechanical sense; systems theory, ecology, and other disciplines have been raided for concepts that are used out of any warranted generalizing context. The disingenuous manner in which such models have often been applied is one of the most disturbing aspects of the last decade of archaeological work." Their solution to the dilemma was, however, only partially correct in our view. They suggested that borrowing conceptual bits and pieces from nonanthropological disciplines would be more successful if no attempt was made to force them into "anthropologically derived archaeological frameworks," and instead archaeology was viewed as merely a technique for testing notions derived from other disciplines (Gumerman and Phillips 1978:186–187).

The critical missing element here is that Gumerman and Phillips provided no guide for selecting analytical tools or concepts from nonanthropological disciplines; such a guide can come only from explanatory theory. The overall lack of a uniquely archaeological explanatory theory—written in explicit archaeological terms—was noted twenty years ago (Dunnell 1978c), and things have not changed much in the intervening years. This is no doubt a result of the fact that the research protocol Binford advocated in the 1960s and 1970s was still being advocated by him in the 1980s (e.g., Binford and Sabloff 1982; Sabloff et al. 1987). However, by that time, he had changed his tune a bit, arguing that archaeologists must develop methods "that will build [high-level] theory about the nature of cultural stability and change over time" (Sabloff et al. 1987:203), yet he focused on strengthening middle-range theory and did not mention the necessity of explanatory theory (e.g., Binford and Sabloff 1982). Perhaps this is why some processual archaeologists (e.g., Spencer 1997) and some behavioral archaeologists (e.g., Schiffer 1996) are now examining the Darwinian archaeology we advocate in this volume to determine if it can serve as an explanatory theory.

DARWINIAN ARCHAEOLOGY: A RETOOLED CULTURE HISTORY

Beginning in the late 1970s and early 1980s, archaeologist Robert Dunnell (1978a, 1978c, 1980; Dunnell and Wenke 1980a, 1980b) and anthropologist David Rindos (1980) suggested that Darwinian evolutionary theory might provide—though not in unmodified form—the explanatory theory necessary to bring archaeology out of the hole it had dug for itself. As might be expected, the reception was at best lukewarm (e.g., Watson 1986; Yoffee 1980). One reason for this resided in a poor grasp by archaeologists in general of significant differences between essentialism and materialism. Another reason was that human intentions seemed not to have an important role in the program (see various comments in Rindos 1980). Even so, some found value in the suggestion. One proponent, though he never developed the idea further, was none other than Binford (1983:203):

> It is my opinion that we need to give much more serious thought to Darwinian arguments, where the driving forces of change lie in the interaction between the environment and the adaptive system being considered. Given such a view, the system of adaptation may enjoy relatively stable periods of varying duration, representing times when it is able to cope successfully with the perturbations of the environment. Selection for change occurs when the system is unable to continue previously successful tactics in the face of changed conditions in its environment.

What, then, is Darwinian archaeology? What is the role of deductive reasoning within Darwinian archaeology? And what, if any, is the role of human intention in such a program?

What *Is* Evolutionary Archaeology?

Anthropologists have long tried to define evolution in terms applicable to cultural phenomena (Carneiro 1973), though the results have been problematic. First, as Gertrude Dole (1973:254, 258) noted, "The tendency to consider any kind of adaptive change as evolutionary neutralizes the term," and "if the term evolution is to be useful in cultural anthropology it must be used to refer to something distinctive." Second, both increasing complexity and the process of adaptation have figured prominently in anthropological definitions of evolution (e.g., Bargatzky 1984; Bennett 1976; Dole 1973), though no mechanism driving evolution is specified, with the exception, following White (1943:339), of internal "urges" of humans. As we suggested in the preceding section, when Americanist archaeologists claim to follow an evolutionary approach, more often than not they are following Morgan, Tylor, White, and Steward much more closely than Darwin.

As should be clear by now, in our view, a Darwinian archaeology has numerous parallels in modern paleobiology. It is, as we have noted, geared toward providing Darwinian explanations of the archaeological record. The approach comprises, first, building cultural lineages—Robert O'Hara's (1988) historical chronicles—and second, constructing explanations of those lineages—O'Hara's evolutionary narratives (see also Szalay and Bock 1991). Both steps employ concepts embedded within Darwinian evolutionary theory such as lineage, selection, transmission mechanisms, innovation, and heritability. The last ensures that we examine change within a lineage rather than merely a temporal sequence or convergence—topics we take up in detail later.

We favor definitions of evolution such as "any net directional change or any cumulative change in the characteristics of organisms or populations over many generations—in other words, descent with modification. It explicitly includes the *origin* as well as the *spread* of alleles, variants, trait values, or character states. Evolution may occur as a result of natural selection, genetic drift, or both" (Endler 1986a:5). Do not be misled by our use of this definition; it does not mean that we equate evolution with changes in gene frequency. Mayr (1991a:2) argues that evolution defined merely as "a change in gene frequencies" is reductionist and that the concept is "described far better as 'a change in adaptation and in biological diversity.' " The evolution of culture may comprise in part changes in frequencies of what we will call, for lack of a better term, cultural traits, but that is not all there is to it. Mayr (1991a:2) indicates that "it is questionable to what extent changes in the frequency of [adaptively] neutral genes [or traits] can be designated as evolution." There is, as we will see in later chapters, an extremely important role for neutral traits in evolutionary archaeology, and others find a similarly important role for such traits in biological evolution (e.g., Gould and Lewontin 1979; Gould and Vrba 1982). Thus, in our view, trait variation and inheritance, but not fitness differences, are required for evolutionary change to occur; differences in fitness are required only for evolution via natural selection. Evolution in a Darwinian sense is "descent with modification"; it is continuity (heritability) with change (differential production of distinct variants). The historical aspect of Darwinian evolutionary theory is meant to address how and why a particular form of phenomena came to be; *that* is its goal (e.g., Mayr 1961)—a fact recognized by some (e.g., Tschauner 1994) but certainly not by all anthropologists who have examined evolutionary archaeology (e.g., Boone and Smith 1998; Schiffer 1996; Spencer 1997).

Evolution comprises change in the composition of a population over time. In Darwinian archaeology, the population comprises artifacts viewed as phenotypic features, and "it is the differential representation of variation at all scales among artifacts for which [Darwinian archaeology] seeks explanations" (Jones et al. 1995:28). Time is treated as a continuous variable, and "change is conceived [of] in terms of frequency changes in analytically discrete variants rather than the transformation of a variant" (Teltser 1995c:53). Such changes *might* be the result of natural selection and thus represent shifts in adaptational state or adaptedness, or they *might* be the result of drift (O'Brien and Holland 1992). The analytical challenge is to determine which is applicable in any given situation—that is, to construct and explain lineages. On the one hand, such a challenge demands the study of immanent properties and processes and the construction of laws concerning them (Gould 1986; Simpson 1963, 1970), as well as the construction of a set of units for measuring and describing a lineage's fossil record (Chapter 5)—that is, for writing a historical chronicle (Dunnell 1992b). Explaining *why* a lineage has the appearance it does, on the other hand, demands that the uniqueness of historical contingencies and configurations be considered (e.g., Beatty 1995; Simpson 1963, 1970).

The Role of Deduction

Those of us subscribing to Darwinian archaeology perceive a conflation of interpretation and explanation within processual archaeology (e.g., Abbott et al. 1996). This results from efforts to "explain" the archaeological record in terms of the anthropological models in use. The type of science many processualists have in mind—a predictive, law-driven science—will not work in an archaeology construed as historical science, except in the important area of determining how things function (e.g., Jaksic 1981; O'Hara 1988; Simpson 1963, 1970; Szalay and Bock 1991). Deductive science of the sort advocated by many processualists—the Hempelian form—is not particularly useful for studying humans, or any other organism, because it fails to distinguish between laws concerning immanent properties and processes and those concerning configurational properties and processes. Once these differences are pointed out, one realizes that there are no general configurational laws in the materialistic realm, since specific phenomena exist only in specific places and at specific times. Laws of immanence are, on the other hand, what allow us to discuss the mechanical and engineering or chemical and physical properties of artifacts; they *are* atemporal and aspatial. Not so with configuration.

Organisms, human or otherwise, are not subject to the same predictive laws that govern inorganic, essentialistic entities. In fact, if they are governed by any law, it is the law of contingency (Beatty 1995; Brandon 1990), which states that what happens at, say, point E is conditioned to some degree by what happened at successively preceding points D, C, B, and A. Thus, we say that what actually happens at point E is contingent on (1) the particular historical context in which it happens and (2) the particular historical sequence that led up to that point. Perhaps the easiest way of thinking about contingency is by considering the anticipated birth of a human baby. Unless we have independent evidence from ultrasound or some other means, there is no way of knowing exactly what that baby will look like. He or she could have any one of several hair colors, weigh any number of pounds, and have one of a wide variety of skin tones. However, experience tells us that a human baby born today will not have purple hair, weigh forty pounds, and have green skin. Those character states currently are impossibilities given the history of the greater human lineage. Thus, there are only so many pathways, or channels, open to an organism at a particular time and in a particular place. If we want a materialistic law—and it is more of a limiting law than one that allows us to predict, then the only possible law is contingency.

Invariant laws, are, however, precisely the kinds of laws some archaeologists want to apply to humans (e.g., R. A. Gould 1978a, 1978b, 1980a, 1980b, 1985; Reid et al. 1974, 1975; Schiffer 1975, 1996). Fritz and Plog (1970:405), for example, were explicit about their definition of law: "A statement of relation between two or more variables which is true for all times and places." One could, we suppose, try and get around the issue by claiming a distinction between "universal facts" and "laws," but this obscures the real issue: Are there invariant laws that govern human behavior? If there are, then the Hempelian notion of science is quite adequate for the goals of archaeology as a historical science. If there are not, then where do we look for explanation? To us, the obvious place to look is Darwinian evolutionary theory. The place to begin the search is with the units used to identify and track variation. We turn to this issue in the next chapter, but before we do so, it is necessary to return to Stephen Jay Gould's argument with which we began this chapter.

Lamarckism Again

Recall that Gould (1987a, 1996, 1997a) argued that because cultural transmission and inheritance are Lamarckian—a reference to the pre–Darwinian view espoused by Jean-Baptiste de Lamarck that the production of biological variation was directed—and thus potentially much more rapid than genetic inheritance, natural selection has at best a very small role to play in the evolution of cultural lineages. Michael Rosenberg (1990, 1994) similarly argues that cultural evolution is Lamarckian and calls explicitly on human "motivations," or "intent," as an evolutionary mechanism. Lamarck's version of evolution holds that variation is "not randomly generated"; culture is "willed [by its human carriers] to ever better-adapted states of being. [Human] behavioral innovation is by definition adaptive, because adapting is what such innovations are *designed* to do" (Rosenberg 1990:400). Lamarckian evolution is therefore progressive. Rosenberg (1991: 696–697) indicates that

> to the degree that cultural evolution embodies a significant Lamarckian element, it makes intuitive sense that (1) the greater the intelligence of the hominid, the greater the capacity and likelihood of purposeful innovation; (2) the more intense the competitive pressures, the greater the incentive to innovate in an effort to circumvent or ameliorate the proximate manifestation of stress; and (3) the greater both intelligence and rate of innovation, the greater the likelihood that one or more innovative behaviors having selective value will appear.

Rosenberg is, no doubt, correct; intentional innovation provides a source of variants upon which selection can act (O'Brien and Holland 1992; Rindos 1989), a point to which we will return. Rosenberg's three points raise a number of questions. For example, is the categorization of cultural evolution as Lamarckian a fatal blow to Darwinian archaeology? Beginning our discussion with an answer to this question leads us into other questions.

There are two important elements to Larmarck's version of evolution: (1) It is goal seeking in the sense that its direction is one of increasing perfection (hence, it is accurately characterized as progressive); and (2) characters acquired by an ancestral generation can be inherited by the following descendant generation (Mayr 1982a:343–362). Darwin acknowledged that the second aspect of Lamarckian evolution—now known as soft inheritance—played a role in his own version of evolution, though he did not ascribe as significant a role to it as Lamarck did. But Darwin found the first aspect of Lamarck's theory to be nonsense. Rosenberg (1994) takes a hard line on the second element—hard and soft inheritance. In his view, on the one hand, biological "inheritance," or "evolution," is only hard (genetic); this echoes Gould's (1996, 1997a) statement that biological inheritance is Mendelian. Cultural evolution, on the other hand, is soft and thus Lamarckian: Cultural or behavioral innovations "established *during* a lifetime can be transmitted to future generations" (Rosenberg 1994:309; emphasis in the original). This, too, echoes Gould's (1996, 1997a) remarks. Is the characterization of cultural evolution as Lamarckian appropriate because such evolution is goal directed—"willed"—by human motivations and intent?

Citing Rindos's (1984, 1985) work, Rosenberg (1990:401, 399) states that the version of evolutionary archaeology we describe in this book "holds not just that biological change is a product of natural selection, but that the variations on which selection acts arise randomly" and that each new variant will appear "largely at random." Charles Spencer (1997:210, 212), too, characterizes our perspective as assuming "that variation in cultural traits is generated in an undirected fashion" and thus the appearance of new variants is "accidental." While Rosenberg (1990:401) is correct in his statement that our approach does not "*require* that variability arise randomly," he does not define what is meant by "random" appearance, and Spencer does not explicitly indicate what is meant by "undirected." Thus, they miss the critical point of this aspect of Darwinism completely. Rosenberg misses it cleanly when he derogates Rindos's explicit exclusion of intent. Rindos (1985:84) indicated that intent is "inherently unverifiable" in the archaeological record and thus should be excluded from an archaeology that desires a scientific status. Rosenberg (1990:403) counters with the notion that there often is a "relationship between [a selective] situation and [a cultural or behavioral] reaction." In his view, the "study of such possible relationships is *certainly* scientific" (Rosenberg 1990:403; emphasis in the original).

Although we agree that the study of empirical generalizations *can* be scientific, we believe that such study is a poor definitive criterion of science because it does not distinguish science from alchemy, religion, or magic. *Science* demands that theories have empirical content; in other words, hypotheses derived from theories *must* have testable implications. Rindos's point was that human intent leaves no archaeologically visible traces; thus, a hypothesized prehistoric instance of human intent cannot be empirically confirmed or refuted. This point was noted early in the history of processual archaeology when one of its most vocal architects, Kent V. Flannery, noted that "individuals *do* make decisions, but evidence of these individual decisions cannot be recovered by archaeologists" (Flannery 1967:122, emphasis in the original; see also Kushner 1970). Rindos (1989) elaborates at length on the problem, beginning with a definition of random and undirected variation. He sees undirected variation as " 'random' only in the sense of having no preordained relation to the 'needs' of the organism" (Rindos 1989:8). Such variation "fuel[s] the engine of evolutionary change by generating new forms that may then be subject to selection" (Rindos 1989:9), thereby ensuring the important role of selection.

Regarding the "random" appearance of new variants, biological variants arise in what appears to be random fashion from a synchronic perspective, but from a diachronic perspective, those variants arise stochastically; that is, what comes later depends at least in part on what came earlier. Metaphorically, an ancestral clay vessel will not beget the space shuttle; it will beget a descendant clay vessel similar to, yet different from, its ancestor. The descendant, whether a clay vessel or an organism, may not be identical to its ancestor because of accident or transmission error (Bell 1997). Similarly, the vessel, perhaps because of the intent of the replicator, may be different from the ancestor. As Dunnell (1982:22) noted, a "*theory* of scientific [evolutionary] archaeology could no more ignore the interactive properties of cultural phenomena than evolutionary biology could ignore those of organisms" (emphasis in the original). So variation is not random over the long term; it is random with respect to immediate need but not with respect to the ancestral form from which it arises. Rosenberg's, Spencer's, and apparently Gould's position is that variation is not even random with respect to immediate need because people are smart. This leaves unanswered and unanswerable the question of how they became smart; we doubt it was because they wanted to or that such was their intent beginning at that ancient time when some unidentified hominid first manufactured a tool.

O'Brien and Holland (1992:45) state explicitly that evolutionary archaeologists assign human intent "the important role of introducing variation into a cultural system ... [but v]ariation, not intent, is the significant component" in evolution. This is quite different from Rosenberg's (1990:402) implication that evolutionists hold human intent and motivation to be "irrelevant" to studies of the evolution of cultures. In other words, individual intent "means little statistically within Darwinian evolution, as evolution is a population phenomenon measured in terms of changing representation of individual actions or traits through time. Furthermore, the goals of individuals vary, and, as we are all too aware, the difference between intent and outcome is often quite wide" (Leonard and Reed 1993:650). As Dunnell (1982:21) noted, humans "have been enormously successful without the slightest knowledge of how culture changes."

Philosophers Edward Stein and Peter Lipton (1989) present an intriguing argument that is quite relevant to the question of the role of intentional innovation. They examined how a scientist produces "a new hypothesis" for testing during the process of the development and evolution of a discipline's theory. Is that production blind, or is it guided? It is guided in the sense that the scientist has antecedent knowledge about which hypotheses failed and which withstood previous tests, but this is merely the historical channeling of available variation—part of the historical contingency of evolution. A new genetic mutation is not chaotic; it depends on what is available for mutating and, thus, what has gone before. If the scientist seeks to produce a hypothesis about the relation between atmospheric circulation and local weather patterns, the hypothesis will not contain statements about the workings of the computer on which this sentence is being written, because it is already established that one has little to do with the other. Furthermore, many possible hypotheses are probably considered before one is (or a few are) selected for testing. Those that are mentally discarded will likely not leave any visible trace in the scientist's lab notes or equipment. Only those subjected to testing might leave such traces; those that fail in testing might not leave any traces. But to return to the key point, the evolution of ideas and the evolution of gene pools both rest on random (with respect to immediate needs) yet stochastic (based on what came before and was retained) variation. Natural selection, not human intent, motivation, or choice, has been the final arbiter (Mayr 1961, 1962) concerning what works and what does not; thus, it greatly influences what is transmitted and replicated over long temporal spans.

Chapter 5

Archaeological Units and Their Construction

> In retrospect, one of the most serious flaws in processual archaeology was its failure to rethink its language of observation, its systematics. Anthropologists and archaeologists disposed toward the use of evolutionary theory seem on the brink of following suit. The issue of units has not generally been addressed; traditional units drawn from the investigators' own culture or from academic lore continue to be used. Even when the importance of units has been recognized, researchers have not bothered to acquaint themselves with modern biological theory in more than a cursory fashion. As a result biology is depicted as having reached a consensus on key systematic concepts when such is not the case. (Dunnell 1995:33)

Dunnell's comments underscore a point we have made repeatedly: Evolutionism, regardless of whether it is used to study organisms or things in the archaeological record, has its own set of requisites regarding the kinds of units that are acceptable for use in analysis. Given the distinctly different perspectives of reality that essentialism and materialism possess, the fact that each requires different units should be nonproblematic. It stands to reason that if reality is viewed as a series of natural kinds, then analytical focus is placed squarely on understanding the differences between those *kinds*. Alternatively, if reality is viewed as a system of endless change, then analytical focus will be placed on trying to understand how and why *things* (not *kinds*) change. Intuition might lead us to suspect that units devised to measure similarity and difference probably cannot also be used to measure change, though to this point, as Dunnell points out, archaeologists interested in applying Darwinian evolutionism have paid little attention to the question of units and instead have focused on programmatic issues. This suggests that evolutionary archaeologists are either indifferent to the entire issue or believe that the matter of units has been settled.

Careful reading of the evolutionary-biology literature should convince anyone that not only is the issue of units important but also it is far from settled. If biologists are still wresting with the concept of a species—clearly the most fundamental unit in evolutionary biology—why should we believe that the matter is less complicated in archaeology? Although programmatic issues such as adaptation, selection, drift, homology, and analogy are important considerations in the development of an evolutionary archaeology, failure to confront head-on the issue of units has led evolutionary archaeology onto thin ice. Culture history witnessed its share of arguments over units, though those arguments were based for the most part not on theoretical grounds but more on the basis of who had a more persuasive commonsensical argument (Lyman et al. 1997b; O'Brien and Lyman 1998). Most cultural historians overlooked entirely the problems inherent in the way they were using some of their units, which most frequently were not being drawn directly from theory. Without a clear statement of why certain units are applicable or nonapplicable to an evolutionary study, as well as clear statements of how to construct such units, evolutionary archaeologists are going to find themselves in the same predicament in which culture historians eventually found themselves, having created a set of units with limited or questionable analytical utility.

Being a materialist is not easy. In both essentialism and materialism, explanation is tied inextricably to observed variation, but there the similarity ends. In materialism, kinds are constantly changing nonempirical configurations that are nothing more than observational consequences (Dunnell 1988a); inferences about the nature of change can be made only after variation has been identified and measured precisely. Thus, the difficulty with materialism is twofold: Not only do we have to create units that allow us to identify and measure *variation*, but we also need access to units that allow us to measure *change*. In materialism, variation is the result of change, which is *not* to say that different variants cannot occur simultaneously, whereas in essentialism, variation is merely the result of essential differences between things. In essentialism, the units await discovery, and we have only to worry about measuring the differences between them; we do not have to worry about measuring change between units, because essentialist units cannot change. Once essentialist units have been identified, one looks for a reality that underlies the diversity, which then serves as the explanation for the discreteness of the units. Conversely, materialism postulates a reality *sustained* by diversity: "This standard [Darwinian] way of viewing evolution assigns a causal role to individual idiosyncrasies. Individual differences are not *the effects* of interfering forces confounding the expression of a prototype; rather they are *the causes* of events that are absolutely central to the history of evolution" (Sober 1980:371; emphasis in the original). In materialism there can be no prototype or archetype; documenting variation between and among specimens has to be, as Richard Lewontin (1974a:5) stated, "the cornerstone of [Darwinian evolutionary] theory."

The question is, how do we study the variation that "is rendered as change" (Dunnell 1988a:16); that is, how do we arrive at explanations that are historical in nature? If, as materialism holds, things are always in the state of becoming something else, how do we make evolutionism more than merely a kind of metaphysical exercise? This is the key question in evolutionary archaeology, just as it is in evolutionary biology, and although it has been answered in both disciplines, the expositions contain epistemological murkiness. The answer is found in the construction of a set of units that allows phenomena to be measured. The analyst selects dimensions—mutually exclusive variables such as color, length, and weight—that appear to be relevant to some problem, and it is those dimensions and their attribute states, the values each variable might take, that are used to sort specimens into internally homogeneous, externally heterogeneous classes. Thus, from the materialist view, historical change is rendered as alterations in the frequencies of *analytical*, not *natural*, kinds. This point cannot be overemphasized.

How this complex exercise differs from traditional archaeological method is the primary focus of discussion here. A portion of our examination stems from Dunnell's (1971a) arguments in *Systematics in Prehistory*, a book that, despite the almost thirty years that have passed since it was published, remains one of the few treatments of archaeological systematics in archaeology. We suspect that there are at least three reasons for a general lack of attention to systematics. First, the subject is a difficult one, largely because the issues are ontological and epistemological. Second, precise terminology is necessary; thus, one must learn a set of terms and their associated definitions. Compounding this problem is the fact that many of the terms are routinely used in English to refer to various kinds of things, and it often is difficult to sweep those other things aside and focus solely on the more restricted definitions. Third, systematics is not as interesting as, say, cultural processes or human behaviors, and it can be tedious to teach in a classroom setting. Learning how to recognize cultural processes will hold a student's interest longer than will learning the epistemological differences between classes and groups. It is our impression that systematics is almost a thing of the past in graduate-program curricula (for a similar assessment of biological systematics, see Mayr 1968). But if we are ever to bring archaeology in line with evolutionary biology, we need to pay more attention to units and especially to determining which units are applicable to the problems encountered in doing materialist science. This is an area, as Dunnell (1995) points out, in which archaeology has something to contribute to biology. Our discussion makes it clear that by carefully distinguishing between two fundamentally different kinds of units, ideational and empirical units (Dunnell 1986), there is a way out of the species conundrum. To date, conflation of these two fundamentally different kinds of units has created massive confusion in biology as well as in archaeology.

IDEATIONAL AND EMPIRICAL UNITS

Dimensions and their accompanying attribute states are *ideational* units as opposed to *empirical* units, meaning that they are not real in the sense that they can be seen or picked up and held. A ballpoint pen is an empirical, or *phenomenological*, unit—we can see and feel it—and it has various properties that can be measured using ideational units. The easiest way of defining ideational units is to say that they are tools used to measure or characterize empirical things. An edge angle on a stone tool is measured, using a goniometer, in ideational units known as degrees. Inches, centimeters, grams, ounces, and degrees do not exist in any empirical sense—we cannot see or feel them—but they do exist in an ideational sense. As such, they are useful for measuring properties *of* empirical units. Note that edge angle itself is an ideational unit with different empirical *manifestations*. Thus, a 60-degree edge angle on a specific tool is an empirical unit. Ideational units are not interchangeable—centimeters, for example, measure length and not weight—but experience rarely allows for confusion here. More often, despite best efforts to keep them straight, empirical units become confused with ideational units. Think back to Phillips et al.'s (1951:66) warning: "Exigencies of language require us to think and talk about pottery types as though they had some sort of independent existence," though we ought to remember that they do not.

Two Kinds of Ideational Units

Ideational units can be *descriptional* units, used merely to characterize or describe a property or a thing, or they can be *theoretical* units, which are created for specific analytical purposes. We can choose to characterize something on the basis of the dimension color and to specify several different attribute states of color, even though that property may have nothing to do with the particular focus of analysis. Still, color does describe empirical units. Alternatively, in light of a proposed causal relation between, say, function and various edge angles of stone tools, edge-angle units (attribute states) such as 1–30 degrees, 31–60 degrees, and 61–90 degrees would be theoretical units. In short, a theoretical unit is an ideational unit that has explanatory significance *because of, and only because of, its proposed relevance*.

Numerous artifact types created in Americanist archaeology are descriptional units. They are purely ad hoc constructs that include any amount of descriptive, often nonredundant, information; type A might include color, weight, and length in its description, whereas type B might include color and width but not weight and length. In contrast, the pottery types that culture historians created, while including a host of information that was ancillary to the main analytical focus—keeping track of time—were also theoretical in that they were formed by the analyst for a specific purpose, their validity was tested, and then they were used for that purpose. James A. Ford's Southeastern pottery types were derived directly from his conception of time as a continuum and from his understanding of the fact that the continuum had to be subdivided in order to get any analytical work done. Early on, Ford developed marker types—index fossils—to accomplish his goal (e.g., Ford 1935a, 1935b, 1935c, 1936a, 1936b); later, his types were constructed around the popularity principle (O'Brien and Lyman 1998, 1999b). Those types that passed the historical-significance test, that is, that showed normal frequency distributions through time, were useful; those that did not were either abandoned or relegated to the status of descriptional units.

Intensional versus Extensional Definitions

Ideational units can be defined either intensionally or extensionally. An *intensional*[1] *definition* comprises the necessary and sufficient conditions for membership in a unit, that is, the explicitly listed distinctive features that a phenomenon must display to be a member of the unit. Importantly, the significant characteristics of the unit are derived from propositions about the way things operate. Thus, the three classes of edge angles mentioned earlier—1–30 degrees, 31–60 degrees, and 61–90 degrees—are derived from our ideas about stone-tool function that indicate some edge angles are necessary for some functions and other edge angles are necessary for other functions. This does not mean that our ideas are correct; perhaps after further investigation we will find that the attribute states 1–25 degrees, 26–65 degrees, and 66–90 degrees are more appropriate units of measurement—meaning that more of the observed variation can be more readily explained—but this possibility is something to determine analytically. Furthermore, finding that particular phenomena do not exist, such as, say, edge angles of 1–10 degrees, has no bearing on how units are constructed.

Units can also be defined extensionally. An *extensional definition* comprises the necessary and sufficient conditions for membership in a unit and is derived by enumerating selected attributes shared by the unit's members; that is, the definition is based on observed attributes of the actual members placed in a unit. Thus, the significant characteristics of extensionally defined units are seldom theo-

[1]Not to be confused with *intentional*. Intensional as used here in the logical sense refers to the properties connoted by a term.

retically informed in an explicit manner. Most units that have been traditionally employed in archaeology are extensionally defined—units formed as the result of an analyst sitting down, looking at a pile of artifacts, and dividing them up into smaller piles based on perceived similarities and differences. The unit definitions are dependent entirely on the specimens originally examined.

As we argue later, intensionally defined units are required to study change. Specifically, one examines alterations through time in the frequencies of units comprising empirical specimens and explains those alterations in terms that theory informs us are relevant to them. Explanation of frequency changes in non-theoretical, extensionally defined units is virtually impossible because the unit definitions are potentially modifiable every time a new specimen is found, thus creating all sorts of problems if one seeks to measure change.

Scale and Field of Observation

Most of us are familiar with the statistician's scales of measurement termed nominal, ordinal, interval, and ratio (Stevens 1946), but here we use the term *scale* differently. By *scale*, we mean a particular degree of inclusiveness or rank (Dunnell 1971a:146). The most obvious scale is that of discrete object—a *thing* that can be moved without losing its integrity and without destroying the spatial relations or structural integrity of its constituent parts. When archaeologists speak of artifacts, they generally mean discrete objects. But things can be of larger or smaller scales than discrete objects. An archaeological "feature" such as a post mold or hearth is a thing but not a discrete object; it cannot be moved without its structural integrity being destroyed. So, too, with an "assemblage" of artifacts or "tool kit," both comprising a set of discrete objects. Scales larger or more inclusive than discrete objects are regularly referred to by archaeologists.

So, too, are scales that are smaller or less inclusive than discrete objects. We speak of attributes of a projectile point, such as the placement of notches or the presence or absence of a stem or shoulders; similarly, attributes of pottery sherds include thickness, temper, and color. Clearly, an "attribute" of a discrete object is not so inclusive as the object itself, the latter including all of the attributes of the object. An apple as a discrete object has attributes such as a particular shape, a stem, some seeds inside, and a skin. But the skin also has attributes such as thickness, color, and texture. We say an apple is red, but actually it is the skin of the apple that is red. One can think of the skin color as being an attribute of the apple rather than of the skin, and this may be acceptable in some situations. But if it is realized that removing the skin changes the color of the apple's surface, then it is clear that the color of an unskinned apple is an attribute of the skin and only indirectly of the apple. The point to realize here is that discrete objects are classified based on their attributes—at the scale of properties of discrete objects. When classifying aggregates of discrete objects, the attributes are typically at the scale of discrete object and are generally the types of discrete objects represented in each aggregate.

The second issue concerns what we call, following Dunnell (1971a), the *field of observation*. By this we mean the material or materials that are to be classified. Only by specifying a field will we avoid considering, for example, the variable "tempering agent" (as in pottery) when we classify only lithic projectile points. It usually is not difficult to determine from a classification what the field comprises, but it is mandatory to know what that field is from the start.

GROUPS AND CLASSES

In discussing the differences between empirical and theoretical units, Dunnell (1986:151) stated that this was "the basis of the distinction that I (Dunnell 1971a) earlier characterized clumsily as the contrast between groups and classes, the former designating empirical entities and the latter encompassing the 'theoretical' units." We would not characterize his earlier distinction as clumsy; rather, Dunnell took two terms that have enjoyed common usage in archaeology and gave each a very distinctive meaning. Dunnell's later distinction—that between ideational units and empirical units—actually heightened the earlier distinction as opposed to replacing it. The point to keep in mind is that we view classes as ideational units and *only* ideational units. As such, they can be either theoretical or descriptional units; our focus will be almost exclusively on the former. As ideational units, classes can be defined intensionally or extensionally. Similarly, groups are empirical units and only empirical units. They also can be defined intensionally or extensionally.

Groups as Empirical Units

Empirical things can occur singly, whereby the thing in question shares particular properties in common only with itself, or there might be multiple objects that share particular properties in common. Collections of empirical things are termed *groups*, defined as aggregates, or sets, of things—of whatever scale—that are either physically or conceptually associated as a unit. The properties that things share in common might be of analytical interest, but there is no stipulation that they have to be. Again, because they contain actual specimens, groups, like the specimens themselves, are empirical, or phenomenological, units. There are two general procedures for producing groups. The first uses grouping methods, from simple visual inspection to complex algorithms, to produce the groups. Such groups are extensionally defined, meaning the definition is derived from the pile of artifacts one actually examines; hence, those groups are idiosyncratic. Once such a group has been defined, it carries that definition until either the original pile of artifacts is divided into two or more piles, at which time new definitions are written, or new artifacts are added to the pile, at which point the original definition might be revised, depending on the amount of variation that is tolerable. The second procedure produces groups based on a set of classes, discussion of which we delay until the section on classes later in this chapter.

Grouping methods take many forms that, for our purposes, we lump under the generic term *clustering approaches*. The objective of a clustering exercise is to produce groups—clusters—of things, each of which is more like the other things in that group than things in other groups. To produce the clusters, objects are taken one pair at a time and scored in terms of their similarity to each other. Similarity is generally measured as the number of shared attributes or characters. Similarity coefficients are calculated in like manner for all pairs of objects, and the coefficients are linked in descending order of similarity, producing the familiar dendrogram pattern of linkage. Clusters then are identified either by visual inspection or by the use of conventional threshold values. This type of approach to object clustering is termed *numerical taxonomy* (Sokal and Sneath 1963), or *phenetics* (Mayr 1981). It developed in modern form in biology (Sokal and Sneath 1963) as an all-purpose method of constructing groups using a large number of morphometric traits.

The various methods available for grouping phenomena can serve as excellent pattern-recognition devices, but they are incapable of explaining anything. The problem is that being the output of an inductive "throw-it-up-and-see-what-sticks" approach, the units created by grouping methods are extensional. This is not to say that inductively derived observations are invalid, but it is illusory to deny the deductive component of observation (Dunnell 1986:189). Importantly, it is this deductive component that links observations to theory. Our feeling is that any theory that does not yield a series of logical implications that can be tested is essentially useless in science. This does not mean that it has to be easy to test the implications through observations, but there has to be some logical means of doing so. As we discuss later, extensionally derived groups might or might not yield logical implications. The problem is in deciding whether they do or do not in any particular case.

Cluster analysis became popular in archaeology (e.g., Cowgill 1982; Doran and Hodson 1975) after publication of David Clarke's (1968) *Analytical Archaeology*, though it was around in varied form much earlier. An influential paper that preceded Clarke's book by two years was "Some Experiments in the Numerical Analysis of Archaeological Data" (Hodson et al. 1966), coauthored by Peter H. A. Sneath, who, along with Robert R. Sokal, was one of the originators of numerical taxonomy in biology. Harold Driver and A. L. Kroeber (1932) may have been the first to argue for its use in archaeological analysis as a method of assemblage comparison, but the approach reached its fullest expression a few years later in the Midwestern Taxonomic Method (MTM). Although its major architect did not intend for it to be numerical (McKern 1939, 1940, 1943, 1944), the MTM operated in a fashion similar to numerical methods such as modern phenetics, and some argued that it should be based on quantitative analyses (e.g., Kroeber 1940, 1942). Recall from Chapter 3 that the MTM arose as a result of Americanist archaeologists having come to something of an impasse over the means and terms used to describe and discuss assemblages of artifacts. The MTM filled this gap by arranging sets of associated artifacts, termed *components*, into successively greater levels of inclusiveness through the use of linked (shared) traits and diagnostic (unshared) traits (Figure 3.10). The taxonomic, or hierarchical, structure of the MTM—several similar components comprise a focus, several similar foci comprise an aspect, and so forth—was adopted apparently because McKern believed this procedure would eventually reveal phylogenetic relations among the included cultural units (Fisher 1997). Despite the branching structure of the MTM, it is distinct from any taxonomic system that shows such relations. Hierarchical units in the MTM show only formal similarity or resemblance (Dunnell 1971a:179), largely because of a failure to consider whether recorded similarities are homologous or analogous.

Classes as Ideational Units

In contrast to groups are units that are nonphenomenological. These are classes—ideational units that have been defined explicitly in terms of their characteristic attributes. More correctly, a class *is* a definition, one that specifies the necessary and sufficient conditions, termed *significata*, that must be displayed by a specimen in order for it to be considered a member of the class. Given that a class is a definition, it is also a measurement unit in that it measures phenomena against its necessary and sufficient criteria for membership. If a phenomenon measures up, so to speak, it is identified as a member of that class and included in the group designated by that class; if not, it is excluded from that group. Importantly, specifying the necessary and sufficient conditions for membership in a class does not automatically mean that there are actual specimens that meet the conditions; that is, specimens that possess the attributes comprising the necessary and sufficient conditions might not exist in a studied collection. Or they might not exist anywhere. Specimens that do meet the conditions and hence are identified as members of a class constitute the *denotata* of that class. One might well ask what the difference is between the denotata of a class—the actual specimens identified as members of a class—and a group. They both contain empirical "stuff," but they are not the same. A set of denotata is a group, but a group is not a set of denotata unless the included specimens have been *previously identified* through intensional classification. As Dunnell (1970:307) put it, "The denotata of a class constitute a special kind of group, one with an explicit meaning." The second procedure for creating groups, then, involves sorting specimens into piles based on one's identification of each specimen as possessing the significata of one class or another.

Because they are ideational units as opposed to empirical units, classes have *distributions* in time and space; because they are empirical—that is, made up of real things—groups have *locations* in time and space; that is, groups are simply inhabited areas of morphospace (Conway Morris 1998:4–5). The locational aspect of groups applies equally to groups produced by clustering approaches and to groups formed on the basis of a set of classes. There is, however, a significant difference in the way we interpret the two kinds of locational information created by groups and classes. The locations of groups formed by clustering methods are completely dependent on the specimens used to create the groups; hence, we may have no way of ascertaining whether like specimens occur in other localities. Groups created from classes are not dependent on the specimens used to create them. Since classes are spaceless and timeless entities, their denotata are free to vary in their distributions across space and time; theoretically, specimens could occur anywhere.

We noted previously that Dunnell's (1986) self-criticism of his earlier attempt to distinguish between ideational and empirical units by using the terms *classes* and *groups* was misplaced. We made that notation to draw attention to the equivalence between such things as degrees and centimeters on the one hand—both of which are commonly recognized measurement units—and, say, a projectile-point class on the other—something that is not so obviously a measurement unit. It seems counterintuitive to think of a class as not containing "things," given the manner in which the term is used in everyday conversation, but classes do not have to contain anything. We cannot overemphasize that classes are ideational units used to measure variation. The significata—the necessary and sufficient conditions for membership in a class—not only define the class, but they *are* the class. A class might have denotata, but it might not. The fact that there could exist a class with no denotata—empirical specimens—should convince us that as a measurement unit, a class cannot have an essence. There can be no equivocation on this point.

Kinds of Classification

There are numerous kinds of classification—the method that creates classes—but we treat only two of them here. One is taxonomic classification, and the other is paradigmatic classification. The former is a hierarchic structure comprising "an ordered set of oppositions or contrasts which amounts to a division of the field of the classification into classes, sub-classes, and so on" (Dunnell 1971a:76). A hierarchical classification involves "the nesting of lower-level entities within those of higher levels. Thus, the units at each rank either include or have parts (subunits) that are units at lower ranks. This architecture produces a distinctive set

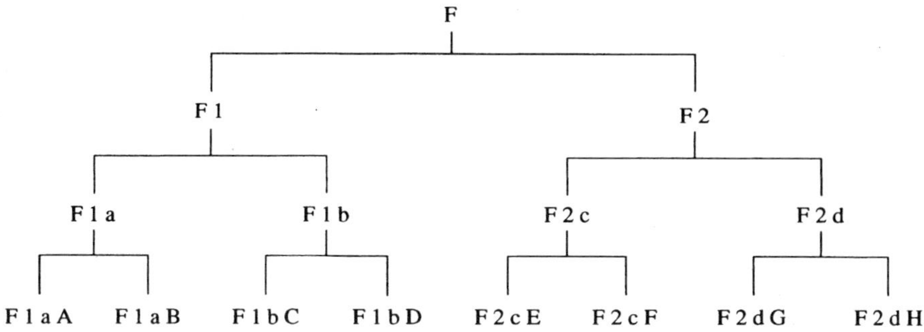

FIGURE 5.1. A symmetrical taxonomy producing eight classes (after Dunnell 1971a).

of properties that is useful in understanding the interrelationships of the units within the hierarchies" (Valentine and May 1996:23). Figure 5.1 illustrates the relations between and among different levels within a simple taxonomy. Importantly, at any level, the resulting units could be termed classes. For example, in Figure 5.1, there is a class F, but if levels of greater exclusivity are used, the classes become Fl and F2. At an even greater level of exclusivity, the classes are Fla, Flb, F2c, and F2d, and at the greatest level the classes are FlaA, FlaB ... F2dH. Thus, the definition of any taxonomic class comprises

> the series of oppositions leading *from* the field [the kind of stuff to be classified] to the class. From the point of view of any class, the definition derives from the inclusion of the class in a series of super-classes at higher and higher levels culminating in the field. The features which make up the *significata* of individual taxons reflect the series of oppositions from field to class as a serial order.... The net effect of this serial ordering of the features of the taxonomic definitions is to restrict the range of the features constituting an opposition to a portion of the classification. (Dunnell 1971a:76–77)

In taxonomic classification, the significata of a class—the features, or attributes, that define the class—reflect a serial order that, as opposed to coming from any inherent essential ordering of the things to be classified, is constructed by the analyst. For example, note in Figure 5.1 that attribute states a and b (themselves classes) are relevant only for distinguishing classes of "superclass" Fl. This does not mean that objects (denotata) in class F2cE, say, do not display attribute states a or b; it means only that since they display the attribute state that makes them fall in superclass F2, attribute states a and b are not considered but c and d are considered. In short, the attribute state that places objects in superclass Fl or F2 is more important in the judgment of the analyst than a–d are. Once the distinction between Fl and F2 is made, the analyst judges a and b to be relevant only to Fl, whereas c and d are judged to be relevant only to F2. Thus, we talk about the

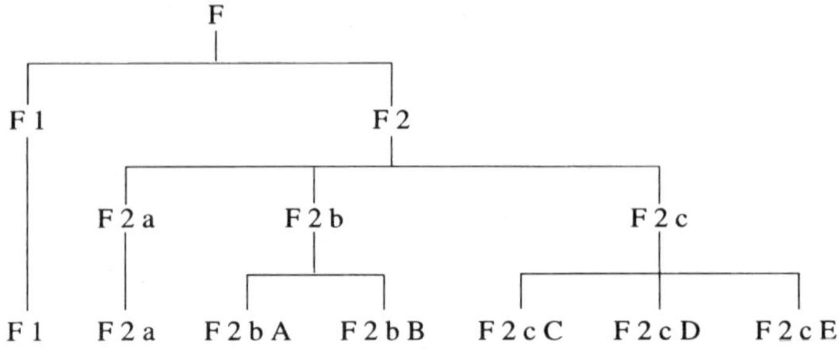

Figure 5.2. A nonsymmetrical taxonomy producing seven classes (after Dunnell 1971a).

attribute states that comprise significata as being weighted, which yields the hierarchical structure of taxonomies. Lest we think that taxonomies have to be symmetrical, consider Figure 5.2, which illustrates a nonsymmetrical arrangement (at the greatest level of exclusivity) of seven classes. In the case of class F1, a single feature—the attribute state designated "1"—distinguishes it from all other other classes. This is not the case with the other classes, where various numbers of features define them. Class F2a is defined by the combination of two attributes signified as "2a," and classes F2bA through F2cE are each defined by combinations of three attribute states.

We know taxonomic classification best through its application in biology—the familiar Linnaean taxonomic system. Although the various levels of that taxonomy can be said to house empirical referents—organisms—it usually is the case that only the lowest levels—species—are treated that way. We normally do not, for example, categorize individual plants and animals by family but by species. However, species are members of genera, which are members of families, and so on. This method of ordering is used to reflect a "degree of structural similarity" (Dunnell 1971a:79) among the taxa being categorized. The Linnaean system illustrates similarity but is also often thought to illustrate ancestry: Species of the same genus are thought to share a common ancestor. This may or may not be so in any particular case, however; "the Linnaean hierarchy has been criticized as a scheme of systematic classification because it does not trace the phylogenetic tree" (Valentine and May 1996:31). Part of the problem is this: "The Linnaean hierarchy was devised to represent God's essentialist plan.... What has dramatically changed from Linnaeus's day is that for him taxonomic inclusiveness indicated degree of similarity (in essential traits), whereas now taxonomic inclusiveness indicates genealogical connectedness" (Ereshefsky 1994:188–189; see also Padian 1994; Renner 1994). As Mayr (1981, 1994, 1995b) remarks, only after publication of Darwin's *On the Origin of Species* was the Linnaean taxonomy perceived as possibly reflecting common evolutionary descent. That there was no "thorough upheaval of existing [biological taxonomies]" was a result of biologists having selected many characters of similarity that were also homologs (Mayr 1995b:424), a topic to which we will return. Biologists have nonetheless sought to perfect their taxonomies in various ways better to reflect phylogeny (Mayr 1981, 1994, 1995b). One way has been to develop a phylogenetic-species concept. Such units are irreducible clusters of organisms distinct from other such clusters and within which there is a pattern of ancestor and descendant (for reviews see Cracraft 1989b; Mayden 1997). The major difficulty with this concept is that it denies polyphyly or hybridization—in short, that evolution is reticulate—so it is of limited utility (Endler 1998) and may not help either biologists (Ereshefsky 1989) or archaeologists solve the problems they face.

In contrast to taxonomic classification, paradigmatic classification uses an unordered arrangement of attribute states to construct classes. Any attribute state belonging to a single dimension *can* combine with any other attribute state belonging to any other dimension. Whether they actually *do* combine on one or more specimens is an entirely separate question. One can conceive of a 12-inch-long red pencil, but one need not have an actual specimen displaying this combination of length and color attributes to realize that such a thing may exist.

Paradigmatic classes exhibit several important characteristics. First, all dimensions are equivalent in terms of importance, as are the attribute states within each dimension. In contrast to taxonomic classification, none is more or less important than any other in forming classes. Second, paradigmatic classes are unambiguous in terms of both internal structure and their utility for creating groups. Since all attribute states of a dimension are mutually exclusive, there can be no internal contradiction. Things, for example, cannot be red or blue at the same time if under the dimension color the attribute state blue and the attribute state red are each listed but the attribute state "blue and red" is not listed. In this case, if objects are red, they are in one class; if they are blue, they are in another class. Third, paradigmatic classes are absolutely comparable with all other classes in the same classification.

A paradigmatic classification is relatively straightforward to apply to archaeological materials *if* the dimensions and the attribute states of each dimension are well defined. One application is detailed later in this chapter and another in Chapter 8. Theoretically, any two archaeologists could, using identical dimensions and attribute states, independently sort a set of specimens into the same groups, and each group should contain the same specimens (ignoring observer bias; Beck and Jones 1989). Even though two of us are faced with the same pile of specimens and use the same classification, we may end up with groups of somewhat different composition. But here the problem is with the observers, not the units, and that is an important point. Many arguments in archaeology over the typological identity of a particular specimen are the result of poorly defined units; interobserver biases merely exacerbate the problem. Classes, especially paradigmatic ones, lessen such problems and leave us with only interobserver bias. We do not consider this bias further, as only recently have efforts been made to explore how pernicious such bias is in archaeology (Whittaker et al. 1998).

BIOLOGICAL SPECIES AS CLASSES

Any event or object is amenable to *classification* if by this term we mean categorization, or sorting into kinds. Organisms regularly are sorted into kinds termed *species*. But in biology, it goes against the very core of the discipline to consider species to be classes, as we have defined the term. Most evolutionary biologists and philosophers of biology agree that under a materialist perspective species do not have essences, but lacking an essence does not automatically impart classness to a set of phenomena. Biologists and philosophers have been largely in agreement on this issue, stating that species are not and cannot be classes. Mayr (1995b:421) remarks, for example, "A species is a very different kind of taxon from a higher taxon. A species, indeed, is always a particular (individual), never a class." The reasons for this view are varied, but in the end they come back to a central point: Species comprise organisms, and organisms (actually, the populations thereof) evolve. If something evolves, it cannot have an essence. Classes, it is said, have essences. A class is a " 'natural kind' ... is spatio-temporally unrestricted and defined by its essence (its definition) (Ghiselin 1981)" (Mayr 1995b:421). Therefore, species cannot be classes (but see Gould 1983c).

This is a belief that in our view leads to various problems that can be escaped by adopting the notion that classes are measurement units rather than kinds with essences, and as such are as capable of measuring variation in organisms as they are of measuring variation in stone tools. The fact that a large segment of those who study the natural world have a difficult time conceiving of species as *classes*—not as we define the term, but as something else—means that the issue deserves closer examination. It has broad epistemological implications. As an entry point, consider how Mayr (1988:337) handles the issue of classes:

> Membership in a class is determined strictly on the basis of similarity, that is, on the possession of certain characteristics shared by all and only members of that class. In order to be included in a given class, items must share certain features which are the criteria of membership or, as they are usually called, the "defining properties." Members of a class can have more in common than the defining properties, but they need not. These other properties may be variable—an important point in connection with the problem of whether or not classes may have a history.
>
> The class concept is widely applicable to inanimate objects. For instance, one can recognize a class of chairs as consisting of pieces of furniture built in such a way that a human can sit on them. By far the most important defining

> property of a class is its constancy, a necessary correlate of its being based on an essence. At the same time, class membership is not spatiotemporally restricted. If two-legged hominoids on Mars were to construct pieces of furniture with the defining properties of chairs, these would belong to the class of chairs. Nor is there any special relationship among members of a class, such as one finds among parts of an individual [read a *species*]. Clearly there is no relation among the members of the class "chair." In other classes, there is sometimes an indication of relationship, but this relationship is not part of the definition.

There is one point in Mayr's statement with which we agree: Class membership is based on "defining properties" that members hold in common. In other words, the denotata of a class, if there are any, share properties in common. But we strongly disagree that Mayr's "defining properties" comprise an essence (Mayr 1995b). If classes are ideational units—measurement units rendered as sets of significata—and are conceived as such, they do not and cannot have essences. Recall that there are two kinds of ideational units, descriptional and theoretical. Neither has an essence, even though part of the definition of one might include attribute states that themselves define essences. We might, for example, list blue as an attribute state under the dimension color. Blueness is an essentialist kind *if* it is used to measure light reflectivity. Similarly, we could build a classification of chairs in part around physical–chemical dimensions and attribute states. The point is, classes can be constructed using attribute states that define essences, but this does not make the classes themselves essentialist units.

Mayr says that the "class concept is widely applicable to inanimate objects." He does not explicitly say that the concept is inapplicable to animate objects, but the implication is there (see also Mayr 1996:266). Toward the bottom of the paragraph containing that line, he goes further, explicitly separating "parts of an individual" from members of a class. By "parts of an individual," Mayr is referring to the actual organisms comprising a species, using terminology introduced by Michael Ghiselin (1966, 1974a, 1974b, 1981), who argued that members of a species function as parts of an individual much like the fifty states function as part of an individual entity, the United States. As Mayr (1987:153) notes, biologists had long abandoned the idea of species as classes, though it was not until Ghiselin (1966) pointed this out that philosophers (e.g., Hull 1976, 1978, 1980, 1981a, 1984; Sober 1984a) took note. We have no problem with Ghiselin's species-as-individuals argument and in fact have supported it (e.g., O'Brien and Holland 1990, 1992; O'Brien et al. 1994). What we do not support are arguments that species cannot be classes as we have defined the latter term.

Mayr's efforts in this regard seem to revolve around two issues. The first is the issue of unit discreteness:

> Individuals are reasonably discrete in time and space, they are bounded rather than being potentially unlimited, as classes are. However, because species

> speciate and occasionally merge, their borders are sometimes "fuzzy" and the point at which one species leaves off and another begins is often arbitrary, as is also true for mountains and many other objects of nature that are individuals. Owing to their discreteness, species are particular things with proper names. (Mayr 1988:344).

To equate the fuzziness of species borders with the fuzziness of borders of mountains "and many other objects of nature that are individuals" does not address the key issue of how boundaries between units are set, that is, how unit discreteness is established. As Mayr (1996:267) later put it, a biologist faced with the problem of placing organisms (and populations thereof) into a species must accomplish two things: First, develop a concept of what a species is such that it fits within the Linnaean taxonomy, and second, "to apply this concept when combining [organisms and] populations into species taxa." Mayr suggested the first goal can be met by using the biological-species concept; in short, "species are groups of interbreeding natural populations that are reproductively isolated from other such groups" (Mayr 1996:264). This definition establishes boundaries based on reproduction and thus allows living organisms and extant populations to be combined into species taxa—the second requirement. Mayr (1970:13) indicated such units were time-specific but emphasized that they had "an evolutionary potential for continuing change." Otherwise, it leaves unspecified, as Simpson made clear, the time dimension (see also Hull 1992).

The second issue around which Mayr's consideration of classes revolves concerns whether the members of a unit are somehow related. In the preceding long quote, Mayr indicates that members of a class do not have a "special relationship" to one another. Because in Mayr's view chairs have no such special relationship to one another, they comprise a class. Although Mayr is not explicit, we hazard the guess that because the individual organisms comprising a species in fact have the special relationships of sharing genes and a reproductive system (they are potential mates), they cannot be members of a class; hence, a species is not a class. Again, Mayr has precluded the properties of a class from species units merely by definition. We find the property of "special relationship"—particularly a genetic-like relationship—between members of a class to be critically important in evolutionary archaeology, and we return to it in later chapters.

We have to exercise care in presenting Mayr's case, because he has argued variously that species *are* classes and that they are *not*. The variation resides in how he defines the concept of class and in what he assigns to it. He lays out the basis of his argument in the following terms:

> A major advance in conceptualization and terminology of the taxonomist is curiously unknown to many philosophers. I am referring to the difference between the species as a category ("species definition") and the species as a taxon ("species delimitation"). A species taxon is an object in nature recog-

> nized and delimited by the taxonomist. The species category is the rank given to a species taxon. That the species category is a class is not disputed by anybody. What is involved is the ontological status of the species taxon. (Mayr 1987:146)

Earlier, Mayr (1980:34) noted that "species taxa are individuals, with each member of the species being a part of the species, while the species category is a class, the members of which are the species taxa." What Mayr is saying is this: On the one hand, the empirical member populations of the species as taxonomic units comprise all the individual species that ever were or that ever will be—what Mayr (1996:263) elsewhere refers to as "the concrete phenomena of nature" and what we would call groups. They do not have essences; rather, the members of each species merely have properties in common, the critically significant one being the potential to interbreed successfully (the "special relationship" noted earlier). This feature, together with the facts that species can "speciate, they can hybridize, and they can become extinct," indicate to Mayr that they are not classes (Mayr 1987: 151, 157). On the other hand, species as a category is what we might call a theoretical unit—our class. It is, in the Linnaean taxonomic system, the rank below genus and is signified by the biological-species concept of reproductive isolation. This comprises the significata of the unit and might be conceived of as the essence of the unit (e.g., Gould 1983c), though this is decidedly not what Mayr believes. In his view, such a position comprises typological thinking (e.g., Mayr 1959a). The reproductive-isolation part of the definition is critical because a "definition of any class of objects must be applicable to any member of this class and exclude reference to attributes not characteristic of this class" (Mayr 1996:263). The species definition of reproductively isolated populations avoids the operational problems of the typological-species concept, as Mayr makes clear in the following:

> In order to determine the species status of a population or taxon, one must attempt to apply to it the species definition and see whether the situation is consistent with it. What is involved has been excellently illustrated by Simpson (1961) in the case of monozygotic twins. Two persons are monozygotic twins not because they are so similar, but they are so similar because they are monozygotic twins. Two populations belong to different species, not because they can tolerate coexistence without interbreeding, but rather they can tolerate such sympatry because they are species. Probabilistic inference is the major method in systematics for determining species status of a population. It can be observed directly only in the nondimensional situation and not always even there. In all dimensional situations (with longitude, latitude, or time added) species status can be determined only by inference. No working naturalist can escape this simple fact of nature. (Mayr 1987:146–147)

Mayr makes an excellent point when he cites Simpson's (1961) clever illustration of monozygotic twins: They *are* identical because they are monozygotic twins; that is, they are members of the same unit of evolutionary descent—a species. But thus far, Mayr has still not bridged the gulf between his biological-species concept and Simpson's chronospecies. If Mayr had followed through to the logical end of his argument, he might have seen that it provided a bridge. Two things are required to build the bridge. First, because members of a species tend to be phenetically similar, one can infer, as Simpson did, that similar fossils comprise a species. The biological-species concept serves to specify a scale of observation—individual organisms comprise a species—and a field of observation—organisms that do or do not interbreed. Second, and following from the first, instead of viewing the taxon *species* as a category and the individual species (Mayr's "taxa") as empirical members of that category, treat *each species* as a class in the sense we advocate and the individuals within it as the denotata. This clears up multiple problems. For example, instead of speaking of chair as a class, we define "chair" in a manner that specifies a field of observation. Mayr's (1988:337) "pieces of furniture built in such a way that a human can sit on them" might be a necessary but certainly insufficient part of that definition because, according to his own rules that such a definition exclude reference to attributes not characteristic of this class, it does not exclude a coffee table or a bed. Once this definition is clarified, then we can define particular classes of chairs—recliner, bean bag, overstuffed, and so on—with individual chairs in each class being the denotata of the class, denotata that have time and space locations.

When we employ classes as measurement units, what we are measuring is variation. Species as classes—theoretical units—do exactly that. Species as classes have precise defintions—more precisely, the definitions *are* the classes—and as such they are, theoretically, timeless and spaceless. By the rules of the game, if we found something on Mars that met all of the necessary and sufficient conditions for the class *Homo sapiens*, then we would have to classify it as a denotatum of that class. We would certainly want to know how it got there, since it clearly falls outside the known distribution of the *group* known as *Homo sapiens*, but that is an empirical and historical problem, not a theoretical one.

Similarly, if we found something on Mars that met the necessary and sufficient conditions of the field of observation "chair," then it is a chair of one class or another. But notice that Mayr states that if "two-legged hominoids on Mars were to construct pieces of furniture with the defining properties of chairs, these would belong to the class of chairs." As far as something meeting the necessary and sufficient conditions, it is totally irrelevant who or what made it. By bringing "two-legged hominoids" into the picture, Mayr is tying chairs to certain members of the superfamily that includes hominids and apes—all of which are themselves classes in our sense of the term. He has inferred that a member of our lineage or a closely allied lineage made the Martian chair. This provides an "indication of relationship," to use Mayr's term. But our point is, for something to be identified as a member of the field "chair," at least as Mayr defined it, in no way requires that a two-legged hominoid made it. Mayr, of course, has consistently argued that biological units such as are represented by the Linnaean taxonomy comprise not only phenetically similar things but also, with greater or lesser accuracy, phylogenetically related ones (e.g., Mayr 1981, 1995b, 1996).

Mayr, in choosing "chairs," picked a bad example to illustrate his conception of a class. In his rush to assign inanimate objects to the world of classes as essentialist units and to save organisms for the world of "taxa," he overlooked the implications of the fact that species fully qualify as both ideational and empirical units—our classes and groups, respectively. Once that duality is recognized, it does away with the paleontological problem of trying to slip assemblages of dead organisms into the neontological definition of a species. The problem reduces to defining the boundaries of a class—Mayr admits that species boundaries are inferred—and then identifying the denotata of that class, which together form a group. Those members—the individual organisms—are empirical units and thus have *locations* in space and time. The class, in that it is timeless and spaceless, has a *distribution* represented by its empirical members. The paleontological problem is trying to understand the hows and whys of the actual locations displayed by the denotata of a class; it is *not* worrying about whether a set of specimens possibly comprising a species actually does or not, and this is why Mayr so dislikes Simpson's concept (e.g., Mayr 1997:132). Simpson understood this distinction, as do most modern paleontologists, though making the distinction between classes and groups part of the issue clarifies the notion of species considerably.

How does one tie classes to evolutionary change? Mayr (1988) addresses this issue, but because he views the species *category* as a nonessentialist class or "individual" as opposed to viewing individual species as *classes*, he misses the critical point entirely:

> To include the capacity to evolve among the evidence in support of the claim that the species is an individual has been frequently criticized. Is evolutionary change a diagnostic difference between class and individual? Is it not also possible for a class to evolve? Has there not been a historical change concerning the class of chairs in the last several hundred years? Indeed, if we compare three sets of chairs, those in use in 1780, those in use in 1880, and those in use in 1980, we may discover considerable change. If it had been part of the defining criterion of the 1780 chair that the supporting structure is of wood, this definition would no longer be appropriate in an age of steel and plastic. But "being made of wood" was never a truly defining criterion of chair. What had changed over the centuries was not the essence of chairness, but rather some accidental properties of chairs that are not part of the defining essence. A class, having a constant essence, cannot evolve. (Mayr 1988:347)

Mayr is correct: A class cannot evolve, but the reason it cannot evolve is not because it has an essence, which it does not in our view, but because it is an *ideational* unit. Such units are used to keep track of variation—in effect, to measure it. Only populations of empirical phenomena—actual things in groups—can evolve. By equating "chair" with an essentialist class Mayr overlooks the fact that there are *classes* of chairs—in fact, an infinite number of chair classes. To make the point somewhat differently yet along the lines we used earlier, the species unit (as Mayr's taxon) can be construed as a class. Mayr would admit that there are species of the taxon rodent, species of the taxon bird, and so forth. Precisely the same can be done with the taxon chair. We could speak of the family "furniture," the genus "chair," and the species "recliner," "beanbag," and "overstuffed," or the species "wooden," "metal," and "plastic." The point is that *the analyst* chooses the definitive attributes for each level in the taxonomy. Mayr does not like the fact that the analyst picks the criteria and instead wants some biologically, particularly evolutionarily, meaningful criterion; he thus opts for interbreeding.

The definitive criteria—the significata—of any class of chair include, if only implicitly, whatever it is that *we* decide will be included within the larger-scale, more inclusive unit or field "chair." Were the larger-scale, more inclusive unit "species," then the latter would entail within-group interbreeding and between-group reproductive isolation. We emphasize that Mayr, not the organisms, specified the species criterion—what breeds with what—and that this criterion was a biological one derived from the specimens observed. Because Mayr's specimens were temporally limited, when talking of evolutionary descent he had to append the phrase that a species was capable of evolving, speciating, going extinct, and so on. Similarly, we specify what makes something a chair, not the thing. The only difference between Mayr's procedure and ours is that we would construct classes of chairs intensionally, perhaps using composition as one analytical dimension. We then would create groups by identifying the denotata in the individual classes and finally plot their time–space positions. If we were interested in evolutionary phylogeny, we would include some notion of transmission in the construction of our units, but those units would still be classes in the sense that we use the term.

The question of whether classes evolve—which we agree with Mayr is an impossibility—overlooks the interesting question of whether *chairs*—the empirical phenomena—evolve. Since only living things are actually capable of evolving, are we reduced to speaking metaphorically if we say chairs—or projectile points—evolve? We would argue that metaphor plays no part here. If we accept that projectile points are parts of the human phenotype in the same way that birds' nests are phenotypic, then certainly the people who made the points evolved. All we have to do is to be careful to state that when we talk about projectile-point evolution, what we really are doing is using a shorthand notation to speak of the evolution of the point makers themselves. Similarly, when biologists talk about the evolution of the beak, what they are really talking about is the evolution of birds, all of which possess beaks. Tracking change in morphological characteristics of the beak, which is a phenotypic feature, is a way to get at the evolutionary history of birds.

ARCHAEOLOGICAL TYPES AS UNITS

The distinction between ideational and empirical units is not difficult to understand, at least conceptually, but biologists are not the only ones who occasionally conflate empirical and ideational units. Archaeologists, despite a few warnings such as that by Phillips, Ford, and Griffin (1951:66), have also been unable to keep the distinction clear. Types, which Phillips, Griffin, and especially Ford viewed as arbitrary constructs, are treated as "real," empirical units that are discovered as opposed to created. They are treated as real in part because they house real stuff, in part because they are extensionally derived, and in part because they are viewed as reflections of cultural norms. During the heyday of culture history, failure to appreciate the difference between theoretical and empirical units resulted in some interesting debates, the most notable of which was over how to construct archaeological types. This often heated exchange, which began in 1953, came to be referred to as the Ford–Spaulding debate, and in it one can see many of the problems inherent in trying to do culture history without the theory to support the end product. The problems exist today in almost unaltered form and bear considerably on efforts to incorporate Darwinian evolutionary theory into archaeology.

Until the early 1950s, most archaeologists were content to worry about the chronological placement of the types they drew from various cultural lineages. It was thought that these units *might* reflect the cultural norms of the people who made the artifacts being placed in the types (e.g., Rouse 1939)—meaning that the units might in some way be "real"—but such suspicions were merely commonsensical rationalizations for the units. They were not empirically testable save in a tautological manner, such as in George Brainerd's (1951a) notion that the popularity of cultural norms produced normal frequency distributions of types through time; thus, empirical manifestations of such frequency distributions denoted norms or customs. In the absence of theory common sense prevailed, and the belief that types were somehow real units grew stronger, helped along with statements such as Walter Taylor's (1948) admonition that archaeology should be anthropological. The issue reached a head in 1953, when Albert Spaulding (1953b) published a paper describing a technique for discovering those real types. James Ford (1954a) responded, Spaulding (1954b) replied to his response, and Ford (1954c) finally produced a more programmatic statement concerning his views on types and their construction.

Ford was always clear on the purpose of his types: "The inadequacy of the procedure of dividing pottery into 'types' merely for the purposes of describing the material is recognized. This is merely a means of presenting raw data. Types should be classes of material which promise to be useful tools *in interpreting culture history*" (Ford and Griffin 1938:3; emphasis added). Here, Ford established a position that would remain essentially unchanged in his mind (e.g., Ford 1935a, 1935b, 1935c, 1936a, 1936b, 1938, 1951, 1952, 1954b; Ford and Quimby 1945; Ford and Webb 1956; Ford et al. 1955). He, like Irving Rouse (1939) and others, figured that the types could, in some way, reflect cultural ideas (e.g., Ford 1936a), but this was irrelevant to the greater problem at hand. Ford was correct in stating that historical types were classes (in our sense of the word), but this was simply a fortunate choice of words. Ford's types were ideational units formed on the basis of trial and error—their definitions were derived extensionally—and then tested to ensure that they measured time. His types thus had a somewhat arbitrary appearance with respect to their cultural significance.

Spaulding, on the other hand, wanted a classification technique that "expressed at one stroke the classifier's opinion of the cultural relationship and the chronological position of an assemblage," as such a technique would allow "a combined presentation of [the] independent units of chronological position and cultural affinity" (Spaulding 1949:5). This was a lofty goal, the creation of a kind of unit that marked not only time but also ethnicity, but the means of creating such a unit were unclear to Spaulding in 1949. That changed as a result of a paper that Brainerd published in 1951. In it Brainerd (1951b) made two observations that undoubtedly influenced Spaulding (Lyman and O'Brien 1997; Lyman et al. 1997b; O'Brien and Lyman 1998). First, he indicated that "archaeological taxonomy [read *typology*] is in itself a generalizing procedure which ultimately depends for its validity upon the archaeologist's success in *isolating the effects of culturally conditioned behavior* from the examination of human products" (Brainerd 1951b:117; emphasis added). His second point was that

> the first step of procedure in artifact analysis is usually the formulation of types, groups of artifacts, each of which shows a combination of similar or identical attributes or traits.... If [the observation quoted above] is acceptable, the systematics used must have cultural validity in that they must mirror the culturally established requirements met by the artisans. In his search for these tenets of the unknown group it behooves the archaeologist as a scientist to work objectively, free of a priori conceptions. The attributes used in sorting artifacts into types should thus be *objectively chosen as those which occur most often in combination in single artifacts*. Criteria based upon subdivisions of an attribute which occurs in a continuous range through the material are preferably used only when the distribution curve of the attribute in the archaeological samples shows binodality, and *the dividing line for sorting should be drawn between the nodes*. By use of the above requirements for type attributes, the archaeologist can *objectively describe the cultural specifications followed by the artisans. Statistical procedures for the formulation of,*

> *and sorting of specimens into, types satisfying these requirements are feasible*, and may in some cases be useful. It seems conceivable also that mathematical studies of attribute combinations may demonstrate more finely cut cultural differentiation without the use of the intermediate concept of types, for types are, after all, simplifications to allow qualitative division of the material into few enough categories to permit inspectional techniques of analysis. (Brainerd 1951b:118–119; emphasis added)

Brainerd (1951b:124) argued that by carrying out his analytical protocol, "it is conceivable that a bridge may be found uniting the objectivity of the taxonomist to the cultural sensitivity of the humanist. Cultural intangibles can, if they exist, be made tangible," and he concluded that "better technique is the solution" to what Spaulding (1949) perceived as archaeology's dilemma.

Spaulding (1953b) used statistics to launch his attack against the perceived arbitrariness of artifact types, arguing that types formed by clustering algorithms had sociobehavioral meaning. His definition of a type as "a group of artifacts exhibiting a consistent assemblage of attributes whose combined properties give a characteristic pattern" (Spaulding 1953b:305) was compatible with earlier definitions (e.g., Rouse 1939), including Ford's (1936b), because of its emphasis on the recurrence of attribute combinations. The difference was that for Spaulding, recurrence was empirically determined on—that is, extensionally derived from—a closed set of materials. His emphasis on attribute combinations meant that his types had class-like properties, but since recurrent attribute combinations were empirically determined, his types also had group-like properties. Spaulding worked with sherds from one site at a time, measuring similarities between them in terms of attribute combinations and creating groups based on statistically frequent patterns. Were the types represented at locations outside of the one that produced the sample that was analyzed? This question was impossible to answer because of the idiosyncratic nature of the types. Adding a single new sherd required that the entire exercise be repeated. In the end, despite their class-like properties, Spaulding's types were nothing more than extensional groups that were tightly space- and, occasionally, time-bound.

Ford's types were also formed from combinations of attributes, but the pottery samples were much more widely distributed in space and time (O'Brien and Lyman 1998). Types were explicitly, if extensionally, constructed to have distinct spatiotemporal distributions. Once formed, they were tested to determine whether they measured time—in Alex Krieger's (1944) words, did they pass the historical-significance test? If not, they were thrown out or refined (O'Brien and Lyman 1998). Ford's orientation was narrow: Types were analytical tools that the archaeologist constructed in such a manner as to permit chronological ordering of assemblages. Unfortunately, his attempts to explain the culture-historical paradigm to which he subscribed were "badly muddled by a conflation [of] the definitional and interpretive meanings attached to [the] type [concept] and by frequent recourse to intuitive rationales necessitated by the lack of any truly theoretical justification for the culture-historical position" (Dunnell 1986:172).

Spaulding's types were real and inherent in the specimens; in other words, they were essentialist, empirical units. Thus, to Spaulding (1953b:305), "classification into types is a process of discovery of combinations of attributes *favored by the makers* of the artifacts, not an arbitrary procedure of the classifier" (emphasis added). Because types are inherent in the data, they must be discovered inductively, and statistical techniques, as suggested by Brainerd (1951b), provide the objective means of determining which attributes regularly, and more often than random chance would allow, co-occur on specimen after specimen. Since artifacts are products of human behaviors, discovery of recurring attribute combinations—types—is simultaneously discovery of that behavior, or so Spaulding's reasoning went.

In his comment on Spaulding's (1953b) paper, Ford (1954a) protested that Spaulding's approach was "amazingly naive" because it only suggested cultural norms; it did not help write culture history. In his response, Spaulding (1954b:393) noted that Ford had not "challenged the validity of the techniques [Spaulding] used to discover [attribute] clusters" and underscored the procedural murkiness in Ford's constructions of "attribute combinations." Here, the debate was over method, but Spaulding failed to see the difference between Ford's types as theoretical units—classes—and his own types as idiosyncratic groups. The latter were obviously different creatures from Ford's, and they were set up to do different work. Spaulding noted that his "attribute clusters" included "inferences as to the behavior of the makers of the artifacts." Here, the debate was over the meaning of types. Faced with Ford's questioning of both the usefulness of his units and the lack of a nontautological test of their interpretive meaning, Spaulding was forced to turn to his method—the identification of statistical modes—for justification. The legitimacy of the claim that discovered attribute combinations reflected human behavior had to come from method, as there was no other place from which it could come. The implicit reasoning was that such combinations existed as human creations and were sortable into recognizable, empirical sets; thus, they *had* to be real. The same kind of reasoning guided Mayr's (1969b:313) assertion that because he and a Papuan tribesman both consistently classified birds the same way, the resulting groups (species) *had* to be real.

Ford (1954c) went on the attack again, and although a careful reading of his paper makes clear the distinction between Ford's materialist position and Spaulding's essentialist position, it also points out the manner in which Ford conflated elements of the two perspectives. Cultural types certainly once existed, Ford thought, but he was not particularly interested in discovering them (just like Simpson with respect to species as reproductively isolated populations of organisms). Rather, he wanted "type groupings consciously selected [by the archaeologist to produce] a workable typology ... designed for the reconstruction of culture history in time and space" (Ford 1954c:52). However, he never specified how such groupings were to be extracted from the flowing, constantly changing braided stream of cultural evolution. Ideational units were called for that allowed one to measure change, but in lieu of theoretically informed unit construction, Ford used trial and error, and this gave his procedure and his units an arbitrary appearance. To Ford, significant formal variation existed at any point in the time–space continuum, and although that variation might "tend to cluster about a mean, which [the analyst] could visualize as the central theme of the type ... [he or she] cannot rely upon the culture bearers to define this theme. They may or may not be aware of it.... The [type], then, is an abstraction made by the [analyst] and derived from cultural activity" (Ford 1954c:45). Ford's view of the temporally continuous nature of artifact variation is well captured in Figure 5.3.

In Ford's view, discontinuities along either the temporal or spatial dimensions of the archaeological record presented the archaeologist with natural seams, at which points the time–space continuum could be broken up into chunks. Importantly, this perspective was not essentialistic, though it sounds that way. Ford simply was proposing that discontinuities presented convenient points at which to insert *arbitrary* breaks in the continuum; if such discontinuities did not present themselves, one could make the cuts at any arbitrary points (e.g., Ford 1935a, 1935b, 1935c). Ford's critics, especially Phillips (e.g., 1970:908–909), never understood how Ford could make temporal breaks when no natural divisions—either stratigraphic disruptions or the appearance of new cultural traits—presented themselves. Ford could do it because in his mind the flow of time, and hence of culture, was seamless, punctuated only by changes in tempo (O'Brien and Lyman 1998). Chunks had to be carved out of the continuum for analytical purposes, but there was nothing particularly real about them. This sounds exactly like the paleontological position of George Gaylord Simpson. Types to Ford were nothing but accidents of the samples available: "The particular locality where an archaeological collection chances to be made will be one of the factors that determines the mean and range of variation that are demonstrated in any particular tradition in the culture that is being studied" (Ford 1954c:49). Chance samples of the continuum would provide discontinuous snapshots of that continuum; hence, types "are easily separable and they look natural [that is, 'real']" (Ford 1954c:52). This, too, sounds a lot like Simpson.

In the end, Ford's strategy for refuting Spaulding's position failed because his allusions to customs and standards gave Spaulding's types a certain reality. Despite Ford's poorly constructed arguments and obtuseness, his basic position stands in stark contrast to that of Spaulding, whose types were "founded in the essentialist view of variation. They presume that significant variation occurs as more or less discrete packages and that variation not assignable to such packages lacks explanatory significance" (Dunnell 1986:181). Comparison of Spaulding's

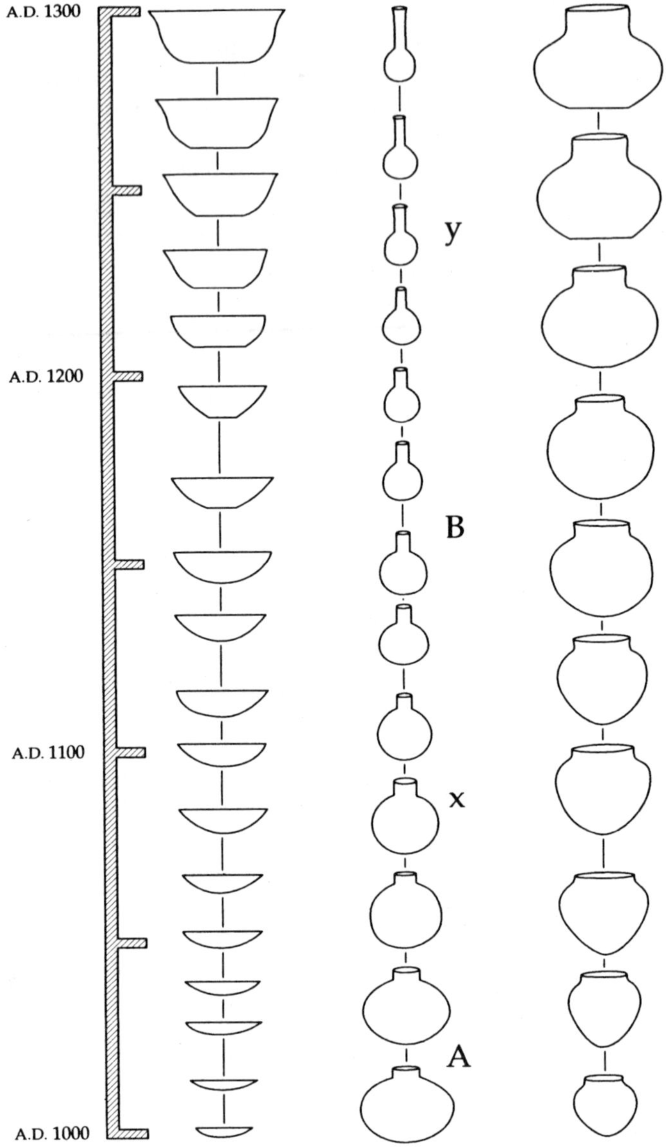

Figure 5.3. James A. Ford's conception of culture change as exemplified in pottery. Note the constant gradation in the three forms through time. Ford also used the diagram to reinforce the notion that types were arbitrary constructs of the analyst. If a type A is set up in the "water-bottle" tradition (center) as shown, then the classifier has to select the next type in this tradition at least as far away as B in order to differentiate among types consistently. The examples midway between these "typical" type examples are the difficult borderline cases. Alternatively, if the first type established had been set up at *x*, then the second type would have to be as far away as *y*. In this arrangement, the forms that were typical for type B of the first system become the doubtful specimens of the second (after Ford 1962).

"real" types must be qualitative and must focus on differences between them. Ford's types stem from a materialistic conception of reality, that is, variation in form is continuous across space and through time. Division of that continuity into chunks through the use of measurement units—classes—is a trial-and-error process, the success of which Ford chose to evaluate with the historical-significance criterion (Krieger 1944).

The Ford–Spaulding debate did little to clear up the confusion over the nature of types, and in fact, some archaeologists (e.g., Cowgill 1963) saw little or no contradiction in the two positions. Given the general belief that types probably have some cultural meaning (e.g., Brainerd 1951a, 1951b; Krieger 1944), it is not surprising that Americanist archaeologists took typology a step further and broke types down into *varieties* in an attempt to get at smaller-scale sociological phenomena. James Gifford (1960) suggested on the one hand that a variety represents individual or small-group social variation within a society—a position taken to its extreme in Phillips's (1970) treatment of pottery from the lower Mississippi River valley (O'Brien and Lyman 1998). On the other hand, a type is "the material outcome of a set of fundamental attributes that coalesced, consciously or unconsciously, as a ceramic idea or 'esthetic ideal'—the boundaries of which were imposed through the value system operative in the society by virtue of individual interaction on a societal level. These ceramic ideas occurred in the brains of the potters who made the ceramic fabric that constitutes a type, and they are not by any means creations of an analyst" (Gifford 1960:343). Gifford's rationalization of his units was the typical one: Potters "tend to conform to the demands of a majority of the norms that are a part of their culture at a particular time in history" (Gifford 1960:343), and cultural phenomena are not randomly distributed in time and space (Gifford 1960:342). Thus, despite a lack of citation, Gifford (1960:342), like Spaulding (1953b) before him, believed that an inherent—essentialist—order in the data was discoverable and that "classificatory schemes ... are in part useful as a means toward this end." Such schemes, however, are of little use in an evolutionary study.

Types as Groups and Classes

Intensionally defined classes—that is, classes whose definitions are imposed on rather than drawn from a set of objects or events—have played only a small role in Americanist archaeology. Of far greater importance have been extensionally defined units, the definitions of which have been drawn from the set of objects or events to be classified. More to the point, we could say that such ideational units have been *described*—sometimes in great detail, where the description borders on a definition, but more often it does not. By the middle of the second decade of the twentieth century, archaeologists working in the Southwest realized that pottery designs and surface treatments had changed over time; hence, sherds carrying traces of those designs and treatments could be used as temporal markers. Gradually, Southwesternists realized that pottery types were becoming idiosyncratic; there often was no standardization in how types were created and named. The Pecos Conference and the Globe Conference were founded in part to alleviate this problem and resulted in a standardization in how types were named and described. The binomial naming system that appeared in the Southwest eventually made its way into the Southeast through the efforts of Ford and Griffin (1937, 1938; see O'Brien and Lyman 1998), who also sought to standardize the manner in which types were created. Despite such efforts, classifications were still inconsistent from region to region and from investigator to investigator.

Dunnell (1986:165–166) points out some of the problems that resulted from the ad hoc creation of pottery types using extensionally derived definitions, often based on small samples of sherds:

> Although the list of dimensions varied somewhat from area to area based on tradition and the variable character of archaeological materials, this approach provided that a type hold some value in each of a standard set of dimensions of observation and measurement. Just as importantly, beyond setting limits on the dimensions that could be used, this approach did not provide any guidance in the *selection* of definitive attributes. They were literally *descriptions of pottery assigned to a particular type*, not definitions of the type. While a value or set of values was required for each dimension (e.g., a particular paste texture, a particular temper, or a particular exterior and surface treatment), which dimensions were definitive varied from type to type and were unmarked in the descriptions themselves. Thus, type descriptions embodied a very substantial amount of operational ambiguity. One had to know, case by case and a priori, which attributes were cause for the assignment of type and which were simply attributes associated typically with the definitive elements. Because the type descriptions were in fact descriptions of particular pottery assemblages (how else could one obtain a "thickness" range, for example), a further and even more far-reaching structural ambiguity was introduced. Were types the creations of archaeologists and, thus, definitionally associated sets of attributes (i.e., analytic tools for the dissection of the archaeological record), or were they empirically associated sets of attributes that "discovered" existential entities? Either interpretation was possible, and both would be pursued. (Dunnell 1986:165–166; emphasis in the original)

As we have seen, both interpretations indeed *were* pursued. After the Ford–Spaulding debate, most archaeologists followed the Fordian tradition of creating types in ad hoc fashion, assuming that the types were useful for keeping track of time and space, but they also assumed that their types were useful for getting at sociocultural issues such as what kind of marriage pattern a particular community practiced (e.g., Deetz 1968; Hill 1970; Longacre 1968). This perceived ability to access the sociocultural past that the early processualists discovered was a legacy of the conflation of units that occurred during the heyday of culture history (see also Tschauner 1994).

What do we do with the myriad artifact types that have been created in archaeology? Do we refer to them as groups or classes? Dunnell (1971a:157) points out that the term *type*, especially when preceded by the term *descriptive*, often is used "for intuitive groups which do not in any respect meet the criteria of classes and is thus employed as a synonym for English 'kind' "—exactly the point we made in Chapter 2, where we substituted usage of "kind" for "type." However, Dunnell (1971a:158) also provides a technical definition of "type" as a "paradigmatic class of discrete objects defined by modes." Both Ford's and Spaulding's types, for example, were of this kind, though the latter's lost their class-like characteristics after the statistically significant attribute clusters were identified. When Spaulding was finished, his types depended on the set of artifacts polled and thus could not be used to classify other materials because they comprised group descriptions. In Ford's system, the analyst selected only those combinations of modes that were useful for keeping track of time and used those combinations as definitions of types. The resulting definitions could be used without modification to identify—to "type"—new materials.

By way of further example, we note that archaeologists have long searched for ways to categorize what normally are referred to as projectile points, recognizing that many of the specimens so labeled could and undoubtedly did serve functions other than as stone tips on the end of spears, darts, and arrows. By far the most common approach to categorization has been to use form-related characteristics as a basis for type creation. Some types that have been created contain narrow ranges of variation around the means (the significata are very precise), and others allow large amounts of variation. Additionally, some types have been based on large samples from several sites, and others are based on small samples, often from a single location.

One of the first efforts to standardize projectile-point categorization on a regional scale was *An Introductory Handbook of Texas Archeology* by Dee Ann Suhm, Alex Krieger, and Edward Jelks (1954), which summarized trait lists of the recognized cultural complexes of Texas as well as descriptions of pottery and projectile-point types; it was subsequently reissued as "Handbook of Texas Archeology: Type Descriptions" (Suhm and Jelks 1962). By 1960, at least two other guides to projectile-point types had been published, incorporating Suhm et al.'s (1954) types and adding new ones (Bell 1958, 1960). Those early volumes were followed by Gregory Perino's (1968, 1971, 1985, 1991) guides and Noel Justice's (1987) single-volume compilation.

How good are named types for doing archaeological work? There is no question that some of the types were described in such minute detail that their descriptions border on being definitions. Some descriptions were based on a large number of specimens from a fairly large geographic range, which can be both a blessing and a curse, because although idiosyncrasies tend to be minimized, overall variation in what was placed in the type increases. Many of these types do their job of marking time, whereas other types encompass so much variation that they are essentially worthless for measuring time. Marie Wormington (1957:105), in commenting on the early nomenclature used with reference to what became known as Paleoindian points, underscored the problems in using extensionally defined types, especially those that are little more than descriptive units:

> Naturally, many difficulties arose because so many widely divergent types were placed in a single category, and various efforts were made to define the term "Yuma" in such a way as to include all of them. One such effort is represented by the definition submitted by the Committee appointed by the Chairman of the round table discussion of Folsom and Yuma points at the International Symposium on Early Man held in Philadelphia in 1937. "A Yuma point is triangular. It runs from triangular through parallel-sided to leaf-shaped. Its base is either straight or convex or concave. It is frequently stemmed but when stemmed has parallel sides.... The sides of the stem are parallel. It is never fluted. It is pressure flaked from both sides, the flakes being parallel."
>
> This definition was virtually meaningless. It made it possible to include a heterogeneous assortment of types under a single term without differentiating between them.

Technically, Wormington was incorrect; it was not that Yuma included "so many widely divergent types," because a type—an ideational unit—cannot include other types. What she meant was that the amount of variation in the points continually placed in the Yuma type ensured that the type description was all encompassing and hence of little value archaeologically. Because the type was extensionally defined, as more and more specimens were called Yuma, the definition had to be expanded to keep the conditions for membership up to date. Finally, when the type reached the point of being "meaningless," to use Wormington's term, it was split into multiple types.

Virtually all archaeological types in use in the United States during the twentieth century are extensionally defined categories that reside somewhere in the realm between classes on the one hand and groups established by grouping procedures on the other. All of them to a greater or lesser extent have group-like properties; a few, those for which the necessary and sufficient conditions are spelled out in detail, also exhibit class-like properties. The key to understanding the nature of these archaeological types is not by identifying them as classes or groups but rather by recognizing that they are extensionally defined. In some cases, this characteristic does not negate the usefulness of the type relative to chronological concerns.

Traditional Types versus Groups Derived from Classes: An Example

Exactly how useful are some other common types? Do they partition variation in a useful way—that is, do the types allow us to document small-scale change through time? To answer this, we discuss the Big Sandy projectile-point type. Side notching had long been a technique used in haft modification; it can be seen in the slight indentations in the haft areas of Dalton points in the Midwest and South and in the slightly later Hardaway points in the East, and it reached its peak in the several millennia following 7000 B.C.[2] The ascendancy of side-notched points is evident throughout much of the southeastern and portions of the eastern United States, and numerous types have been devised to document the variation in such specimens. For example, Otter Creek (Ritchie 1971) is a commonly used type name in the Northeast, and Big Sandy (Kneberg 1956) is by far the most common type name used in the Southeast and Midwest. Based on stratigraphic positioning in several deep sites and on radiometric determinations, Big Sandy points were originally viewed as Early Archaic (ca. 7000–5000 B.C.) in age, but in the four-plus decades following the initial type formulation, the age range of Big Sandy points has been expanded considerably. This is attributable in large part to the incorporation of points into the type that do not match the original definitive attributes as set forth by Madeline Kneberg (1956), based on specimens from the Big Sandy site in Henry County, Tennessee. In subsequent decades, Big Sandy has become little more than a default category.

Kneberg (1956:25) described Big Sandy points as follows: "The basic shape is trianguloid, with excurvate side edges and incurvate or straight basal edge. The blade is usually thin and flat, and well retouched with pressure chipping. The notches, which are perpendicular to the long axis of the blade, are usually narrow and short. The base is usually incurvate, either definitely or slightly. The portion of the blade between the notches and the base represents the unmodified basic shape of the blade before the notches were made." Three decades later, Noel Justice (1987:60) expanded the definition considerably:

> The typical blade shape is a narrow and elongated triangle. Cross sections vary from biconvex or plano convex to rhomboid or median-ridged depending on the degree of blade resharpening. Beveling and serration are common resharpening features produced using pressure flaking techniques. The shoulders of these forms may be parallel with the basal ear margins, representing the unmodified preform sides, or nearly exhausted from resharpening.
>
> The haft element of Big Sandy specimens exhibits side notches that tend to be shallow, in that the notch width may be greater than notch depth. The notching techniques exhibited vary from indentations produced bifacially following a single inward direction, to those exhibiting a Y-pattern resulting in two notching directions in the interior of the notch [similar to that on Thebes points]....
>
> The basal edge ... varies from nearly straight to deeply concave and is sometimes nearly bifurcated in appearance. Basal thinning is a conspicuous attribute.... The basal ears are usually squared but may be slightly rounded. Full haft grinding is common, although specimens lacking grinding are not atypical.

[2]All dates in this section are in uncalibrated radiocarbon years.

Notice in Justice's description that the cross section of a Big Sandy point varies considerably, the basal edge varies from nearly straight to deeply concave, and the haft element may be ground or unground. Now, there is nothing technically wrong with such a *description* provided the analyst does not mind having considerable variation within the resulting group, but if used as a definition, it does little good when one is after small-scale morphological change—the kind of change that might be tracking functional change as well.

Justice also included ontogenetic (developmental, or life-history) factors in his description of Big Sandy points, noting that cross-sectional shape depends on the degree of blade resharpening. This is important information, since without it, the analyst could erroneously assign a resharpened point to another type. This was not an uncommon occurrence in early biological taxonomy, where, because they looked different, puerile forms were assigned to a different species than parental forms. Whether things start out small and get larger through time, as organisms do, or whether they start out large and get smaller through time, as projectile points do, it is ontogenetic development. We need to keep this in mind when we set up classes that are based on morphological dimensions.

Our illustrations of Big Sandy points (Figure 5.4) are based on specimens from the Big Sandy site that T. M. N. Lewis furnished Robert Bell for incorporation into the second volume of his projectile-point guide (Bell 1960). Comparison

FIGURE 5.4. Big Sandy projectile points (based on specimens from the Big Sandy site in Henry County, Tennessee) that T. N. M. Lewis furnished Robert Bell for incorporation into the second volume of his projectile-point guide (from Bell 1960; drawings courtesy of the Oklahoma Anthropological Society). All specimens at same scale; specimen at the left is 5.9 centimeters long.

of (1) published descriptions of Big Sandy points and photographs of specimens that accompanied those descriptions with (2) more recently published photographs of so-called Big Sandy points suggests that few archaeologists are familiar with the earlier work. Carl Chapman (1975:242), for example, stated that the

> suggested dating of the type is 5000–500 B.C. The specimens in central and southwestern Missouri date primarily in the period 5000–3000 B.C., and those in eastern Missouri and Illinois date as late as 500 B.C. The type was manufactured in the Middle Archaic period in Graham Cave [Missouri] as a smaller and less precisely made version of Graham Cave Notched, which appears to be the progenitor. It continues throughout the Middle Archaic period and is the most distinctive type of the period in Missouri. It appears again in the Early Woodland period [1000–600 B.C.] associated with the ... Black Sand phase in Illinois.

Chapman was incorrect; Big Sandy points did not "disappear" and then resurface several thousand years later. There was indeed a small resurgence in side notching in the centuries just before the Christian era, a clear case of evolutionary convergence, but the points do not fall in the Big Sandy type unless we decide that all side-notched points by default fall in the ideational unit Big Sandy. Chapman's error is typical of the confusion that can surround the placement of specimens in extensionally defined types. If we follow the earlier type descriptions (e.g., Cambron and Hulse 1960, 1964; Kneberg 1956) and more modern updates by authors who have studied variation in side-notched points (e.g., Cambron and Hulse 1975; DeJarnette and Knight 1976), it appears that Big Sandy points are Early Archaic in age, probably on the order of 7000–6000 B.C.

Graham Cave side-notched points (Scully 1951) (Figure 5.5), are morphologically similar to Big Sandy points. The morphological relation between the two types is difficult to miss: If you take a Big Sandy point, lengthen it considerably, and turn its feet slightly downward while keeping the notches oriented perpendicular to the long axis of the point, you have created a Graham Cave side-notched point. This might be taken as evidence that there is a phylogenetic relation between Big Sandy points and Graham Cave points, though it is equally possible that the forms are analogous. Wilfred Logan (1952:30) noticed considerable variation in side-notched points from levels 5 and 6 in Graham Cave, particularly in base shape, and he created three types, one of which, his type A, contained specimens that exhibit the characteristics of what might be termed prototypical Graham Cave side-notched points:

> This is a long, side-notched, concave-based point. The side notches are small and precise. The base is sharply concave to the extent that the corners of the base present a barbed appearance. The length ranges between 4 and 6 inches. The point is three to four times as long as wide. The chipping is carefully controlled in small, shallow flakes. The edges may be slightly serrated, and the

FIGURE 5.5. Graham Cave side-notched points from Missouri. Both specimens at same scale; specimen at the left is 10.5 centimeters long.

> blade may make an "S" shaped curve from shoulder to tip. In all but one point, the cross-section is thick and biconvex, almost to the point of being diamond-shaped.

Graham Cave points are common over most of Missouri and also occur in western Illinois and northern Arkansas, but with lesser frequency. Justice (1987: 62) noticed this geographical separation: "Big Sandy forms may extend into Missouri and other areas to the west. Once again, however, published information would indicate an absence of Big Sandy forms beyond the middle Mississippi Valley. In Missouri (C. H. Chapman 1975:242) side notched materials are primarily morphological variations of the Graham Cave Side Notched and Raddatz types rather than Big Sandy." This is a correct statement and one that archaeologists working in the Midwest should carefully note. Perino (1985:36) made more or less the same point: "Too often, side-notched points of various time periods and in other localities have been called Big Sandy points when, in fact, they are not." He limited the distribution to western and central Tennessee and adjoining parts of Kentucky and Alabama.

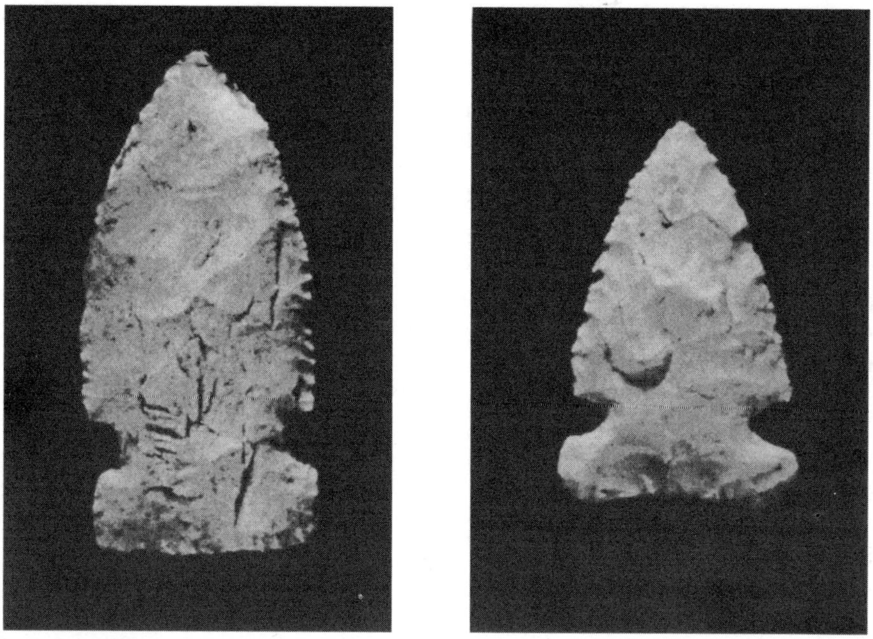

FIGURE 5.6. Raddatz (Hemphill/Godar) points from Missouri. Both specimens at same scale; specimen at the left is 6.8 centimeters long.

All available evidence indicates Graham Cave points date to the Early Archaic period and probably ceased to be made by 5500 B.C., if not slightly earlier (O'Brien and Wood 1998). Their successor was Raddatz points (Figure 5.6), which in the literature are consistently misidentified or equated with Big Sandy points. For example, Chapman (1975:158) stated that

> the Forager Tradition of the Middle Archaic period had associated with it a chipped-stone technique that was used to produce a characteristic form of side-notched point or knife of different proportions from that of Graham Cave Notched, an Early Archaic period type. This side-notched form has been given various designations, such as Raddatz, Black Sand, Big Sandy, White River Archaic, and has previously been considered to be a variety of Graham Cave Notched (Chapman and Chapman 1964). It is a distinctive type, and the basic form is similar to and may have been developed from Graham Cave Notched. It is smaller in size, and the technique of manufacture involved heat treating, which may be a distinctive characteristic of the Middle Archaic period.... The several varieties of the form can be separated typologically on the basis of size and heat-treating technique employed in their manufacture. I have selected the term *Big Sandy Notched* to designate the side-notched points

from the Middle Archaic period sites in order to reduce the number of names for this particular type.

Chapman clearly recognized that Graham Cave side-notched points were Early Archaic in age, and he, like others, recognized morphological differences between Graham Cave points and other, later side-notched points. His error was in selecting the type name Big Sandy to use for the later Middle Archaic points. Chapman certainly has not been alone in this error, and one of us (MJO) made the same mistake in equating type names with some of the side-notched-projectile-point classes from the central Salt River valley of Missouri (Curry et al. 1985). The reason for the error lay in a blind acceptance of type designations by others and an unfamiliarity with original type descriptions and the context of the archetypical specimens.

The distinguishing characteristics of Raddatz points are their small, U-shaped notches and their square feet, though tremendous variation exists among points described in the literature as being Raddatz. Bases usually are straight, but variation in base shape also exists among illustrated specimens. Justice (1987:68), like Chapman, noted that there are numerous other types—including the deeply side-notched points usually placed in the Godar type (Perino 1963) and the concave-base Hemphill points (Scully 1951)—that exhibit the general characteristics of Raddatz points and which can be collapsed within a single type. But unlike Chapman, Justice did not include Big Sandy among them, instead reserving that type for "large Side Notched cluster forms which possess a range of haft and resharpening characteristics unlike those typical of Raddatz projectile points" (Justice 1987:68). Based on inspection of published photographs and drawings, Raddatz points appear to have a distribution that extends from western Missouri and Iowa, north to the Great Lakes, east to Pennsylvania, and south as far as southern Kentucky—more or less the area mapped by Justice (1987:69). Raddatz (Godar/Hemphill) points appear to date from the tail end of the Early Archaic period well into the Middle Archaic period.

Our venture into point typology demonstrates some of the attendant problems in using extensionally defined types. Sometimes the definitions are so precise that actual specimens placed in the type are clones, but in most cases, the definitions tend to be general descriptions that encompass significant variation. Paradigmatic classification, however, allows us to observe variation in objects; to do so objectively and repeatedly, thereby suppressing observational bias; and, when the chronological ordering of variants has been completed, to track that variation through time. In short, paradigmatic classification gives us an *objective* means of examining variation. We *know* what a particular class is simply by looking at the definition and do not have to worry about whether what we are calling a Raddatz point is what someone else is calling Raddatz (or Godar, or Hemphill). Classes are replicable by anyone using the classification scheme—a feature that is not present in most typological schemes. If it *were* present, we would not have different type names being assigned to more or less identical specimens; we would not have wide ranges of morphological features designating types (as classes); and we would not have a wide range of variation within the types (as groups).

Recall that in a paradigmatic classification, the investigator, before classifying, selects the dimensions that he believes are required for analysis and then chooses the attributes that he expects to encounter. Any additional attributes that later appear in the sample can be added as they become apparent during classification. As an example of the different results produced by typical categorization versus paradigmatic classification, we use side-notched projectile points from the Early Archaic– and Middle Archaic–period levels at Pigeon Roost Creek, a deep site in northeastern Missouri (O'Brien and Warren 1982, 1983, 1985) that was occupied intermittently throughout much of prehistory (Figure 5.7). Points from the site originally were classified by Marianne Curry and her colleagues (Curry et al. 1985) as part of a larger effort to classify over two thousand projectile points from the central Salt River valley. Table 5.1 lists the eleven dimensions used during the exercise and the attribute states of each dimension. Each individual class is defined by a specific combination of attribute states—one from each dimension—and the resulting eleven-attribute class carries a number string as its name. Each number corresponds to the specific attribute state of a particular dimension, with each dimension being the place number in the string. For example, in class 538562(03)43(12)3, the first number, 5, is attribute state 5 of dimension 1; the second number, 3, is attribute state 3 of dimension 2; and so on. Dimensions 7 and 10, because they contain double-digit attribute states, are set off by parentheses. Thus, by looking at Table 5.1, we see that class 538562(03)43(12)3 is defined as "side notched; slightly concave base; expanding curved stem; the widest point of the blade is at the distal end; obtuse shoulders; slight shouldering; right-angle rounded shouldering; notching that occurs as broad, shallow arcs; perpendicular notch orientation; large, parallel tangs; and medium in length."

Theoretically, there is an infinite number of classes that could derive from a classification of objects; here, using the closed array of eleven dimensions shown in Table 5.1, there are slightly over one billion classes possible. Obviously, given the number of points classified, very few classes actually contained denotata; those that did were given group numbers. For example, the group of points extracted from class 538562(03)43(12)3 was labeled group 25.12. Figure 5.8 illustrates representatives of the side-notched groups from the Early Archaic and Middle Archaic levels of Pigeon Roost Creek. If we had used the existing point typology for the Midwest, points in groups 25.1, 24.0, 23.0, 18.1, 18.2, 14.0, and 25.10 would be placed in the Raddatz type (Table 5.2). The advantages of paradigmatic classification is evident by noting that within the groups of points that normally would be placed in the Raddatz type there is considerable variation in such

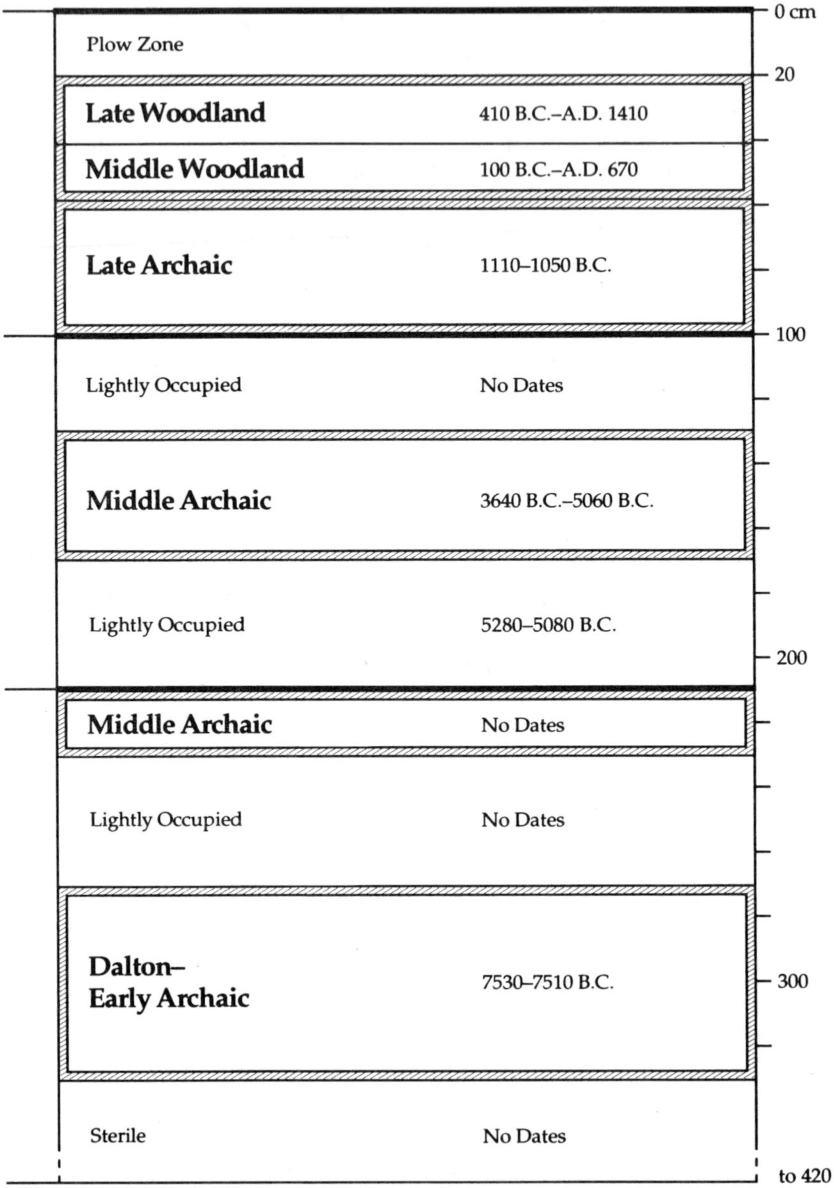

Figure 5.7. Generalized stratigraphic profile from a vertical section at Pigeon Roost Creek, Monroe County, Missouri, showing analytical strata and radiocarbon dates (after O'Brien and Warren 1985).

TABLE 5.1. Paradigmatic System Used to Classify Projectile Points from Pigeon Roost Creek, Missouri

Dimension Attribute state	Dimension Attribute state
I. Primary hafting preparation	VII. Shape of shouldering
1. No discernible haft modification	1. Not applicable
2. Grinding	2. Right-angle squared
3. Grinding and flaking	3. Right-angle rounded
4. Lateral edge flaking	4. Right-angle modified rounded
5. Side notching	5. Acute pointed
6. Corner notching	6. Acute rounded
7. Basal notching	7. Acute modified rounded
8. Stemmed	8. Obtuse pointed
II. Shape of stem	9. Obtuse rounded
1. Convex	10. Obtuse modified rounded
2. Straight	VIII. Notching proportions
3. Slightly concave	1. Not applicable
4. Slightly concave–angular	2. Deep, narrow arc
5. Deeply concave	3. Equal arc
6. Deeply concave–angular	4. Broad, shallow arc
7. Pointed	5. Short, shallow arc
III. Shape of stem	6. Broad, shallow rectangle
1. Not applicable	7. Angular
2. Contracting linear	IX. Notch orientation
3. Contracting curved	1. Not applicable
4. Contracting incurvate–excurvate	2. Acute angle
5. Parallel	3. Perpendicular
6. Expanding linear	4. Parallel
7. Markedly expanding linear	5. Asymmetrical
8. Expanding curved	X. Shape of tangs
9. Markedly expanding curved	1. Not applicable
IV. Widest point of blade	2. Rounded
1. At proximal end	3. Parallel
2. In the proximal third	4. Straight expanding
3. In the middle third	5. Contracting
4. In the distal third	6. Pointed
5. Distal	7. Footed
V. Angle of shoulder	8. Expanding rounded
1. Lacking (contracting blade)	9. Modified rounded
2. Lacking (parallel blade)	10. Expanding, modified rounded
3. Lacking (expanding blade)	11. Large, straight expanding
4. Acute	12. Large, parallel
5. Perpendicular blade	XI. Overall size (length)
6. Obtuse	1. Micro (< 3 cm)
VI. Degree of shouldering	2. Small (3–4.5 cm)
1. Not applicable	3. Medium (4.45–8 cm)
2. Slight	4. Large (> 8 cm)
3. Moderate	
4. Pronounced	

Source: O'Brien and Warren 1985.

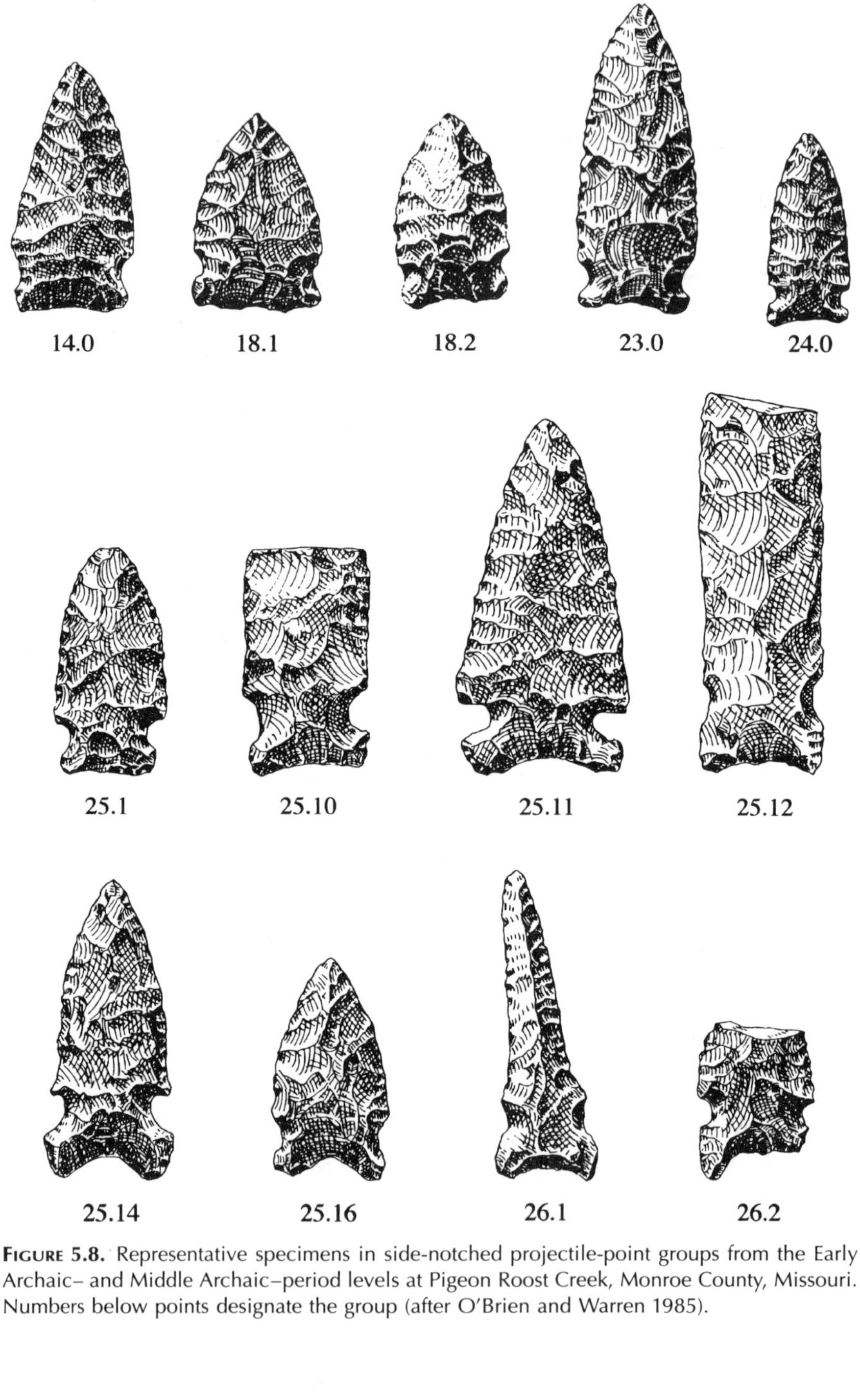

Figure 5.8. Representative specimens in side-notched projectile-point groups from the Early Archaic– and Middle Archaic–period levels at Pigeon Roost Creek, Monroe County, Missouri. Numbers below points designate the group (after O'Brien and Warren 1985).

Table 5.2. Classes, Class Definitions, and Frequencies of Side-Notched Points from Early Archaic and Middle Archaic Levels at Pigeon Roost Creek, Missouri

Class	Class definition	Number of specimens	Established type
25.12	538562(03)43(12)3	1	Graham Cave
25.11	539154(03)33(04)0	2	Graham Cave
25.10	538563(03)33(04)3	1	Raddatz
25.16	558162(09)33(02)3	3	Graham Cave
14.0	538162(10)45(09)3	6	Raddatz
18.2	538162(10)53(09)3	4	Raddatz
18.1	528162(10)53(09)3	5	Raddatz
23.0	538262(10)33(02)3	2	Raddatz
24.0	538153(04)33(09)2	2	Raddatz
25.1	528163(03)33(09)3	1	Raddatz
26.1	548163(10)33(03)3	2	Graham Cave
26.2	568153(03)33(12)3	1	Graham Cave
25.14	559153(03)33(04)3	1	Graham Cave

dimensions as shape of shouldering (dimension 7) and tang shape (dimension 10). Base shape (dimension 2) is fairly consistent, usually slightly concave, though two groups—18.1 and 25.1—exhibit straight bases. Specimens in groups 26.1, 26.2, 25.14, 25.16, 25.11, and 25.12, because of their deeply concave bases and downturned feet, would typically be placed in the type Graham Cave side-notched, but again, the paradigmatic classes allow recognition of the fact that considerable morphological variation is present in the more frequently used type unit.

Groups have specific time–space locations, and we can examine the vertical locations of point classes as evidenced by the various groups. Table 5.2 lists the frequencies of specimens by group, the latter of which are arranged in stratigraphic order with specimens from the deepest (and presumably oldest) levels at the bottom. Figure 5.9 illustrates the vertical distribution of specimens by group and level. Based on radiocarbon assays (O'Brien 1985; O'Brien and Warren 1983; O'Brien and Wood 1998), the 270- to 330-centimeter levels date to the Early Archaic period and the 100- to 230-centimeter levels to the Middle Archaic period. No points were found in the 230- to 270-centimeter levels.

Despite the low frequencies in the groups, several trends are noteworthy. The nine specimens in groups 18.1 and 18.2 share identical distributions, being confined to the 130- to 160-centimeter levels. Notice from the definitions of the two classes that produced those groups that the only difference between the two is in base shape, with points in group 18.1 being straight based and those in group 18.2 being slightly concave based. The two specimens in group 24.0 occur on either side of the 40-centimeter-thick layer (170–210 centimeters) that produced no projectile points and which is assumed to represent a dramatic decrease in site-

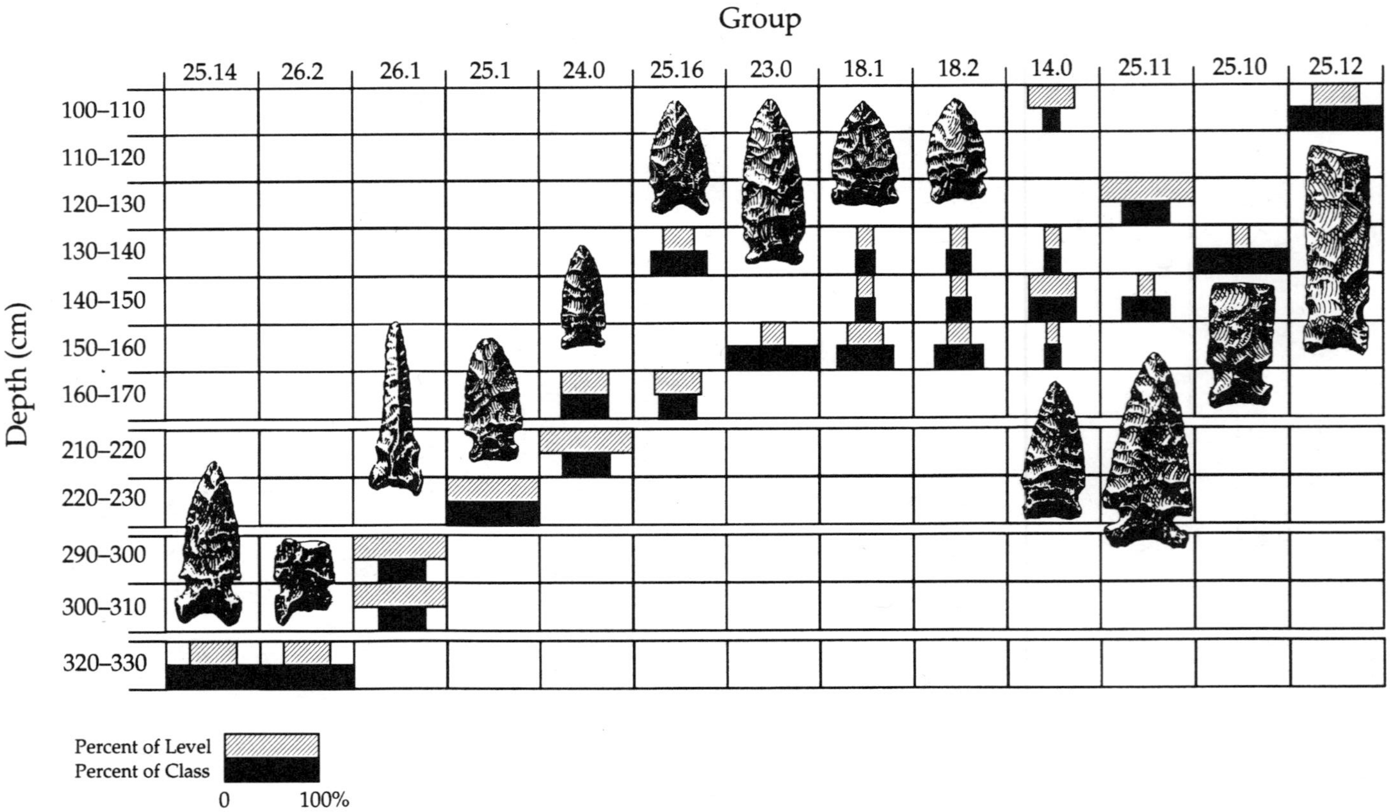

FIGURE 5.9. Relative projectile-point group frequencies for side-notched specimens from Pigeon Roost Creek, Monroe County, Missouri. The top bars show the percentage contribution made by each group to the total point assemblage in that level, and the bottom bars show the percentage contribution of points in each level to the total number of points in each group. For example, points in group 18.1 occur in three of the 10-centimeter levels and are most frequent in the 150- 160-centimeter level (revised from O'Brien and Warren 1985).

occupation intensity between about 4500 B.C. and 4000 B.C. One specimen in that group and the single specimen in group 25.1 are the only points from the earlier of the two Middle Archaic components (Figure 5.7). Not surprisingly, those two groups are very similar in terms of attribute states.

If we plot the vertical distribution of specimens by midwestern point types, the results are all over the board in terms of where various specimens fall. Graham Cave points fall anywhere between the 120-centimeter level and the 330-centimeter level, a block of sediment that accumulated over some 3000 years or so (Figure 5.7), and Raddatz points fall anywhere between the 100-centimeter level and the 230-centimeter level, a block of sediment that accumulated over perhaps as much as 2000 years. In defense of this typology, one might argue that in general terms, most Raddatz points *were* stratigraphically superposed over Graham Cave points—exactly where they were supposed to be. One might even argue that the stratigraphically higher Graham Cave points—specimens in groups 25.11, 25.12, and 25.16—were "misidentified"; that is, they belong in another type, since they are so clearly out of the time range of Graham Cave points. This kind of argument, however, is fraught with peril. If we can pick and choose which specimens to include or not to include in a type based on how we feel about their suspected age, then we have destroyed any chance at objective categorization. In essence, if we work hard enough at it—that is, if we carefully select the specimens—we can make any artifact type pass the historical-significance test. We might, if we need to, even throw out certain radiometric determinations as being "bad," perhaps on the grounds that the samples were somehow contaminated.

What have we learned from all of this? For one thing, extensionally defined types, if properly constructed, can serve as excellent chronological markers. Projectile points placed in the Clovis type appear to date to a short time span, between roughly 9250 B.C. and 8950 B.C. Other types are not as tightly defined as Clovis and thus are not particularly useful as time markers except in a very general sense. It would appear, however, that if we are interested in tracking small-scale change in things such as projectile points, most types, no matter how tightly defined they might be, are inadequate for the job, primarily because they lump too much variation into a single normative category and often are difficult, if not impossible, to replicate from sample to sample and from investigator to investigator. It could be argued that types are still legitimate categories to use; if one wanted to look at the evolution of a single feature of projectile points—say, haft-element grinding—one could simply arrange the points, by type, in chronological order and then examine that feature relative to each group of points. This might be acceptable if that is all that one wanted to do, though we still have the problem of replicability, but this rarely is the case. More likely, we would want to know how grinding is associated with other features, if it indeed is associated. It seems to us that the only way to measure such co-occurrence in objective fashion is to use paradigmatic classification.

FOSSIL SPECIES AS GROUPS AND CLASSES

Before moving on to how we might go about examining lineages from an evolutionary perspective, the topic of Chapters 6–8, we think it is important to point out some of the similarities between archaeological systematics and paleontological systematics. Recall our previous discussion of how traditionally minded paleontologists—phyletic gradualists—such as George Gaylord Simpson and Philip Gingerich identify chronospecies, and how Niles Eldredge, Stephen Jay Gould, and their followers—those subscribing to punctuated equilibrium—identify species. For the former, chronospecies are more or less arbitrary chunks of the evolutionary, that is, morphological, continuum (compare Figure 4.2 with Figure 5.3). Boundaries between such units can be signified by convenient stratigraphic boundaries, but they need not be. The critical point is that definitions of chronospecies are extracted from the fossils available and thus are extensional (e.g., Maglio 1971; Rose and Bown 1986). Punctuated equilibrium uses differentia—what we would call significata—but what has gone unremarked by paleobiologists of either stripe is that the differentia are extracted from the fossils available. Thus, a species for Gould, Eldredge, Elisabeth Vrba, and others is an extensionally defined unit, much as is a Graham Cave side-notched point or a Raddatz point. As with projectile points, one result of the extensional derivation of species differentia is a reinforcement of the perception that species are real, empirical units. Another result is an effective masking of the fact that one has merely shifted to a larger, more inclusive scale—species—rather than remaining at the scale of individual organisms.

The central difference between the systematics of phyletic gradualists and that of proponents of punctuated equilibrium is epistemological, a point some paleobiologists have recognized (e.g., Eldredge 1982). *Arbitrarily* defined chronospecies are so defined because of the view that evolutionary change is gradual and continuous, whereas viewing species as evolutionarily and thus morphologically static is argued to be a better conception (e.g., Eldredge and Gould 1977). Variation *within* a species might denote evolutionary change, but the species is the more fundamental unit of such change for proponents of punctuated equilibrium, so they focus on change from species to species, usually referred to as taxonomic turnover (the scale change). But to conclude that "for the most part, taxonomic rates were invented and continue to be compiled because they are a handy way of estimating transformational rates: morphological rates pose practical problems of measurement" (Eldredge 1982:340) results, in our view, in precisely the problems we find with archaeological types. This is not to deny the utility of using taxonomic rates—what could be construed in archaeology as rates of type replacement or typological turnover, for as Simpson (1944:16) himself noted, "Inasmuch as it seeks to divide phyla into generic and specific stages, representing roughly equivalent amounts of total morphological change, the taxonomic system is a rich source of data." Even Gingerich (1985:38), a staunch phyletic gradualist, admits that "inclusion in the same species of organisms differing in size or form by factors greater than those characterizing living species reduces the functional comparability of [fossil] species." But the keywords in Simpson's statement should be noted: Each taxon represents "roughly equivalent amounts of total morphological change." This *must* be an assumption with extensionally defined units: "The assumption that two successive or related genera do cover equivalent amounts of evolution [read *morphological variation*] is obviously uncertain in any one instance. It becomes more reliable and useful when taxonomists involved are of equal and great skill, when one student with extensive first-hand knowledge has revised all the genera, and especially when a large number of genera based on more or less comparable criteria can be averaged" (Simpson 1944:16).

Interestingly, the preceding quote echoes the opinion of some archaeologists who claim the only way to evaluate another's work is to consider that person's expertise—for example, archaeologist Raymond Thompson's (1956b:332) statement that "the final [judgment] of any archaeologist's cultural reconstruction must ... be based on an appraisal of his professional competence." More to the point, Simpson's remark indicates that paleontological taxonomic units are extensionally derived: They and their contents can be altered by looking at more or different specimens. As paleobiologist Thomas Schopf (1982:1148–1149) noted in one of his critiques of punctuated equilibrium, the use of "new characters ... to distinguish taxa [based on] new methods" results in new characterizations of taxa, and "the greater number of available traits in paleontological material, the more readily are species discriminated" (see also Maglio 1971; Rose and Bown 1986). Our point here is that Simpson's assumption—that a similar degree of morphological variation exists within each taxonomic unit—can be tested if one uses intensionally defined units of equal magnitude. Theoretically informed construction of ideational units has other benefits as well, as we will see.

Is Species Stasis an Artifact of Systematics?

In Robert Schoch's (1983:360) view, there are three questions—the answers to which, it should be noted, are time, space, and lineage specific—that must be investigated given the punctuated-equilibrium versus phyletic-gradualism debate:

1. What is the tempo of evolutionary change?
2. Are instances of rapid change the times when we find most diversification (cladogenic events)?
3. If the answer to the second question is yes, is cladogenesis (rather than anagenesis) the dominant mode of evolutionary change?

Worded in such fashion, the second and third questions demand units equivalent to species or higher-level taxa such as genera. That the first question does not necessarily demand such units is clear from the observation that the "problem of the origin and duration of species is not the same as the problem of stability in form of morphological 'hard parts' " (Schopf and Hoffman 1983:439). This is so because when one studies morphological stability or change *within* taxonomic units, "one is no longer referring to a theory about biological species, which was the focus of punctuated equilibrium" (Schopf and Hoffman 1983:439). The question asked by punctuated equilibrium is, as Schopf (1981:158) put it, "Does the same biologic *species* persist [over long time spans]?" or "Is stasis a common factor in *species* evolution?" (emphasis added). From the phrasing of these questions, it should be clear that the central issue under punctuated equilibrium is recognizing units termed *species* in the fossil record. Schopf's (1981:161) most critical point in this respect is that the "philosophy of systematics which makes one ascribe variation to the same biologic species ... is a severe limitation on independently evaluating whether or not stasis occurs in fossil species."

Richard Fortey (1985) presents a cogent discussion of how systematics influences perceptions of stasis and rates of change. He argues that only four specimens are required to empirically show an instance of rapid change (Figure 5.10). Two specimens from different times, and representing the same species, that is, the specimens are morphologically and thus taxonomically identical, represent one period of stasis; the other two specimens are morphologically and thus taxonomically identical but represent a species morphologically distinct from the first. This model leads to the observation that hypotheses of rapid change "are not falsified by the discovery of intermediate morphologies" (Fortey 1985:18). This is so because, as Fortey (1985:18) continues, "The discovery of a chronologically and morphologically intermediate form AB between two species A and B, suggests only that there were *two* [rapid] events of change" (see also Fortey 1988; emphasis in the original). This is well illustrated in Figure 5.11a–d, which begins with the model in Figure 5.10 and then adds a first, then a second, and finally a third form, all of which are variously chronologically and morphologically intermediate to the original two forms. The root of the problem is that such difficulties arise simply because the differentia, or significata, of units A, AAB, AB, ABB, and B are extensionally derived.

Recall that Gould and Eldredge argued that ancestral species should be contemporary with daughter species. This would give the graph in Figure 5.11d the appearance of an asymmetrical candelabra, a not unreasonable model according to Gould (see Figure 4.3). As Fortey (1985:19) makes clear, such a result is easily derived given how species are extracted—extensionally derived—from paleontological data. Morphology may vary more or less continuously through time, as shown in Figure 5.12a. If "species" are extensionally derived from morphology alone, that is, without respect to time, then the vertical stringers in Figure 5.12b

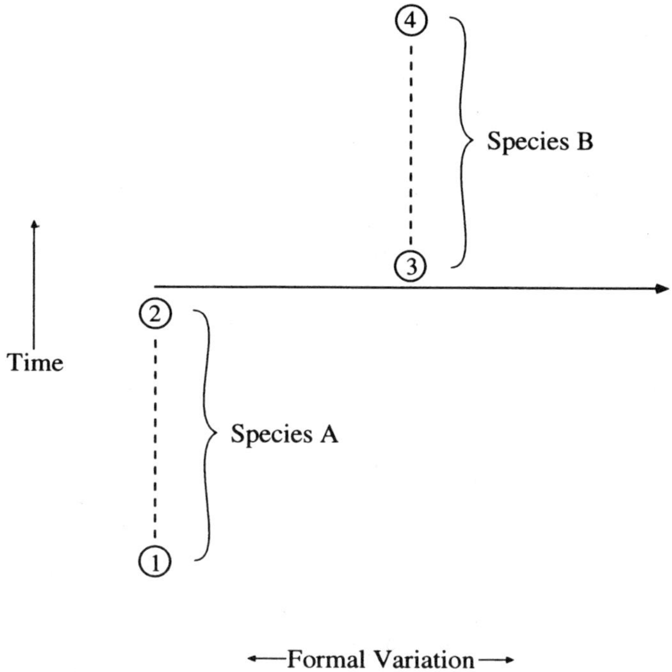

FIGURE 5.10. Diagram showing that only four specimens are needed to support an example of punctuated equilibrium. Species A is represented by two specimens, 1 and 2, that demonstrate stasis and are stratigraphically separated; the same applies to species B, represented by specimens 3 and 4. The speciation "event" producing species B appears as the solid line connecting specimens 2 and 3 (after Fortey 1985).

will result because a species is viewed as a morphologically homogeneous unit. Conceiving of each stringer as representing a species based on morphological homogeneity results in the asymmetrical candelabra of Figure 5.12c. That model not only masks potentially significant morphological variation *within* the species units, but it also suggests cladogenesis and contemporaneity of ancestral and daughter species, two key features of the punctuated-equilibrium model. Fortey (1985:26) suggests a change in graphing conventions might avoid such problems. As we indicate later, we agree.

Measuring Stasis and Change

Gould (1983a), at least, apparently was initially unimpressed by this most fundamental issue of systematics, though now he is arguing, to alter the order of

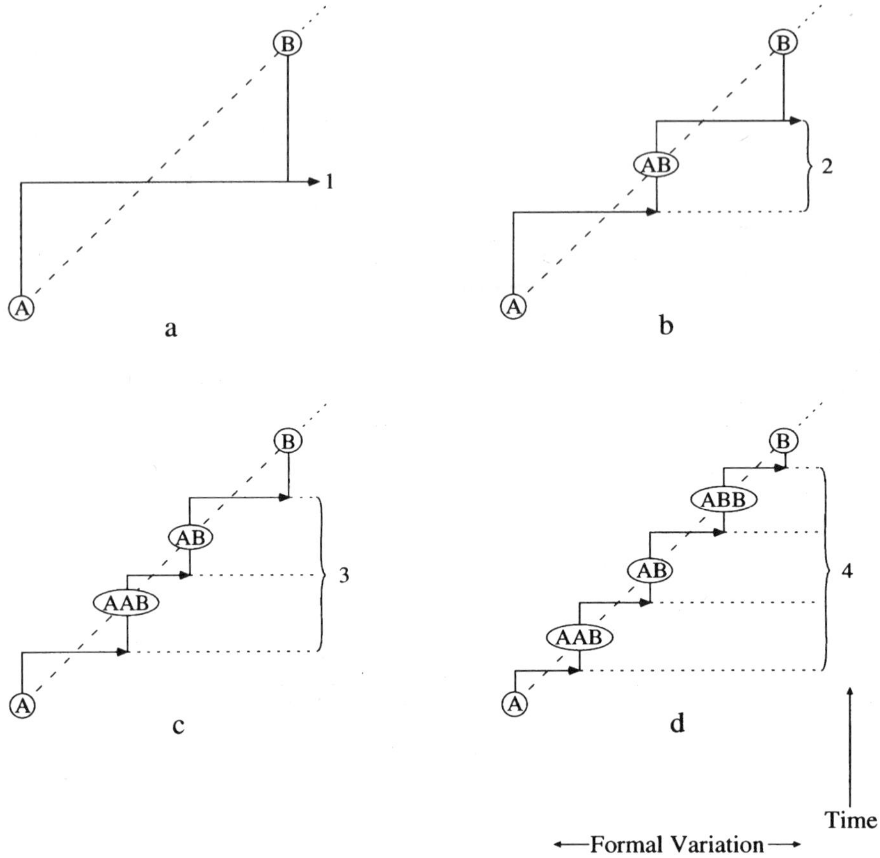

FIGURE 5.11. Diagram showing gradualistic and punctuational hypotheses competing against data on equal terms. In a, two "species," A and B, are discovered, separated by stratigraphic time and morphological distance; they can hypothetically be related by either gradualistic (dotted line) or punctuational (solid line) mechanisms, here having an equal chance (1). In b, intermediate form AB is found at an intermediate horizon; the gradualistic explanation still applies, but now two "invisible" speciation events are required to defend the punctuational view. In c, another intermediate form, AAB, is discovered, requiring three speciation events. In d, a third intermediate, ABB, raises the odds further, to four to one over the gradualistic view (after Fortey 1985).

his wording a bit, that "we need to define a full range of the abstract (and richly multivariate) space into which all organisms may fit (the morphospace). We must then be able to characterize individual organisms and plot them within this encompassing space.... A morphospace [is] seemingly idiosyncratic for each kind of organism" (Gould 1991b:420). We find it interesting, but predictable, given that

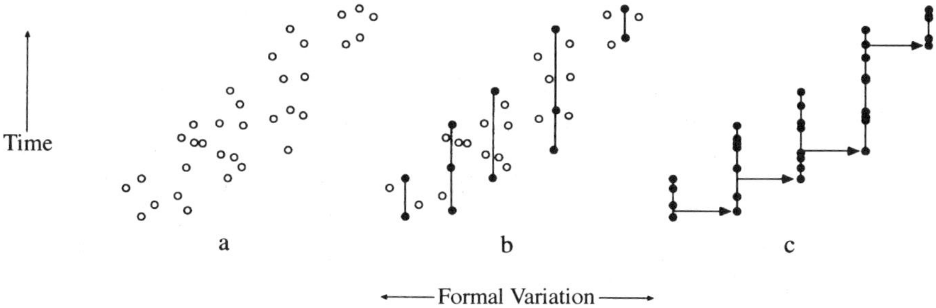

FIGURE 5.12. Diagram showing the possibility that punctuated equilibrium could be an observational artifact. In a, a gradational series of morphological variation through time (a chronocline) is shown. In b, the observer may see the gradation as a series of bars and interpret the distribution as the result of punctuated equilibrium. In c, the points have been moved no more than 5 percent of the total morphological distance; the resulting distribution is what the observer believes he or she is really seeing in the distribution, namely, a clear case of punctuated equilibrium, minus the "sampling error" shown in b (after Fortey 1985).

fossil taxa have consistently been defined extensionally, that at first Gould (1991b:420) did "not pretend to have a solution [to the measurement problem]." He wrote, for example,

> If we define morphospace empirically [read *extensionally*] by a range of organisms, how do we plot other creatures not of the original set and outside its bounds (by dimensionality, not just extrapolation of distance)? How do we deal, theoretically, with the fact that potential measurements are infinite? How do we treat correlation, covariation, the simultaneous consideration of state, meristic [shape-related], and metric variables? How do we express quantitatively the vague but vital concept of relative importance or developmental depth, when superficial characters may display vast metrical differences among species, while small changes in deeper characters may trigger much more fundamental divisions?

We suggest that the answer to just such questions has been available in the archaeological literature since 1971 (Dunnell 1971a)—the construction of paradigmatic classes. We note that Gould's phrases "not of the original set" and "outside its bounds" indicate that his taxonomic units are extensionally derived. Ideational units will allow the plotting of nonoriginal specimens. That the variables to be measured and the values each might take are potentially infinite is readily acknowledged under paradigmatic classification, but theory will suggest which variables and values to start with, a point recognized by some biologists and philosophers of biology (e.g., Hull 1970; Lewontin 1974a). Paradigmatic classification also readily accommodates "state," or nominal-scale variables; "meristic," or shape-based variables; and "metric," or interval-scale variables of size because of its dimensional structure. The "relative importance" of attributes can be accommodated by weighting variables and attributes, if necessary, to produce a hybrid classification incorporating elements of paradigmatic and hierarchical, or taxonomic, classifications. Finally, we interpret the related notions of "developmental depth," "superficial characters," and "deeper characters" to comprise the issue of scale. As archaeologist J. O. Brew (1946) noted decades ago, multiple classifications can and must be built to solve multiple problems, to which we would add, they can be at various scales in order to explore the importance of "developmental depth" and related notions.

We can illustrate the way to solve Gould's problem of how to measure morphospace by returning to the lesson of Figure 5.11—how to handle specimens that are morphologically "transitional," or intermediate, to established units (e.g., Maglio 1971). To do this, we use, with some modification, an example described by biologist Evan Hazard (1998). Hazard's example concerns (fictitious) extant birds on an island, but it is quite applicable to fossil mammals, projectile points, or pottery sherds. We will simply call the material specimens and note that (1) we use the words *type* and *species* as synonyms and *genus* and *ware* as synonyms, and (2) the only requirement is that all specimens be of one kind, such as birds, fossil mammal teeth, or sherds. To begin, the first researcher to study a set of the specimens of interest collected a sample in 1950 and distinguished and described types 1–5 in a 1953 publication. Types 1–3 seemed to be similar enough to consider them members of the same genus, and types 4 and 5 were considered to represent a second genus, as indicated in Figure 5.13a. Here, the types are numbered in order of their description, and the genus/ware to which they belong is indicated by an X or Y.

Another researcher collected a second sample in 1960 and found that some specimens fell in types 1, 2, and 5, but none seemed to fall in types 3 and 4, *if* the original type descriptions were retained and used as (extensional) definitions. Furthermore, some specimens fell in none of the previously identified types, so new types—6–11—were created and described in a 1964 publication, in which it was noted that type 8 was rather similar to types 1–3, type 6 was similar to type 3, and type 11 was similar to types 2 and 3 (Figure 5.13b). Furthermore, type 7 was similar to types 4 and 5, type 9 was similar to type 4, and type 10 was similar to type 5. Although all the new types variously seemed to fall under the two previously identified genera, they extended the range of morphological variation encompassed by the original genera. Finally, the new types seemed to form a partial bridge across the morphological gap between the two more inclusive units or genera, as indicated in Figure 5.13b. That bridge was completed when a 1970 expedition recovered specimens that resulted in the identification of type 12 (Figure 5.13c).

As Hazard (1998) emphasizes, one has several options upon the "discovery" of type 12. Does it belong to genus/ware X or Y or to a new genus/ware Z? Does

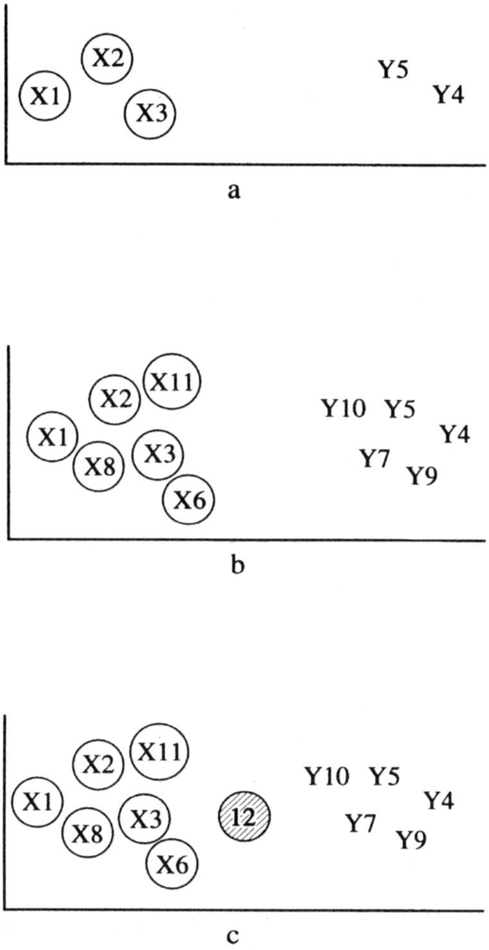

FIGURE 5.13. Results of extensional definition of classes. In a, types 1–5 were identified from available specimens and placed in two genera or wares, X and Y. In b, types 6–11 were identified from new specimens and placed in the existing genera or wares. In c, type 12—an intermediate, or transitional, type—is identified from new specimens. Should it be placed in genus/ware X or Y, or should a new genus, Z, be created? Should genera/wares X and Y be collapsed into one? (after Hazard 1998).

type 12 suggest that genera/wares X and Y should actually be considered a single genus/ware? If so, which one? Hazard's point is simple. Regardless of which of these options is chosen, the analyst has "redefined X and Y. Every time you define something new, here species 12, you implicitly redefine all other categories as 'not 12' " (Hazard 1998:359). Hazard is correct *if* the original X and Y were extensionally defined, which they were. If all types/species and wares/genera were intensionally defined from the start, then one has not redefined anything by adding type 12. Rather, one has merely found members (denotata) for what was previously an empty class, that is, for a class that previously had no empirical members.

Visualizing time as flowing from left to right along the horizontal axis of the graphs in Figure 5.13 should make clear the kinds of problems in measuring morphospace that Gould noted. In case simple imposition does not make it clear, consider paleontologist Vincent Maglio's (1971:371) remark that "where a lineage is better known and morphological overlap is great, it is perhaps better to set up informal stages along the sequence [of forms, rather than to assign species or subspecies names to various chunks]. The latter may be designated by successive numbers or by stage names." Maglio's point is well taken, for what he construes as paleobiology's ultimate goal is "not the recognition of fossil 'taxa' and the establishment of a formal terminology for ever smaller segments of a phyletic continuum. On the contrary, it is the establishment of evolutionary units that can be traced through long periods of time and with which broad evolutionary phenomena can be studied" (Maglio 1971:372). We agree.

Gould (1995a) notes that a way out of the conundrum of measuring morphospace seems to be available. It involves a technique that uses theoretical units (classes) and is described in a series of papers by paleobiologist Mike Foote (1993a, 1993b, 1994, 1997b, and references therein). This technique for measuring morphospace, concomitant with retention of taxonomic designations of the measured specimens, allows insights into the way what is variously filled or empty in multidimensional morphospace—morphospace "disparity"—can be compared to taxonomic "diversity" (actually, "richness"). It has allowed new insights into the evolutionary process because variation both within and between taxonomic units can be monitored simultaneously (Foote 1993a, 1993b; Gould 1993; for an overview of the arguments that led to this insight, see McShea 1993). Thus, one can detect and measure (1) decreases in taxonomic diversity that are coincident with stability in morphological disparity; (2) adaptive radiation expressed as increases in morphological disparity and perhaps, but not necessarily, increases in taxonomic diversity; and (3) other such phenomena. For example, finding that low taxonomic diversity is coincident with high morphological disparity, and that taxonomic diversity increases as morphological disparity decreases within taxa, may suggest increased constraint and less plastic development (e.g., Foote 1994). We return to precisely these issues later.

Classes and Scale

The use of classes construed as theoretical units allows the measurement of change at multiple scales. One can monitor change in the frequency of a particular class of attribute state regardless of whether it forms part of one or multiple classes of discrete objects; similarly, one can monitor change in the frequency of a particular class of aggregate of discrete object classes. Archaeologists seem fairly aware of the former but perhaps less so of the latter. We suspect this results from their using anthropologically conceived large-scale units such as tribes, chiefdoms, and the like. Because many archaeologists seek to explain cultural evolution, we are not too surprised that some of them have borrowed the biological concept of *Bauplan*, which signifies a unit larger than a species (e.g., Rosenberg 1994, 1998; Spencer 1997). In biology, a *Bauplan* is a basic structural design, what "comparative anatomists now call the morphotype" (Mayr 1988:109). The concept appeared early in the nineteenth century, when it was recognized that "animals cannot be seriated in a smooth, continuous chain from the simplest to the most perfect.... Instead, a limited number of discrete types can be recognized, such as vertebrates, insects, and mollusks" (Mayr 1988:405–406; see also Hall 1996). "Within the mammalian *Bauplan* ... such strikingly different functional types evolved as whales, bats, moles, gibbons, and horses, without any essential change of the mammalian plan" (Mayr 1982a:468). Rosenberg (1994:320) defines a culture's *Bauplan* as its "essential *organic design*," its "ideational system," and its "central superstructural elements" (emphasis in the original). Given this notion, he argues that "we can say that a culture remains *that* culture so long as the systemic integrity of its *Bauplan* is maintained" (Rosenberg 1994:320). Spencer (1997:235) suggests that "the categories 'chiefdom' and 'state' are examples of sociopolitical *Baupläne*" and that many anthropologists agree that "the existence of the chiefdom form of organization is a precondition for the emergence of the primary or pristine state in a particular developmental trajectory."

We perceive two potential problems with this borrowing. The first consists of two parts and hinges on the question: What are the archaeological differentia of sociopolitical *Baupläne*? Without a clear and theoretically informed answer, we suspect archaeologists will find themselves open to the criticisms leveled at units such as chiefdoms and states by Robert Leonard and Tom Jones (1987). Their points were that sociocultural *Baupläne* are units extensionally defined on the basis of extant cultures and which encompass so much variation as to be of little use analytically. The first demands the use of configurational analogy to reconstruct the archaeological record into something like an ethnography, and the second underscores the atheoretical nature of the units used to measure variation.

The second problem we perceive with borrowing the concept of *Baupläne* rests on the fact that Darwin's "theory of common descent [is] a strictly hierarchical theory" (Mayr 1989:153). A *Bauplan* is a large-scale hierarchical unit and can be one of several levels of inclusiveness; that is, the taxonomic level of a *Bauplan* can vary. Archaeologists who have borrowed the concept have done so without considering the appropriate scale or level of *Baupläne*. Thus, we fear that such units may become little more than essentialist kinds, given that archaeologists seem to be lost in this ontology and intent on building analytical units extensionally (inductively) from specimens at hand rather than from theory (Lyman and O'Brien 1997, 1998; Lyman et al. 1997b). Under a materialist ontology and the ideational units this demands, units not unlike *Baupläne* may be useful as variously inclusive hierarchical units in a taxonomy of cultural units. This is precisely the purpose for which the concept was invented in biology, and thus one can speak of the vertebrate *Bauplan*, the less inclusive mammalian *Bauplan*, the even less inclusive rodent *Bauplan*, and the like (Hall 1996). But the critical point to keep in mind is that a *Bauplan* at whatever level of inclusiveness is an ideational unit much like an inch or a centimeter; it is a unit of measurement, nothing more, and must be constructed from theory rather than extensionally derived from phenomena. Paleobiologists in particular find such units useful (Foote [1996] and references therein), and there is no reason archaeologists might not find them equally useful *if* they keep in mind that *Baupläne* are ideational units. Failing to do so may result in cultural *Baupläne* being little more than a modern expression of the Midwestern Taxonomic Method.

Other Kinds of Units

Various critics of what we are outlining in this book have suggested that we and those of like mind too often fail to take into account cultural transmission as a significant driving force behind the evolution of cultures (e.g., Boone and Smith 1998; Schiffer 1996; Spencer 1997; see particularly the discussion in Lyman and O'Brien 1998). Such criticisms are easily dispensed with by adopting a line of reasoning developed by David Hull (1980, 1981b, 1982, 1988), who was concerned with what are variously termed units of selection or units of evolution—a contentious issue in biology (e.g., Sober and Wilson [1994] and references therein). Richard Lewontin (1970) had more or less argued that biological units of any scale—genes, organisms, colonies, populations, species, communities—could be the units of selection and evolution. Hull (1980:311, 1981b:41) detected a problem with this "traditional organizational hierarchy" because of debates over group selection (e.g., Williams 1966): "The current majority view is that genes mutate, organisms are selected, and species evolve" (Hull 1981b:23). Hull wondered if a different set of units might work better, because it was clear that genes seemed to be only one of several units of transmission—chromosomes or large portions thereof, for example—and there also seemed to be cases in which a phenotype—typically construed as *the* unit upon which selection works (e.g., Mayr 1963: 184)—could be an organism or a population thereof. Could these kinds of problems be resolved by rethinking the issue?

Evolution requires heritability and thus replication, and it also requires differential replication if it is driven by selection. The latter will occur as entities are more or less successful at dealing with their environment and thus at surviving and reproducing. "Both processes are *necessary* for evolution by natural selection to occur. In most cases, entities at *different* levels of organization perform these two functions" (Hull 1981b:26; emphasis in the original). The problem with units of selection and evolution, then, Hull (1980, 1981b) reasoned, was one of scale. To escape that problem, he distinguished between two basic sorts of units without specification of scale; he defined a *replicator* as "an entity that passes on its structure directly in replication" and an *interactor* as "an entity that directly interacts as a cohesive whole with its environment in such a way that replication is differential" (Hull 1980:318; see also Hull 1981b:33). These two units, then, represent "two quite distinct functions" (Hull 1981b:30). From the perspective of evolutionary change, "the structure of replicators is differentially perpetuated because of the relative success of the interactors of which the replicators are part" (Hull 1981b:41).

Hull went on to note that (1) "replicators replicate themselves directly but interact with increasingly inclusive environments only indirectly. Interactors interact with their effective environments directly but usually replicate themselves only indirectly" (Hull 1980:319), and (2) only the structure of replicators "persists, and that is all that is needed for them to function as replicators" (Hull 1980:320). Here, Hull (1981b:31) noted that fidelity in replication concerned only *structure*, as "neither material identity nor extensive material overlap is necessary for copying-fidelity." Also, (3) *selection* could be defined as "a process in which the differential extinction and proliferation of interactors cause the differential perpetuation of the replicators that produce them" (Hull 1980:318), and (4) a *lineage* is "an entity that changes indefinitely through time as a result of replication and interaction" (Hull 1980:327). Here, "for selection to take place, spatiotemporal sequences of replicates are necessary. Similar entities alone won't do; neither will spatiotemporal sequences of entities alone" (Hull 1981b:31–32). Finally, (5) replicators and interactors could each be specified at a multitude of scales (for an exemplary list, see Brandon 1990:97): "A replicator must be small enough to retain its structural pattern through numerous replications, yet large enough to have a structural pattern worth preserving" (Hull 1981b:32). Later, Hull argued that in order to study evolution with useful units, the "appropriate levels [read *scales*] are not genes, organisms, and species as they are traditionally conceived, but replicators, interactors, and lineages. Evolution needs to be viewed as a hierarchical process. The issue is the character of this hierarchy. My claim is that the regularities that elude characterization in terms of genes, organisms, and species can be captured if natural phenomena are subdivided differently: into replicators, interactors, and lineages" (Hull 1988:26).

Hull's (1980, 1988) idea has been accepted by others (e.g., Brandon 1990; Mitchell 1987), but whether or not it proves to win the day in biology and biological philosophy, it certainly has utility in evolutionary archaeology. Replicators and interactors are both represented in the notion of a culture trait. They are equally important, we think, in paleobiology and archaeology. Paleobiologists assume a change in genes (replicators) is represented by late Pleistocene fossils (interactors) of bison assigned to *Bison antiquus* and middle Holocene fossils assigned to *Bison bison*. They do not, however, *know* this to be the case, although they do not argue about it, nor do they worry about identifying a precise correspondence between one or more replicators and the particular formal attributes of the interactors used to identify fossils as belonging to a particular species. They do not worry about it because of such things as pleiotropy—one gene influencing multiple phenotypic traits—and polygeny—multiple genes influencing one phenotypic trait. Thus, we find the change from one "type" of projectile point or pottery to another to represent a change in replicators—some would call them memes (e.g., Dawkins 1976; Hull 1982)—as manifest by interactors (the points or pottery) (Lyman and O'Brien 1998). Such changes may or may not have been coincident with a change in the gene pool of the point makers (Neff 1998) because the replicator units that result in the production of projectile points and pottery are separate and distinct from the replicator units that produce organisms. This does not mean that the fitness of the persons manufacturing the points or pots *was not* affected; rather, it means that we need to examine the case further (Chapter 8). There no doubt are cultural processes analogous to pleiotropy and polygeny (e.g., Heyes and Plotkin 1989), so we cannot expect a particular cultural attribute to reflect precisely a particular replicator. All we need to know is that replicators and interactors are two distinct kinds of units and that they have different purposes in evolution.

The utility of the concepts of replicators and interactors resides in the fact that without specification of a particular scale, they distinguish between units of transmission and units upon which selection can act directly, respectively. Selection will influence the relative frequencies of classes of interactors within a population over time; the vagaries of transmission will influence the relative frequencies of replicators within a population over time. Of course, sometimes the two may comprise the same entity (Brandon 1990; Hull 1980, 1981b), but we need not worry about this either. One might protest, then, that given that inheritance is required, inheritance need not result in perfect fidelity yet is identified as similarity. In other words, "How far can one bend the fidelity requirement and still have replication?" (Heyes and Plotkin 1989:150). Identifying replication and its degree of fidelity are topics that we take up in the next chapter; here, it suffices to note that archaeology has possessed the analytical tools for doing this for eighty years, although the majority of archaeologists have not recognized those tools as such. The units of interactors and replicators are important conceptually and, in slightly modified form, critical to using those particular analytical tools. Identifying particular classes of artifacts as interactors, for example, allows us to hypothesize causes of natural selection (Wade and Kalisz 1990). Identifying other classes as more reflective of replicators allows us to identify transmission pathways.

BEYOND UNIT CONSTRUCTION

The construction of paradigmatic classes and the creation of groups with them gives us an objective means to measure variation—Gould's "morphological disparity" *and* "taxonomic diversity"—and to plot that variation in time and space, but we do not construct classifications simply to measure variation. The ultimate purpose is to stack that measured variation up against time so that we can begin to understand not only how changes occurred but also why they occurred when they did. To us, the end result of any evolutionary study in archaeology is explanation, but to explain something in evolutionary terms first requires detailed histories of organisms and their phenotypic features. In archaeology, those phenotypic features are artifacts—stone tools, ceramic vessels, and the like. By constructing lineages of artifacts, we are simultaneously tracking the effects of evolutionary processes—for example, selection and drift—on the possessors of the objects.

Again, we see no difference between this approach and what any paleontologist does. Paleontologists, though, might argue that they are studying more or less complete organisms and not simply a single feature of an organism, and we would make the same argument. Archaeologists interested in taking an evolutionary approach do not examine only single features, though any particular study might focus exclusively on a single feature, just as a paleontological study might focus exclusively on the evolution of teeth in the family Equidae (e.g., MacFadden 1992). But like the paleontological record, the archaeological record usually preserves only the hard parts of the phenotype; thus, neither discipline really ever has access to complete organisms. If the central goals of paleontology and archaeology are the same—writing histories of organisms through the analysis of phenotypic features—then there is no reason not to believe that the means to achieve the goals exhibit considerable similarity.

Chapter 6

Building and Testing Historical Lineages

> A classification is an arbitrary procedure; the grouping or categorizing of phenomena reflects the attitudes of the classifier toward his data rather than any inherent "truths" in the materials themselves....
>
> In establishing the culture "periods" as the major classificatory device of this report we are operating with the basic assumption that culture changes through time. The culture period is the means of measuring and describing cultural forms, both material and non-material, as these have existed in a time continuity.... [W]e have defined these culture periods by their own content. (Willey 1949:3)

One does not have to pledge allegiance to the materialist epistemology in order to understand the simple fact that time is a continuum. Archaeologists have always recognized this, just as most of us have realized that the classifications we use are, to one degree or another, arbitrary constructions. Some archaeologists have argued on occasion that certain units were not arbitrary—Albert Spaulding's pottery types, for example—but these were, for the most part, minority opinions. This is not to say that all archaeologists have been as opinionated on the subject of the arbitrariness of units as, say, James Ford, but by and large, most would find nothing wrong with the above-cited comments by Gordon Willey in his introductory remarks in "Archaeology of the Florida Gulf Coast," written between 1942 and 1948 (Willey 1949). Time is, as Willey pointed out, a continuum, but to make sense out of it, that is, to make archaeology more than an intellectual exercise, we need to carve it up into units that can then be studied. Continua cannot be studied except from afar, for the simple reason that there are no natural entry points into them. We can, though, metaphorically dam the flow of time at certain points and thus create artificial segments. But there is a troubling aspect to this: Archaeological usage makes the segments appear as if they are real. Although most of us know and understand the difference between things that are arbitrary and things that are real, we continually conflate the two when it comes to getting archaeological work done.

Look at what Kent Flannery (1986:506–507) had to say on the subject, with reference to his own work at the preceramic site of Guilá Naquitz in the Valley of Oaxaca, Mexico. The curves to which Flannery refers are graphs of efficiency of a simulated collector/forager group before and after the appearance of cultivated plants in the archaeological record:

> As anthropologists, what we claim we most want to know about are the processes underlying these curves. In our grant proposals we talk about "preagricultural adaptation," about agriculture "reaching the takeoff point," about ancient cultures "achieving a new adaptive plateau." One would therefore expect that our preceramic chronologies would be based on the major landmarks of these sigmoid curves. We might expect to hear statements such as, "Guilá Naquitz E lies near the top of the curve's upswing, just before it levels off, while Guilá Naquitz D lies on the stable plateau formed after the upswing levels off." ...
>
> Do we hear such things? We do not. Instead, we have living floors assigned to archaeological phases that are based on projectile point styles....
>
> Thus, the whole sequence of plant collecting, incipient cultivation, and gradually developing preceramic agriculture in the valleys of Oaxaca and Tehuacán has been broken down into time segments based on stylistic changes in deer-hunting equipment.... We are therefore confronted with a paradox: the processes we wish to document proceed as a series of logistic curves, while our chronologies are composed of linear phases based on stylistic changes in artifacts that may have little or nothing to do with these processes.... [S]o long as our evolutionary sequences are tied to stylistic phases, we have an unsolved dilemma.

Without using the term, Flannery summed up the *materialist paradox* perfectly: Archaeologists want to explore evolutionary issues, but they resort to using the wrong kinds of units to do it. When we hear terms such as *preagricultural adaptation* or see references to agriculture "reaching the takeoff point" and ancient cultures "achieving a new adaptive plateau," we might expect to find some discussion of evolutionary mechanisms such as selection and drift, because things such as adaptive plateaus fall under the purview of evolutionism. Even if archaeologists are not as well versed in evolutionary theory as biologists, we might expect that somewhere along the way at least one archaeologist would have adopted

Sewall Wright's (1931, 1932) notion of adaptive peaks, especially after George Gaylord Simpson (1944) popularized the concept. If someone *had* adopted it, he or she might have realized that the body of theory underlying Wright's and Simpson's work had some application to archaeological instances of "adaptive plateaus." But when humans are viewed simply as goal-directed organisms, such things as agriculture become tagalongs—ideas that are carried subconsciously in people's heads, to be hatched when needed. Thus, the search for agricultural origins becomes an exercise geared toward unraveling the orthogenetic pathways that particular crops took on their inexorable journeys toward eventual domestication by humans (and vice versa) (O'Brien 1987; O'Brien and Wilson 1988; Rindos 1984).

Without undermining Flannery's position, we need to point out that there is nothing wrong with tying such things as logistic curves to stylistic changes in artifacts. In fact, this is a necessity if we ever expect to construct historical sequences. We doubt Flannery would disagree with this point. Rather, he is implying that this is where things usually stop archaeologically. Most chronologies are nothing more than a vertical series of units—phases, cultures, stages, and so forth—that themselves are variously based on perceived but rarely measured similarities and differences among artifact assemblages. These units might in some cases be useful shorthand devices, but they have no place in an evolutionary study because they are the wrong units with which to measure change.

How do we do better? One way is by employing units designed to measure change—units such as paradigmatic classes and the groups derived from them. By now we should have made it clear that groups identified through the use of classes contain specimens that are identical in terms of significata. Hence, we know with 100 percent certainty that any time–space location mapped as carrying a specimen of a particular group actually exhibits a member of that group—a denotatum—as opposed to a specimen that sort of looks like other specimens in the group. This, as we have seen, is virtually impossible using extensional types. Most archaeologists would agree that if there is a superior method of keeping track of variation temporally and spatially, then they probably would adopt it, but the problem is getting everyone to adopt the same system so that we are not constantly reinventing the wheel every time we want to examine variation. Without consistency in measurement, results are incomparable. One of the reasons types were constructed in the first place was to ease communication among investigators, but the extensional nature of types only adds to the lack of comparability except at the most general level. Regardless of how widespread or localized the classificatory efforts, identifying some way to measure variation objectively and replicably is the first step in building models of change. We refer to these models of change as historical lineages because the concept of *change* as opposed to simply *difference* entails the notion of heritable continuity. These are concepts that we had best be clear on from the start.

BIOLOGICAL LINEAGES AS HERITABLE CONTINUITY

We have defined a *lineage* as a temporal line of change owing its existence to heritability. To borrow Darwin's (1859) famous phrase, such lines of evolution represent "descent with modification." Similarity of descendant phenomena to their ancestors is the result of a genetic connection—transmission—between the two. There is replicative fidelity, and like begets like as a result. Thus, the more similar two things are to each other, the more character states they hold in common, the more closely related they are thought to be. Conversely, the more dissimilar they are, the more distantly related they are thought to be. But as we noted earlier, Simpson (1961) and others (e.g., Mayr 1981) indicated that a degree of care is required here; we want to avoid the pitfall represented by the interpretive axiom that two things are related because they are similar. Rather, we must keep in mind that what underpins such assessments is the notion that two things are similar because they are related phylogenetically. Thus, Gould (1991b:414), correctly in our view, indicates that "temporal branching order and morphological disparity are separate issues (empirically related only by the high probability of a rough correlation between cladistic and phenetic distance [read *formal similarity*]). Branching order is the cladistic component of 'relationship' among taxa; disparity is a phenetic component of the same ambiguous term."

In short, we refer to the two issues that must be dealt with as *heritable continuity* and *similarity*, and we consider them in turn. But before we do that, we need to introduce two concepts that will be useful in our discussion. Biologists who write phylogenetic histories assume heritable continuity at two scales. First, given genetic transmission between individuals comprising the species, each fossil specimen belonging to a particular species is hereditarily continuous with every other fossil specimen identified as a member of that species. We refer to this as the *type/species* scale of heritable continuity (continuity resulting from heredity). Second, congeneric species—that is, distinct species belonging to the same genus—denote heritable continuity, again given genetic transmission and also in this case evolutionary divergence from a common ancestor (the same applies for confamilial genera, and so on). We refer to this as the *tradition/lineage* scale of heritable continuity.

Heritable Continuity in Organisms

Similarity of phylogenetically related organisms is a result of replication through genetic transmission. During reproduction, organisms transmit genetic material, creating either an offspring that is an exact copy of the parent (asexual reproduction) or an offspring that has characteristics of both (sexual reproduction). Reproduction is more or less faithful, and the offspring is either very similar or identical genetically and phenotypically to the parent or parents. Over time, because of transmission errors, mutation, and/or recombination, the organisms comprising a population (or species) change. These changes might not be detectable from one generation to the next, but after sufficient time, we notice that the two ends of the lineage comprise dissimilar individuals, whereas individuals adjacent to one another in the lineage are virtually identical. As we saw in Chapter 4, if the individuals at any two points along the continuum are sufficiently different, they would be assigned to two separate (chrono)species.

Paleontologists have no way of determining whether the fossil organisms they are studying could have interbred with taxonomically similar fossil organisms a few centimeters higher or lower in a stratum (Pearson 1998b; Rose and Bown 1986). In other words, they cannot tell how many biological species might be represented. Simpson (1943:174) did not worry about this a great deal and instead indicated that the heritable continuity represented by a temporal sequence of genetically related forms—a lineage—comprised a chronocline. Some paleobiologists refer to the arbitrary chunks of a chronocline as "chronospecies." If the cline is documented across geographic space, it is a "chorocline" in Simpson's (1943:174) terminology. So, how do we trace lineages among sets of fossils thought to comprise several species?

Day-to-day changes in the composition of the species rendered as turnover in the individuals comprising its population(s) not only drive the evolution of that species but leave traces—manifest as morphological variation—that allow us to reconstruct its evolutionary pathway. Piecing together lineages, measuring rates of evolutionary change, and trying to understand the processes that create change are part and parcel of the world of paleobiology. At their core, the specific methods and techniques that have been devised to trace historical lineages revolve around identifying similarities and differences in character states of organisms. A *character state* (we also use the term *attribute state*) is the condition of a *character*, or dimension, of variation. At any particular time, all organisms in a species might have, for example, five digits, but in a succeeding generation, some might have six digits. Such an appearance constitutes a change in the state of the character "number of digits." This simple example of a character gradient should make it clear that measurable changes in characters—a shift from one state to another—are useful for both documenting the passage of time and correlating strata temporally based on their fossil content (e.g., Eldredge and Gould 1977).

But paleobiologists are interested in creating an order of specimens that, when viewed overall, measures the passage of time *within a lineage*; that is, they do not want simply a historical sequence of forms, which could be provided by a superposed sequence of specimens in a stratified column of sediments. They want not only a temporal *sequence* but also a *line of heritable continuity*. In other words, the line must comprise specimens that are related to one another as a result of transmission. Because paleobiologists cannot track change in genes, they distinguish between *analogous* characters and *homologous* characters in order to assess relatedness. The former, characters that two or more organisms possess that, while they might serve similar purposes, did not evolve because of any common ancestry. Birds and bats both have wings, and those characters share properties in common, yet we classify birds and bats in two widely separate taxonomic groups because birds and bats are only distantly related; those two large groups diverged from a common vertebrate ancestor long before either one of them developed wings. Thus, we say that bird wings and bat wings are analogous structures, or analogs for short. They are of no utility in reconstructing lineages because they evolved *independently* in the two lineages after they diverged. The character of having wings is held in common by birds and bats, but the *state* of the character—the details of its osteological composition and anatomical structure—differs between the two groups.

Homologous characters, or homologs for short, *are* useful for tracking heritable continuity because they are holdovers from the time when two lineages were historically a single lineage. For example, all mammals have a vertebral column, as do animals placed in other categories. The presence of vertebrae is one criterion that we use to place organisms in the subphylum Vertebrata. The vertebral column is a homologous character shared by mammals, birds, reptiles, and some fishes, and it suggests that at some remote time in the past these organisms shared a common ancestor. Biologists did not always view homologs this way, and we need to comment briefly on the history of this concept before proceeding. Knowing that history helps us understand Simpson's caveat that two things are similar because they are related rather than the converse.

The published definition of homologous structures often cited today preceded Darwin's *On the Origin of Species* by sixteen years. In 1843, Sir Richard Owen defined a homolog as "the same organ in different animals under every variety of form and function" (Panchen 1994:40). Detailed analysis of the early nineteenth-century literature indicates that Owen's conception of homolog originated much earlier and that rather than denote historical or phylogenetic affinity, it was taken to mean representative of a common or archetypical plan (Boyden 1943; Brady 1985; Hubbs 1944; Padian 1997; Panchen 1994; Rieppel 1994). By archetypical plan, or archetype for short, Owen meant basically an ideal, generally primitive, form (for historical details, see Desmond 1984). Owen later distinguished between "homotypy"—the "same" organ in various organisms—and "general homology"—resemblance of an organism to an archetype (Donoghue 1992:170), thereby explicitly including variation in scale in his concept.

The important thing to note is that the early nineteenth-century biological concept of homolog "did not imply anything more than purely formal relation" (Szarski 1949:124). In short, resemblance—of some unspecified degree—in form denoted homologous structures, regardless of scale. Darwin's insight provided a new and logical causal explanation as to *why* there would be formal similarities between organs and organisms and also why the Linnaean taxonomy was hierarchical. Darwin 1859:206) argued that "by unity of type is meant that fundamental agreement in structure, which we see in organic beings of the same class, and which is quite independent of their habits of life. On my theory, unity of type is explained by unity of descent." In two short sentences, Darwin clearly distinguished between analogous and homologous characters and provided the first explicitly scientific and theoretical explanation for the existence of homologs: Homologous similarity is historical for the simple reason that it results from heritable continuity. Homology in this sense denotes what post–Darwinian biologists interested in evolutionary descent have identified as "special," or "phylogenetic," homology; similarities are the result of common ancestry and heritable continuity (e.g., Haas and Simpson 1946; Moment 1945; Patterson 1982; Sluys 1996). With such a definition, a biologist after 1859 could answer the following questions: Why are the wings of robins and crows similar, yet the two organisms are different species? Why are the wings of little brown bats and fruit bats similar, yet bird wings and bat wings are superficially similar but structurally different?

Similarity of Organisms

Darwinian evolutionary theory provides the explanation for special homology—an explanation found in the definition of the concept. But the definition does not tell us how to identify instances of special homology. Not surprisingly, then, there is disagreement on procedure within biology, though it is generally agreed that the results should be testable (e.g., Brady 1985; McKitrick 1994). As one should guess from Simpson's caveat, operationalizing the concept in biology does not rest solely on formal similarity: "Homologous structures may be extensively similar or very dissimilar. Therefore similarity is to be considered something quite apart from considerations of homology.... Similarity does indeed suggest homology, but dissimilarity is also abundantly evident in homology. The concept of homology cannot logically be tied into any definitions of similarity or dissimilarity" (Smith 1967:101). Or as Ernst Mayr (1981:511) put it, "So-called similarity is a complex phenomenon that is not necessarily closely correlated with common descent, since similarity is often due to convergence." These statements do not mean that mere morphological or formal similarity is unimportant. To the contrary, such "similarity is the factor that compels us to *postulate* homology" (Cracraft 1981b:25). As Colin Patterson (1988:604) indicates, "Morphologists would agree with [the] point [that similarity] can be factual, whereas homology 'must usually remain an hypothesis.' " Of course, if it is a hypothesis, then it is testable.

Whatever the procedure for recognizing fossil species, it assumes heritable continuity at the type/species scale. Given that designating a species also designates membership within all higher levels of the Linnaean taxonomy, heritable continuity is assumed at those levels as well. This is heritable continuity at the tradition/lineage scale. One might wonder, then, why we have distinguished between these two scales. It is simply because the validity of the assumption—at either level—depends on the underlying systematics. As Joel Cracraft (1981a:461) well noted, "The classificatory methodology underlying the taxonomy of the groups may have an influence on the number of [taxa or groups] recognized at various stratigraphic levels." If a set of units low in the taxonomic hierarchy, say, genera, is incorrectly subsumed under a single unit higher in the taxonomy, say, an order, when in fact they should be subsumed under multiple orders, then clearly any measurement of the richness of orders will be influenced. Furthermore, if a taxon is assigned to the incorrect rank, then an incorrect measure of taxonomic richness at other ranks will result. As Mayr (1981:514) put it, "Rank determination is one of the most difficult and subjective decision processes in classification." Paleobiologist Mark Norrell (1993:414) put it this way: Assignment of fossils to various high taxonomic ranks are "most of the time ... subjective [assignments] independent of a detailed knowledge of phylogeny." Mike Foote (1996:5) added, "Factors that go into assigning taxonomic rank, from morphological divergence to taxonomic convenience, are quite varied." In short, one's systematics must contend with both phylogenetic history *and* position within the taxonomy hierarchy. We suggest part of this problem can be alleviated by careful definition of one's field of observation.

As should be clear from previous discussion, use of different differentia (significata) will result in the definition of different groups of specimens and potentially in an assignment to different units within the taxonomic hierarchy. Among all the potential differentia, which should be chosen in order to contend with phylogenetic history and position within the taxonomy hierarchy? If we are interested in phylogenetic relations and thus heritable continuity, it is clear that we must start with homologs. But there is much more to it than that. Prior to the late 1960s, paleontologists spoke of ancestral (primitive) and derived traits. Evolutionary innovations comprise derived traits, which by definition occur later in time than their ancestors. Derived traits are the critical ones and comprise a particular kind of homolog, what is called a *shared derived* character, or *synapomorphy*; shared ancestral characters, or *symplesiomorphies*, are another kind of homolog, but they are not used when building phylogenies. The latter characters, such as the vertebral column mentioned earlier, are indeed homologous and suggest the relatedness of all vertebrates, but they do not help in determining the phylogenetic relations between the various groups of Vertebrata, because they are shared by all members of those groups; we suggest such character states can be used to define fields of observation and *Baupläne*. Synapomorphies are critical because they comprise "*the* way to recognize branches of the phylogenetic tree"; they provide "an explicit methodology that allows us to evaluate hypotheses of monophyly," or common ancestry (Eldredge and Novacek 1985:67–68).

The use of synapomorphies to determine phylogenetic relations is founded in the notion of evolutionary descent with modification. In short, synapomorphies are homologous traits and as such "are nested. From the point when a feature [character state] first appears in phylogenetic history, it will be passed along in some guise (i.e., in the same or transformed version) to descendent taxa. [By] mapping the distribution of such attributes[,] monophyletic taxa—branches of the tree of life—are delineated, defined, and recognized" (Eldredge and Novacek 1985:67). Such taxa comprise "actual historical units" (Eldredge and Novacek 1985:66). How, then, does one recognize synapomorphies in the paleobiological record?

Cladistics and Synapomorphies

There are two basic approaches in biology to identifying instances of special homology, each founded in a unique epistemology. Criteria typically used by those following a cladistic approach are *positional*, or what is referred to as topological, identity; *typological*, not necessarily functional, identity; *ontogenic* identity, which seems to be the least frequently used criterion; and *congruence*—multiple traits, especially traits independent of one another (e.g., Mayr 1981, 1995b), should display the same ancestor–descendant relations between organisms. Although there are potential flaws in these criteria (e.g., Striedter and Northcutt 1991), they do not seem to be fatal, meaning they allow at least an inference of homology to be made. Cladists agree that the validity of the inference must be tested in various ways, such as by using each criterion as a test of every other criterion (Brower and Schawaroch 1996; Forey 1990).

Noncladistically inclined biologists—those usually referred to as evolutionary taxonomists—prefer to identify characters as homologous "if they can be traced back to the same state in the common ancestor" (Mayr 1969a:85; see also Bock 1977; Mayr 1981, 1995b; Simpson 1975). Evolutionary taxonomists point out that the positional and typological criteria used by cladists are necessary but insufficient because "homologous structures are by no means necessarily similar" in terms of these attributes, as in the case of mammalian ear ossicles and the jaw bones of lower vertebrates (Mayr 1969a:85). Evolutionary taxonomists argue that one must be able to trace through "transitional" specimens the shift in position and form of, for example, a bone from serving as a part of the chewing apparatus to serving as a part of the hearing apparatus (for a lucid account of these particular homologous structures, see Crompton and Parker 1978). This is a valid point and one that occupies our attention in Chapter 8.

Some cladists argue that the temporal polarity of two apparently related traits—such as might be suggested by their order of appearance in a stratigraphic column—may be misleading as a result of the way in which the fossil record was formed or sampled (e.g., Stevens [1980] and references therein). Thus, they argue that one should completely ignore the apparent temporal positions of traits and instead focus on the distribution of the traits in question across different taxa. Depending on which taxa are compared, a particular trait may be either ancestral or derived. Mark Ridley (1985:83) provides an excellent example: "If we compare amphibians with fish, the pentadactyl limb is the derived state and fins the primitive state of this trait; but if we compare an amphibian with a horse, the five-toed state becomes primitive and the one-toed state in the horse is now the derived. Similarly if we compare the hand of a human with the front foot of a horse, the pentadactyl human hand is in the primitive state relative to the single-toed equine foot."

What is important to realize is that when attempting to determine phylogenetic relations, "derived traits are selected because shared derived traits [synapomorphies] indicate common ancestry, whereas shared primitive traits [symplesiomorphies] do not" (Ridley 1985:83). Continuing with Ridley's (1985:83) example,

> Suppose that we wish to classify a five-toed lizard, a horse, and an ape in relation to each other. The ape and the lizard share the trait "five-toed," but this does not indicate that they share a more recent common ancestor than either does with the horse: the trait is primitive and does not indicate common ancestry within the group of horse, ape, lizard. Whenever there is an evolutionary innovation, it is retained (until the next evolutionary change) by the species descended from the innovator species: shared derived traits [synapomorphies] do indicate common ancestry. That is why they are used to discover phylogenetic relations.

Of course, as Ridley recognizes, this example assumes a particular evolutionary course. How can we use synapomorphies—how can we in fact recognize them—when we do not know the course of evolution, the very thing we are trying to discern? The solution many cladists use is called outgroup comparison: "The outgroup should be a species which is not more closely related to one of the two species than the other, that is why it is an *out*group: it is separated from the species under consideration" (Ridley 1985:84; emphasis in the original). If, say, we want to determine between a mammal and an amphibian whether lactation is an ancestral or a derived trait, we could compare them with a fish, the outgroup: "By the method of outgroup comparison, that trait is taken to be primitive which is found in the outgroup" (Ridley 1985:84). Thus, the absence of lactation is the ancestral trait, whereas the presence of lactation is the derived trait. The problem with outgroup comparison is convergence. What if the taxon comprising the outgroup has evolved in such a manner as to entail a derived trait that is similar to the trait of interest in the two compared taxa? Comparing the body shape of dolphins and dogs and using fish as the outgroup would lead one to suppose that the dog had the derived body shape, when in fact it is clear on the basis of historical evidence that the dog's body shape is ancestral and dolphin body shape merely converged—for functional reasons—on the same body shape as that of fish.

A degree of circularity is built into the cladistic technique of outgroup comparison because it presumes that we *know* a priori that the chosen outgroup is no more closely related to one group than it is to any other group under consideration. In other words, it assumes something about what it is we are trying to discover. This has resulted in observations such as that of Eldredge and Novacek (1985:68) that "cladistics is no panacea. It is hard to do a good piece of systematics research using cladistics—indeed, harder than ever before. The system is logically much more rigorous, requiring every available item of evidence." As Gould (1986:66) remarked, "Cladistics is the science of ordering by genealogical connection, *and nothing else*" (emphasis in the original). Its difficulty, he continued, is that it has often been applied to "inappropriate data," by which we believe Gould means nonhomologous attributes. The available items of evidence now regularly used to lessen the circularity include the congruence criterion (prompting the development of computer programs such as MacClade™ for manipulating multiple variables), an assumption of parsimony—that particular innovations will appear a minimum number of times (Sober 1988)—and independent temporal data such as that derived from superposition to test indications of trait polarity suggested by cladistic analysis (e.g., Donoghue et al. 1989).

Both cladists and evolutionary taxonomists pay attention to the number of characters that might denote homologous similarity, believing that many are better than a few (the test of congruence [Rieppel 1994]), and they agree that the "kind, extent, and amount of similarity provide the primary data of historical science" (Gould 1986:66). A growing awareness among cladists that space and time are important has resulted in what is called stratocladistics (e.g., Fisher 1994; Smith 1994); this involves the use of temporal data, such as from superposition, to help construct a cladogram of evolutionary divergence rather than constructing a cladogram based solely on intrinsic formal attributes (e.g., Benton and Hitchin 1997). Evolutionary taxonomists are studying character states more closely and in terms of concepts developed by cladists rather than merely examining the spatio-temporal distribution of formal characters (e.g., Jernvall et al. 1996; Szalay 1994).

An example will show that what Gould (1991b) would probably term the phenetic–cladistic distinction is a serious problem. Figure 6.1 illustrates three possible historical patterns—two are termed *phenograms* and one is termed a *cladogram*—for four taxa (A–D) and five characters (1–5). The first two measure similarity only; the last is an attempt to measure phylogeny. As shown in the matrix at the lower left, for each taxon the five characters are in one of two states, designated by a lowercase letter with and without a prime. The matrix in the middle shows the number of shared character states between pairs of taxa. Which pattern is correct historically, that is, phylogenetically? Perhaps phenogram 1 is correct, since it denotes a close phenetic relationship between taxa A and C, which

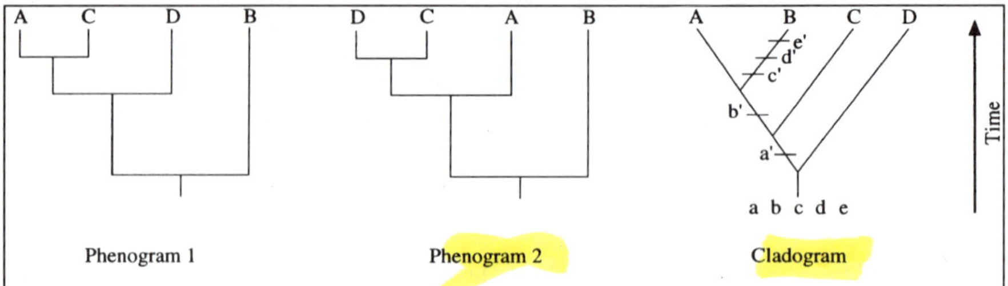

Taxa	Characters 1	2	3	4	5
A	a'	b'	c	d	e
B	a'	b'	c'	d'	e'
C	a'	b	c	d	e
D	a	b	c	d	e

	A	B	C	D
A	–	2	4	3
B		–	1	0
C			–	4
D				–

	A	B	C	D
A	–	2	1	0
B		–	1	0
C			–	0
D				–

FIGURE 6.1. Comparison of two phenograms (upper left and middle) with a cladogram (upper right), based on the hypothetical states of five characters in four taxa (lower-left matrix). The phenograms are constructed from the total number of character states shared by any pair of taxa (lower-middle matrix), whereas the cladogram is constructed from only the derived character states (marked by primes) shared by pairs of taxa (lower-right matrix). The character–state transitions are marked on the cladogram. Note that the evolutionary tree implied by the use of shared derived characters differs from those implied by the use of all shared characters (after Futuyma 1986).

have four character states in common, but then again, it splits out taxon D, which has three states in common with taxon A and four with taxon C.

Maybe phenogram 2 is better in that it minimizes the phenetic distance between taxa D and C and shows that taxon A is phenetically closer to D and C than any of them is to taxon B. Taxon B shares no states with taxon D—thus, we maximize the distance between them; one state with taxon C—thus, we place C closer to B; and two states with taxon A—thus, we minimize the distance between them. On the face of it, phenogram 2 is far superior to phenogram 1—a result that we probably would have gotten by plugging the data into most cluster-analysis programs. Such algorithms search all the data to find the most equitable solution in terms of minimum–maximum distances (similarity rendered as shared attributes) between pairs. But does phenogram 2 configure the historical evolutionary relationships among the four taxa correctly? We would bet heavily against the correctness of any phenogram that purportedly illustrates phylogenetic relationships. We would make this bet because in the Figure 6.1 example, as is true of most phenograms, we made no attempt to discriminate between synapomorphies—shared derived states—and symplesiomorphies—shared ancestral states.

The pattern at the upper right in Figure 6.1—the cladogram—illustrates the evolutionary history of the four taxa based solely on an analysis of shared derived characteristics, shown in the matrix at the lower right. In other words, here we have identified the ancestral (primitive) state of each of the five characters—lowercase letters without a prime—and we ignore them in favor of the derived states—lowercase letters with a prime. Now taxa A and C share little in common, whereas before they were placed closer together because of the total number of character states they had in common. But three of them were ancestral states, not derived states. Analysis of synapomorphies alone indicates that taxon C split off from the ancestral form that produced A and C at some point when taxon B had not yet come into existence. Cladistical analysis produces what are called *monophyletic groups*, or groups that include all descendants of a particular ancestor plus that ancestor. In the Figure 6.1 cladogram, there are three monophyletic groups, or clades (not counting single taxa): AB, ABC, and ABCD (each would also include the here-unidentified common ancestor). Any other group, ACD, for example, is polyphyletic and thus is disallowed.

HISTORICAL CONTINUITY, HERITABLE CONTINUITY, AND THE STUDY OF ARTIFACTS

Archaeologists have long used changes in artifact form rendered as changes in character states to measure the passage of time, the idea being that stone tools or ceramic vessels, for example, evolve in terms of form. Ford's (1962) illustration of culture change is an excellent heuristic device in this respect (Figure 5.3) because it clearly and concisely displays heritable continuity in vessel form. Such a neat, orderly evolution of form through time could only be reconstructed through detailed research. For example, Sir John Evans (1850) produced what we elsewhere (Lyman et al. 1997b, 1998b; O'Brien and Lyman 1999a) term a *phyletic seriation* of British gold coins, and Sir William Matthew Flinders Petrie (1899, 1901) created "sequence dates" for the Egyptian pottery he phyletically seriated. Similarly, Clark Wissler (1916c) discussed the "genetic relations" of decorative motifs on the moccasins of North American Indians, and A. V. Kidder (1917) phyletically seriated Southwestern pottery (Figure 3.5). All of this research was based on the notion that formal similarity denoted heritable continuity and thus phylogenetic relations.

Heritable Continuity in Cultural Lineages

The notion that formal similarity of cultural phenomena denotes heritable continuity originated with the use of the comparative method in linguistic studies of the late eighteenth and early nineteenth centuries (Leaf 1979; Platnick and Cameron 1977). As Franz Boas (1904:518) observed, "Owing to the rapid change of language, the historical treatment of the linguistic problem had developed long before the historic aspect of the natural sciences was understood. The genetic relationship of languages was clearly recognized when the genetic relationship of species was hardly thought of.... No other manifestation of the mental life of man can be classified so minutely and definitely as language. In none are the genetic relations more clearly established." Boas was, of course, speaking metaphorically about the "genetic relationship of languages," but his point was solid; linguistic similarity was the result of cultural transmission and heritable continuity.

Boas (1904:522) commented on a "theory of transmission" but was contrasting it with another popular theory of the day—the psychological unity of mankind—as two opposing explanations for cultural similarities. He did not explore the differences and similarities of cultural transmission and genetic transmission explicitly. A. L. Kroeber, Boas's student, highlighted the differences between these two forms of transmission when he equated the term *heredity* solely with genetic transmission—because it was from parent to offspring only—and what we today think of as cultural transmission with "the non-biological principle of tradition" (Kroeber 1923a:3). He distinguished between the two because the latter could be from parent to offspring, from offspring to parent, or between siblings. On the basis of this distinction, Kroeber (1923a:8) remarked that " pure Darwinian anthropology would be largely misapplied biology." Such remarks contributed to the eventual discard of any model of cultural evolution constructed with the biological model of evolution and transmission as a blueprint (Lyman and O'Brien 1997). Was there an alternative?

We pointed out earlier that Kroeber (1931a) argued that culture traits were analogous to species and that cultures, which comprise suites of traits, were analogous to faunas and floras. We suspect Kroeber chose to equate species with culture traits because he was thinking that both sorts of units evolve monophyletically, and thus could not be reduced to smaller evolutionarily significant units, whereas cultures are polyphyletic (comprising a constellation of traits). But Kroeber left the critical term *culture trait* undefined, as did virtually all of his anthropological and archaeological colleagues. It seems to have denoted any unit of culture that had empirical manifestations but which could also be culturally transmitted. Teltser (1995c) suggests, correctly we think, that for Kroeber and his contemporaries, traits comprised ideas.

Although culture historians and anthropologists regularly referred to general processes of cultural transmission such as enculturation and diffusion, they were hardly explicit about what was transmitted or why such might be important. In the middle of the twentieth century, Gordon Willey (1953:368) remarked that "theories of culture change and continuity are fundamental to [Americanist] archeological studies" and "the treatment of archeological assemblages in any historico-genetic system has a basis in theories of continuity and change." He did not, however, elaborate on what those "theories" were or point out that the notion of continuity had to rest on heritability and transmission. Albert Spaulding provided a brief but revealing insight into the "theories":

> If we view the ultimate task of archeology as the development of the ability to explain the similarity or lack of similarity of any two [archaeological manifestations], the significance of [classifying those manifestations according to their positions in time and space and their resemblance] is easy to state. All [three—time, space, and form—] can be related to the proposition that culture change is systematic rather than capricious and to the auxiliary proposition that an important basis for the systematic behavior of culture is its *continuous transmission through the agency of person to person contact.* The observed degree of similarity between the two [archaeological manifestations] sets the problem to be solved, and time and geographical position are two discrete systems, each of which offers an independent measure of closeness. (Spaulding 1954c:14; emphasis added)

Spaulding was merely making explicit what had been implicit in the reasoning of early twentieth-century Americanist archaeologists. Irving Rouse (1939: 14), for example, had written about the historical processes of diffusion, persistence, origination, extinction, and replacement; the first two implied heritable continuity, but Rouse did not comment on them in those terms. Alex Krieger (1944: 272) characterized archaeological types as merely "identifying distinct patterns of behavior or technology which can be acquired by one human being from another, and thus serve as tools for the retracing of cultural developments and interactions." Americanist culture historians in the early and middle twentieth century followed their predecessors in the social sciences and seldom referenced Darwin's theory, though they believed at least implicitly that if formal similarity was taken to signify a historical relation between compared items, then heredity or transmission of some sort must be involved. Their belief was the result of using ethnologically documented mechanisms such as diffusion and enculturation to account for typological similarities in the archaeological record. But for want of an explicit and well-developed theory, they tended to put the cart before the horse. Thus, Willey's (1953:363) statement that "typological similarity is an indicator of cultural relatedness (and this is surely axiomatic to archeology), [and thus] such relatedness carries with it implications of a common or similar history" was cause for little concern within the discipline. It should have caused *some* concern because it is clear that Willey's axiom falls prey to Simpson's caution: Two things are not related because they are similar; rather, they are similar because they are related.

Similarity among Cultural Traits

Kroeber well understood the difference between analogs and homologs. He implied that a "true homology" denoted "genetic unity" (Kroeber 1931a:151), arguing that

> there are cases in which it is not a simple matter to decide whether the totality of traits points to a true [genetic, homologous] relationship or to secondary [analogous, functional] convergence.... Yet few biologists would doubt that sufficiently intensive analysis of structure will ultimately solve such problems of descent.... There seems no reason why on the whole the same cautious optimism should not prevail in the field of culture; why homologies should not be positively distinguishable from analogies when analysis of the whole of the phenomena in question has become truly intensive. That such analysis has often been lacking but judgments have nevertheless been rendered, does not invalidate the positive reliability of the method. (Kroeber 1931a:152–153)

Kroeber was suggesting that there are two forms of similarity—one homologous and the other analogous. The former results from shared genetic ancestry and the latter from evolutionary convergence, such as when two genetically unrelated populations of organisms reach a similar adaptive solution. How are the two distinguished? Kroeber (1931a:151) suggested that identifying "similarities [that] are specific and structural and not merely superficial ... has long been the accepted method in evolutionary and systematic biology." Kroeber also recognized the problem of independent invention—analogous similarity—and noted that anthropologists had too often not ascertained if the traits they deemed similar were the result of common ancestry—homology—or the result of evolutionary convergence—analogy. Although Edward Sapir (1916) had spelled out in general terms the analytical procedure and criteria for rendering such ascertainment, they were summarized in some detail by Julian Steward in 1929.

When a cultural trait was found in two or more localities, the criteria were the "uniqueness" or quality of the trait, the presence of a probable ancestral trait in the same geographic area, the quantity of other shared traits, and the geographic proximity of the localities (Steward 1929). Only by close study of each of these criteria could one determine whether a cultural trait in an area had originated there or elsewhere. The uniqueness criterion was often expressed as a trait's complexity; the more complex a trait, the less likely it was independently invented multiple times and the more likely it was a homolog. The issue of scale—the shape of a decorative design, the elements and color(s) it comprised, its placement on a vessel, or some combination of these—never arose. The quantity of shared traits and the geographic proximity of compared localities were thought to correspond directly to the probability of contact. The more traits shared and the closer in space the cultures sharing them, the more likely the traits were homologs. The criterion of the presence of a probable ancestor comprised at least indirectly the inference desired. Not surprisingly, then, rather than indicate how ancestral traits were to be identified, Steward fell back on the other three criteria to help determine if an ancestral trait was present. This probably contributed to Kroeber's (1931a) lament two years later that anthropologists had failed to borrow biological procedures for distinguishing between homologous and analogous traits.

Spaulding was a bit more explicit than his predecessors about what "similarity" meant. What he referred to as "resemblance" meant "actual physical similarity of tools, utensils, remains of houses, and so on. It means in addition quantitative similarity, a concept which includes not only the form and other physical properties of the artifacts but also the relative popularity of artifact types in the [archaeological materials] being compared" (Spaulding 1954c:12). Rouse (1955) pointed out that there were three steps to determining the historical relatedness of archaeological units. First, determine the extent or degree of their similarity or resemblance, but he did not elaborate on how this should be done. Second, determine their degree of proximity in time and space; contiguity in both denotes the potential for contact or interaction and thus the potential for transmission and heritable continuity—in short, a phylogenetic relation. To determine if contact had taken place required the third step, which comprised the distinction between analogous and homologous similarities. From this third step, one could determine the phylogenetic history of the units. But Rouse, like Steward before him, was not explicit about how the third step was to be accomplished, nor did he identify the properties by which similarity or resemblance was to be determined. Spaulding (1954c), as we have seen, mentioned Rouse's first two steps but not the third. Within the discipline generally, the first and third steps of Rouse's procedure were accomplished with Willey's (1953:363) axiom that "typological similarity is [an] indicator of cultural relatedness," which, as one should correctly guess by now, did little to resolve the problem because not only was it typically operationalized as mere formal or typological similarity but it also was susceptible to Simpson's caveat.

Similarity and Heritable Continuity in Artifacts

In 1968, Lewis Binford (1968a:8) identified what he characterized as "a basic, unsolved problem" in archaeology—culture historians lacked a method to "distinguish between homologous and analogous cultural similarities." We agree with Binford's conclusion that to write culture history requires "an overhaul of method and theory [because] traditional methodology and analytical procedures are inadequate for the successful achievement of the stated aims" (Binford 1968a:11). As we have indicated, the overhaul effected by processual archaeologists was not the appropriate one, largely because Binford and his followers did not attempt to design a solution to the problem he identified. Rather, he suggested that a focus on function, or analogous similarity, rather than on homologous similarity would allow access to cultural processes, by which was meant the processes identified a decade earlier by Willey and Phillips (1958)—diffusion, migration, and the like. At the same time that Binford identified the problem, an archaeologist on the other side of the Atlantic came close to solving it.

Although he used rather obtuse terminology, did not list the word evolution in the index to his classic *Analytical Archaeology*, and only mentioned Darwin in the context of his discussion of numerical taxonomy, David Clarke had some significant insights into the problems of measuring similarity and detecting heritable continuity. He well understood the importance of transmission to maintaining heritable continuity, and he anticipated the arguments of evolutionary archaeologists a decade later when he remarked that "it is the artefact maker who feeds back into the phenotypic constitution of the next generation of artefacts the modified characteristics of the preceding population of artefacts, and it is in this way that the artefact population has continuity in its trajectory and yet is continuously shifting its attribute format and dispersion" (Clarke 1968:181, 1978:198; we cite both editions of *Analytical Archaeology* to aid the reader who wishes to track Clarke's discussion in greater detail). Importantly, he explicitly identified Gould's (1991b) phenetic–cladistic distinction when he defined *phenetic relationship* as "relationship by ancestry; transform types from single multilinear time-trajectory, or tradition" (Clarke 1968:229, 1978:493). Here, the term *tradition* had its basic archaeological meaning—it was an evolutionary lineage—and the term *trajectory* was basically a synonym for tradition, with explicit recognition that it could vary in scale; a trajectory is "the successive sequence of states of an attribute, entity, or vector generated by successive transformations" (Clarke 1968:82, 1978:495). The term *transform type* meant "the relationship existing between successive and collateral type-states from a single multistate artifact-type trajectory" (Clarke 1968:229, 1978:495). In short, a transform was a transition or change, and a transform type was any state of phenomena at a particular time *within a lineage*; transform types "are descent related and are really successive or multilineage type-states" (Clarke 1968:211, 1978:228). Clarke (1968:148, 1978:165) was keenly aware of the reticulate nature of cultural evolution:

> The taxonomic assessment of affinity between entities will suggest the limited number of possible transformation trajectories which might link the network of particular entities in passing time. Great care must then be taken to avoid the danger of interpreting affinity relationships simply as descent relationships—a condition further complicated by the peculiar nature of branch convergence and fusion found in cultural phylogeny [here, Clarke referred to a version of our Figure 3.8]. This problem can only be controlled by providing an adequate chronological frame and by postulating multiple alternative hypotheses of development to link the established degree of affinity between sets of entities under investigation.

The model of change Clarke developed was couched in terms of systems theory, but it was remarkably similar to Ford's model (Figure 5.3). Whereas Ford used the metaphor of a flowing, braided stream, Clarke used the metaphor of a braided cable:

> [W]e have a static model expressing the structure of an artefact-type population as a nucleated constellation of attributes arranged in clustered complexes

> and secondary nuclei in terms of the attribute intercorrelation in *n*-dimensional space. We now wish to develop some model of the kinematic trace or time-trajectory "behavior" of successive generations or phase populations of these artifacts—the phenetic output of one phase being the input of the succeeding phase.... The arbitrarily expressed system trajectory of the developing artefact-type population may be arbitrarily expressed as a single overall integration of such subsystems and lineages within a single multilinear and mosaic development. The archaeological record provides sporadic and successive sections of strands within this continuous cable of development and it is the relative ordering of these sample phase sections in relation to the orientation of the tradition cable that most exercises the archaeologist's researches. (Clarke 1968:210, 1978:225–226)

Clarke (1968:211, 1978:226) then turned to the problem of concern here:

> One of the fundamental problems that the archaeologist repeatedly encounters is the assessment of whether a set of archaeological entities are connected by a direct cultural relationship linking their generators or whether any affinity between the set is based on more general grounds. This problem usually takes the form of an estimation of the degree of affinity or similarity between the entities and then an argument as to whether these may represent a genetic and phyletic lineage or merely a phenetic and non-descent connected affinity.

Clarke then basically reiterated the traditional criteria of Steward (1929), Rouse (1955), and others: The more similar two phenomena are, the more characteristics they share, and the more correlations between "idiosyncratic attributes" they share, the stronger the hypothesis of "phyletic relationship" (Clarke 1968:211, 1978:228). His real contribution to the issue, in our view, was how he illustrated his model of culture change. A version of that model is shown in Figure 6.2.

Interestingly, Clarke presented his model in the context of a discussion of frequency seriation. He correctly indicated that the direction of time's flow could move "from bottom to top, or vice-versa" in Figure 6.2. It does not take a great deal of imagination to conceptualize each column of dots in the figure as comprising an ideational unit—that is, in fact, their symbolic meaning in other graphs Clarke produced—and that the frequencies of each of those units describe a unimodal curve. In the figure, Clarke's "type states" comprise assemblages of classes (our term) of some unspecified scale—attributes of discrete objects, types (attribute combinations) of discrete objects, or assemblages of particular types of discrete objects. Each assemblage of material is designated by an X combined with one or more primes. In our terms, the primes designate a particular lineage; the bottom of the graph comprises lineage X′ and the two branches lineages X″ and X‴. The Arabic numbers denote the sequence of assemblages 1–4 within each of the two branches. Each "type state" comprises, then, a particular cultural unit within a lineage.

We think that Clarke was signifying the ideational and extensional nature of cultural units with his "imagined phase assemblage" and was distinguishing them

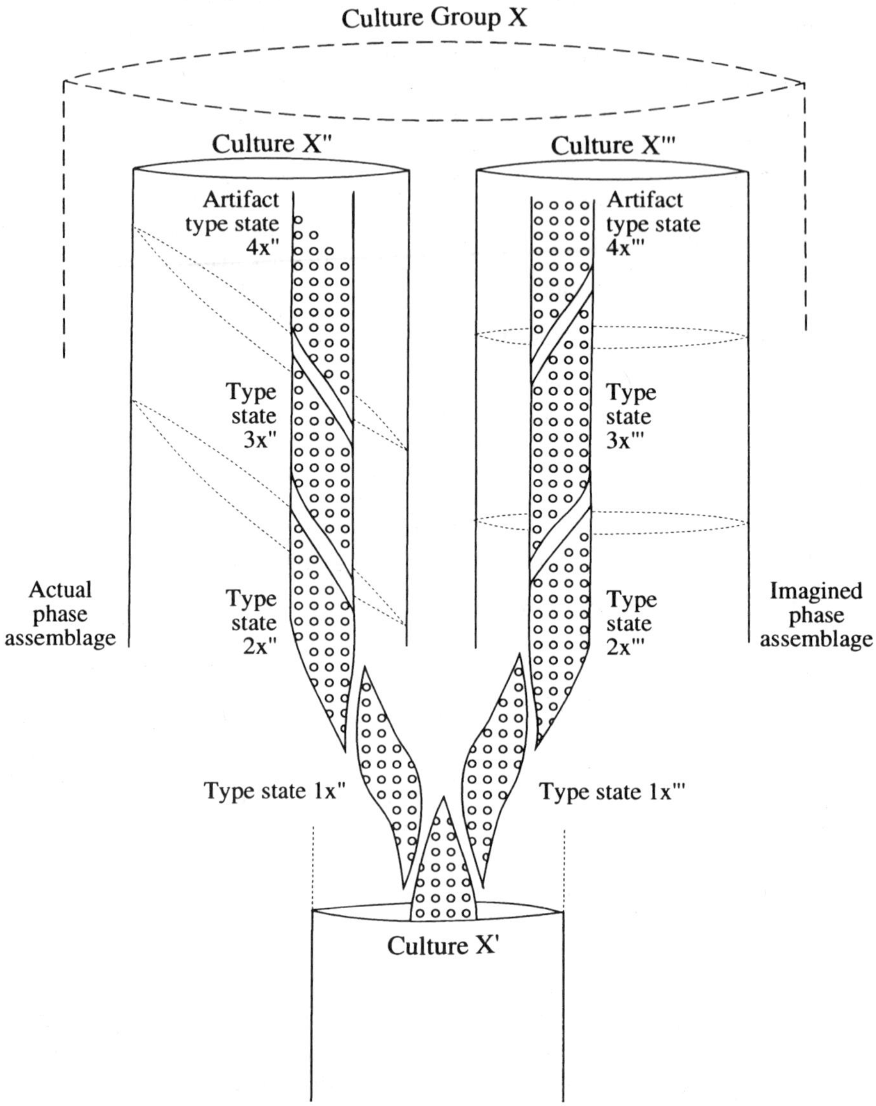

FIGURE 6.2. David L. Clarke's model of culture change. Time may be passing from bottom to top, or top to bottom. Each branch is a lineage, and a "type state" is a cultural unit within a lineage representing an assemblage of classes of unspecified scale. The "actual phase assemblage" spans a duration of time, whereas the "imagined phase assemblage" occupies a point in time suggesting it was extensionally derived (after Clarke 1968, 1978).

from the empirical reality of his "actual phase assemblage." We find this reasonable because he described variation in artifacts as multidimensional, or polythetic, and constantly changing, and Figure 6.2 shows the "actual phase assemblage" encompassing a time period—the cylinder section is slanted—whereas the "imagined phase assemblage" encompasses a single point in time—the cylinder section is horizontal. He wrote that a cultural "system is dynamic and continuous, with the attributes or entities [artifact types] having specific values or states which vary by successive transformations" (Clarke 1968:46, 1978:45). In short, a combination of phyletic seriation and frequency seriation (see below) produced precisely the desired result—the indication of a phylogeny, or lineage—at both the type/species scale and the tradition/lineage scale, though Clarke was not explicit about this. Probably because he was not explicit about it, archaeologists failed to appreciate the significant implications of his model, which rested on the related notions of cultural transmission and heritable continuity.

The Role of Transmission

Kroeber (1919:239) suggested that what would become known as artifact styles did not "vary [from each other] in purpose"; a decade earlier, he had suggested that changes in styles comprised merely "passing change of fashion" (Kroeber 1909:5). Most commentators on his famous hem-length study (Kroeber 1919) have failed to recognize these critical aspects of his reasoning. What Kroeber had done, without using modern terminology and without sufficient explication, was to define styles as selectively neutral. The significance of this is that it implies that styles diffuse over space and through time merely as a result of transmission rather than for functional or adaptational reasons. They are, then, proxies for replicators. Kroeber failed to develop this notion further, however, and did not refer to it in his seminal frequency seriations of Zuñi potsherds (Kroeber 1916a, 1916b), nor in his later pleas that anthropologists should follow evolutionary biology and pay more attention to the distinction between homologous and analogous structures (Kroeber 1931a, 1943).

In a seldom-cited paper, Binford (1963), with an unappreciated bit of insight, reiterated Kroeber's implicit distinction between style and function. He assumed "that the range and stability of individual variations in the execution of stylistic norms between parent and daughter communities is a function of the generational continuity in learning and enculturation between the populations" (Binford 1963: 91–92). This addressed the transmission and heritable-continuity issue and the concept of replicators. After considering the phenomenon of genetic drift, he stated that cultural drift "implies a process of formal modification in culture content, particularly within classes of *functional equivalents* or in the relative frequencies of stylistic attributes which may crosscut functional classes. This process is dependent upon the operation of probability factors in sampling variations between the generations of any given social unit" (Binford 1963:93; emphasis added). This underscored the fact that styles—fashions—would have spatiotemporal distributions dependent only on the vagaries of transmission. Binford (1963:95) examined variation in a "single projectile point type," the specimens of which he and others believed to be "functionally isomorphic." He analytically determined which attributes of the points covaried, as well as the extent of within- and between-site sample variation. He concluded that such studies may reveal "processes of interaction and change within a region [and] serve as the basis for the definition of interaction zones and sub-regional style zones" (Binford 1963:106). He did not, however, follow up this insight to the analytical utility of artifact styles as proxies for replicator units in later studies and instead focused on functional variation in artifacts (e.g., Binford 1965; see also Shennan 1989, 1991). His analysis, like Kroeber's (1919) before him, did, however, illustrate that there were two critical dimensions of variation to artifacts.

Style and Function: A Necessary Dichotomy

The analytical potential of artifact styles of which Binford spoke has only recently been explicitly realized, probably because until recently it had little explicit basis in a theoretically informed model of cultural transmission and change (Lipo et al. 1997; Neiman 1995). In 1978, Robert Dunnell laid the foundation for that model when he published a paper entitled "Style and Function: A Fundamental Dichotomy." He took as his starting point an explicitly Darwinian perspective. Dunnell (1978c:199) defined stylistic similarity as "homologous similarity; it is the result of direct cultural transmission once chance similarity in a context of limited possibilities is excluded." By the last, Dunnell meant that the behavior of style in the archaeological record—that is, the spatiotemporal distributions of various artifact styles—was independent of natural selection and thus was the result only of transmission. Thus, styles could be used to build chronologies, to measure the interaction of spatially separate populations, and to determine and document the phylogenetic histories of artifact lineages. Styles were, in short, reasonable proxies for replicator units. Dunnell, correctly in our view, indicated that the definition of style he proposed was "quite close to its usage in archaeology, particularly as employed by culture historians" (Dunnell 1978c:199).

Dunnell's discussion appears to have had minimal impact on Americanist archaeology, probably because anything remotely related to the culture-history paradigm had been vilified a decade earlier by processual archaeologists (e.g., Binford 1968a; Flannery 1967). Thus, today there is little about "style" upon which modern archaeologists seem to agree (e.g., Braun 1995; Carr and Neitzel 1995; Conkey and Hastorf 1990; Hegmon 1992; Hill 1985). In fact, there was more agreement on archaeological style between 1915 and 1960 than there is today. Prior to 1960, common sense provided the understanding of how and why artifact styles displayed the spatiotemporal distributions they did; it was generally, if implicitly, agreed that styles were types of artifacts that denoted historical—that is, evolutionary—relations because they were the result of transmission mechanisms such as enculturation and diffusion.

Importantly, Dunnell differentiated between style and function. Style "denotes those *forms that do not have detectable selective values. Function* is manifest as those *forms that directly affect the Darwinian fitness of the populations in which they occur.* In an archaeological context, the term *form* should be limited to artificial attributes without any specification of scale" (Dunnell 1978c:199; emphasis in the original). Elaborating, forms "that have discrete selective values over measurable amounts of time should be accountable by natural selection and a set of external conditions [the selective environment]. [Forms] identified as adaptively neutral will display a very different kind of behavior because their frequencies in a population are not directly accountable in terms of selection and external contingencies. Their behavior should be more adequately accommodated by stochastic processes" (Dunnell 1978c:199). Given these definitions of style and function, Dunnell predicted that these two kinds of forms, rendered as classes, could be analytically distinguished because they would display different spatiotemporal distributions. For example, functional forms would covary directly with variation in the selective environment, whether arrayed spatially (e.g., Dunnell 1978b) or temporally (e.g., Braun 1987). Styles, also rendered as classes, would not covary with the selective environment and would display stochastic distributions over space and time because they owed their existence to the vagaries—that is, historical contingencies—of transmission.

We emphasize that by "stochastic" we do not mean chaotic or statistically random. Rather, the term means simply that what comes later depends in part on what came earlier, precisely because of transmission and heritable continuity. As Ernst Mayr (1988:108) put it, "Evolutionary pathways, for this reason, are largely channeled." In referring to the behavior of styles as stochastic, all that is meant is that there is a degree of infidelity of self-replication within heritable continuity. Novelties arise simply because "self-replication is always to some degree imprecise, and variation is a property of self-replicating systems that does not require any special explanation" (Bell 1997:6). Bell's "infidelity of self-replication" is not only a sufficient source of variation for natural selection to act on but also a *necessary* part of Darwinian evolution. From an anthropological perspective, the *unintended* mistakes—Boyd and Richerson's (1985) "transmission errors"—of replicating pottery, arrowheads, houses, and other artifacts may well be a major source of variation in their evolutionary histories. To us, the terms *invention* and *innovation* typically used by anthropologists and archaeologists carry too much of a connotation of *conscious*, *thoughtful*, or *anticipatory intent*. We prefer the term *novelty* and its derivatives (Lyman and O'Brien 1998).

Do not be misled by the dichotomy between style and function. Functional similarity may and often does result from transmission: "Because selection presupposes inheritance, the investigation of selection [of functional variation] cannot be separated from the investigation of descent" (Neff 1993:36). The critically important point to remember is that functional similarity can also result from adaptive convergence—finding similar solutions to similar adaptive problems, such as the evolutionarily independent appearance of agriculture in different geographic locations. The analytical challenge, as should be more than clear by now, is to determine whether two phenomena are similar because of common ancestry (similarity is homologous) or adaptive convergence (similarity is analogous).

Archaeologists with interests in applying Darwinian evolutionary theory to archaeological problems have devoted some effort to exploring the ramifications of transmission. Hector Neff (1992, 1993), particularly, discusses the importance of the transmission of "recipes" (replicators) for making pottery, though his remarks apply equally well to recipes for making projectile points, building houses, planting fields, or taking mates. Parts of his discussion are worth quoting at length:

> Inter-individual transfer of pottery-making knowledge must produce historical phenomena that [can be referred] to as "ceramic traditions." Shared information dictates where to find clay, how to prepare clay, how to form and decorate pots, how to fire pots, and other aspects of behavior related to pottery-making. Just as the techniques employed by an individual to produce his/her distinctive pots are determined by pottery-making information that individual carries, so the distinctive, collective phenotypic expressions recognizable in particular regions, during particular time intervals, are determined by information shared among individuals working within a tradition. The importance of ceramic traditions to ceramic evolution is that traditions, like individuals, are partitions of information concerning how to produce pottery-making aspects of potters' phenotypes; because evolution results from any disruption of information flow through time, ceramic evolution will result both from disruptions of inter-individual transmission of pottery-making knowledge and from disruptions that simultaneously affect the ability to transmit pottery-making knowledge of all individuals working within a ceramic tradition. (Neff 1992:152–153)

Neff (1992:153) suggests that "the archaeological phenomena recognizable as 'types' or 'wares' may represent [heritable] continuity of phenotypic expression, a result of the flow of information on pottery-making within traditions." And while he notes that although "individual pots constitute a phenotypic expression of [a ceramic] tradition; and archaeological 'types' are analytical constructs that, in some cases, represent an archaeologist's attempt to group together phenotypic expressions pertaining to a single tradition," he also indicates that types may be constructed "for purposes completely unrelated to the problem of recognizing shared information" (Neff 1992:153–154). Style is included in types because archaeologists tend to focus "on ceramic collections from a particular site or region, [and this] increases the likelihood that any intuitive groups will represent the work of potters among whom there was cultural transmission" (Neff 1993:31). Function is included when archaeologists distinguish between utilitarian and elite wares and decide whether the former were produced for internal or external consumption. Neff (1993:28) reasons that "traits vary in how tightly their expression is controlled by selection." Furthermore, because some traits may covary with other traits, yet selection may operate on only a few of them, "the configuration as a whole may be strongly influenced by selection, even though particular traits have negligible selective impact on the configuration, and would behave stochastically in other contexts" (Neff 1993:28). He notes that ceramic "traits" may be subject to selection during only part of their existence, and thus history is important to the style–function distinction (Neff 1992, 1993). We agree with all these points (e.g., O'Brien and Holland 1990).

Neff's (1993:27) observations prompted him to suggest that Dunnell's style and function distinction does not characterize "two poles of a dichotomy" but rather "may be more practically seen as defining a continuum." Recognizing that artifacts, like organisms, are subject to engineering constraints, which can affect our ability to distinguish between functional and stylistic traits, we still view style and function as representing a dichotomy that is not only fundamental but also necessary to developing an evolutionary theory for archaeology. Styles are homologs, which, as we have seen, are critical to writing phylogenetic history; functional forms may also be homologous, but they need not be. Neff's (1992) argument that the two concepts specify a continuum rather than a dichotomy is little more than a problem of scale, and it is readily dealt with by adopting the concept of sorting and the concepts of interactors and replicators we described earlier. Similarly, his argument that attributes or types may at one time be stylistic but at another can become functional is readily accounted for with the concepts of spandrel—in the metaphorical sense of Gould and Lewontin (1979) (Chapter 4)—exaptation, interactor, and replicator. The important question is, how are style and function to be distinguished archaeologically?

Sorting Out Style and Function

John Terrell (1986:188) identifies four "patterns" of phenomena typically mentioned by anthropologists and archaeologists: "(1) *chance* or *random variation*; (2) *adaptation*; (3) *historical inheritance*; and (4) *teleological design or purpose*" (emphasis in the original). He subsumes the fourth pattern within the second because that seems to him to be the way archaeologists treat it, and we agree (e.g., O'Brien and Holland 1992). That leaves the first three, which he characterizes as "three basic types or 'ideal' patterns of variability" (Terrell 1986:189) and defines as follows:

> RANDOM VARIATION: When the state of a system at time T_{n+1} cannot be inferred (predicted) from knowledge of the prior state of that system at time T_n or from knowledge of its present environment.
>
> HISTORICAL INHERITANCE: When the state of a system at time T_{n+1} can be fully inferred (predicted) from knowledge of the prior state of that system at time T_n.
>
> ADAPTATION: When the state of a system at time T_{n+1} is totally dependent on the environment of that system at that time and totally independent of the prior state of that system at time T_n.
>
> Phrased so basically, it seems clear that no instance (i.e., pattern) of variability in language, customs, or biology is likely to qualify as solely 100% chance, 100% history, or 100% adaptation.... [E]xplaining human variability over space ... is a matter of deciding how much sense it makes to say that the variation observed is the product of chance, historical inheritance, or adaptation.

We find Terrell's characterization of patterns in variation insightful, but we also note two things. First, he bases his discussion on a genetic model and finds that he cannot sort out the three "patterns" with respect to the phenotypes of human populations because of the lack of tight links between particular genes (replicators) and particular phenotypic traits (interactors). Second, the "random variation pattern" is analogous to genetic mutation and thus comprises a process as well as a pattern; similarly, although "adaptation" is often construed by archaeologists as a pattern rather than a process, it is used by them to signify both (O'Brien and Holland 1992). The "historical inheritance pattern" is clearly only a process and entails the transmission of mutations that are either selectively neutral or selectively functional, both of which can be produced randomly. We find Terrell's discussion illuminating because it underscores the necessity of keeping patterns and processes that create patterns conceptually and analytically separate when attempting to sort out style and function.

Archaeologist David Meltzer put the discussion in the preceding several pages in clear and concise language:

> At one level, style can be defined as those forms that do not have detectable selective value: those forms are adaptively "neutral." On another level, however, because style, like all elements of cultural baggage, has a cost in energy, space, and matter, it remains a part of the total selective picture of the cultural system. So in a general sense style has a function because, by virtue of its independence from its environment, it can be employed to delineate spatial interaction and demarcate cultural boundaries. In other words, while *particular styles* are not directly accountable by natural selection, *style in general* has a role in the cultural system.
>
> The difference is primarily one of scale, and can be illustrated as follows. In many instances, the choice between certain kinds of design elements on ceramics is not a functional consideration, but rather is historically determined and selectively "neutral," because there is no inherent advantage

> between one element and the next. The actual presence of the design, however, has a selective value, because that particular design serves to mark a certain individual or group boundary (or whatever other function it may serve).
>
> Archaeologists are most familiar with the former instance of stylistic behavior because it is the historical and "neutral" basis of stylistic attributes that causes them to take on in a temporal dimension that peculiar *Markovian* ("Battleship Curve") pattern that makes [frequency] seriation work. Because stylistic similarity, with the exception of fortuitous circumstances, is necessarily homologous similarity, the result of cultural contact or common descent, and because particular styles are independent of external [selective] conditions, one can obtain the purely historical, non-repetitive classes used to tell time.
>
> Function, on the other hand, includes those elements that directly affect the Darwinian fitness of the populations in which they occur. These elements are traits that can be accounted for by selective factors. For this reason, functional similarity, while it may be homologous [the result of transmission], may also be analogous, the result of similar activities or adaptations in similar environments. The replication of functional traits is not dependent on direct character transmission. (Meltzer 1981a:314; emphasis in the original)

Meltzer's discussion suggests avenues for analytically distinguishing between stylistic and functional *forms*. Meltzer, like Dunnell (1978b), and David Braun (1987), states that comparative analyses of groups of specimens of a form—that is, a class—from widely separate spatiotemporal contexts will reveal no significant between-group variation *if* the specimens are functionally similar. The separation in time and space ensures against heritable continuity between the specimens (Meltzer 1981a). O'Brien and Holland (1990:47–48), following biologists (e.g., Mayr 1983), suggest that one begin with the hypothesis that a class (of whatever scale) is functional and thus subject to selection. The hypothesis is tested, and one concludes the class is stylistic only after failure to determine via engineering and mechanical analyses that it has selective value within the context in which it appears. Neff (1993:28) agrees. Another way to separate the two analytically is to perform a frequency seriation; we turn to why this should indicate whether a class is stylistic or functional in the next section. We advocate all of these methods for sorting out stylistic from functional classes at all scales; each not only serves as a test of the others, but also each provides detailed data relevant to answering other questions of interest to evolutionists.

SERIATION

Along lines similar to those indicated by Neff, Hartmut Tschauner (1994:84) remarked, "It is the ideational elements that are cultural, not the artefacts themselves which are studied by the archaeologist as the material manifestations of cultural instructions [Neff's 'recipes']. Artefacts are the products of culturally transmitted information [read *replicators*] working upon material found in the environment, with phylogenies existing in the transmission of this information and selection working on the information and/or the product." Echoing Mayr (e.g., 1981, 1995b), he argued that "of the two evolutionary processes, transmission and selection, our classifications should primarily focus on measuring transmission—that is, establishing cultural phylogenies or descent" (Tschauner 1994:86). Thus, he stated that a "basic task for a Darwinian archaeological systematics" is "to trace the flow of cultural instructions both through time at any one site (transgenerational transmission) and across space (oblique transmission), beginning at the intrasite level and distinguishing homologous from analogous variation" (Tschauner 1994:87). The last, of course, brings us back to where we started this chapter: How are homologs and analogs to be distinguished?

In what is sometimes referred to as the seminal paper on evolutionary archaeology (e.g., Schiffer 1996), Dunnell (1980:38) remarked that

> evolution is a particular framework for explaining change as differential persistence of variability.... It is exceedingly important to note, given the tenor of archaeological discussion, that the subject is *change*, not simply similarities and differences. The continuity implied in the terms change and persistence bespeaks a fundamental assumption: the phenomena being examined are historically and empirically related to one another [by heritable continuity].... Variability is conceived as discrete. Change is accomplished by alteration of the frequency of discrete variants rather than alterations in the form of a particular variant (emphasis in the original).

Dunnell's remarks provide important clues to our interrelated problems of distinguishing between homologs and analogs, of sorting out stylistic and functional variation, and of writing and explaining phylogenetic history. First, variation must be measured in discrete units, meaning that ideational units, or classes, are necessary. Second, persistence is denoted by stability over time in the groups created by the classes, and change is denoted by alteration in the frequencies of members of classes and/or turnover in the classes represented. Third, because we are using classes, heritable continuity at the type/species level is hypothetically possible. And fourth, by using the seriation method and meeting its requirements, we are ensuring that heritable continuity at the tradition/lineage level is also hypothetically possible. In fact, the seriation method is founded on exactly all these points. It is critical, then, that we describe the seriation method in some detail, as it comprises the only patently archaeological method for testing hypothesized heritable continuity. In a later section, we discuss the theoretical underpinnings of seriation; here, our interest is only in the method itself.

The seriation method involves placing objects or sets thereof in an order based on their formal similarities—Spaulding's (1954c) "resemblances" mentioned earlier. The more attributes two phenomena share, the closer they are to one another in the order; the fewer they share, the more distant from one another they are in the order. The implicit but key assumption behind the procedure—regardless of how similarity is measured—is that propinquity in formal properties denotes propinquity in time. We refer to this as the assumption of *historical continuity*. One might well ask why formal similarity should denote temporal similarity. In short, the assumption of historical continuity rests in turn on the assumption that formal similarity is the result of *heritable continuity*; that is, the seriation method as a chronological tool rests on the assumption that things produced at any particular moment resemble in significant respects things of the same kind produced at slightly earlier moments, just as they will closely resemble—in different ways—those produced at slightly later moments because of a genetic-like connection between them. Assuming a temporal sequence of moments 1–5, for example, the seriation model cast in phyletic terms holds that objects of kind A produced at moment 3 will most resemble those produced at moments 2 and 4 and will resemble less those things produced at moments 1 and 5. "Kind," as used here, is without reference to scale.

One might also ask why we have distinguished between historical continuity and heritable continuity. We have done so for two basic reasons. First, most archaeologists implicitly hold to a definition of historical continuity such as "shared tradition" or "the result of information sharing within some definite span of space and time" and agree that it is "a basic assumption that archaeologists must make about human societies" (Neff 1993:25). Such a definition implicitly assumes at the start that formal similarity is the result of "shared tradition or information." Second, we make the distinction because we want to make the implicit—and most critical—aspect of the traditional definition of historical continuity explicit. That term *could* be taken to signify that two phenomena are merely formally and temporally similar. Thus, the phylogenetically independent appearance of multiple cultural traits in multiple lineages in approximately the same order—such as hunting and gathering followed by horticulture followed by agriculture, or the often accepted "universal" sequence of band, tribe, chiefdom, and state—would denote historical similarity but not necessarily heritable continuity. It is the latter that is critical to the analytical construction of lineages on the basis of historical continuity.

The assumptions of historical continuity and heritable continuity, though related in important ways, are separate and distinct. Historical continuity denotes a sequence of *similar* forms that measures the passage of time; heritable continuity underpins the assumption of historical continuity when the latter is rendered as formal similarity and used to order phenomena in what is *inferred* to be a temporal sequence. That it *is* an inference is clear from the early literature: Kroeber's (1916a, 1916b) noting that stratigraphic data would prove that his seriation was a temporal ordering is the most notable example (see also Rowe 1961). Heritable continuity, or "relatedness," is one explanation for the observed similarities; convergence is another. As we have seen, archaeologists have long been aware of precisely these sorts of problems, but they have chosen to deal with them analytically in different ways than have paleobiologists. Some archaeological approaches to sorting analogous similarity from homologous similarity are reasonable, but the general failure conceptually to distinguish between historical and heritable continuity has resulted in conflation of the two. We want to avoid such conflation. How is that accomplished with the seriation method?

Let us say we have an array of objects of a specific kind, all of which were produced in one locale over some span of time. Furthermore, let us assume that each sequential object resembles its predecessor, as it will its successor, in numerous ways. The problem is that the set of objects is jumbled, and our job is to put the objects in order based on their formal similarity. The principle of ordering in seriation dictates the procedure: Place the objects most similar to one another adjacent to one another; as similarity between objects decreases, increase the distance between them. The *inference* that the order is chronological is explicitly founded on the principle of historical continuity and implicitly founded on the notion of heritable continuity. Whether or not the order we produce is in fact a chronological ordering of forms—a temporal sequence—and thus measures time is an inference that must be tested with independent data such as finding the same ordering of forms in a sequence of superposed sediments. If the ordering is in fact chronological and thus denotes historical continuity, then the next problem is to determine if the *sequence* is also a *lineage*—a line of heritable continuity. As we have indicated, the two—heritable continuity and historical continuity—are not necessarily one and the same. If heritable continuity is indicated, and we discuss how this is done below, then historical continuity follows automatically. If only historical continuity—a mere temporal sequence of forms—is indicated, heritable continuity is possible and perhaps even highly probable, but it is not assured. In other words, historical continuity is founded in the similarities of the ordered phenomena; *why* the similarity exists may be the result of heritable continuity or some other reason (e.g., temporally coincident convergence).

The assumptions underpinning the seriation method—historical continuity and heritable continuity—should sound familiar to anyone with a background in biology or paleontology, as they are the backbone of those disciplines. Biologists and paleontologists are vitally interested in historical and heritable continuity because they are central to understanding organismic evolution. But we do not need a background in evolution to understand these concepts and their value as tools for building chronologies and phylogenies of artifacts. Most of us would not have much trouble correctly ordering examples of a particular line of automobile. We believe that if ten people were shown twenty cars of the same line produced in Detroit between 1978 and 1998, most of them would put most of the cars in correct chronological order, regardless of how much or how little they knew about automobiles. They might not know which end of the sequence was the most recent (although we suspect they would), but they would get the basic sequence correct. The key to successful ordering would come from noting features that were exactly the same or nearly the same on different vehicles and putting them next to or close to each other in the sequence. This is simply the principle of historical continuity in action, and it works because—note that we said "a particular *line* of automobile"—of the principle of heritable continuity.

The seriation method has an interesting history in archaeology (Lyman and O'Brien n.d.a, n.d.b; Lyman et al. 1998b; O'Brien and Lyman 1998, 1999a). Our view of that history suggests there are three basic techniques of operationalizing the method: *phyletic seriation*, *occurrence seriation*, and *frequency seriation*. These can be subsumed under what are elsewhere referred to as similiary and evolutionary (Rowe 1961), or developmental (Rouse 1967), seriation (Figure 6.3). Evolutionary seriation rests on the assumption that phenomena in a lineage will evolve according to a rule, such as from simple to complex, and thus should not be confused with Darwinian evolution, which follows no such rules. We do not consider it further here because the techniques we describe assume nothing about the direction of change. What Rowe (1961) labeled similiary seriation includes the three techniques of interest, and we devote the remainder of the chapter to them.

Phyletic Seriation

In our view, what we have termed *phyletic seriation* involves the most commonsensical way to order objects based on their similarities. It is what A. V.

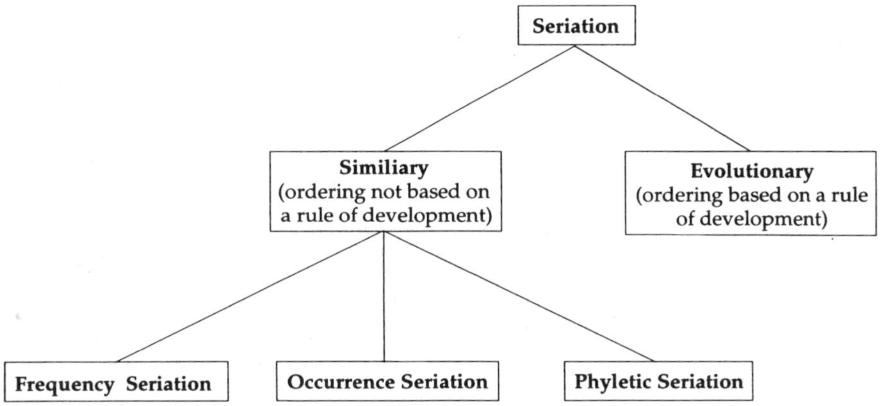

Figure 6.3. A categorization of seriation techniques. Seriation comprises techniques of ordering based on formal similarities (after Lyman et al. 1998b).

Kidder did with his pottery designs from Pecos (Figure 3.5). Heritable continuity, and thus historical continuity, is represented by a line of forms believed on theoretical grounds to comprise a lineage. Each form in the line is succeeded by a slightly different form, one that differs in one or several character states, such that change occurs continuously and gradually over time. The earliest phyletic seriation using pottery of which we are aware was that carried out by the aforementioned British archaeologist Sir William Matthews Flinders Petrie, who was faced with making chronological sense out of some 4,000 predynastic burials from several localities along the Nile River north of the Valley of the Kings in Egypt. Petrie (1899:297) believed that artifacts could be ordered chronologically because they occurred in a "series of development and degradation of form." Once established, a phyletic sequence enables "a long period to be arranged in approximate order, and serves as a scale for noting the rise or disappearance of other types" (Petrie 1899:297).

Petrie's (1899) drawings make it clear how the phyletic seriation technique works. In Figure 6.4, the sequence runs from oldest at the top to youngest at the bottom, though this is irrelevant. Petrie could have had the sequence reversed without disrupting the validity of the technique because, as we noted above, nonevolutionary seriation assumes nothing about the direction of change. In the diagram, Petrie illustrated various ceramic-vessel forms that were in the sample of roughly 900 burials he used initially (he added more than a thousand additional burials once the sequence had been established). Vessel forms are arranged by what Petrie termed "sequence dates," from a point in the distant past—sequence date 30—up to more modern times—sequence date 80. Given that this is a much simplified drawing, Petrie combined many of the periods into larger units. Note that through time, some vessel forms continued through several periods and others did not. For example, the outflaring cylindrical vessel form shown in the upper left continued from sequence date 30 into sequence date 34. The bowl form in the upper right did the same thing. This overlapping of forms from one sequence date to the next allowed Petrie to work his way up or down through time to create a sequence.

The key to Petrie's analysis was that he used an attribute that changed states through time and thus allowed him to construct a temporal sequence of pottery forms. This attribute was vessel handles: "At the left ends of the five lower rows is the wavy-handled type, in its various stages; the degradation of this type was the best clue to the order of the whole period" (Petrie 1899:300). Petrie suspected that the handles were functional on the earlier jars, which tended to be large and bulky, but that through time they had become less functional and more decorative, such that by the time of sequence date 63 they were simply adornments (Figure 6.4): "The most clear series of derived forms is that of the wavy handled vases.... Beginning almost globular, with pronounced ledge-handles, waved (as in stage 35 to 42), they next become more upright, then narrower with degraded handles, then

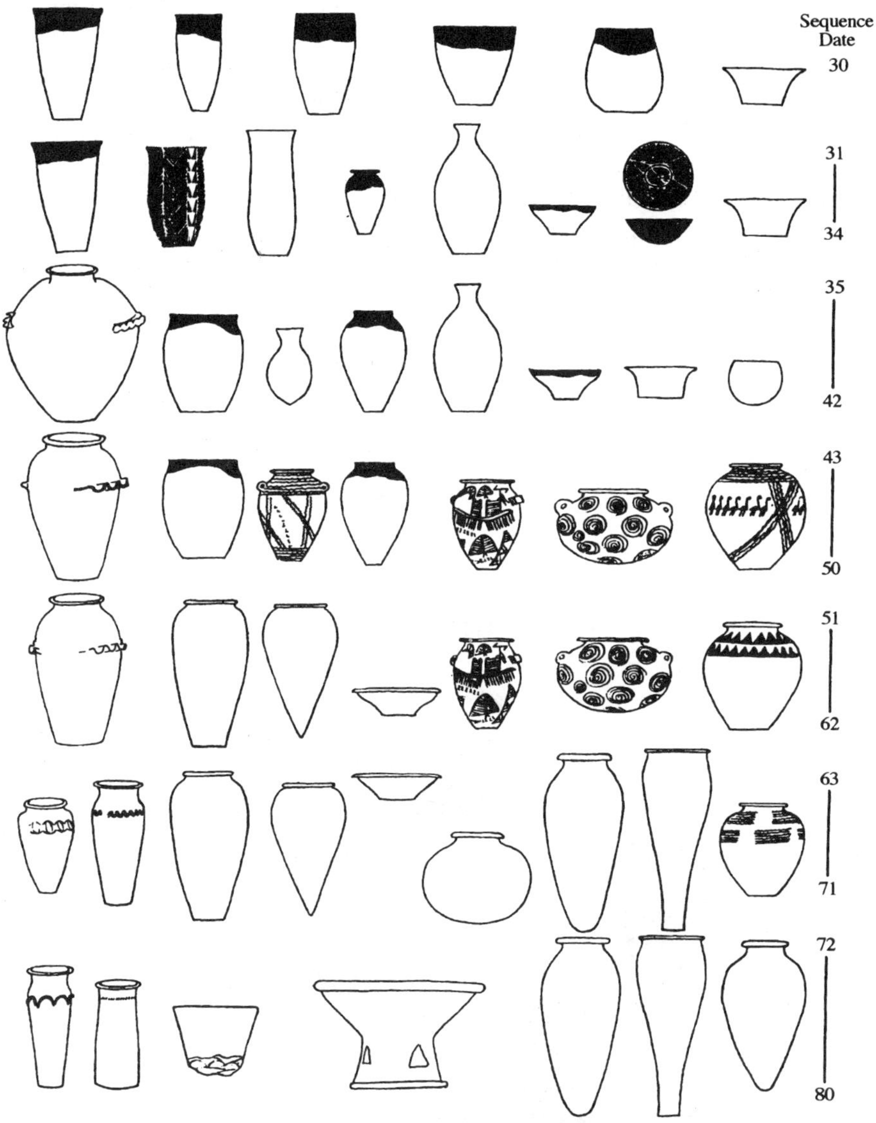

FIGURE 6.4. W. M. Flinders Petrie's chronological ordering of ceramic vessels recovered from burials in three localities in Egypt. Periods from 30 to 80 are shown at the right; vessels from each period are placed in rows. Notice the wavy handles on vessels at the far left of the lower five rows; the shape of the handle was the first clue Petrie had as to the sequential ordering of the burials (after Petrie 1901).

the handle becomes a mere wavy line, and lastly an upright cylinder with an arched pattern or a mere cord line around it" (Petrie 1901:5). This is phyletic seriation.

Petrie (1899:300) referred to portions of his temporal sequence of pottery forms as "genealogies." Nowhere did he expand this notion, which, because of its evolutionary aspects, is unfortunate. For example, recalling that low numbers fall early in the sequence and high numbers late, as shown in Figure 6.5, the jar form at sequence date 38 appears to be a "hybrid" of two different forms, as does the form at sequence date 70. If a correct interpretation, it illustrates that the evolution of the cultural trait of vessel form is, contrary to Kroeber's observation decades later,

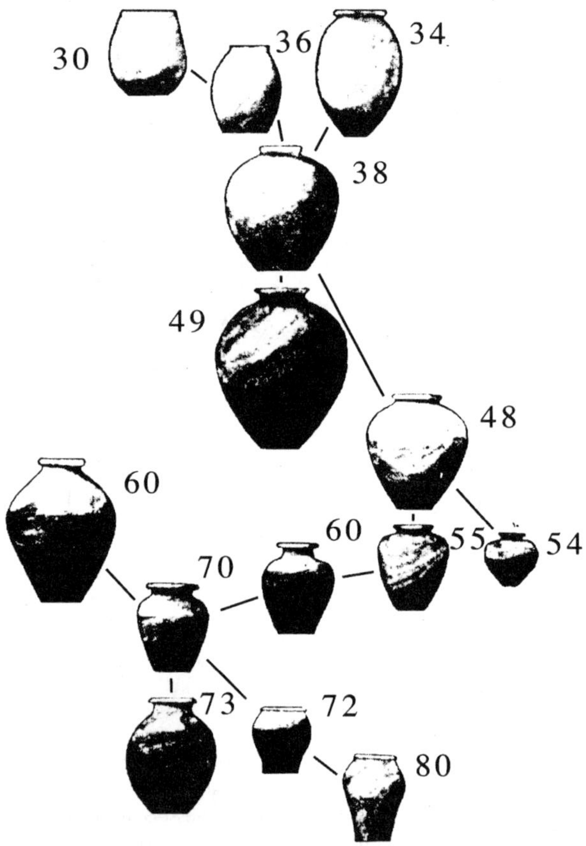

FIGURE 6.5. W. M. Flinders Petrie's genealogy of ceramic-vessel forms recovered from burials in three localities in Egypt. The numbers refer to periods (see Figure 6.4) (after Petrie 1899).

polyphyletic. But we do not know if this is a correct interpretation because we do not know if vessel form is a homologous or an analogous character.

Petrie's use of phyletic seriation as the basis of chronological ordering had precedence in the work of several British archaeologists, including A. L. Pitt-Rivers (1870), who seriated copper and bronze axes, and Sir John Evans (1850), who seriated gold coins from Great Britain that were minted prior to and after the Roman invasion of Britain in 54 B.C. Only Pitt-Rivers (1875a) was explicit about why heritable continuity provided an explanation for phyletic seriations. He "selected and arranged [artifacts] in sequence, so as to trace, as far as practicable, the succession of ideas"; thus, his arrangement illustrated "the development of specific ideas and their *transmission* from one people to another, or from one locality to another" (Pitt-Rivers 1875a:294, 295; emphasis added).

Evans (1850) used changes in two dimensions of variation—weight and design—to seriate his British coins. A third dimension, die size, did not produce chronologically useful results. Not visible in Evans's seriation (Figure 6.6) is the decrease through time in coin weight. For example, type 2 coins on average weighed 103.5 grains, type 3 coins 91.5 grains, and type 4 coins 87.25 grains. Highly visible, though, is the change in design on both sides of the coins. The sequence begins with the natural-looking laureated bust of Phillip II of Macedon on the obverse and a horse-drawn chariot on the reverse. Through time, the designs on both sides became successively more stylized, until a point was reached at which they again became naturalistic. Evans perceptively identified typological "creep"—the problem noted by Phillips, Ford, and Griffin (1951) a century later—as occurring when one moves away from the "centers" that produced the analytical archetype:

> Thus far I may observe at present, that the coins generally recede farther from the prototype as the places of their discovery recede from the southern coast—as, for instance, the Yorkshire and Norfolk types Nos. 24 and 16; and that in the southwestern counties the workmanship of the coins appears continually to have deteriorated; while in the southeastern and eastern, after declining for a time, it again improves, probably through the introduction of foreign artists, till, under Cunobeline, it attains its highest perfection. (Evans 1850:137)

The work of Petrie, Evans, and Pitt-Rivers was founded in the commonsensical notion that similarity of form denoted historical continuity. Only Pitt-Rivers (1875a) offered an explicit rationale for why this should hold for cultural phenomena: Heritable continuity and thus historical continuity, especially as indicated by phyletic seriation, reflected empirically the transmission of ideas. The simultaneous if implicit use of spatiotemporal propinquity of the seriated items increased the probability that the phyletic seriations produced by Petrie, Evans, and Pitt-Rivers were founded on homologous similarity and thus comprised heritable continuity, though they did not guarantee this. In fact, such similarity was of minor

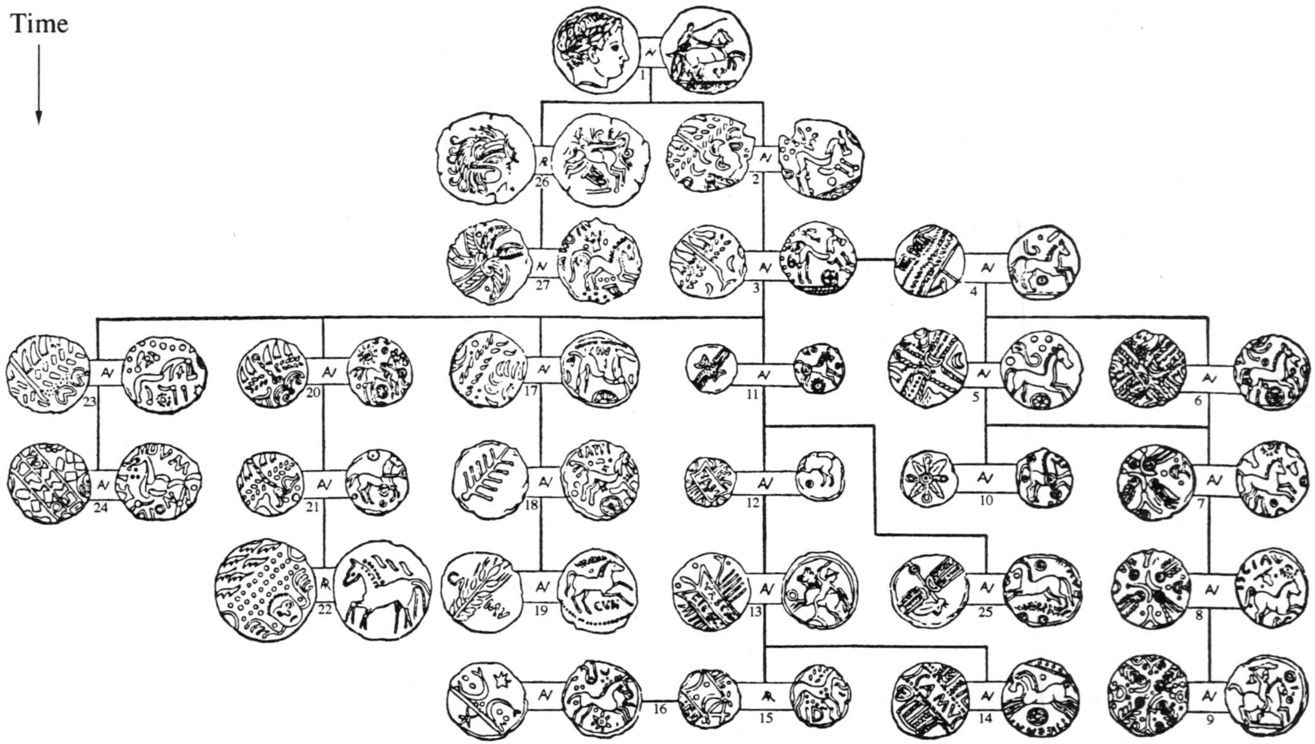

Figure 6.6. Stylistic changes in British (Roman period) coins proposed by John Evans. The sequence begins with the natural-looking laureated bust of Phillip II of Macedon on the observe and a horse-drawn chariot on the reverse. Through time, the designs on both sides became successively more stylized until a point was reached at which they again became naturalistic (after Evans 1850).

concern relative to the more immediate problem of constructing an ordering of artifact forms that was chronological.

A. V. Kidder probably learned the phyletic-seriation technique from George Reisner, who had worked in Egypt (Browman and Givens 1996:86) and undoubtedly was familiar with Petrie's work. Kidder's (1915) original pottery sequence (Figure 3.5) was based on his suspicions regarding the evolution of various design and technological attributes. In a brief paper published in 1917, he demonstrated how such an evolutionary and thus temporal sequence could be worked out. The chronological sequence of attributes or character states was *confirmed* in 1917 and later by study of superposed collections from Pecos Pueblo (Kidder 1924). The suspected order had been constructed two years earlier (Kidder 1915) using phyletic seriation of the character states of pottery decoration. Kidder proposed that through time, the pottery design became less intricate, changing from a stepped motif, to a pair of stepped motifs, and finally to a series of ever-larger white spaces. Kidder here bypassed the notion of type (though it was incorporated into his later work at Pecos) and concentrated strictly on a single character, using changes in character state to tell time.

Kidder termed the results of his phyletic seriation a "series," and his discussion here and elsewhere (e.g., Kidder 1917; Kidder and Kidder 1917) implies that for him, a series was not only a chronological sequence (denoting historical continuity) but also an evolutionary, or phylogenetic, one (denoting heritable continuity). The former was an inference that had to be tested with superposition, and Kidder (1916, 1924, 1936a) performed such tests. The latter was an interpretation that could either be derived *after* a chronological sequence had been built, or be used, as Kidder did, to construct a chronological sequence. Kidder was not explicit about this, nor were his contemporaries or intellectual followers. The result was disastrous, as chronological method and interpretive algorithm were conflated. Sometimes the confirmed chronology (historical continuity) was used to substantiate phylogenetic inferences (heritable continuity); other times, the suspected phylogeny was the basis of the chronological inferences.

The process of ordering that comprises phyletic seriation rests on measuring and evaluating the *similarity* of the phenomena to be ordered. Phyletic seriations are typically constructed on the basis of the similarity of artifacts at the scale of the attributes they share (e.g., Deetz and Dethlefsen 1971; Rowe 1959). As we have noted, the more attributes shared by two objects, the closer they are placed to one another in an ordering and the closer they are thought to have occurred in time. We believe that the phyletic-seriation technique is not only a good chronological tool but also a reasonably good, but not the best, tool for building lineages and writing phylogenetic histories. One major problem that we perceive with it is identified by Evans's (1850) recognition of typological creep. Because one begins by ordering specimens, any resultant types or class definitions (significata) are extensionally derived. Typological creep is a predictable result. The second major problem relates closely to the first. Because types are extensionally derived, there is no effort to ensure heritable continuity in the type/species sense by testing. Each class is more or less unique from every other class, and one is merely stacked on top of another. Historical continuity might be confirmed by an independent dating technique, and the probability of heritable continuity might be raised by regular use of the criterion of spatial propinquity, but heritable continuity cannot be confirmed this way.

Classical archaeologist William Biers (1992:25) characterized well what we perceive to be the issue of units in phyletic seriation when he wrote, "The particularly characteristic or distinctive way an object appears to the eye can be said to be its style. A change in its appearance, or details of its appearance, or attributes, is seen to be a stylistic change or development. Stylistic change can be related to time, but is not necessarily always caused by the passage of time, and can be slow or almost nonexistent." Although he correctly identified stylistic variation as occurring across time and space, Biers's discussion indicates that the units used in phyletic seriations are not explicitly distinguished as empirical or ideational. His discussion also indicates that the units in phyletic seriations are at least initially treated as empirical—one form blends into another (e.g., Deetz and Dethlefsen 1971). These units might eventually become theoretical, but they begin as empirical units. Heritable continuity in the type/species sense is not assured, nor can it be tested. Contributing to the appearance that the units are empirical and the lack of a test of heritable continuity is the fact that no overlapping (see below) is acknowledged, and change thus appears to be transformational, that is, from one form to another. These are precisely the problems we want to avoid. The seriation techniques left to be described—occurrence and frequency seriation—avoid them. But before we turn to these techniques, we need to discuss the theoretical underpinnings of the seriation model and to outline why successful—that is, tested and confirmed—occurrence and frequency seriations actually comprise lineages.

Seriations as Lineages

Americanist archaeologists working early in the twentieth century found various ways to measure similarity. Kidder's phyletic seriations (Figure 3.5) comprise one, but there were others as well. Basically, these involve either or both a shift in how units are viewed and a shift in scale from the attributes of artifacts to the attributes of aggregates of artifacts. The shift in scale that occurs with occurrence and frequency seriation, although not mandatory for such seriations (Chapter 8), results in the ordering of aggregates of artifacts we here term *assemblages*, or *collections*. Similarity is measured by first noting the presence or absence of types of artifacts or the frequencies of types of artifacts in each aggregate, and then either subjectively (visually) or objectively (statistically) ascertaining how similar various aggregates are to one another in terms of the types they share or in the frequencies of the types. The more types shared by two assemblages, the more similar they are and thus the closer in time the two are thought to be; likewise, the more similar the relative frequencies of shared types in two assemblages, the closer to one another the two are placed in an ordering and the closer in time they are thought to be. These comprise occurrence seriation and frequency seriation, respectively. Importantly, because the similarity of assemblages is measured in terms of shared types or similar type frequencies, the types *must be* theoretical units. Only then can they have distributions *over* time and space rather than have locations *in* time and space as empirical units do. Specimens, or empirical units, have locations; classes have distributions that are indicated by the summed spatiotemporal locations of the specimens—the denotata—included within them. Empirical specimens are not mapped in time and space; rather, particular theoretical units capable of occurring in more than one place simultaneously are mapped (Dunnell 1970).

Lest we be misunderstood on a key point, we are not saying that groups extracted from paradigmatically formed classes (Chapter 5) are the only theoretical units that can be used in a seriation. If this were so, then none of the types developed to date would do the job many of them were created to do, which is to measure the passage of time. Because historical types were formed precisely to keep track of time, when establishing the types, archaeologists selected traits that experience showed changed over time. Those traits became the defining criteria of the types, which ensured that the types did what they were intended to do. The types, then, were extensionally derived by trial and error.

Occurrence and frequency seriation employ the notions of historical and heritable continuity, particularly the latter, in somewhat different fashion than does phyletic seriation. This is because the former two entail certain expectations about how change will appear—not its direction—and thus how time and heritable continuity can be tested and then measured. We therefore need to outline the various assumptions that guide the workings of occurrence and frequency seriation before presenting the details of how they work. Together, the assumptions comprise what we refer to as the *seriation model*, which is actually little more than a set of statements regarding the expected spatiotemporal distribution of specimens classified as historical types.

For any seriation technique—phyletic, occurrence, or frequency—to work, the rate of change within a lineage must be gradual. Although this cannot be assumed literally, it is precisely such an assumption that allows the positioning of like adjacent to like and unlike to be placed some distance away in an ordering. Under the assumption of heritable continuity, replicative fidelity is rather good; were it not, we could not seriate materials. But do not be misled by the word *gradual*. Characterizing the rate of change as "gradual" does not preclude rapid change or fluctuation in the rate of change during the history of a lineage, nor is stasis precluded (Chapter 7); rather, only abrupt or sudden change of great magnitude is precluded. By assuming change is "gradual," one is saying that time manifest as variation in material objects is a continuum rather than a discontinuum. George Cowgill (1972:384) suggested that rather than assume change is gradual and continuous, "for seriation to be useful as a basis for chronology, it seems to me that all that is required is that there never, among the set of units being seriated, be a break in the sequence so abrupt and catastrophic that units immediately following the break bear no (or only accidental) resemblance to units before the break." Cowgill was being critical of James Ford's (1962) work. Early in his career, Ford (e.g., 1938:11) displayed a sophisticated awareness of the necessity of change being gradual, as we have construed it, when he noted that the act of ordering via seriation must include "overlapping." By this term, he meant that type (class) A may fall within periods 1–3, type B within periods 2–5, and type C within periods 4–6. The various *overlapping* temporal occurrences of the types are what connects the sets of materials being seriated, ensures heritable continuity between collections (at the type/species level), and allows an ordering to be derived. This is precisely what Cowgill meant when he indicated "abrupt and catastrophic breaks" cannot be so great as to preclude resemblance of units before and after the break. If overlap does not occur, then one or more of the conditions or requirements of seriation (see below) has not been met, and the seriation model will not be approximated.

The Seriation Model

Historical types are often, but not always, *styles*, and by definition occur in only one portion of the spatiotemporal continuum and thus have what we refer to as a "continuous distribution." Each will also display during that period of occurrence a unimodal frequency distribution through time relative to the abundances of other historical types. The relative abundance of a historical type will initially be rare, eventually rise to a single peak abundance, though not necessarily at the midpoint of its temporal duration, and finally decrease in abundance until it no longer occurs. Irving Rouse (1939:15) put it this way: "A [class] may have come into existence at a site by origination on the spot, or by diffusion from a neighboring tribe. It will then persist for a certain length of time, at first (perhaps) increasing in popularity, later decreasing in popularity. Finally, the [class] will die out, either by becoming extinct or by replacement by another [class]."

George Brainerd (1951a:304) later wrote that

> if a series of collections comes from a culture changing through time [read a *tradition*], their placement on the time axis is a function of their similarity; collections with closest similarity in *qualitative* or *quantitative* listing of types lie next to each other in the time sequence. This ... allows a "seriation" or ordering of the collections to be formed which, if time be the only factor involved, must truly represent the temporal placing of the collections, although determination of the direction early to late must be obtained by other means (emphasis added).

By "qualitative," we suspect Brainerd meant the presence or absence of types in collections, though this is not clear in his discussion; by "quantitative," we suspect he meant the relative frequencies of the types in collections. The former is referred to as occurrence seriation and the latter as frequency seriation. Brainerd was correct when he said that "the direction early to late must be obtained by other means." Cowgill (1972:382) phrased it this way: "[F]or any specific sequence of entities, another sequence that is its exact reverse is an equally good seriation."

Both occurrence and frequency seriation assume transmission and heritability, and they do so at two levels (Rouse 1939). First, each artifact identified as a member of a particular class is assumed to be related phyletically to every other specimen within that class, given their properties in common and, typically (though not necessarily), their spatiotemporal propinquity (e.g., Phillips et al. 1951; Rouse 1955). Thus, Brainerd (1951a:304) observed that a historical type "must be of sufficient complexity ... that the presence of an artifact belonging to [a historical type] suggests that its maker lived in the same cultural milieu as that of makers of all other artifacts classified into the same sorting group [read *historical type*]." This is what we have referred to as the type/species sense of heritable continuity. Second, the multiple classes that are seriated—whether by their occurrence or their frequency—are assumed to be related phylogenetically given the requirement of seriation that all seriated collections derive from a single cultural tradition—Brainerd's "cultural milieu." Willey and Phillips (1958:37) defined *tradition* as "a (primarily) temporal continuity represented by persistent configurations in single technologies or other systems of related forms." As such, a tradition reflects transmission, persistence, and heritable continuity (Phillips and Willey 1953; Willey and Phillips 1958)—the same point made by Raymond Thompson (1956a:39) in his definition of cultural tradition: "a socially transmitted cultural form which persists in time."

Recall that Nels Nelson (1916:163) wanted to find multiple types stratigraphically mixed together, because this would show "one gradually replacing the other [and thus account] for the otherwise unknown time that separated the merely superposed occurrences of types and from the point of view of the merely physical relationships of contiguity, connected them." The connection was evidence of heritable continuity. We suspect Nelson got this remarkable notion from his academic adviser, A. L. Kroeber. The *overlapping* (e.g., Ford 1938; Meighan 1959) temporal occurrences of the types connects the sets of materials being seriated and indicates heritable continuity between collections (Dunnell 1970; Lipo et al. 1997). Because traditions can be conceived of and constructed at the scale of an attribute of a discrete object, a type of discrete object or attribute combination, or particular combinations of multiple types of discrete objects (e.g., Neff 1992), we have referred to this as the tradition/lineage sense of heritable continuity to signify the potential for a diversity of units—of whatever scale—within a tradition or lineage. As we will see, both occurrence and frequency seriations that meet the requirements of the model indicate heritable continuity in both the type/species and tradition/lineage senses (Lyman and O'Brien n.d.a).

Requirements and Conditions of Seriation

If archaeological seriation is taken to be a method of comparing phenomena so that they may be ordered in such a manner as to reflect both the passage of time and heritable continuity, then the phenomena to be compared and ordered must be alike—Willey and Phillips's "persistent configurations"—except for their position in time. This means in part that the phenomena must be measured with theoretical units, termed *historical types*, or *classes*. It also means that the phenomena must meet certain requirements if a seriation is to successfully produce a chronological ordering of phenomena comprising a lineage. Determination of whether the phenomena actually meet the requirements necessitates sorting them into an order. Although the roots of the requirements are deep (e.g., Ford 1938), one of the earliest and most detailed statements on them is found in Phillips et al. (1951:219–236). The requirements were variously expanded, amended, and clarified in later years (e.g., Cowgill 1972; Ford 1962; Rouse 1967; Rowe 1961). Dunnell (1970) summarized them concisely, and we follow his discussion closely here.

First, assemblages of artifacts to be seriated must be of similar duration. Meeting this requirement ensures that the positions of particular assemblages in an ordering are the result of their age and not their duration. What duration should be represented by the assemblages? As Ford (1962:41) noted, "Each collection must represent a short period of time—the shorter the better. A sampling of the [artifact] population representing an instant in time would be ideal but, of course, is never achieved." In other words, the shorter the duration, the finer the temporal resolution in the final seriation. Although it often is difficult to determine if in fact each set of material spans a similar duration of time, Rouse (1967:162) indicated that "one can design [units] in such a manner that all will represent roughly equivalent periods of time." Ford (e.g., 1935c, 1962) merely collected sufficient artifacts to ensure against variation in sample size per assemblage influencing the final order; this may or may not have resulted in units that spanned similar chunks of time given that the rate of change may have varied over the total time span represented. Is there a way to meet this requirement? Dunnell (1970:312) suggested that without absolute chronological control, which when available "obviates the need for a seriation," the best procedure is to attempt a seriation and determine if there are any sets of material that are "at substantial variance with the model stipulated by seriation." If there are, then chances are that those variant sets are of different durations than the others. We provide an example later.

The second requirement is that all assemblages to be ordered must come from the same local area. This requirement is Julian Steward's (1929) geographic-propinquity criterion and is meant to ensure that what is being measured is variation in time rather than difference in geographic space (see Quimby 1943). It attends the fact that diffusion over geographic space can influence the results of a seriation (e.g., Phillips et al. 1951:223). As Dave Davis (1981:57) noted, the use of seriation to measure time "relies upon general homogeneity in patterns of [artifact] change within a geographic area," and changes in "the nature and rate of inter-community contact" would obscure the patterns of artifact distribution expected by the seriation model (see also Deetz and Dethlefsen 1965; Dethlefsen and Deetz 1966; Dunnell 1981). Thus, Rouse (1967:178) defined a "local area" as a chunk of geographic space "within which it is reasonable to suppose that there has been little, if any, geographic variation in culture." Exactly what a "local area" is varies across space and through time (Dunnell 1981); it is, in short, historically contingent. Only recently have techniques been developed that allow an archaeologist to analytically determine what a "local area" is in any given position along the spatiotemporal continuum (Lipo et al. 1997). We describe these techniques below.

In our view, meeting the second requirement increases the probability of, but certainly does not ensure, meeting the third requirement (following, particularly, Rouse's [1955] reasoning), which is that the assemblages to be ordered in a seriation all belong to the same tradition. Given the definitions of a cultural tradition quoted earlier, *if* one meets this third requirement, then heritable continuity is assured, and phylogenetic affinities between the seriated assemblages are guaranteed. As Dunnell (1970:311) noted, the third requirement means that the seriated assemblages "must be 'genetically' related" (see also Ford 1938). Dunnell (1970:311) indicated that using theoretical units such as are demanded by both occurrence and frequency seriation attends this requirement; that is, the use of theoretical units ensures heritable continuity at the type/species level and, Dunnell contends, without using this terminology, at the tradition/lineage level as well. With respect to the latter, the phylogenetic implications of the hierarchical structure of the Linnaean taxonomy in biology are transferable to a similar hierarchical alignment of artifacts. Thus, "pottery" might be aligned with a biological family, "types" of pottery with biological genera, and "varieties" of pottery with biological species, or the like. Pottery units can be seriated, as they comprise a pottery tradition, or monophyletic group, and projectile points comprise a different tradition. The two "families" of artifacts evolve independently of one another, and each, therefore, can serve as a test of the correctness of the ordering produced by the other (Dunnell 1970).

The roots of the seriation model as they pertain to occurrence and frequency seriation in Americanist archaeology contain insights as to why these seriation techniques map heritable continuity and why archaeologists failed to distinguish between historical and heritable continuity. Rouse (1939:14), for example, remarked that his postulated process of change, termed *replacement*, "should not be likened to the concept of 'evolution.' The latter implies development of artifact types, somewhat like the growth of plants. Replacement, however, expresses merely a change in [classes]. Each [class] is assumed, by definition, to be an immutable pattern [an ideational unit]. Hence, it cannot grow or evolve like a plant. It merely comes into existence, persists, and dies out. It is analogous to a culture trait which diffuses, not to an artifact type which evolves." Rouse here confused biological evolution with biological ontogeny, a confusion that was not Rouse's alone. Biers (1992:26), for example, remarked that an "evolutionary way of viewing stylistic development is common for ancient art, and is perhaps an influence from the natural world in which biological principles of birth, growth, and death can be observed." Biers's description is of ontogeny, not of evolution. Rouse's most important point was that change could be monitored by tracking shifts in the relative frequencies of classes; this was his "replacement," and it was at a larger scale than the ontogeny of an individual plant. A plant species evolves as the variant members change; thus, the composition of a population of them changes over time; *this* is replacement. It is the individual plants—and artifacts—that have ontogenies and life spans; it is the classes of plant—and artifact—forms or variants that have distributions and are replaced by other forms or variants. Rouse did not, however, make this clear, nor did he comment on the heritable continuity at the type/species level that was signified by classes.

Alex Krieger (1944:258), too, was a bit confused when he referred to Rouse's remarks and then wrote that some types in different cultures might be demonstrably

> related to one another because continuities of one kind or another show that the cultures are related. This circumstance involves the problem of genetic affinity in the products of human activity. The postulation of such affinity is based upon the principle that objects of a generally similar appearance found in related cultures are the products of changing tastes or values wrought upon some original ancestral form. Arguments for genetic affinity and evolution in artifact types must, however, be regarded with great caution....
>
> [Following Rouse (1939)], the proof of evolution in cultural products by means of exhaustive historical study is one thing, its assumption on the basis of visual impressions of similarity quite another. Thus basic relationships between specimens cannot be *assumed* to exist in any form, however close their superficial resemblances may be (emphasis in the original).

Krieger was here distinguishing between historical continuity and heritable continuity and stating in other terms George Gaylord Simpson's caveat. For want of a theory of cultural transmission and change, he did not realize how or why the two kinds of continuity might be interrelated and failed to explore the significance of his insight. Things improved a bit when Ford (in Phillips et al. 1951:220) remarked that historical types—as theoretical units—were "measuring units for a continuous stream of changing cultural ideas," but he, too, lacked a well-developed theory for such statements. Darwinian evolution provides just such a theory, and computer modeling founded on this theory indicates that Rouse's "replacement" and his and Ford's measuring units will display the expected spatiotemporal distributions in an occurrence or frequency seriation precisely *because of* transmission and heritable continuity (e.g., Lipo et al. 1997; Neiman 1995). The requirements of occurrence and frequency seriation underpin the seriation model. Only by meeting the conditions will it be reasonable to infer that a seriation represents a chronology and heritable continuity (Dunnell 1970). One has to know, then, if the conditions have been met. Fortunately, failure to meet the conditions of the model will result in an ordering that does not match the theoretically expected result. We turn to occurrence seriation first because it is the simpler of the two and its basic analytical principle is also part of frequency seriation.

Occurrence Seriation

Occurrence seriation was suggested in the late 1950s and early 1960s as an alternative to frequency seriation, which had been developed in the second decade of the twentieth century (Lyman et al. 1997b). The first extended discussion of occurrence seriation of which we are aware is found in a paper by Paul Dempsey and Martin Baumhoff (1963), though John Rowe (1959) mentioned it several years earlier. As William Lipe (1964:103) noted, the major difference between what Rowe, Dempsey, and Baumhoff proposed and what previous researchers such as Kroeber (1916a, 1916b), Ford (e.g., 1949), and Brainerd (1951a) had done was that "with the former [occurrence seriation], collections are compared in terms of presence or absence of types, whereas in the latter [frequency seriation], the relative frequencies of types within the respective collections are employed in the comparisons." Rowe (1959:321) argued that "it is preferable to avoid relying on frequencies [of types] for making chronological distinctions and depend instead on observations of presence and absence," because frequency data were subject to sampling problems. Dempsey and Baumhoff (1963:498) also believed that presence–absence data would be more sensitive to chronological issues because "types that occur with low frequency may be among the best time-indicators [and] the presence of single specimens of certain types may be crucial in establishing chronologies." They also found that "weighting" a type's importance in a seriation on the basis of its relative abundance, though objective, resulted in "gross differences in the contribution of the various [types] to the final ordering" and that "the chief effect of [including rare types] is to increase the amount of busywork" (Dempsey and Baumhoff 1963:498).

To circumvent such problems, Dempsey and Baumhoff (1963:501) recorded each type "merely as being present or absent" in each of the collections they were ordering. Table 6.1 is a simple example of the occurrence-seriation procedure. It

TABLE 6.1. An Example of an Occurrence-Seriation Procedure

Assemblage	Historical type 1	2	3	4	5
Unordered					
A	+		+	+	+
B			+	+	+
C	+		+		+
D			+	+	
E	+	+			+
F			+	+	
Ordered					
E	+	+			+
C	+		+		+
A	+		+	+	+
B			+	+	+
D/F			+	+	

consists of six assemblages, the artifacts of which have been classified as belonging to one of five historical types. The assemblages are unordered in the top half of the table. The seriation model stipulates that "the distribution of any historical or temporal class is continuous through time" (Dunnell 1970:308). This is, in short, the principle upon which occurrence seriation rests; theoretically, each historical type will occur once in time; that is, its distribution will occupy one and only one chunk of the spatiotemporal continuum. Failure to produce such a result suggests that one of the requirements has not been met. The procedure of occurrence seriation is to sort the collections—the rows—such that each type—each column—displays a continuous occurrence (signified by the column of "+" marks). The order resulting from meeting the expectations of the seriation model is given in the lower half of Table 6.1. Note that it makes no difference if the ordering from top to bottom is "E, C, A, B, D/F" or "D/F, B, A, C, E," because the direction of time's arrow is unknown. That knowledge must come from other data independent of the seriation, such as knowing that types 1 and 2 occur late in time and types 3 and 4 occur early in time based on associated radiocarbon dates or stratigraphic excavations. Note that assemblages D and F are identical in terms of the types they contain. They cannot be sorted and must, in this example, be considered *contemporaneous*. In saying this, we mean only that they cannot be distinguished *in terms of the units we have used* (see also Patterson 1963). Other units might have allowed the two assemblages to be separated.

Frequency Seriation

As we discussed in Chapter 3, A. L. Kroeber invented archaeological frequency seriation, and it is clear, despite claims to the contrary (e.g., Browman and Givens 1996; Praetzellis 1993; Willey and Sabloff 1993), that phyletic seriation played no role in Kroeber's development of the technique (Lyman et al. 1997b, 1998b). Part of the reason for the clarity of this distinction is that Kroeber used classes—ideational units—of discrete objects to seriate his collections. Although his classes were extensionally derived, he did not use the empirical units of phyletic seriation, nor did he use shifts in the states of attributes of discrete objects typical of phyletic seriation. Finally, he used frequencies of classes, not the presence–absence of particular attribute states, to derive the ordering. What Kroeber *did* was to invent frequency seriation.

In addition to the stipulation that "the distribution of any historical or temporal class is continuous through time" (Dunnell 1970:308), the frequency-seriation model specifies an additional principle, elegantly stated by Phillips, Ford, and Griffin (1951:220) with respect to pottery types: "If our pottery types are successful measuring units for a continuous stream of changing cultural ideas, it follows that when the relative popularity of these types is graphed through time, a more or less long, single-peak curve will usually result. Put in another way, a type will first appear in very small percentages, will gradually increase to its maximum popularity, and then, as it is replaced by its succeeding type, will gradually decrease and disappear." This is the popularity principle. Phillips et al. (1951:221) noted that if "a complex of cultural materials representing a space–time continuum of culture history is classified in a consistent manner, the popularity curves of the various constituent types will form a pattern. Each portion of this pattern will be peculiar to a particular time and area." Note the term "replaced" in Phillips et al.'s statement; it has the same meaning as Rouse's (1939) use of the term described earlier: Classes will co-occur over time and thus will connect assemblages by their overlapping occurrences. The important thing to note is that with frequency seriation, not only must each historical class be continuously distributed over a single chunk of the spatiotemporal continuum, but also its relative frequency must fluctuate unimodally. Failure of the classes within a frequency seriation to display these properties subsequent to sorting and ordering indicates that one or more of the requirements of the seriation model have not been met.

Phillips et al. (1951) illustrated their concept of a frequency seriation as the history of transportation in Ohio, shown in Figure 6.7. The history is fictional, which they admit, but that is irrelevant to the graph's heuristic purpose. Modes of transportation are arranged vertically, the direction in which time is running. Horizontal bars represent percentages of each mode of transportation at ten-year intervals. Summing percentage bars across any row yields 100 percent. In 1940, automobile transportation accounted for 51 percent of all transportation in Ohio,

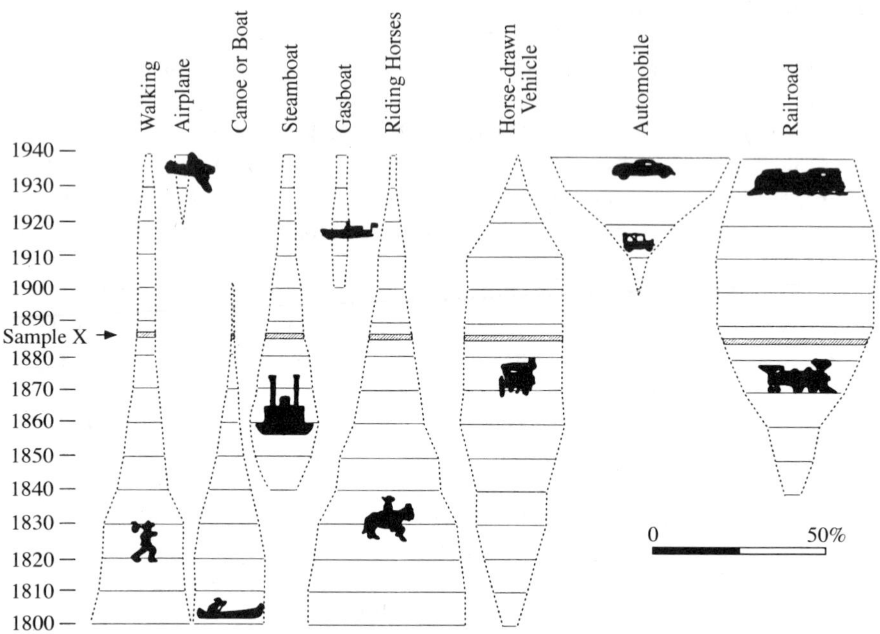

FIGURE 6.7. Fictional percentage-occurrence graph of transportation types of Ohio from 1800 to 1940 (after Phillips et al. 1951).

rising from 42 percent in 1930. Similarly, horse-drawn vehicles accounted for only 2 percent of transportation in 1940, down from 6 percent in 1930. The relative frequencies of all transportation modes differ by ten-year period; that is, the quantitative picture of any particular block of culture history differs from all others. Thus, one could determine the correct chronological position of "Sample X" from Ohio in Figure 6.7 based on the frequencies of each type of transportation. Importantly, Phillips et al. noted that the pattern for Ohio at any particular point in time will differ from patterns obtained elsewhere, a fact later corroborated by a study of New England gravestones (Deetz and Dethlefsen 1965).

The unimodal frequency-distribution model applies to all artifact styles, including archaeological specimens. The example in Table 6.2 shows historical types aligned vertically. Each row represents a single collection of specimens, each from a distinct spatial position on the landscape (not to preclude multiple collections from different areas of a single site), and as in the example of transportation modes (Figure 6.7), the percentages of the various types in each collection sum to 100 percent. Our example begins with the collections in no particular order—the top half of Table 6.2. The seriation procedure involves two

Table 6.2. An Example of a Frequency-Seriation Procedure

	Historical type				
Assemblage	1	2	3	4	5
Unordered					
A	10		30	10	50
B			50	30	20
C	20		15		65
D			40	60	
E	30	25			45
F			20	80	
Ordered					
E	30	25			45
C	20		15		65
A	10		30	10	50
B			50	30	20
D			40	60	
F			20	80	

steps. First, the rows are rearranged such that each type displays a continuous distribution. This has already been done in the lower half of Table 6.1, as the Table 6.2 data comprise the same collections as the former but with frequency data rather than presence–absence data. The second step involves sorting the collections such that each column of frequencies defines a lenticular frequency distribution—the bottom half of Table 6.2. Note that as with occurrence seriation, it makes no difference if the ordering from top to bottom is "E, C, A, B, D, F" or "F, D, B, A, C, E," because the direction of change is unknown. Again, knowledge of the direction of change must come from data independent of the seriation.

In the example in Table 6.2, assemblages F and D are not contemporaneous, such as they are in Table 6.1. The relative-frequency data allow these two assemblages to be separated and ordered along with the other assemblages. Although such increased resolution in temporal ordering may not always be found with frequency data relative to presence–absence data for the same collections, we suspect it often will, simply because of the greater information content in the frequency data. As Steven LeBlanc (1975:23) noted, "Chronologies based on the presence or absence of types or attributes will never be as accurate as those based on their relative frequencies. This is because technological and stylistic changes are rarely instantaneous in inception or momentary in duration"; that is, change will occur as shifts in the particular combinations of variants or the frequencies of variants—whether attributes of discrete objects or types of discrete object—rather than variant A occurring only prior to one point in time and variant B occurring only after that particular point. In short, change is gradual and involves overlap.

As we document elsewhere (Lyman et al. 1998b; O'Brien and Lyman 1999a), archaeologists spent considerable time during the first half of the twentieth century trying to determine the best way to present both the data and the results of a seriation. The "best way" was found to comprise a centered-bar graph that showed not only the relative frequencies of the types but also how those frequencies changed through time. Informally known as the "Ford technique" (O'Brien and Lyman 1998), this manner of presenting data and results simultaneously "presents far more information than do the comparable [and more modern] statistical techniques. Not only is the degree of similarity between two units indicated, but also the actual form and source of similarity is shown. A [statistically based] seriation can always be constructed from the information contained in a Ford seriation, but graphic seriations cannot be constructed from a matrix of similarity coefficients" (Dunnell 1970:306).

Before computers were available, archaeologists did frequency seriations by hand. Ford (1962) suggested using long strips of paper containing bars of length proportionate to the percentage of a particular pottery type in a particular collection. Each strip shown in Figure 6.8 is a separate collection, with bars showing the percentage of each type of pottery contained in that collection. Eleven pottery types are shown, though no collection contains sherds of all eleven. Once each collection is graphed in terms of type percentages, the strips are moved up and down until a best fit is found, meaning that there are as few violations of the continuity and popularity principles as possible; that is, the resulting lenticular type-frequency curves—what Ford (1952:344) referred to as " 'battleship' frequency curves"—are as close as possible to a unimodal frequency distribution.

The Conditions of the Seriation Model Produce Lineages

Our examples in Table 6.1 and 6.2 are simple; in reality, archaeologists often have more than five historical types and more collections than six. There are numerous computer programs that seriate collections, but we do not delve into them here (see references in Johnson 1972a; Marquardt 1978; see discussion in Lipo et al. 1997). Rather, our attention is focused here on understanding the principles of seriation. Why is it that seriations produce lineages? In particular, why should the analytical exercise of seriation result not only in a rendition of historical continuity but also in heritable continuity? These questions can be answered by noting that seriation produces lineages if it meets the requirements of the seriation model. Such an answer demands, however, that we know when the requirements are met. Earlier, we suggested that we may conclude the conditions have *not* been met if the classes in an ordered set of collections fail to (1) occupy

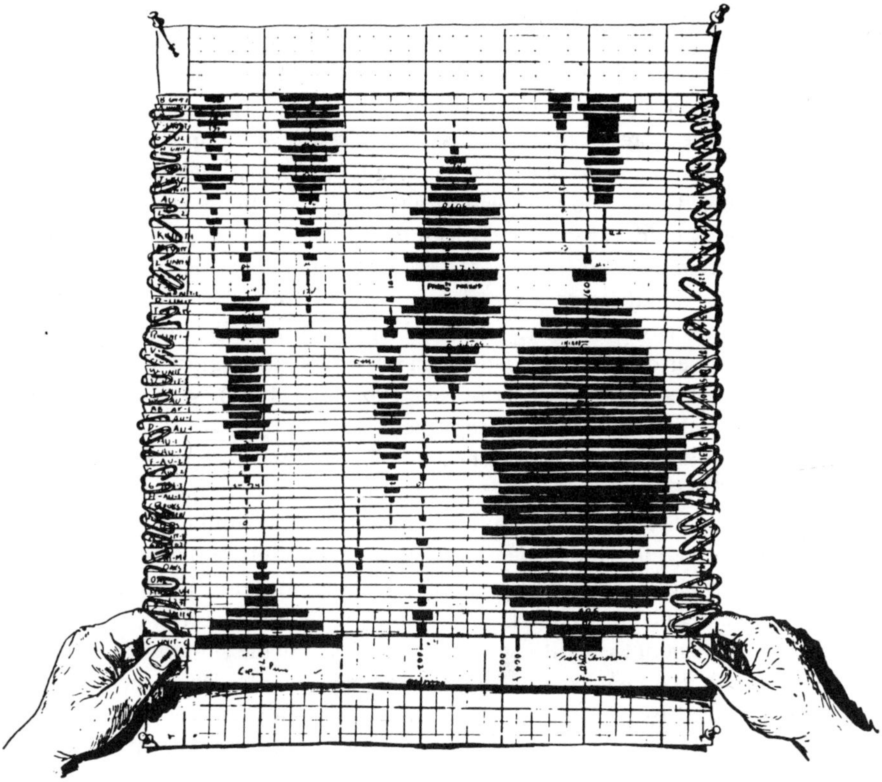

FIGURE 6.8. James A. Ford's thumbs-and-paper-clips method of seriating collections. Each strip of paper represents a surface collection or excavation level; on each strip, bars have been drawn to indicate the percentage of each pottery type. The strips are then moved up and down until the best fit is attained (from Ford 1962).

one chunk of the spatiotemporal continuum and (2) display unimodal frequency distributions.

Archaeologists have long realized that not all collections of artifacts meet the conditions of seriation. With respect to occurrence seriation, Cowgill (1968:518) indicated that it made sense to perform such a seriation only "when we feel sure that the absence [of a type] in the collection means that the [type] was *really* absent" and is not absent from seriated collections as a result of sampling error (emphasis in the original). Slight deviations from the ideal battleship-shaped curves, a few of which are noticeable in Figure 6.8, might be explained by sampling error. Ford (1962:41), writing about frequency seriation, indicated that "each collection must be unselected and large enough to give reliable percentages. A total of 100 or more sherds is desirable, but occasionally collections as small as 50 can be used." Ford's imposition of the 50- to 100-sherd sample was based on his experience in the Mississippi Valley, which had shown that samples of that size often yielded adequate results but that the size of an adequate sample will probably vary from place to place. He was right on the mark with his comments about "unselected" samples being desirable, since anything less than a near-random sample would introduce bias into the analysis. There are now available many ways that archaeologists have for assessing sample adequacy (see Leonard and Jones [1989] and references therein). Suffice it to note that when performing either an occurrence or a frequency seriation, the analyst should check for sample-size effects; such efforts have, to date, been rare in archaeological applications of the seriation method (for an exception, see Lipo et al. 1997). What of the other, more formal conditions of the seriation method?

Cowgill (1968:517) noted that seriation can be conceived of as having two "central tasks": (1) determining the correct chronological sequence of a set of archaeological units and (2) ordering units based on their similarity. Importantly, Cowgill emphasized that the best ordering in terms of unit similarity may or may not also be a chronological ordering; that is, both historical and heritable continuity, as we have defined them, are not necessarily entailed in a seriation. Thus, for the accomplishment of Cowgill's second task to result in the simultaneous accomplishment of his first task, because seriation is merely an ordering technique, an ordering produced by occurrence or frequency seriation can be inferred to be chronological only if the conditions of the seriation model have been met. How do we determine if those conditions have been met? The first condition is that all collections must be of comparable duration. Dunnell (1970:312) suggests that for frequency seriation, if a unimodal frequency distribution for all included types cannot be obtained, then perhaps the collections vary in duration. Furthermore, collections of overly long duration—sometimes termed "mixed" assemblages—may contain types from rather different vertical positions in the seriation. An example of such a collection is given in Figure 6.9, in which both the principle of continuous occurrence and the principle of unimodal frequency distribution are violated by assemblage H. Such deviations from the seriation model cannot be tolerated, particularly when they are found in the middle of the duration of a type: "Variation from the [seriation] model can be tolerated only where it is predictable, namely at the beginnings and ends of distributions and in sparsely populated classes" (Dunnell 1970:313).

Another requirement of the seriation model is that the seriated collections have to come from the same local area. This condition is the most difficult one to meet (Dunnell 1981). As James Deetz and Edwin Dethlefsen (1965) showed, artifact form can vary more or less continuously across space, so the problem becomes one of where to draw a boundary line around a "local area" such that

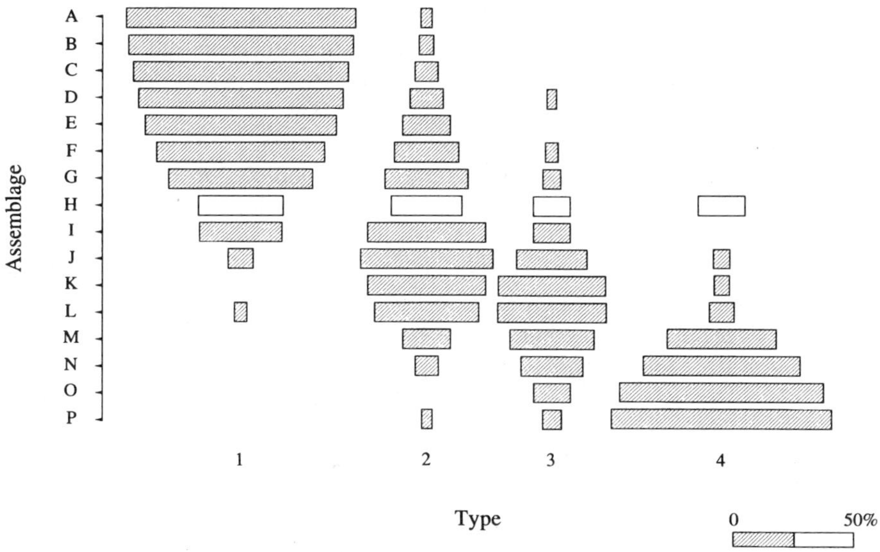

FIGURE 6.9. A seriation of 16 assemblages and four artifact types showing the effect on the expected frequency pattern of an assemblage (H, unshaded) of different duration (longer) than all other assemblages in the ordering (after Dunnell 1970).

formal variation within the bounded area is minimal and variation between such areas is maximal. Another way of saying this is that both within-spatial-unit homogeneity of form and between-spatial-unit heterogeneity of form are maximized. There are at present two ways to meet this requirement. Cowgill (1972: 384) suggests that "it is only necessary that [nontemporal] sources of variation, within the particular set of units being studied, make for differences that are small relative to differences reflecting the smallest time intervals one hopes to reliably distinguish." Thus, one might construct historical classes that vary greatly along the temporal dimension and minimal along the spatial dimension; that is, the classes consist of attributes that "show little variation in space and much variation in time" (Dunnell 1970:315).

The other way to meet the "same local area" requirement is empirical and rests on the notion of heritable continuity in the tradition/lineage sense; that is, it is explicitly founded in a Darwinian notion of heritability effected by transmission, in this case, of ideas (replicators) manifest as artifact styles. Under such a conceptual umbrella, one can argue that a "local area" comprises a "community or localized group of communities [that] produces a distinctive style of pottery that is distinguished easily from the products of other centers of communication and [artifact] production" (Neff 1992:151). Following this notion and using the principles of frequency seriation and the transmission model of Fraser Neiman (1995), Carl Lipo and colleagues (1997) simulated trait transmission and mixture over time and across space. They found that perfect lenticular curves could be generated, with sufficient control of time, using collections recovered from particular limited pieces of geographic space. Applying their findings to archaeological data from the Lower Mississippi Alluvial Valley, they identified instances of prehistoric community interaction and structure; that is, they identified a set of "local areas" on the basis of similarities among the collections, measured as relative type frequencies. Of course, because cultural-transmission pathways can alter direction over time—in other words, cultural transmission and thus evolution is variously reticulate over not only time but geographic space as well—the analyst needs to determine if the "local areas" identified for one portion of the temporal continuum change their boundaries over time. Given our understanding of cultural transmission, we suspect they will. This in turn demands that the spatial dimension be monitored and controlled—so-called "local areas" identified—throughout a seriated set of collections.

The final requirement of the seriation model is that the collections all belong to the same tradition. The easiest way to recognize a failure to meet this condition is when, in a completed seriation, there is marked discontinuity or a lack of overlap among the collections. An example is shown in Figure 6.10. This result may be produced by the representation of more than one "local area" by the collections that are being seriated; this possibility can be checked quickly by determining if, in the case of Figure 6.10, assemblages A–D come from one area and assemblages E–G from another, nonoverlapping area. If the two areas identified by the assemblages do not overlap in space, then it is possible that the two sets of assemblages represent a distinct break in heritable continuity and thus represent two separate traditions. The example in Figure 6.10 is a clear case of a violation of the principle

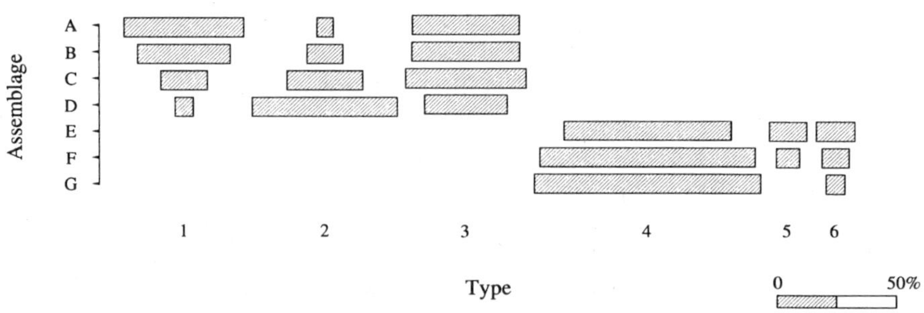

FIGURE 6.10. A seriation of seven assemblages and six artifact types showing the effect on the expected frequency pattern of including two traditions (one containing types 1–3 and the other types 4–6) between which there is no heritable continuity or overlap. Assemblages E, F, and G could legitimately have been placed on top rather than at the bottom (after Dunnell 1970).

of heritable continuity, the principle of historical continuity, and Cowgill's (1972:384) requirement that there never "be a break in the sequence so abrupt and catastrophic that units immediately following the break bear no (or only accidental) resemblance to units before the break." In other words, there is no overlap—historical or heritable—connecting the two sets of assemblages.

A Final Word on Heritable Continuity

The seriation method provides a way to build phylogenetic histories because it is a method that tests hypothesized historical, specifically phylogenetic, relations among artifacts (Dunnell 1970; Teltser 1995c). Underpinning frequency seriation is the notion that a historical type will (1) occupy a single chunk of the temporal continuum and (2) its frequency distribution will be unimodal through that chunk of time. This notion of a historical type in turn rests on the theoretical axiom of heritable continuity (e.g., Lipo et al. 1997; Neiman 1995; Teltser 1995c). That homologous similarity is demonstrated in successful frequency seriations — ones that result in all seriated types displaying unimodal frequency distributions—is so for two reasons. The first relates back to the issue of the nature of types: Are they real or are they arbitrary constructs—what we would term ideational units, or classes? Frequency seriation demands that they be classes so that their distributions as opposed to locations can be determined (Dunnell 1970). Furthermore, because frequency seriation uses classes typically termed *styles*, typological identity is ensured by its being imposed on the artifacts (that is, intensional units are used) rather than being derived from them (as in extensional units). Each artifact identified as a member of a particular class is related phyletically to every other specimen within that class, given their properties in common—Spaulding's (1954b) notion that typological resemblance is the result of transmission—and their spatiotemporal propinquity. This is what we refer to as the type/species sense of heritable continuity (Lyman and O'Brien n.d.a; O'Brien and Lyman 1999a, 2000b).

The second reason that successful frequency seriations denote transmission, heritable continuity, and homologous similarity is equally simple to grasp conceptually. A successful frequency seriation results in *multiple* associated historical types of a particular category of artifact—pottery or projectile points, for example—displaying unimodal frequency distributions. This is the congruence test of phylogeneticists and cladists and is also Julian Steward's multiple-trait criterion. With respect to the congruence test, the multiple classes that are seriated are also assumed to be related phylogenetically, given the requirement of seriation (Dunnell 1970; Lipo et al. 1997) that all seriated collections derive from a single cultural tradition, which by definition reflects transmission, persistence, and heritable continuity. We refer to this as the tradition/lineage sense of heritable continuity (Lyman and O'Brien 2000; O'Brien and Lyman 1999a, 2000b) to signify the potential for a diversity of units—at whatever scale (e.g., Neff 1992)—to exist within a tradition or lineage. As we noted earlier, the phylogenetic implications of the hierarchical structure of the Linnaean taxonomy in biology are transferable to a similar hierarchical alignment of artifacts.

It perhaps is not surprising that occurrence seriation, frequency seriation, and to a lesser extent, phyletic seriation should reflect heritable continuity. This is precisely the notion—if poorly developed theoretically and seldom stated explicitly—upon which these analytical techniques were built in the late nineteenth and early twentieth century (Lyman and O'Brien n.d.b; O'Brien and Lyman 1999a). But the discipline at large seems to have forgotten this. For example, just when discussions over the importance of heritable continuity were reaching a peak in intensity, Clement Meighan (1959:203) indicated that "overlapping of similar traits [read *classes*] in different finds" was critical to frequency seriation, but he did not say why. Two things seem to have happened that exacerbated disciplinary forgetfulness regarding heritable continuity while simultaneously retaining seriation merely for dating purposes, that is, as an indication of historical continuity only. First, stratigraphic excavation, followed quickly by dendrochronology and later by radiocarbon dating, obviated any need to explicate the theoretical underpinnings of the seriation method. Stratigraphic excavation meant relative time was readily visible when an excavated site was stratified, and dendrochronology, and particularly radiocarbon dating, meant absolute dating was possible. Seriation was unnecessary. Second, archaeologists in the 1940s discarded any notion that a Darwinian-like model of evolutionary phylogeny was applicable to cultural phenomena (Lyman and O'Brien 1997). This exacerbated the discipline's failure to develop theory appropriate to the seriation method. Instead, various archaeologists suggested how to distinguish between analogs and homologs, but their suggestions were incomplete and did not provide testable results. Seriation, although derived from commonsensical notions of Darwinian "descent with modification," incorporates those suggestions and provides an analytical means to test results.

As philosopher Marc Ereshefsky (1992:90) points out, the distinctive aspect of Darwinian evolutionary theory resides in the fact that it requires transmission—heredity—and this in turn "requires the different generations of a population to be connected by reproductive ties, and such ties require those generations to be spatio-temporally connected." Control of the formal, spatial, and temporal dimensions is necessary to the production of historical chronicles comprising cultural phylogenies. To ensure that any produced chronicle is a phylogeny and not simply a temporal sequence demands that homologous similarities be identified and used to demonstrate heritable continuity. Dunnell's explicit definition of archaeological style as comprising homologous similarity was a necessary step to building an archaeological theory founded in Darwinism. In the next chapter, we turn to how the seriation method can be used to construct lineages in such ways as to lead to their explanation.

Chapter 7

Tempo and Mode in Evolution

On two topics, in particular, the paleontologist enjoys special advantages.... The first of these general topics has to do with evolutionary rates under natural conditions, the measurement and interpretation of rates, their acceleration and deceleration, the conditions of exceptionally slow or rapid evolutions, and phenomena suggestive of inertia and momentum.... [T]hese problems are meant to be suggested by the word "tempo." The group of related problems implied by the word "mode" involves the study of the way, manner, or pattern of evolution, a study in which tempo is a basic factor, but which embraces considerably more than tempo.... The purpose is to examine the figurative outline of the stream of life and the circumstances surrounding each characteristic element in that pattern. (Simpson 1944:xxix–xxx)

One of George Gaylord Simpson's great contributions to paleontology was the manner in which he separated two variables of evolution: the rate at which a population of organisms changes and the means by which it changes. Biologists and paleontologists before Simpson clearly recognized the separation, but his exposition was unprecedented in terms of clarity. Although time cannot be unlinked from either mode—the process (usually rendered as a central tendency) that takes place over time and leaves patterns—or tempo—the rate of change over time—these two dimensions of evolution must be examined independently. The two can be difficult to distinguish conceptually. For example, it might appear at first glance that punctuated equilibrium is a model about tempo (see Chapter 4), and indeed in the early stages of formulation it *was* largely restricted to tempo (Gould and Eldredge 1977). Later, however, it became

> a theory ... about [both] the tempo and mode of evolution. It holds, speaking of mode, that significant evolutionary change arises in coincidence with events of branching speciation.... [T]he proper geologic scaling of speciation renders branching events as geologically instantaneous and that, following this rapid origin, most species fluctuate only mildly in morphology during a period of stasis that usually lasts for several million years. (Gould 1982b:83)

Relative to tempo, punctuated equilibrium holds that the rate of change will fluctuate; a long period of morphological and taxonomic stasis within a lineage will be followed by a brief burst of speciation (cladogenesis) and concomitant morphological diversification that in turn will be followed by another long period of stasis, followed in turn by another brief burst of speciation, and so on. The mode of change is cladogenesis effected by peripatric speciation. According to the primary architects of the punctuated-equilibrium model, construing evolutionary tempo and mode in this fashion eliminates the "insoluble" problem of identifying biological species in the fossil record—they are signified by their morphometric stasis (e.g., Kelley 1983)—and instead focuses on the evolutionary implications of such units (Eldredge and Gould 1972:93). By shifting the scale from individually variant fossils to populations of fossils that contain some variation (but no more than extant species) and treating those populations *as if they were biological species*, all sorts of evolutionary implications can be found in the fossil record that cannot be observed or conceived of by neontologists. Thus, "speciation is the raw material of macroevolution"; this "represents no departure from Darwinian mechanisms, but only the previously unrecognized mode of operation for natural selection at hierarchical levels higher than the local population"; and the "differing geometry of change is simply an 'allometric' phenomenon of scale: the same process works in differing ways at different levels of complexity and organization" (Gould and Eldredge 1977:139). New evolutionary concepts that were developed as a result of this shift in scale include species selection (Stanley 1975; Vrba 1984), the "effect hypothesis" (Vrba 1983), and sorting (Vrba and Gould 1986; for a recent review of the first two, see Grantham 1995). In our view and that of the architects of punctuated equilibrium, the significant implication of these results is that evolution must be conceived of at various scales, not just at the scale indicated by the definition of evolution as genetic change. The latter is microevolution; change at larger, more inclusive scales in the taxonomic hierarchy—from species on up—is macroevolution.

Phyletic gradualism emerged from the definition of (micro)evolution as genetic change that was adopted by the Synthesis. Its tempo is included in its name—gradual—yet no one, particularly Simpson (1944, 1953), denied that the tempo of change could vary from virtual stasis to relatively rapid, though it was paleobiologists in particular who recognized and studied this aspect of evolution (e.g., Kurtén 1960, 1965). Thus, Simpson explicitly distinguished among three tempos of evolution: *horotely* comprised normal, or modal, rates of change; *bradytely* comprised slower than modal rates; and *tachytely* comprised rates that were faster than the modal rate within a group of organisms. And although Darwin and the architects of the Synthesis never explicitly restricted mode to anagenesis, that tended to be the mode around which the Synthesis hardened (Gould 1983b). This is one reason why punctuated equilibrium was proposed (Eldredge and Gould 1972)—to take more complete account of evolutionary modes. Anagenesis, however, is antithetical to punctuated equilibrium. But phyletic gradualism focused on generation-to-generation change and ignored taxon-to-taxon change. This is another reason that punctuated equilibrium was proposed: Change visible in the fossil record would not be readily reduced to neontologically visible microevolutionary change. The temporal resolution of the fossil record seldom permitted perception—and thus analysis and interpretation—of microevolutionary processes and patterns (Eldredge 1985, 1995a, 1999).

In previous chapters, we described several methods and techniques useful not only for measuring the flow of time but also for constructing lineages, but we were not too concerned with measuring the tempo and mode of change. Here we focus attention primarily on methods of graphing change at various scales that can be used to examine the tempo and mode of evolution. Our discussion underscores the theme of this book—that systematics is critical to analysis. We begin with a brief consideration of why frequency seriation often may not work well with biological units, and then illustrate how a paleobiological technique can be of use to archaeologists if particular sorts of archaeological units are employed.

GRAPHING FINE-SCALE BIOLOGICAL CHANGE

Much debate in paleobiology in the 1980s concerned identifying the tempo and mode of change within lineages of organisms (Chapter 4). Although those debates have died down in the 1990s as researchers have gone back to identifying various tempos and modes (Novacek 1996), the questions raised by the debates are just as important today as they were a decade ago. Is change gradual and phyletic, is it saltational, is it cladogenetic? To simplify discussion, we could build a paradigmatic classification with two dimensions—tempo and mode. The tempo, or rate, dimension could have two attribute states—rapid and gradual. Similarly, the mode dimension could also have two attribute states—anagenetic and cladogenetic. The resulting four *classes* of change, in no particular order, would be gradual anagenetic, rapid anagenetic, gradual cladogenetic, and rapid cladogenetic. The first was termed *phyletic gradualism* by Eldredge and Gould (1972); it comprises gradual, *continuous* change within a lineage and without cladogenesis, or branching. The second—rapid anagenetic—has been termed saltation (e.g., Eldredge 1985, 1999), punctuated gradualism (Malmgren et al. 1983, 1984), punctuated anagenesis (MacLeod 1991), or punctuated stasis (Springer and Murphy 1994) to denote brief periods of rapid change within a lineage, without cladogenesis, followed by relatively long periods of gradual change, stasis, or a combination of the two. Gradual cladogenetic change was perhaps what Darwin (1859) meant to signify with the single figure in *On the Origin of Species*, given his adherance to gradualism and his desire to explain the diversity of life (for a brief, yet cogent discussion of this diagram, see Eldredge 1999:80–89). Change was gradual, in Darwin's view, yet if lineages branch and diversify, then the genealogical hierarchy of life reflected in the Linnaean taxonomy is accounted for readily *and theoretically*.

The last class of change, rapid cladogenetic, might be considered punctuated equilibrium, though we would have to add the significatum that there be long periods of stasis between cladogenetic events. How long? Gould (1982b) suggested the burst of diversification should comprise less than 1 percent of the lineage's duration, meaning the period of stasis would comprise no less than 99 percent of that duration. This suggestion, along with the necessity of admitting stasis, raises other problems with our hypothetical classification of tempo and mode. How rapid is rapid, and how gradual is gradual? Tempo implies a *rate*—recall that Simpson distinguished three classes of rate—which means we must first specify some unit of change per some unit of time and then specify how many units of change per how many units of time differentiate gradual from rapid change. Mode is perhaps less troublesome biologically because change either is within a lineage—from chronospecies 1 to chronospecies 2, and so on—or involves speciation—species A gives rise to daughter species B and C. But identifying mode demands that sister species—those sharing an ancestor—and daughter species be distinguished. Given that fossil species are identified based on similarities and differences in morphometry, and ignoring the fact that interbreeding or the lack thereof cannot be detected in the fossil record, the importance of systematics to identifying a particular mode of change should be evident. How have paleobiologists dealt with change in the fossil record?

Traditional Graphs of Change

The difficulties we have identified with constructing a classification of tempo and mode underscore the fact that change is historically contingent, or as Simpson (e.g., 1963) would have said, configurational. Two decades after the punctuated-equilibrium model of evolution was proposed, sufficient lineages had been studied to show that empirical evidence could be mustered for all four of our hypothetical classes of change plus variants of them (see virtually any issue of *Paleobiology* published in the last twenty years). How are variations in tempo and mode to be detected? It is abundantly clear that, as Philip Gingerich (1980:419) so eloquently put it, "There are three principal components of evolutionary space: morphology, time, and geography," and all must be tightly controlled if any success is to be achieved in identifying kinds of tempo and mode. Given Richard Fortey's (1985, 1988) observations that every class of change should be tested on equal terms against available data (see also Vaupel Klein 1994), how do paleobiologists sort out which class applies in any given case? Part of the problem resides in how paleobiological units—species—are constructed and how they are graphed.

Take a look at Figure 4.2, which shows change in the morphology of a tooth in several taxa of early Eocene primates. In a way, this graph is analogous to an archaeologist's phyletic seriation (see Figures 3.5 and 6.4–6.6), the major difference being that the paleobiologist uses ideational units, typically metric units such as length and width, whereas the archaeologist uses empirical units and largely ignores within-unit variation. Were the archaeologist to use theoretical units like those of the paleobiologist, graphs such as those in Figure 7.1 could be produced. This set of graphs meets Fortey's criterion of illustrating the data—perhaps the differentia of Eldredge (1979)—and is modified from a similar set shown by paleobiologists Robert Martin and Anthony Barnosky (1993:7), who refer to each graph as illustrating a "pattern of morphological change," where a pattern is a representation of one or more morphological traits arrayed as a statistical summary against time. The graphic standard is that each horizontal bar represents the range of some measure of variation within a temporally distinct sample, and each vertical bar represents the mean of that measure for the sample. Although ideational units are used to measure specimens, these specimens are grouped not by their membership within a morphometrically defined class but rather on the basis of the temporal association of specimens. As Figure 4.2 demonstrates, one can modify the graph to include the standard deviation if that is desired. One can, in fact, plot each individual specimen (e.g., Schankler 1981), though such a depiction occurs but rarely in the literature with which we are familiar.

The patterns of change illustrated in Figure 7.1 are labeled with descriptive terms based on the behavior of the means over time. We say "descriptive" for two reasons. First, although the temporal axis is labeled, it is not indicated whether that axis is ordinal or interval in terms of scale. Only by presuming that it is interval scale is it reasonable to label graph D as "variable rate" and graph C as "saltational." We return to this later. Second, we have graphed ten samples for each pattern of change, but is this a sufficient number? In other words, how long must the mean of samples fluctuate back and forth to be considered a case of stasis? Similarly, recalling that a simple best-fit regression line can be drawn through two data points, how many samples are required to represent stasis or change? Certainly at least three, but is that enough? Despite these difficulties, the graphs in Figure 7.1 are useful heuristic devices for illustrating some of the possible tempos and modes of change. The concept of "mosaic change" is a critical one, as it

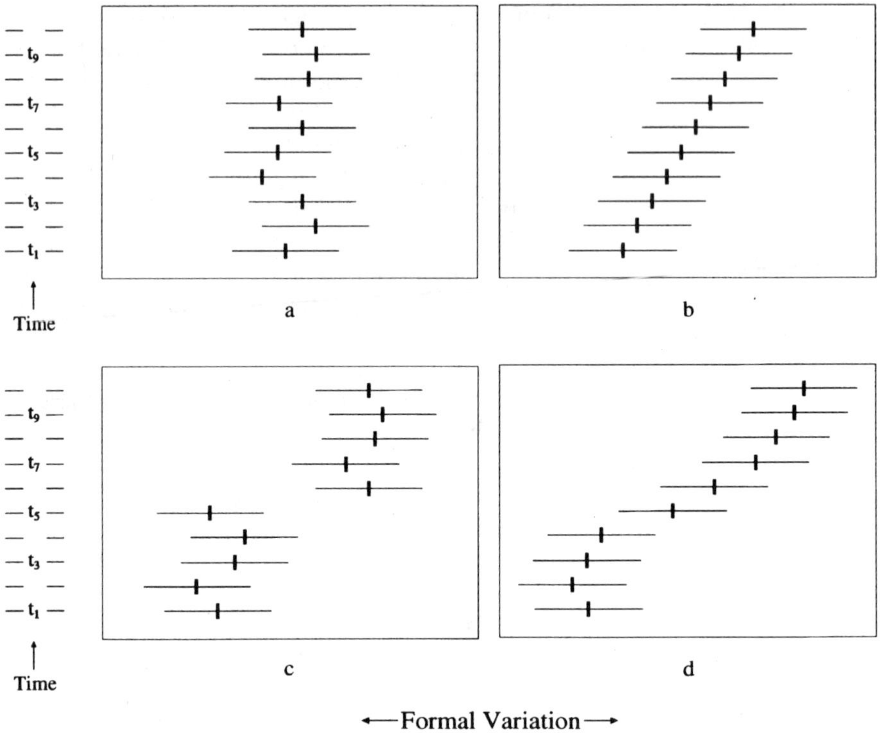

FIGURE 7.1. Patterns of morphological change. Each horizontal line displays the range of variation in one or more morphological traits in a sample; the vertical line shows the mean of the sample. Time (vertical axis) could be either ordinal or interval scale, but is presumed to be interval scale when patterns a–d are labeled. In a, stasis occurs when the mean does not move in one direction over time; there is no change. In b, gradual change occurs when the mean shifts uniformly in one direction over time. In c, saltation occurs when the mean shifts abruptly from one static state to another. In d, a variable rate of change occurs when tempo varies within a directional trend. Mosaic change occurs when one trait displays b, c, or d pattern and a second trait displays b, c, or d pattern but not the same pattern as the first trait (after Martin and Barnosky 1993).

signifies that length, for example, may change in one way, whereas width may change concurrently but in a different way. As Martin and Barnosky (1993:7) indicate, "The combination of stasis in one trait and change in another does not indicate a mosaic pattern, because most traits show stasis; only two or more traits changing in different ways will indicate a mosaic pattern."

The graphing technique exemplified in Figure 7.1 is a handy way to summarize data critical to the construction of lineages, assuming the requirement of heritable continuity has been met. But has it? Perhaps the best way to illustrate the significance of this requirement is to consider another set of graphs, this one originally presented by paleobiologists Kenneth Rose and Thomas Bown (1986: 120). Their set is shown in modified form in Figure 7.2. Whereas Martin and Barnosky's graphs focus more closely on what might be considered within-lineage and perhaps even within-taxon patterns of change, Rose and Bown's set illustrates the problem of identifying lineage diversification (cladogenesis), or between-taxon variation. The graphs in Figure 7.2 mimic those in Figure 7.1 in important ways: formal variation is plotted on the horizontal axis; time is plotted on the vertical axis, but no distinction is made between ordinal and interval scale; each horizontal bar represents the range of variation in some morphological trait,

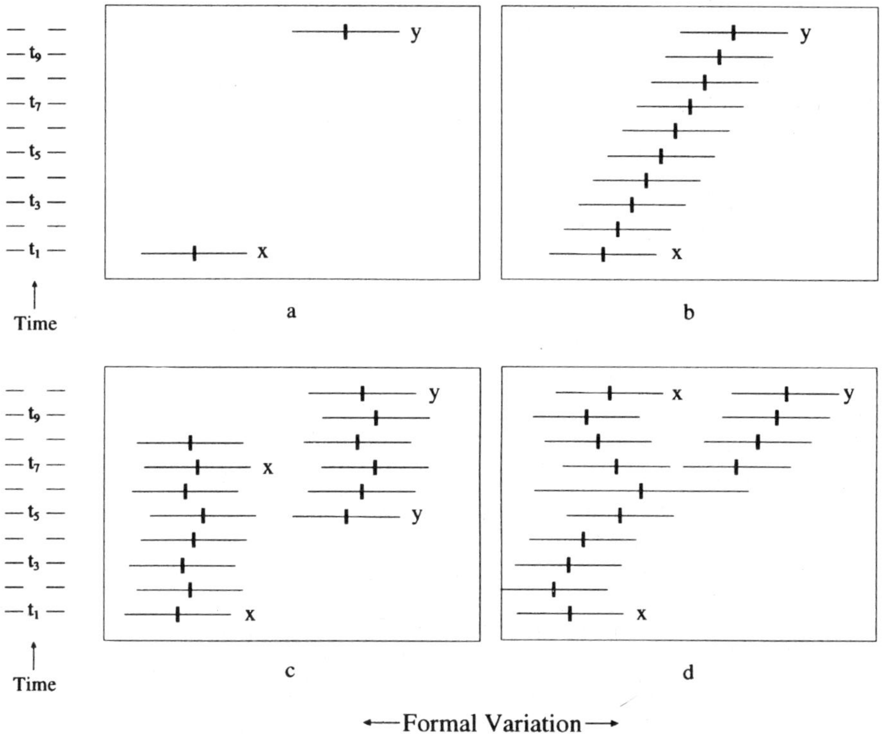

FIGURE 7.2. Change as difference in species. Graphing conventions are the same as in Figure 7.1, but *x* is one species and *y* is another species. In a, two species (*x* and *y*) are separated in morphology and time. In b, two species (*x* and *y*) are connected by intermediate specimens. In c, two species overlap in time, but species *y* could be a descendant of species *x* or an immigrant. In d, two species overlap in time, and intermediate specimens indicate cladogenesis (after Rose and Bown 1986).

and the vertical bar represents the mean of that trait within the measured sample signified by the horizontal bar; and measured specimens are grouped together based on their temporal similarities rather than on their formal resemblance. The important difference between the two sets of graphs is that in Figure 7.1 there is no indication of taxon, whereas in Figure 7.2 there is. This is why the issue of heritable continuity is critical.

In Figure 7.2a two fossil species—*x* and *y*—are indicated. They are well separated in time and morphology; the taxonomic significance of the latter could have been determined on the basis of the magnitude of morphological distance between extant taxa related to the fossil species—for example, all being members of, say, the same taxonomic family. The separation of species *x* and species *y* over time could be the result of stratigraphic discontinuities in the sampled exposures or of taphonomic factors, or it could be the result of sampling error. Over time, however, the temporal gap is filled with specimens that turn out to be morphologically intermediate between species *x* and species *y* (recall Figure 5.13). When arranged against their temporal position, the result could be that shown in Figure 7.2b. As illustrated, change is gradual *and* also phyletic; thus, cladistically species *x* and *y* should be considered "one species, although they are no less distinct [than in Figure 7.2a] simply because [our knowledge of] the fossil record has improved" (Rose and Bown 1985:120). Here is the first problem, and it arises because specimens are grouped not on the basis of their formal similarities but on the basis of their temporal similarities. Other problems arise when, as in Figure 7.2c, a new form seemingly appears suddenly. We say "seemingly" because there is no apparent morphological connection or continuity between species *x* and species *y*, yet they are somewhat similar. Should species *y* be considered an immigrant or a daughter species resulting from cladogenesis? Here is where detailed knowledge of large regions is necessary, as such data will demonstrate if species *y* originated elsewhere at an earlier time. This consideration brings the spatial dimension into sharp focus. Similarly, increasing the size of available samples might, as in Figure 7.2d, eventually produce specimens that are morphologically intermediate to species *x* and species *y*. Yet, here again, the problem is the same as in Figure 7.2b: Given their shared ancestor and morphological continuity, where does one draw the line between *x* and *y* (recall Figure 2.6)? The problem, it seems, is with the units employed in the graph—units that are thought to comprise species. We have argued in previous chapters that frequency seriation allows one to test for heritable continuity between units and that frequency seriation demands the use of theoretical units. That our argument holds for artifacts and for fossils is easily demonstrated.

Seriating Species

There is no reason biological units cannot be seriated in order to ascertain if they reflect heritable continuity. There are, however, epistemological reasons why a successful seriation may not be produced. We chose as an example a set of fossils comprising early Eocene mammals of Wyoming largely because the published data were conducive to purposes of illustration. David Schankler (1981) plotted the frequencies of five species (representing two genera) against their stratigraphic provenience by arbitrary 10-meter-thick increments. Because the sizes of many of his 10-meter-increment samples were small (eighteen increments contained no fossils, and four others produced fewer than ten fossil organisms), we lumped individual samples together until each sample contained a minimum of twenty specimens. This converted Schankler's original sixty-nine samples into eleven samples, the stratigraphic thickness of which ranged from 10 to 100 meters. We then calculated the relative abundance of each taxon per sample and graphed those abundances against time (Figure 7.3). Although the graph looks like a frequency seriation, it is more correctly characterized, in the terminology we use elsewhere (Lyman et al. 1997b, 1998b; O'Brien and Lyman 1998), as a percentage-stratigraphy graph.

The frequency-distribution curves per species in Figure 7.3 do not conform to the lenticular shape of historical classes used in frequency seriations. This is because of a failure to meet the requirements of frequency seriation. It is doubtful that the temporal increments represented by each of the eleven graphed assemblages represent similar durations of accumulation. Furthermore, the graphed specimens come from "over 240 fossil localities" dispersed over 650 square kilometers (Schankler 1981:135), suggesting that spatial variation is contributing to the fluctuations in frequencies. Schankler (1981:137) suggested that environmental change might be driving the changes in taxonomic frequencies, variously increasing or decreasing interspecific competition and prompting immigration or emigration. Although the graphed species likely comprise heritable continuity in both the type/species and the tradition/lineage sense, we contend that species as typical analytical units of paleontologists are *functional* units. The differentia of the species represented in Figure 7.3, for example, comprise differences in tooth morphology and size, both undoubtedly functional attributes. Furthermore, species not only differ in such attributes, but they also alter their geographic ranges in response to environmental change. Such units do not comprise the historical types—often selectively neutral *styles*—demanded of frequency seriations (*sensu* Dunnell 1978c; Lipo et al. 1997; Teltser 1995c). This means that we should not expect units such as are represented by typical paleontological species to display a normal frequency distribution over time; as analytical units, species are not constructed to do so.

We emphasize that the preceding statement does not mean that frequency seriation of biological organisms is precluded. Rather, it means that one must explicitly consider how variation is measured among such units if one wishes to produce a frequency-seriation–like graph of biological variation. Paleobiologists have tended to focus on change in the mean of one or more morphometric dimensions of a taxon over time (Figures 7.1 and 7.2) rather than on shifts in the

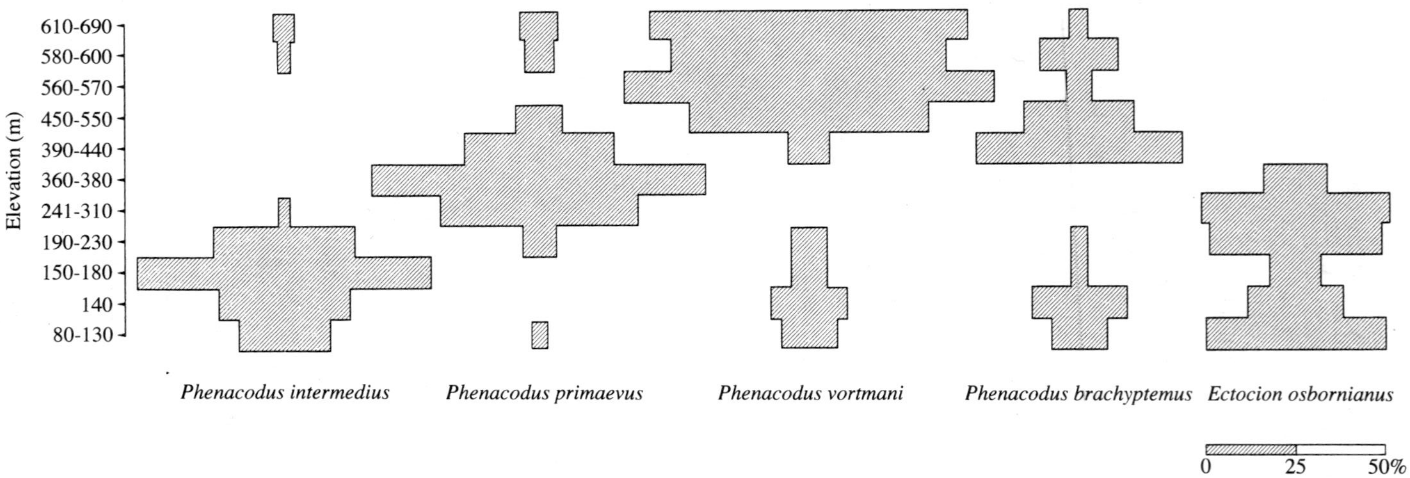

FIGURE 7.3. Percentage-stratigraphy graph for five species of Eocene mammals (data from Schankler 1981). Note that the frequency-distribution curves are not battleship shaped.

frequency of specimens within a particular morphometric class over time. The former comprises a chronocline (e.g., Koch 1986). The data on which these are based, if plotted as frequency distributions of morphometric classes over time, may well approximate the lenticular curves of a frequency seriation, particularly if the data comprise selectively neutral features. This may, in fact, be much of what is presently being measured and plotted against taxonomic diversity by Mike Foote (e.g., 1993a, 1993b, 1996) that so excites Gould (e.g., 1991b, 1995a).

Species as Units

The fundamental reason why Schankler's data do not produce unimodal frequency distributions for each species is easily understood. Species as units do not seriate for the simple reason that they are not historical types in the sense understood by most archaeologists. The spatiotemporal distribution of a species can be modeled as shown in Figure 7.4 (after Pearson 1998a). If the total real

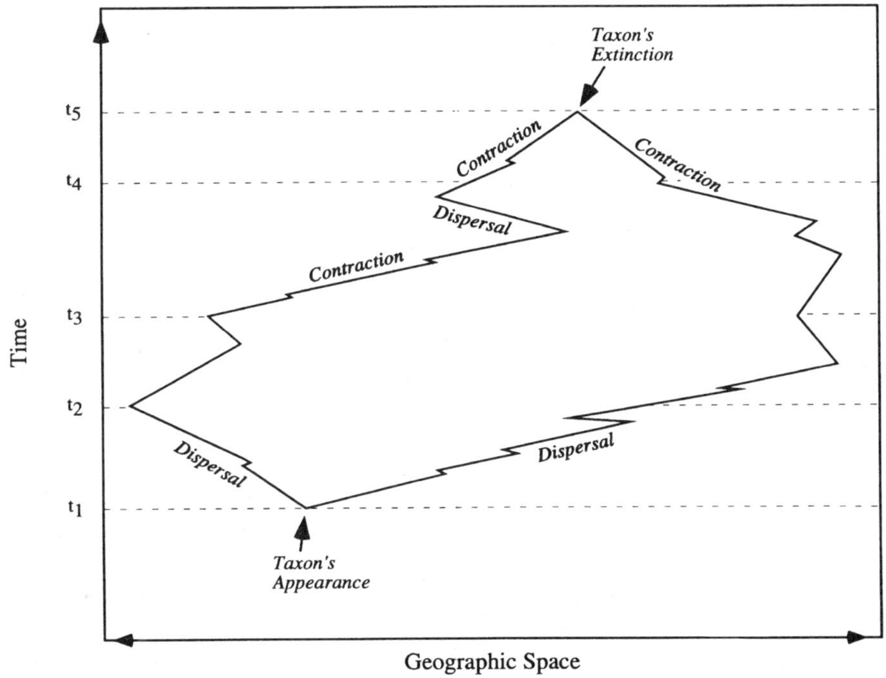

FIGURE 7.4. A model of the spatiotemporal distribution of a biological taxon (after Pearson 1998a). Were a sample of fossils to span only one portion of the spatiotemporal distribution of this taxon, then the spatiotemporal distribution of the taxon would be incompletely known.

spatiotemporal range of a species is not represented in available fossil samples, as in the case when only the left half of the distribution in Figure 7.4 is known, then the times of appearance and extinction of an extinct taxon will influence the shape of curves in a frequency seriation based on such data. Culture historians worried about this problem, and it prompted them to account for the spatial dimension when they constructed the units (types) they intended to seriate. Kroeber, Nelson, Spier, and Kidder viewed their pottery types as analytical tools rather than as "real" entities (Lyman et al. 1997b, 1998b, n.d.b; O'Brien and Lyman 1998, 1999a); they were ideational and theoretical units specifically constructed to measure the passage of time. In contrast to the biological concept of species, the units archaeologists plotted in their frequency seriations and percentage-stratigraphy graphs were "historical types" (e.g., Krieger 1944; Rouse 1939).

Darwin showed that the immutable nature of species was incorrect, although the empirical reality of species lives in the modern biological-species concept precisely because it has this biological meaning and therefore entails particular implications for biological evolution, despite the problems of recognizing interbreeding populations of organisms among inanimate fossils (e.g., Fox 1986; Trueman 1979). When such units are graphed in a graphic seriation or percentage-stratigraphy graph, the fossil species' real spatiotemporal distribution (Figure 7.4) is effectively converted to a rectangle. This conversion brings with it the two problems graphically depicted in Figure 7.5. First, the real distribution may be much more complex than the relatively simple one displayed in Figure 7.5; the more complex the real distribution, the greater the number of samples necessary to approximate accurately that distribution. Even the relatively simple real distribution shown in Figure 7.5 will be poorly approximated if only samples A, C, and E in the figure are available. Second, the more complex the real distribution, the greater the discrepancy between the perceived and real distributions.

Kroeber, Nelson, Kidder, and Spier escaped these problems in archaeology by constructing ideational units of a particular kind—an analytical unit that would allow them to measure time while simultaneously controlling the spatial dimension. Given the view that artifact form varies more or less continuously both over time and across space, the analytical units were built to have limited spatiotemporal distributions that approximated rectangles, such as those shown in Figure 7.6, column B; that is, they monitor the passage of time rather than difference in geographic location. This is the analytical unit that came to be known as a historical type, and it had to be built by trial and error (Rouse 1939), a point rarely acknowledged explicitly. Given such a mode of construction, the utility of the types for measuring the passage of time had to be tested—a significant point made explicit by Krieger (1944) when he indicated that archaeologically useful types must pass the historical-significance test. The test implications were that a useful historical type had to have a distribution like one of those shown in Figure 7.6, column B. Types that had distributions like those in Figure 7.6, columns A and C

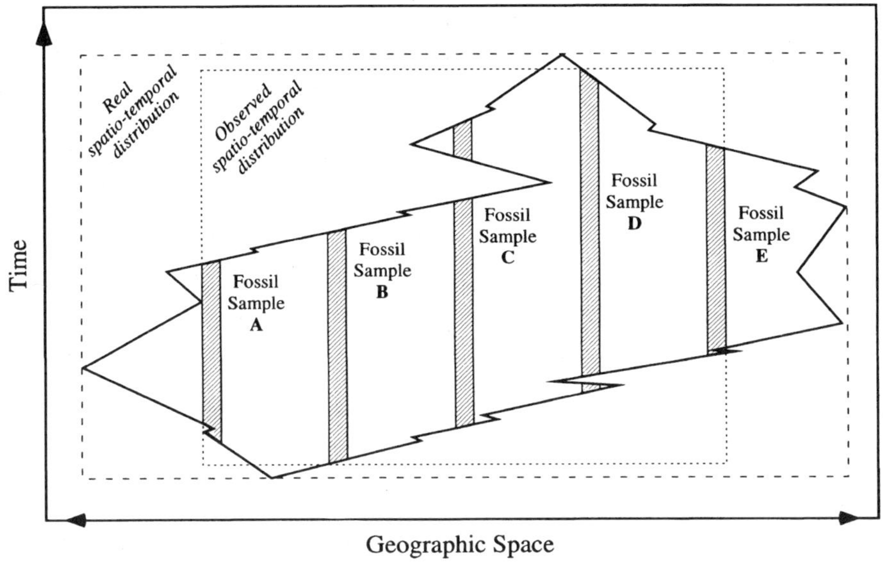

FIGURE 7.5. A model of what happens when the real spatiotemporal distribution of a biological taxon is estimated with fossil samples. The distribution becomes rectangular, and will, if samples are inadequate, be smaller than the rectangle defined by the real distribution. The rectangle that would be defined if only fossil samples A, C, and E were available would be smaller than that defined by the five samples indicated.

could be used, but they were less satisfactory because they measured variation in space is well as variation in time. If the constructed types did not pass the historical-significance test, they were discarded and new types were erected. This trial-and-error, classify–test–reclassify, process continues to this day (e.g., O'Connell and Inoway 1994; Thomas 1981) as archaeologists attempt to construct analytical units that allow them to reliably and validly measure the passage of time.

Species and artifact types might display distributions such as those signified by the rectangles in Figure 7.6, column B, but that is unknown when these units are first constructed. Whether or not the constructed units have such distributions is what is determined by the historical-significance test. Species units often have spatiotemporal distributions like that shown in Figures 7.4, 7.5, and 7.6, column C, and thus are potentially less reliable and valid measures of time. The worldwide distribution of the taxon must be known such that the time of that taxon's appearance and extinction can be determined accurately. This is not the case with the analytical units used by culture historians who built their units to have limited spatiotemporal distributions such that they were useful for measuring time.

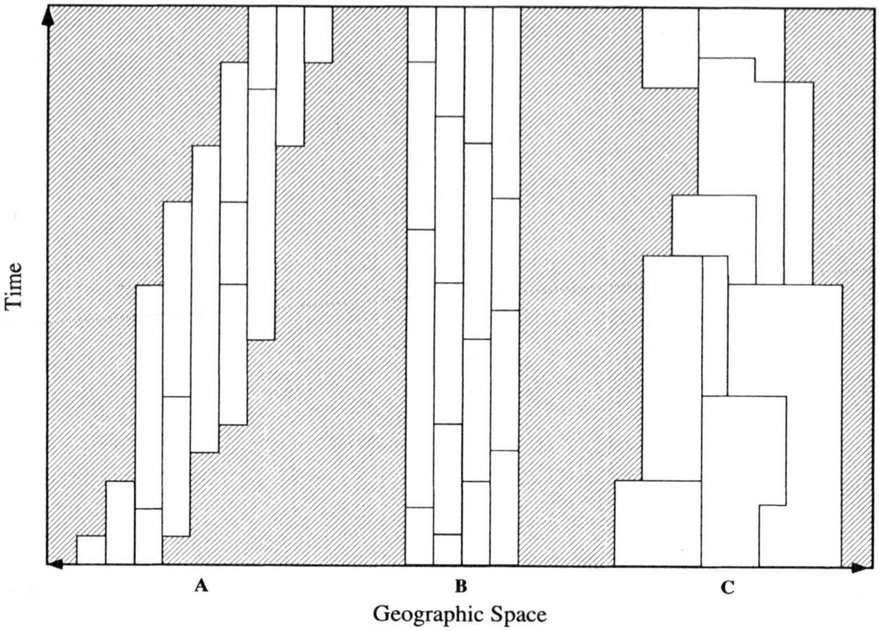

Figure 7.6. A simplified model of ideational units used by archaeologists to measure time. Artifact form varies continuously along both axes, but there is no absolute scale on either axis. Each polygon represents an ideational unit used during analysis to measure variation; shaded areas represent formal variation not measured by those units. Each column of polygons (A–C) denotes a set of analytical units comprising a typology: A, analytical units overlap through time but include spatial variation as well, meaning that change over time as well as variation over space is included; B, analytical units overlap through time but do not include a change in spatial variation, meaning that only time is measured; and C, variation in time and space varies from unit to unit, meaning that although units measure time and thus change, they also measure spatial variation.

The model in Figures 7.4 and 7.5 comprises the spatiotemporal distribution of an empirical unit termed a *biological species* (Pearson 1998a). Clearly, such a unit has significant analytical constraints, the most important one in terms of measuring time being that the unit's total spatiotemporal distribution must be known for it to be a useful chronometer. This underscores the fact that the kinds of units required by frequency seriation are theoretical units that have spatiotemporal distributions like those in Figure 7.6, column B. By recognizing the utility of such theoretical units, paleobiologists could begin to use such units in frequency seriations and interdigitation to develop faunal chronologies of greater resolution than many currently available (for initial examples, see Lyman and O'Brien [n.d.a]). We suspect paleobiologists have used frequency seriation only rarely (e.g., Brower and Burroughs 1982; McKee et al. 1995) because species are the units seriated and paleobiologists are well aware of the problems with seriating such units exemplified in Figures 7.4 and 7.5. Furthermore, some paleobiologists (e.g., Gould et al. 1987) have derogated frequency seriation because they do not understand the ontological differences between species and the theoretical units upon which frequency seriation depends (Lyman and O'Brien n.d.a). But here is where each discipline benefits the other.

From the perspective of punctuated equilibrium, Gould (1997c) would label such fine-scale temporal and morphological resolution as is illustrated in Figure 7.2d "the paradox of the visibly irrelevant." Although this stance is admissible given that it comprises merely a change in scale, we find it unsatisfactory. It forces variation into extensionally derived cubbyholes termed *species* and, as Fortey's (1985) graphs, reproduced in Figure 5.12, show, renders it invisible. But do not misinterpret this. What we mean is that starting and ending an analysis with such large-scale units as "species" can potentially mask important variation and ultimately mask the operation of selection and other sorting mechanisms; that is, within-unit variation—measured at less-inclusive scales—is equally important to monitoring between-unit variation—measured at more-inclusive scales such as species, genera, and families. As Steven Stanley (1979:64) notes, "By studying chronospecies, we can assess rates of evolution *within* lineages" (emphasis added), but we would omit the term *chronospecies* and just focus on available samples—samples, we hope, with fine temporal resolution that do not span large periods of time. We would do this for the simple reason that it is easier to lump than to split; one can always lump data that have been measured at overly fine scales, whereas data measured with inclusive scales such as are represented by species-type units cannot be split into finer scales. But large-scale, inclusive units *are* useful. We turn next to graphing what we think of as coarse-scale change, or change at a level of inclusive units higher in the taxonomic hierarchy—*Baupläne*, if you wish—than individual variants. As we will see, it too rests on unit construction. The method we describe was developed to address paleobiological questions but is readily applied to archaeological data.

Clade-Diversity Diagrams

Given that species are valuable units for studying the tempo and mode of change, how might we graph them for study? Elliott Sober's (1992:18) definition of evolution as change in "the frequency of some characteristic in the population" provides a place to start. Enlarging the scale, we might ask: How do frequencies of taxa change over time? Paleobiologists have long constructed graphs showing the frequency of lower-level taxa within a particular higher-level taxon in order to display the history of life (for a brief history and references, see Signor 1990). John Phillips published one of the first such graphs in 1860, showing the fluctuating but generally increasing diversity of life through time (reprinted in Rudwick 1976: 240). Similar graphs were produced by others as the fossil record became better known (e.g., Simpson 1949). These later graphs—now referred to generally as clade-diversity diagrams (e.g., Raup et al. 1973), where "diversity" denotes richness, or the number of taxa at some specified level in the Linnaean taxonomy—are plots of the frequency of taxa within a particular monophyletic lineage, or clade. Some clade-diversity diagrams take the form shown in Figure 7.7, what we term a centered-bar graph, though their form varies and includes histograms, with diversity on the vertical axis and time on the horizontal axis, and broken-stick graphs, which show lines connecting a series of points plotted against diversity on either the vertical or horizontal axis and time on the opposite axis (Figure 3.4 is a broken-

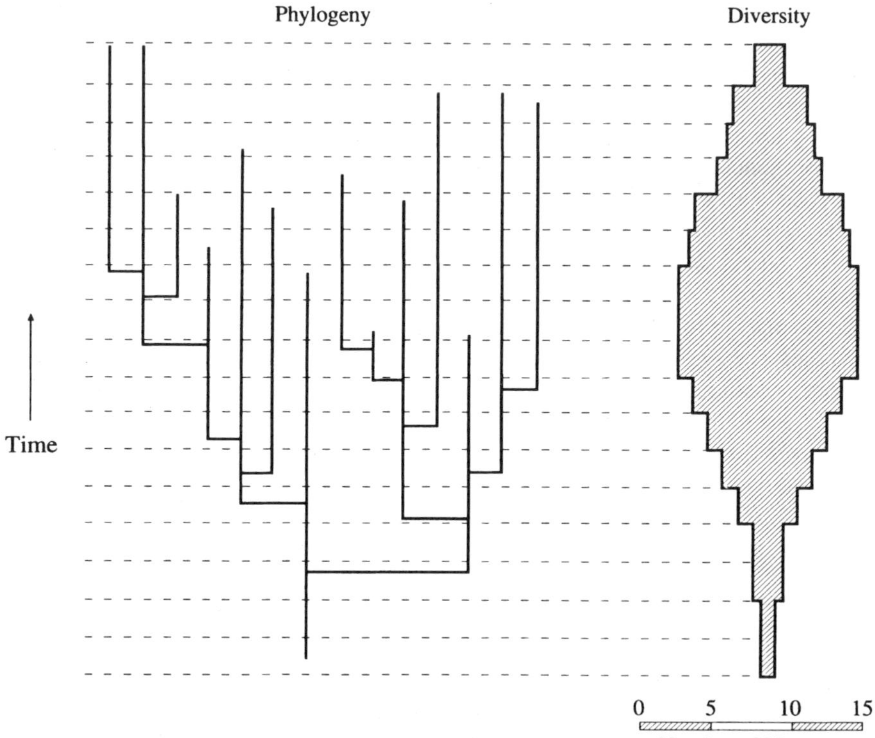

FIGURE 7.7. A model for producing a clade-diversity diagram. The phylogenetic history of taxa is shown on the left, and the resulting clade-diversity diagram is shown on the right. Once the phylogenetic history of related taxa has been worked out, including the placement of their starting and ending points, the data are summarized as at the right, which shows the waxing and waning of the number of taxa (after Raup et al. 1973).

stick graph, but is not an illustration of clade diversity.) Because a clade-diversity diagram shows change in the number of extant lineages (taxa) over time, it comprises a history of the origination and extinction of taxa within a clade: "If the rate of branching exceeds the rate of extinction ... the number of coexisting lineages increases. If extinction exceeds the rate of branching, diversity decreases. If the two rates are equal over a period of time, diversity remains constant" (Raup et al. 1973:528).

Efforts to better understand the shape of and the reasons for the history of life have resulted in comparisons of the taxonomic richness of empirical monophyletic lineages, or clades, with those generated randomly (e.g., Gould et al. 1977; Raup and Gould 1974; Raup et al. 1973; Uhen 1996). A monophyletic lineage arises only by branching (cladogenesis) and is defined as all taxa deriving from a common ancestor plus the common ancestor. Comparisons involve examination of clade-diversity diagrams rendered as centered-bar graphs. Each stack of bars represents a clade, and each bar's width represents the absolute taxonomic richness within a clade at a particular time. Graphs are generated according to the model in Figure 7.7. Often, as well, phylogenetic connections between clades are shown (Figure 7.8). One result of these efforts has been the recognition that the richness and duration of a clade is influenced by the taxonomic level—family, order, class—used to quantify clade diversity (Stanley et al. 1981).

Clade-diversity diagrams have been compiled for a number of taxa and time periods and have revealed various evolutionary events that took place throughout the history of life (e.g., Prothero 1985; Raup 1972, 1976a; Sepkoski 1981; Sepkoski et al. 1981) and the potential causes of those events (e.g., Miller and Sepkoski 1988). As well, variations in the computer programs used to generate random clade-diversity diagrams have granted insights into possible evolutionary processes (e.g., Benton 1995; Gilinsky and Bambach 1986; Uhen 1996). Paleobiologists developed the various computer programs to simulate the diversity history of a clade because they "wished to predict what phylogeny would look like if it were determined by random processes and then to compare this with the real world, to enable us to separate random elements from those that require interpretation in terms of specific and perhaps nonrecurring causes" (Raup et al. 1973:527). Using more explicit wording, Raup (1977:53) indicates that "the first question for the paleobiologist faced with an evolutionary trend should be: Does the trend represent a statistically significant departure from chance expectations?" This question must be asked, as Gould et al. (1977:24) note, if explanations are to result: "Until we know what degree of apparent order can arise within random systems, we have no basis for asserting that any pattern in the history of life implies a conventional [typically functional or adaptive] cause for its generation."

Gilinsky and Bambach (1986:251–252) summarize possible conclusions if the question were answered negatively: "Beyond the orderliness necessarily imposed by the system, major features of the history of diversity are truly (onto-

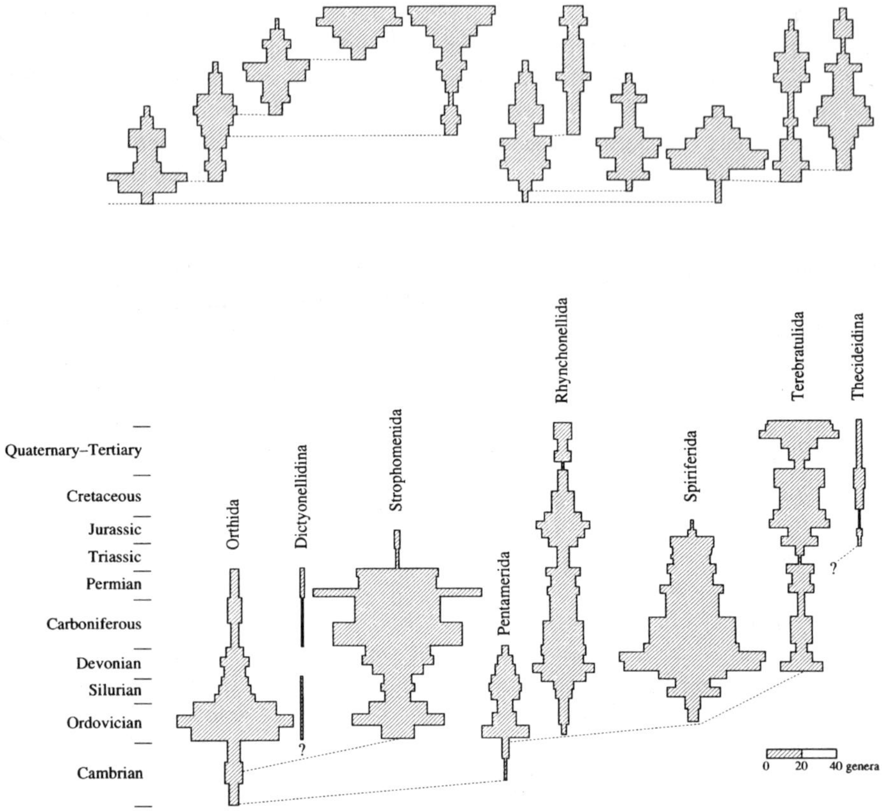

FIGURE 7.8. Clade-diversity diagrams of randomly generated (top) and real (bottom) clades. The Markovian structure of such graphs depends on heritable continuity (after Gould et al. 1977).

logically) random; or ... major features of the history of diversity result from large numbers of complexly intertwined causal factors none of which predominates, such that [clade histories] often *appear* to be random." The "orderliness necessarily imposed by the system" can be summarized as follows:

> Initial diversity is always zero (before the taxon's inception); final diversity is always zero (after extinction); diversity never swings below zero; and diversity is Markovian (diversity at any time $t + dt$ depends in part upon the diversity at previous time t). These features limit the range of possible diversity histories and cause patterns of diversity change through time to appear more orderly than in independent-events processes, which have no Markovian memory and have no constraints at their beginnings and ends. (Gilinsky and Bambach 1986:251)

Gould et al. (1987:1437) posed the question, "[D]oes any asymmetry exist, statistically defined over large numbers of lineages, in the vertical [temporal] dimension of clade diversity diagrams?" If so, one might measure time with the diagrams or detect a recurrent pattern in the history of changes in life's diversity. They found such an asymmetrical pattern in the form of "bottom-heavy clades" and suggested that there indeed was a particular direction to biological evolutionary time (but see Uhen 1996). Many of the real clades they examined appeared to be taxonomically richest early in their history and to become progressively less rich over time (but see Kitchell and MacLeod 1988, 1989). This and related observations resulted in greater interest in extinction events (e.g., Kitchell and Pena 1984; Raup 1986; Raup and Sepkoski 1982, 1984) and the detection of a possible pattern in mass-extinction events (Jablonski 1986; Raup 1976b; Raup and Boyajian 1988; Raup and Sepkoski 1986). Massive vacating of adaptive zones—ecological niches, if you prefer—may have allowed rapid diversification and what are perceived as bottom-heavy clades (Benton 1995).

Clade-diversity diagrams are built to display the multiplication and extinction of species or some higher taxonomic unit within a clade and are founded on the assumption of phylogenetic continuity, or heredity, between taxa included in the graph. Clade-diversity diagrams assume heritable continuity in two ways. First, each fossil specimen belonging to a particular species is, given genetic transmission, hereditarily continuous with every other fossil specimen identified as a member of that species. This is the type/species sense of heritable continuity. Second, congeneric species—that is, distinct species belonging to the same genus—denote heritable continuity, again given genetic transmission and evolutionary divergence (the same applies for confamilial genera and so on) within a monophyletic group. This is the tradition/lineage sense of heritable continuity. Of course, what is measured is fossil morphometry, or formal variation, not genetic similarity (Pearson 1992, 1998b). The latter is inferred on the basis of the former, and this demands the separation of homologs and analogs and the recognition of synapomorphic character states to ensure that common ancestry informs the classification of fossils into particular species. Otherwise, one ignores both Simpson's caveat about twins (Chapter 5) and Gould's (1991b) phenetic–cladistic distinction (Chapter 6) and adopts Gordon Willey's axiom of culture history: Historical continuity comprises heritable continuity.

But is this statement really accurate? Perhaps not, for the simple reason that the "Markovian structure [of a clade-diversity diagram] is insured by inheritance" (Dunnell 1980:58); that is, what comes later depends in part on what came earlier. However one recognizes species in the fossil record, the procedure automatically assumes heritable continuity in the type/species sense. Given that designating a species also designates membership within all higher levels of the Linnaean taxonomy, continuity may be assumed at those levels as well—the tradition/lineage sense. However, the validity of the assumption depends on systematics.

Recall Joel Cracraft's (1981a:461) remarks cited earlier: "The classificatory methodology underlying the taxonomy of the groups may have an influence on the number of [groups] recognized at various stratigraphic levels, and this obviously can bias, to some unknown degree, a mathematical characterization of changes in diversity through time." If a set of units low in the taxonomic hierarchy, say genera, are incorrectly subsumed under a single unit higher in the taxonomy, say, an order, when in fact they should be subsumed under one or more other orders, then, clearly, any measurement of the richness of orders will be influenced. One's systematics must contend with both phylogenetic history *and* position within the taxonomy hierarchy. Techniques for analyzing clade-diversity diagrams allow the identification of departures from the random background generation of variation in artifacts and grant insights into evolutionary modes.

MEASURING THE TEMPO AND MODE OF CHANGE IN PREHISTORY

It should be clear by now that a good ordering of collections (Cowgill's [1968] task 2) may not be a good chronology (Cowgill's [1968] task 1). One reason why the best ordering may not reflect the passage of time involves the fact that the units within a seriation may not be temporally sensitive (Cowgill 1968:519). But presuming that a good ordering can be shown empirically to be a good chronological ordering—such as with, say, stratigraphic information—and that not only historical continuity but also heritable continuity is represented by the seriation, then the next issue concerns identifying the tempo and mode of change displayed by the lineage represented by the ordering. This issue has seldom been addressed by archaeologists. Although Phillips, Ford, and Griffin (1951) and others sometimes referred to what appeared to be variation in the rates of change as depicted in their centered-bar graphs, they could not accurately estimate the rates (O'Brien and Lyman 1999a). To determine rates of change requires an interval-scale dating method that produces absolute, or calendrical, dates. Efforts to identify modes of change—and hence to derive explanations—have typically been cast in terms of ethnographically documented processes of culture change such as innovation, migration, trade, and diffusion.

Inferences about tempo and mode can be derived only with the correct kind of data, in this case, theoretical units that group specimens based on formal rather than temporal similarities. Thus, not surprisingly, we prefer Ford's centered-bar graphic technique because, unlike a matrix of similarity coefficients, it allows one to *see* the tempo and mode of change. Similarity coefficients may, of course, be used to help produce the ordering of changes such as those depicted in Figure 6.8, but as we indicated earlier, time is best plotted on an interval scale if one's goal is to monitor tempo and mode. Given that the seriation method by itself is ordinal scale and provides only a relative chronological sequence rather than calendrical ages, it is mandatory that we be aware of how absolute dates can be aligned with an ordering produced by seriation. The literature on this is extensive (e.g., Braun 1985a; O'Brien and Lyman 1999a; Read 1979), and review of it is beyond our scope here. Frequency-seriation graphs can, however, tell us something about the tempo of change, and we discuss that in the following section.

The Tempo of Change

The degree of temporal resolution obtained by a seriation will depend on the relation between the rate of change within the lineage or tradition represented by the seriated materials and how one chooses to measure that change. Clement Meighan (1977:628) noted that "if there is rapid [cultural] change, any orderly seriation method will recognize short time periods; with data showing little observable change over long periods, no seriation method can identify short time periods." The key point is that sortability—how easily collections can be distinguished one from another—rests on the dimensions and attributes used to measure formal variation within the collections. In the case of frequency and occurrence seriation, the variables forming the basis for sorting and ordering are either the frequencies or the presence–absence of classes, respectively. The problem thus reduces to the construction of classes; which variables or attributes should be used? Meighan (1977:629) correctly noted that "sometimes the quality of the data [used in a seriation] can be improved merely by improving the definition of the ceramic types ... so that the types more closely reflect chronological change." Steven LeBlanc (1975:24) suggested that "too few types do not distinguish short intervals of time," so one might think more types would be better. But as George Cowgill (e.g., 1968) and others have noted, the more types (and assemblages) included in a seriation, the more difficult it becomes to produce an ordering that does not violate the continuity and unimodal-frequency principles specified by the seriation model.

LeBlanc (1975) suggests that in some cases, frequency seriation of particular attributes—rather than attribute combinations typically termed types—may provide an ordering that more closely approximates the model of seriation and that also represents finer temporal resolution. He refers to this as "micro-seriation." In the sense that temporal resolution may in fact be finer, we agree with his term. But "a combination of a larger number of [attributes] is likely to cover a shorter span of time than a combination of a smaller number" (Rowe 1959:322); that is, each attribute—as a class—will have a particular duration, and there is no reason to suspect a priori that each attribute will occupy a portion of the temporal continuum that overlaps perfectly with any other attribute. Thus, a combination of attributes that we choose to call a type will have a shorter duration than any one of its constituent attributes if the *combination* of attributes is established independently of the duration of any given attribute. This would be easy to determine empirically in any given situation: merely seriate collections based on types, then based on attributes, and compare the two. Individual attributes may be found to span longer durations—occur in more assemblages—than individual types. Such comparative seriations would allow one to identify the *mosaic* pattern of change (Figure 7.1).

Similarly, Frank Hole and Mary Shaw (1967:96) indicate that "a comparison of different data sets from the same series of sites gives dramatic evidence of relatively different rates of change among data sets. These data, graphically expressed, could go a long way toward making the nature of change more understandable"; that is, projectile points, say, will have a history of change different from the history of change in associated pottery even though both belong within the same *cultural* lineage. This is so because, as Hole and Shaw (1967:86) note, types that produce good seriations "are likely to have a relatively neutral adaptive value in a culture. Changes in them will derive more from fashion or from random drift than from necessity" (see also Dunnell 1978c; Lipo et al. 1997; Neiman 1995). In other words change in projectile points may be slow (or fast) relative to change in pottery; the two will change independent of one another. Such a (fictional) example is shown in Figure 7.9. Rapid change in the ceramic tradition

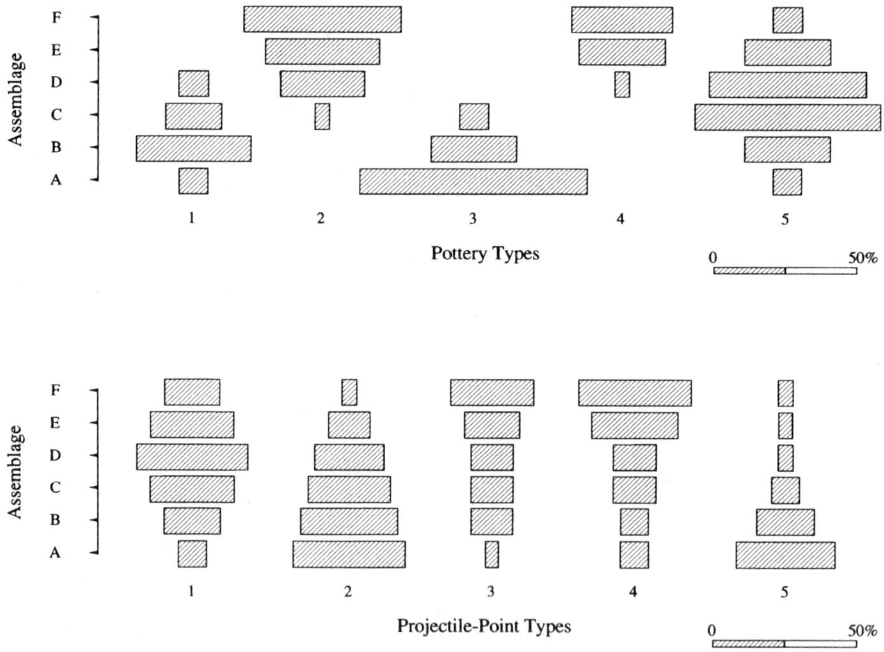

FIGURE 7.9. Frequency seriations of pottery types (top) and projectile-point types (bottom) in six assemblages. This fictional example illustrates how rapid change in pottery might compare with slow, gradual change in projectile points.

in Figure 7.9 is indicated by the turnover of types from the bottom to the top of the graph as well as by the lack of stability of frequencies of some types across the graphed collections. In contrast, the projectile-point tradition shown in Figure 7.9 includes all types throughout the graph, and the relative abundances of projectile-point types 3–5 do not change across two or three assemblages. Thus, seriation can tell us something about the tempo of change, and it could also prove valuable for revealing the mode of change (e.g., Figures 7.1 and 7.2).

Modes of Change

Culture historians typically called on ethnographically documented mechanisms such as innovation, diffusion, and migration as explanations for what they perceived as evolutionary changes in artifact lineages. Such explanations—a better word is *interpretations*—often were chosen on the basis of the magnitude of a perceived difference between temporally distinct assemblages of artifacts. Large differences, as in aggregates of types, were attributed to immigration events, with complete discontinuity ascribed to such things as invasion and total and abrupt population displacement. Smaller differences, as in one or a few attributes of one or a few types, were attributed to diffusion (or contact) if a nearby (spatially and temporally) potential ancestor were known, and to within-lineage innovation if no such ancestor were known (e.g., Lathrap 1956; Thompson 1958a). Seldom, however, were there clear illustrations of the lineage histories being compared. Ford's (1952) "Measurements of Some Prehistoric Design Developments in the Southeastern States" was a notable exception, but that study was rather naive theoretically and simplistic analytically, and as a result it was discounted by many (e.g., Spaulding 1953a), though not all (e.g., Rouse 1954), archaeologists.

The concern with "culture contact" as represented by diffusion and migration was a nod to the fact that cultural evolution was "polyphyletic," as noted by A. L. Kroeber, Julian Steward, and others, or in more modern terms, it was reticulate rather than merely branching (Figure 3.8). Difficulties arose because it was unclear precisely how similar was similar enough to represent contact of whatever magnitude. Discussions of the mode of change were, therefore, plagued by ambiguity and debate. Failure to keep clear the distinction between homologs and analogs and between ideational and empirical units only magnified the problems. Modes of change were inferences, derived via induction and without test implications. In hindsight, the contentious results that occurred were predictable. How *should* the mode of change have been measured? By using theoretical units that allowed one to track change at various scales of inclusiveness. How should kinds of change have been categorized? By using theoretical units that allowed one to unambiguously identify the kind of change being measured.

Earlier, we indicated that few archaeologists have used seriation to order collections continuously through time and then used that order to study change in other variables evident in the ordered collections—a prerequisite to explanation.

Janet Rafferty's (1994) study of the development of sedentary settlement patterns in northern Mississippi is a noteworthy exception. She ordered some forty collections based first on an occurrence seriation of projectile-point types and then on a frequency seriation of pottery types. Collections that represented exceptionally long durations of time were omitted. She then compiled data on seven variables thought to be "indicators of sedentariness" that were associated with the points and the pottery (Rafferty 1994:410; see also Rafferty 1985). These data spanned some 5,500 years and were plotted in the temporal order suggested by the seriations. A period of relative stasis, or gradual, unidirectional change spanning the initial 4,000 years or so was followed by rapid change over the last 1,500 years. Other details of her analysis are unimportant to the discussion here. What *is* important is that Rafferty used theoretical units to produce the temporal ordering and other theoretical units to measure adaptational, or functional, change *continuously* over time. This is what allowed her to use as part of the title of her paper "Gradual or Step-Wise Change"; she could answer the question implied by this phrase with empirical data and thereby discuss the tempo of change. What about the mode of change?

Artifact-Clade Diversity

Paleobiologists generally compare clade-diversity diagrams to graphs of morphological disparity in order to gain new insights into evolutionary tempos and modes (e.g., Foote [1996] and references therein). Precisely the same sort of analyses can be accomplished with artifacts. The key point to remember is that the conventions of graphing clade diversity and frequency seriation produce graphs that look alike but which actually present decidedly different data (Lyman and O'Brien 1999b). Clade-diversity diagrams and frequency-seriation graphs are similar in three ways: Time passes from the bottom to the top, having typically been determined for both kinds of analysis based on stratigraphy or radiometric dating; frequencies of kinds of things are denoted by vertically stacked, horizontally centered bars of various widths; and the things graphed are thought to be phylogenetically related. But clade-diversity diagrams plot the richness of classes (believed to represent descendants of a common ancestor) against time, whereas frequency seriations plot the relative frequency of individual entities of each of multiple classes against time (also probably, but not necessarily, descendants of a common ancestor). A clade-diversity diagram can be easily generated from a frequency-seriation graph. One has but to tally up how many types occur in each temporal interval and then generate the appropriate graph of centered and stacked bars.

To illustrate artifact-clade-diversity diagrams and how they differ from frequency seriations of artifacts, we turn first to some historical data. Michael Schiffer (1996) graphed changes in the frequency of companies in the United

States that manufactured vacuum-tube radios for the home market between 1920 and 1955. We assume, as does Schiffer (1996), that the number of classes of vacuum-tube radios in existence during a time period is correlated directly with the number of companies in existence during that same period. Schiffer's data are summarized in Table 7.1 and are graphed in a form similar to a clade-diversity diagram in Figure 7.10. Schiffer (1996:658) also discussed the frequencies of "different portable radio models manufactured and sold in the United States [between 1920 and 1955]." His data concerning the frequencies of models of portable radio are summarized in Table 7.1 and graphed in Figure 7.11 as if the data represent a clade. Both Figures 7.10 and 7.11 may suggest things about the history of what Gould (1991b, 1995a) would term taxonomic diversity. But do they?

Recall that clade-diversity diagrams are meant to display the taxonomic richness of a *clade*—a group of taxa sharing a single common ancestor. We assume Schiffer's data reflect heritable continuity, as more-recent radios descended from that first radio. It is clear that radio manufacturers shared technological information; thus, the two graphed groups—vacuum-tube radios and portable radios—probably do comprise clades. But our point in asking if the two diagrams actually comprise clade-diversity diagrams rests on several matters of classification. For example, what, precisely, are the classes, and at what taxonomic level do they reside? We return to these matters below; there are other issues that are better explored first. The graph of vacuum-tube-radio manufacturing companies (Figure 7.10) appears to be "bottom heavy" (Gould et al. 1987); that is, there appear to be more companies early and progressively fewer later in time. This suggests that many companies—and probably types of vacuum-tube radios as well—arose quickly and then slowly died out. The graph of models of portable radios (Figure 7.11) appears to be "top heavy," suggesting more models were made late in this clade's history than were made early. What is the significance of such observations?

In paleobiology Gould et al. (1987) sought to determine if clade-diversity diagrams were symmetrical or asymmetrical through time; either would say something about the course of life's history and perhaps help identify its causes. Precisely the same question can be asked of artifacts. Clade-diversity diagrams for artifacts would also suggest something about the mode of culture change. Such diagrams are, after all, good ways to measure the rates of multiplication and extinction of kinds of things, whether biological taxa or artifact types (Leonard and Jones [1989] and references therein). But how do we decide if such a diagram is symmetrical or asymmetrical? Gould et al. (1977) developed a simple way to determine if a clade-diversity diagram was symmetrical or not, and if not, whether the diagram was bottom or top heavy. To do this, they determined the "center of gravity," or what they referred to as the "relative position in time of the mean diversity" of a clade (Gould et al. 1977:26; 1987:1438). The duration of a clade is measured on a scale from zero (the time period immediately prior to the period

Table 7.1. Frequency Data for Vacuum-Tube-Radio Manufacturing Companies and Models of Portable Radios Made in the United States between 1920 and 1955

Year	Number of companies manufacturing vacuum-tube radios	Number of models of portable radios
1955	39	55
1954	44	55
1953	43	54
1952	45	47
1951	47	22
1950	49	69
1949	57	41
1948	86	98
1947	111	75
1946	99	34
1945	0 [89.8]	0 [27.25]
1944	0 [80.6]	0 [19.5]
1943	0 [71.4]	0 [11.75]
1942	0 [62.2]	4
1941	53	55
1940	58	49
1939	60	103
1938	58	4
1937	61	3
1936	67	8
1935	75	2
1934	87	2
1933	87	7
1932	88	3
1931	105	0 [4.5]
1930	99	6
1929	97	4
1928	81	0 [4][a]
1927	101	4
1926	105	5
1925	156	18
1924	141	22
1923	99	24
1922	36	4
1921	8	1
1920	4	0

[a]Values in brackets were derived by interporlation (see text for discussion).
Source: Data from Schiffer (1996).

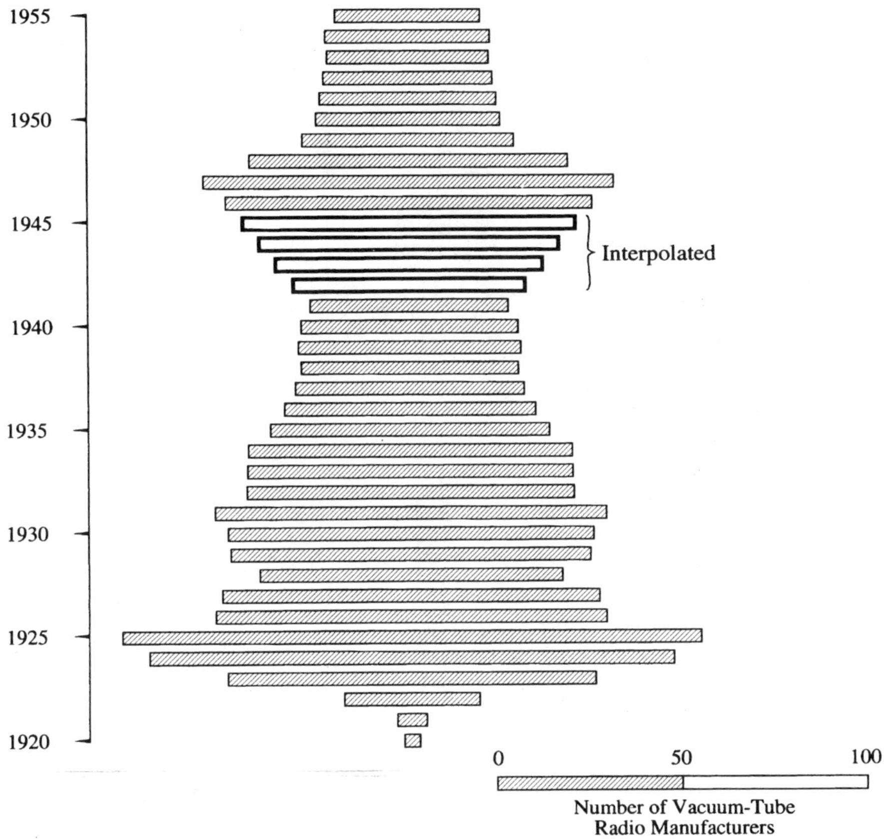

FIGURE 7.10. Clade-diversity diagram of the frequency of companies in the United States that manufactured vacuum-tube radios between 1920 and 1955 (after Schiffer 1996). Data from Table 7.1.

when the clade first appears) to one (the time period immediately following the period of clade extinction). An equilateral, diamond-shaped clade-diversity diagram is symmetrical and has a center of gravity of 0.5; a tear-drop-shaped, or bottom-heavy, diagram is asymmetrical and has a center of gravity of less than 0.5; an inverted tear-drop-shaped, or top-heavy, diagram has a center of gravity of greater than 0.5. The center-of-gravity (CG) value is calculated with the formula

$$CG = (\Sigma N_i t_i)/(\Sigma N_i)$$

where N is richness per time interval and t is the scaled temporal position of the richness measure. For example, the center of gravity of a fictional clade is

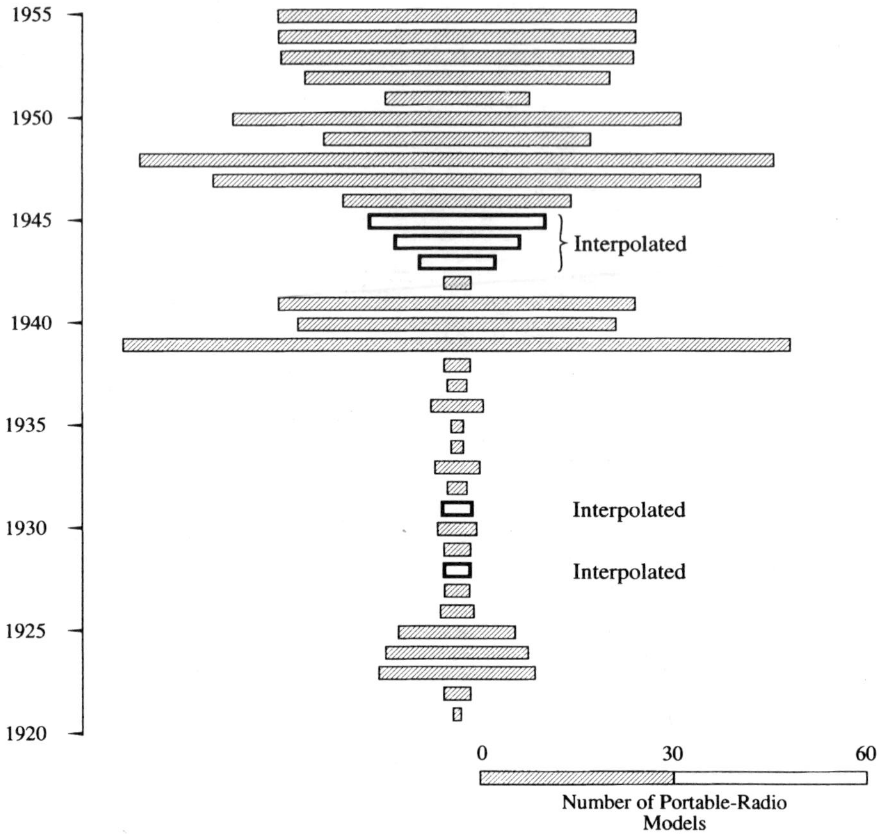

FIGURE 7.11. Clade-diversity diagram of models of portable radios manufactured in the United States between 1920 and 1955 (after Schiffer 1996). Data from Table 7.1.

calculated as shown in Table 7.2. We assume for purposes of illustration that the clade began after the 1840 period and ended before the 1960 period, and use the data in Table 7.2 merely to illustrate the technique of calculating CG.

Kitchell and MacLeod (1988, 1989) note that the categories "symmetrical," "bottom heavy," and "top heavy" are nominal scale. Only a CG of 0.5 denotes the first category, CG values less than 0.5 denote the second, and CG values greater than 0.5 denote the third. Using random simulations, they determined that statistically significant ($p < .05$) bottom-heavy asymmetry was found only in clades with a CG of less than 0.428 and top-heavy asymmetry only in clades with a CG of greater than 0.578 (Kitchell and MacLeod 1988:1192). The statistically significant values indicated were derived from a simulation of the history of 1,000

TABLE 7.2. Mathematics Used to Calculate the Center of Gravity Statistic for a Fictitious Clade-Diversity Diagram

Time period	Richness	Scaled time	Richness × scaled time
1960	0	1.000	
1950	2	0.916	1.832
1940	1	0.833	0.833
1930	2	0.750	1.500
1920	3	0.666	1.998
1910	3	0.584	1.752
1900	4	0.500	2.000
1890	4	0.417	1.668
1880	3	0.333	0.999
1870	3	0.250	0.750
1860	3	0.167	0.501
1850	2	0.083	0.166
1840[a]	0	0.0	0.0
Σ	30	—	13.999

CG = 13.999 ÷ 30 = 0.466

Note: CG, center of gravity.
[a]Each time period spans a ten-year increment, the median of which is listed.

clades over 63 time intervals. Other simulations with similar parameters produce similar mean CG values and statistically significant minimum and maximum values (for details, see Kitchell and MacLeod 1988). In contrast, Gilinsky et al. (1989) argued that one could, using Student's t, determine if differences are statistically significant between a CG value and the mean CG of 0.500 derived from Kitchell and MacLeod's (1988) simulations. These and additional simulations by others (e.g., Uhen 1996) indicate that the number of time intervals included in a simulation combined with the particular parameters of each simulation (particularly the probability of the appearance of a new variant and the probability that an existing variant will disappear during a time interval) influence the confidence intervals bracketing the mean CG. There is much that could be done with simulating the history of artifact diversity, but rather than pursue that here, we compare CG values observed among several sets of artifacts with Kitchell and MacLeod's (1988) simulated-sample mean CG of 0.5 and standard deviation of 0.032 ($n = 1{,}000$); all p values are for one-tailed tests. Thus, for the CG of 0.466 in Table 7.2, $t = 1.062$ ($p > .1$). This clade is effectively symmetrical.

Schiffer's (1996) clade-diversity diagram for vacuum-tube radios (Figure 7.10) appears to be bottom heavy; the CG value of 0.443 ($t = 1.78$, $p < .05$) confirms this observation. More companies manufactured vacuum-tube radios (and presumably produced more kinds of such radios) prior to the middle of the time period considered than after that point. The diagram for portable-radio models (Figure 7.11) appears top heavy, and the CG value of 0.686 ($t = 5.81$, $p < .01$) confirms that it is. But a practical point concerning calculation of the CG value needs to be made here. The clade-diversity diagrams in Figures 7.10 and 7.11 contain gaps for the years of World War II. Schiffer (1996:656) indicates the war "caused [a] hiatus in the manufacture of home radios." We know the reason for this particular gap, because we know the precise historical contingencies influencing the manufacture of radios at the time. How might such gaps be dealt with were they to appear in prehistoric materials for which the historical contingencies are unknown? One obvious response to such gaps—whether one be a paleobiologist or an archaeologist—is to blame them on deficient samples. For example, Eldredge and Gould (1977:27) observe that "it is unfortunately true that we cannot be sure that a collection of fossils is a truly representative sample of a biological population." Among paleobiological treatments of clade diversity there is acknowledgment of sample-adequacy issues, and efforts have been made to ascertain if the available samples are influencing measurements of diversity (e.g., Signor 1982).

Gould et al. (1987) accept as a fact that their samples *are* deficient, evidenced by the manner in which they deal with uneven distributions of fossil samples across temporal periods. They caution that if the durations of periods used to construct a clade-diversity diagram vary—say, if the first sample of fossils is a million years or less in age, the next sample dates between 7 and 8 million years ago, the next between 10 and 11, and the last between 16 and 17 million years ago—the CG value will be skewed if one simply uses the median age of each sample. Their preferred method of compensating for samples assumed to be deficient is to determine the desired duration of chronological periods and then to derive via interpolation a richness value for each temporal period between those periods with known richness values such that each period is of similar duration.

In the example Gould et al. (1987:1438) describe, seven fossil samples are unevenly distributed over a period of 26 million years. Because they desire periods of equal duration to compensate for the effects of varied period duration on the calculation of the CG value, they derive by interpolation 18 richness values plus a zero at both ends of 26 million years (7 observed values, 18 interpolated values, and a zero at each end), such that a value is available for year zero and every subsequent 1 million years of the 26-million-year span. All richness values—observed or interpolated—are then included in the calculation of the CG value. Gould et al. (1987:1439) favor this method because in their view it is the "least biased" of the three they consider. They note that in the example they discuss, the CG value is 0.516 ($t = 0.5$, $p > .5$) without interpolated richness values, and 0.448 ($t = 1.624$, $p > .05$) with the latter values included. The addition of the interpolated values does not change the CG value in a statistically significant manner, though Gould et al. (1987) find the difference important and alter their interpretation of the graph from being more or less symmetrical to being bottom heavy.

One might well ask why a best-fit polynomial equation is not used to determine expected values for all temporal points rather than just those with richness values of zero—a procedure suggested in archaeology (e.g., Kintigh 1984; McCartney and Glass 1990). But were one to pose this question, we suspect it would quickly become obvious that even the interpolation procedure favored by Gould et al. (1987) presumes at least in part precisely what clade-diversity analysis is attempting to determine—fluctuation in the richness of types over time. As David Rhode (1988:711) observes, such an analytical step implies "that a great deal already is known about the nature of the archaeological record." That is, one is presuming—via the interpolation process—exactly what one is attempting to determine—*changes* in diversity over time. Precisely this point has recently been recognized by paleobiologists (Kirchner and Weil 1998).

We calculated CG values for the radio data in Table 7.1 after substituting interpolated richness values for all zero values. We refer to these as CG_{int} to indicate that interpolated richness values are included in their calculation. The interpolated values are listed in Table 7.1 and are shown in Figures 7.10 and 7.11. The CG_{int} value for Figure 7.10 is 0.468 ($t = 0.9995$, $p > .1$), and the CG value is 0.443 ($t = 1.78$, $p < .05$); the CG_{int} value for Figure 7.11 is 0.682 ($t = 5.685$, $p < .01$), and the CG value is 0.686 ($t = 5.81$, $p < .01$). Richness values derived by interpolation do not create statistically significant changes in the CG values for models of portable radios, but they do result in such changes for vacuum-tube radio manufacturers. Given our cautions in the preceding paragraph, we suggest that calculating both CG and CG_{int} values may be advisable.

There does not appear to be a statistically significant difference between the history of vacuum-tube radio diversity and randomly generated background variation within a clade ($p > .1$ for CG_{int}, but $p < .05$ for CG). Although Schiffer's research indicates that the Figure 7.10 graph should not be considered symmetrical, the statistical test of CG_{int} alone would not allow us to identify that graph as such. Statistical tests of the clade-diversity diagram in Figure 7.11, on the other hand, suggest this graph is not symmetrical ($p < .05$ for both CG and CG_{int}). But recall Gilinsky and Bambach's (1986) remark: Perhaps so many diverse and somewhat conflicting "causal factors" are included that any statistical indication of lack of symmetry is muted. We note as well that the diagrams in Figures 7.10 and 7.11 lack their tops; thus, we do not know what the complete history of diversity for either clade was like. A CG value should not be calculated for an incomplete clade-diversity diagram, because if either end is missing, the CG value will be skewed. Any effort to calculate a CG value for the radio data is spurious because that value will automatically suggest a top-heavy graph (Gould et al. 1977, 1987). We have used Schiffer's radio data merely for purposes of illustration. What about archaeological data?

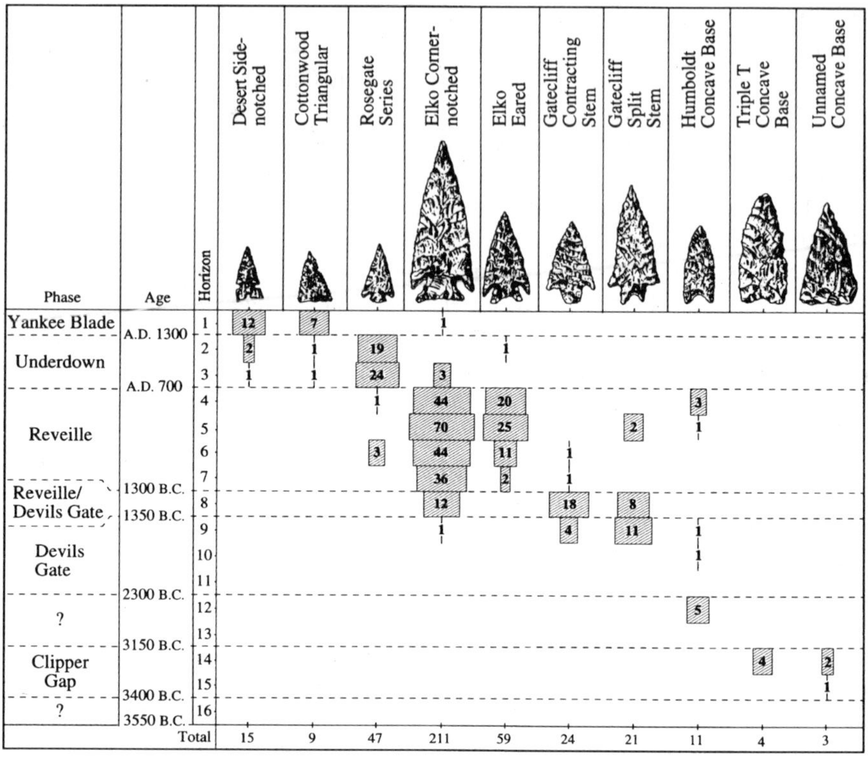

FIGURE 7.12. Frequencies of projectile points by type and stratum at Gatecliff Shelter, Nevada (after Thomas and Bierwirth 1983).

The history of types of projectile points from Gatecliff Shelter in Nevada (Thomas and Bierwirth 1983) is presented in Figure 7.12; observed richness and chronological assignments (after Thomas 1983a:174) are summarized in Table 7.3. A clade-diversity diagram for these materials is shown in Figure 7.13. Only two richness values—for Horizons 11 and 13—require interpolation. The CG_{int} value was calculated by assigning each assemblage of projectile points to the midpoint of its temporal range—a procedure that Gould et al. (1987:1439) indicate will put "too much weight on [temporally] more closely spaced samples." However, if the duration of Horizon 2 is assumed to be 50 years, and the durations of Horizons 4, 5, and 6 are assumed to be 650 years each, then the duration of the horizons is not correlated with the sequence of horizons (Spearman's $rho = 0.11, p > .6$). Furthermore, lumping Horizons 4–6 into a single unit of 1950 years' duration, and assigning Horizon 2 a duration of one year, as is implied in Table 7.3, the correla-

Table 7.3. Temporal Distribution of Projectile-Point Richness at Gatecliff Shelter, Nevada

Horizon	Age	Observed richness (interpolated)	Scaled time
—	post–1500	0	1.0
1	1300–1500	3	0.9375
2	1300	4	0.875
3	700–1300	4	0.8125
4	AD 50[a]	4	0.750
5	—	4	0.6875
6	1250 BC	4	0.625
7	1300–1250	3	0.5625
8	1350–1300	3	0.5
9	1450–1350	4	0.4375
10	2100–1450	1	0.375
11	2300–2150	0 (1)	0.3125
12	3050–2300	1	0.25
13	3150–3050	0 (1.5)	0.1875
14	3300–3150	2	0.125
15	3400–3300	1	0.0625
16	3550–3400	0	0.0

[a]Horizons 4, 5, and 6 span the period 1250 B.C.–A.D. 50.

tion between horizon duration and horizon sequence is no different from zero (*rho* = 0.01, $p > .9$). Gould et al. (1987:1439) indicate that such a result suggests period duration will not influence CG values. The CG_{int} value for the clade-diversity diagram in Figure 7.13 is 0.5995 (t = 3.108, $p < .001$) and indicates that the diversity of projectile-point types increases through time.

Foote (1991:116) indicates that the CG of a clade-diversity diagram "inherent in the time scale" may not be 0.5 simply because of variation in the duration of the periods of which it is comprised. Given the time scale—the set of temporal units and their durations denoted by the cultural horizons—for the Gatecliff Shelter projectile points, the CG inherent in the time scale is 0.436. This value is calculated as the proportion of scaled time since the initiation of the clade indicated by the median shared temporal boundary between horizons (or periods). Thus, the median boundary is that between Horizons 8 and 9; it falls at 1350 B.C., which is 0.436 of the total time since the clade's initiation at 3550 B.C. (with the clade terminating at A.D. 1500). We ignore this problem here and note that were this inherent CG used for statistical comparison rather than 0.5, the clade-diversity diagram for Gatecliff Shelter projectile points would still be significantly top heavy.

In the following, we consider the clade-diversity diagram in Figure 7.13 and

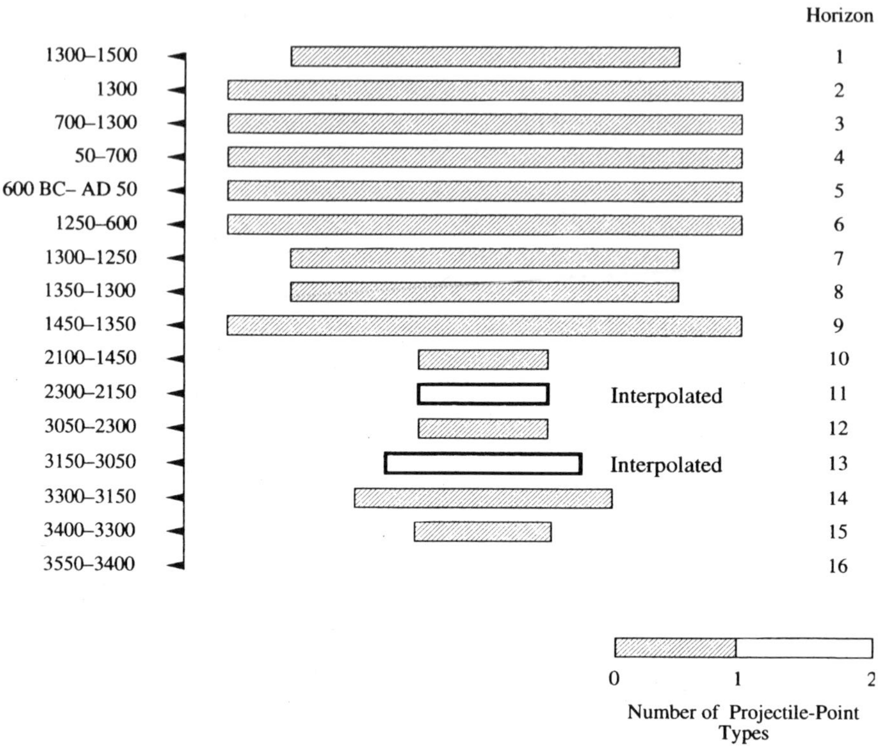

FIGURE 7.13. Clade-diversity diagram for Gatecliff Shelter projectile points. Data from Table 7.3.

its associated CG_{int} value. It is important, therefore, to point out that were we to follow Gould et al.'s (1987) preferred procedure of designating temporal periods of equal duration, those periods each might be of 100 years duration, as that is the smallest temporal increment that can be extracted from Table 7.3. That procedure results in many more interpolated than observed richness values; thus, we do not pursue it here except to note that it produces a top-heavy clade. Alternatively, we might lump various horizons to produce collections of more or less equal duration. One possible result would be the following: Horizons 16–13 (500 yr, 2 types), Horizon 12 (750 yr, 1 type), Horizons 11–10 (850 yr, 1 type), Horizons 9–6 (850 yr, 6 types), Horizon 5 (650 yr, 4 types), Horizon 4 (650 yr, 4 types), and Horizons 3–1 (800 yr, 5 types). This procedure, too, produces a top-heavy clade. Thus, we find the clade illustrated in Figure 7.13 satisfactory for discussion purposes. What can we make of it?

First, Thomas (1983b:425–431) documents in detail that the richness of stone-tool types per cultural horizon at Gatecliff Shelter is a function of sample size: Larger samples produce more types. Sample size and observed projectile-point richness data summarized in Figure 7.12 are correlated (Pearson's $r = 0.744$, $p < .001$), suggesting that caution would be warranted were one to interpret the CG value derived from those data. For sake of discussion here, we assume the effects of sample size on the CG_{int} are insignificant. We also assume that the clade-diversity diagram (Figure 7.13) is not truncated at either end. Granting these two assumptions allows insights into the history of Great Basin projectile-point diversity that can be phrased as hypotheses to be tested with additional data.

Each projectile-point type displays a more or less unimodal frequency distribution when plotted against vertical provenience, but those distributions are imperfect (Figure 7.12). Various explanations could be mustered to account for this kind of distribution, including stratigraphic mixing (e.g., Burch 1959) and recycling and reworking (e.g., Flenniken and Wilke 1989). Most recently, it has been suggested that at least some of the variation in the distributions of the types is attributable to functional causes (Beck 1995, 1998). None of these arguments invalidates the implied temporal ordering of the types, as they rather consistently fall in the indicated temporal order (e.g., Bettinger et al. 1991; O'Connell and Inaway 1994). But these arguments underscore the fact that some of the definitive attributes of the types measure time—are potentially stylistic—and some attributes measure functional change (see also Hughes 1998). The two sets of attributes need not be, and likely are not, mutually exclusive. It is likely that the point "types" at Gatecliff Shelter are combinations of both kinds of attributes and hence produce less than perfect unimodal frequency distributions. Beck (1995) found that side notching occurred frequently early (on large points), then dropped in frequency, and finally became more frequent again late in time (on small points). This is not unusual for a functional trait. She also noted that the proximal shoulder angle of all points was the single attribute that correlated significantly with time, as a stylistic attribute should. This does not mean that the proximal shoulder angle is stylistic, because functional traits can also be transmitted or inherited and thus may mark the passage of time.

Others have suggested that change in the delivery system—lance or javelin, to atlatl dart, to bow and arrow—was the driving selective force resulting in change in Great Basin projectile points (e.g., Beck 1995; Musil 1988). Recalling that the CG_{int} value is significant ($p < .01$), change in the diversity of Gatecliff Shelter points might, then, comprise an instance of stimulated variation. We say "might" because of the kinds of variation measured by the projectile-point types; some of the attributes appear to be functional and others stylistic. Furthermore, some of the definitive attributes of the Gatecliff Shelter projectile-point types may have been merely sorted rather than subjected to selection. Some of the attributes of the hafting elements used to designate point types probably changed as the engineering-design requirements for efficient hafting of a point changed from that for a lance or javelin, to that for an atlatl, then to that for an arrow (Hughes 1998).

We hypothesize that the increasing diversity of projectile-point types comprises an instance of an increase in the number of sets of design constraints that resulted from an increase in the number of weapon-delivery systems.

What is design constraint? "Constraint is a theory-bound term for causes of change and evolutionary direction by principles and forces outside an explanatory orthodoxy" (Gould 1989c:519; see also Carroll 1986; Stearns 1986). The suggestion that change in the delivery system of projectile points was the selective force resulting in change in projectile-point form is orthodox functionalism (in Gould's and Stearns's terms). But note that minimally two scale shifts are made when one calls on weapon-delivery systems as a functional explanation for change in the attributes of projectile-point haft elements. The scale shifts are from the complete weapon to the point (type) and from the point (type) to attributes of the hafting element. The particular history of delivery systems drove the increase in diversity of the hafting element. Each weapon-delivery system posed a set of unique engineering design requirements—constraints—on the sorts of hafting element that would work efficiently; as new delivery systems were added, new constraints on projectile-point haft elements came into play. Attributes of haft elements were sorted as a result of design constraints; they were not the direct focus of selection (e.g., Vrba 1989). That honor hypothetically resides with the total weapon-delivery system, a much larger-scale entity than the scale at which projectile-point types are distinguished. This is not the place to delve further into this issue. Our discussion is meant to outline the kinds of factors that must be considered if one wishes to explain the history of novel variants—their appearance, replication, and disappearance. Schiffer's radio data and data on Gatecliff Shelter projectile points appear to comprise both stylistic and functional variation. What about data comprising only stylistic variation?

The "seriation graph" produced by Phillips et al. (1951: Fig. 21) for ceramic sherds from the St. Francis River area of the Lower Mississippi Alluvial Valley provides an example of changes in the richness of stylistic variants. Because that graph was originally published as a large foldout, we reproduce only a representative portion of it in Figure 7.14. The complete graph, founded in part on superposed collections and in part on surface collections, presents an ordering of 58 assemblages; the 47,413 sherds included represent 18 types and span five "periods." Our discussion includes all 58 assemblages. Phillips et al. (1951) used surface treatment/decoration as the basis for most types. There were few types for the early periods and more for the later periods—a reflection of the increasing variation in decoration in the later periods. The "seriation graph" thus appears to be top heavy, as does the clade-diversity diagram (Figure 7.15) derived from it; the CG value of 0.575 ($t = 2.343$, $p < .01$) calculated from clade-diversity data confirms this appearance.

Not only does the richness of ceramic types increase through time, so, too,

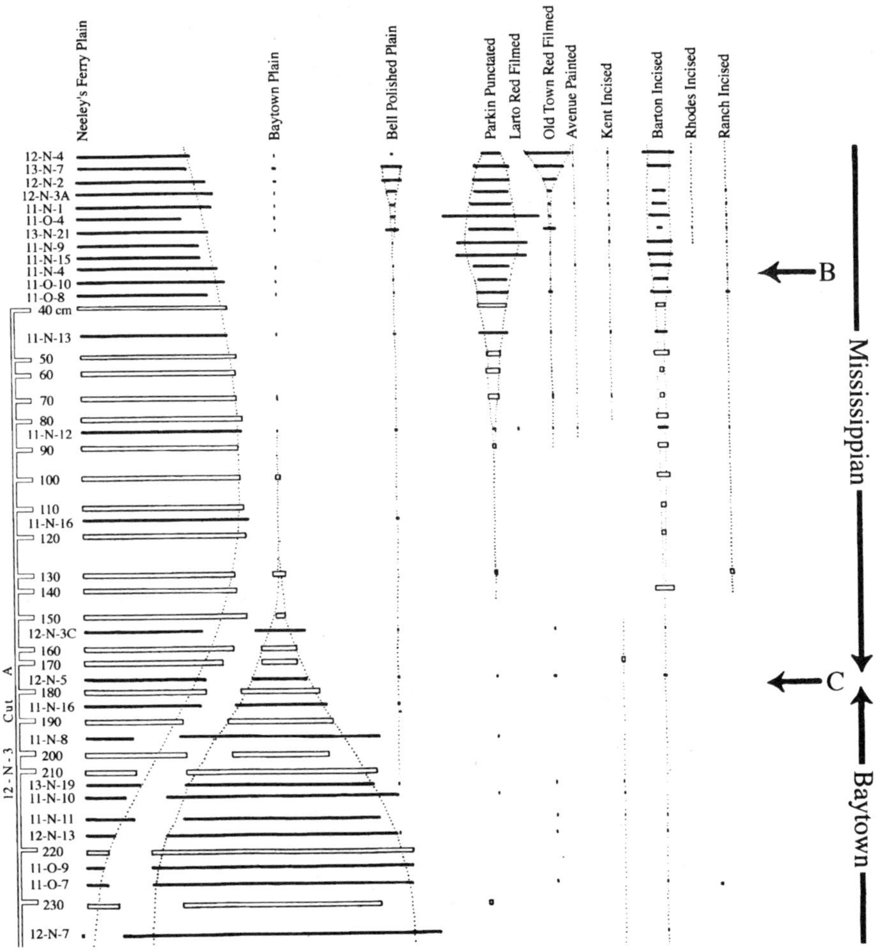

FIGURE 7.14. Chronological ordering of sherd assemblages from excavated and surface-collected sites in the St. Francis River area of the Lower Mississippi Alluvial Valley. The ordering was based primarily on the interdigitation of excavated levels (open bars along left margin) at Rose Mound in Cross County, Arkansas. Other sherd assemblages (solid bars) were placed in the chronology based on percentages of pottery types, denoted by bar lengths. Only 46 of 58 assemblages are shown here, representing the late half of the ceramic chronology; assemblages dating to the early half of the chronology have been deleted. Nine rarely represented types are not included. Boundaries between periods are denoted by the bold horizontal arrows and names on the right (after Phillips et al. 1951).

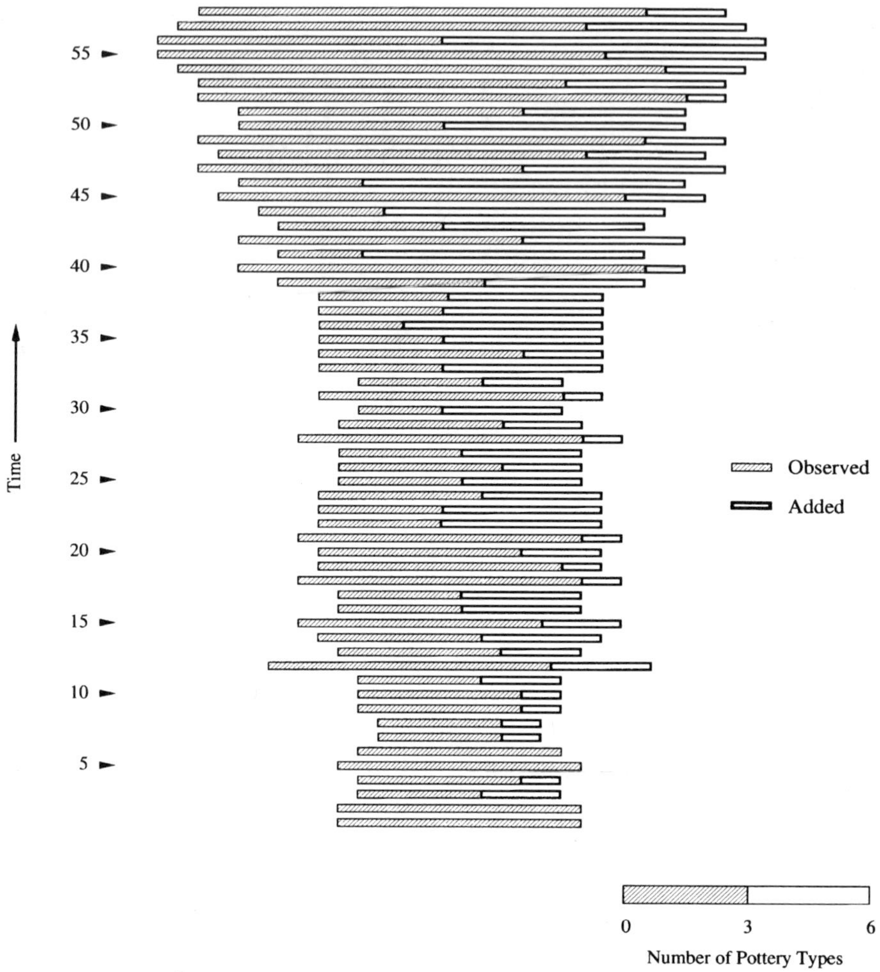

FIGURE 7.15. Clade-diversity diagram for St. Francis River area ceramic types. "Added" richness values are based on Phillips et al.'s (1951) beliefs regarding the temporal span of types (see Figure 7.14). Data from Table 7.4.

does sample size (Table 7.4). These two variables are correlated when all 58 assemblages are included (Pearson's (r = .69, $p < .002$) and when samples are lumped by chronological period (Pearson's (r = .93, $p < .05$). Caution may be warranted when interpreting the history of ceramic diversity represented by the data, but for the sake of discussion we assume sample-size effects are insignificant. We also assume that the ends of the clade are not truncated. The temporal scale of the "seriation graph" is ordinal; thus, we do not know how the assemblages vary in duration. But recall that one requirement of the seriation method is that the assemblages seriated must be of equivalent duration. The "seriation graph" suggests this requirement is at least approximated, else the frequency distributions of the graphed types would not approximate unimodal curves (Figure 7.14). Therefore, we assume that the temporal duration of each assemblage is more or less equivalent to the duration of every other assemblage.

The "seriation graph" for the St. Francis River area represents Phillips et al.'s beliefs regarding the temporal continuity of various pottery types that occur discontinuously in this particular empirical record. Their beliefs are signified in the graph by dotted lines connecting various samples within individual type columns (see, for example, the column under the type "Bell Polished Plain" in Figure 7.14). Tallying richness values for each assemblage according to the combined observed types as well as the types believed to be present—what we term, somewhat inaccurately, interpolated presences—increases the richness values of 54 of the 58 assemblages. The four assemblages with unaltered richness values are among the six oldest assemblages in the ordering (Table 7.4). The richness of each of the oldest 27 assemblages increases by an average of 1.8, whereas the richness of each of the 31 youngest assemblages increases by an average of 3.7. This suggests that the CG value of 0.575 may increase when calculated as a CG_{int} value; the latter is 0.589 ($t = 2.624$, $p < .005$). Both values indicate a top-heavy clade, though the latter is a bit more top heavy than the former.

Richness increases via interpolation in virtually all assemblages throughout the period graphed, but interpolated richness values per assemblage increase progressively more rapidly as assemblages become younger in age, as shown in Figure 7.16, thereby increasing (making more recent in time) the center of gravity of the clade-diversity diagram when it includes interpolated values. The small difference in the CG and CG_{int} values suggests to us that Phillips et al. (1951) knew a great deal about the ceramic chronology of the Lower Mississippi Alluvial Valley. It was precisely this additional knowledge that informed their adding dotted lines to various of the type columns—what we termed interpolations. Their proposed chronology has stood the test of additional data and analyses. What, then, can we make of the increasing richness of ceramic types?

Because attributes of decoration form the basis of the types, we believe the types are stylistic. The relatively smooth unimodal frequency distributions of the types lend credence to this belief but do not confirm it. Later analyses of Phillips et al.'s (1951) types suggest that they are, in fact, largely stylistic (Lipo et al. 1997). Why, then, does the richness of types increase late in time, producing a top-heavy clade? We find Braun's (1995) and Lipo et al.'s (1997) discussions particularly helpful. The richness of decorative types is correlated with social dynamics, particularly the frequency of group interaction or intergroup transmission (Braun

TABLE 7.4. Observed Ceramic-Type Richness, Number of Additional Ceramic Types Believed to Be Present, and Sample Size for the St. Francis River Area, Lower Mississippi Alluvial Valley

Assemblage	Observed type richness	Number of additional types believed present	Sample size	Scaled time
—	0	0	—	1.000
58	11	2	731	0.983
57	10	4	674	0.966
56	7	8	254	0.949
55	11	4	1416	0.932
54	12	2	6898	0.915
53	9	4	722	0.898
52	12	1	2030	0.881
51	7	4	2590	0.864
50	5	6	686	0.847
49	11	2	7287	0.831
48	9	3	4723	0.814
47	8	5	151	0.797
46	3	8	142	0.780
45	10	2	4925	0.763
44	3	7	117	0.746
43	4	5	74	0.729
42	7	4	164	0.712
41	2	7	134	0.695
40	10	1	3063	0.678
39	5	4	106	0.661
38	3	4	65	0.644
37	3	4	60	0.627
36	2	5	103	0.610
35	3	4	113	0.593
34	5	2	99	0.576
33	3	4	59	0.559
32	3	2	83	0.542
31	6	1	273	0.525
30	2	3	73	0.508
29	4	2	69	0.492
28	7	1	698	0.475
27	3	3	56	0.458
26	4	2	193	0.441
25	3	3	87	0.424
24	4	3	179	0.407
23	3	4	103	0.390
22	3	4	146	0.373
21	7	1	266	0.356
20	5	2	211	0.339
19	6	1	298	0.322
18	7	1	1187	0.305

TABLE 7.4. (*Continued*)

Assemblage	Observed type richness	Number of additional types believed present	Sample size	Scaled time
17	3	3	127	0.288
16	3	3	71	0.271
15	6	2	1478	0.254
14	4	3	98	0.237
13	4	2	488	0.220
12	7	2	117	0.203
11	3	2	106	0.186
10	4	1	109	0.169
9	4	1	223	0.153
8	3	1	158	0.136
7	3	1	47	0.119
6	5	0	629	0.102
5	6	0	987	0.085
4	4	1	181	0.068
3	3	2	285	0.051
2	6	0	713	0.034
1	6	0	288	0.017

Source: Phillips et al. (1951).

1995). Decreased intergroup transmission and increased intragroup transmission results in increased localization of decorative styles (Lipo et al. 1997), much like peripatric speciation results from decreased gene flow between reproductively isolated populations. Increasing stylistic diversity may well reflect, then, a decrease in the panmictic characteristics of the human populations in the St. Francis area. Dethlefsen and Deetz (1966:508) state that it "appears that [novel variation] is primarily initiated by a small segment of the population and then spreads to the majority." This is an effective statement of what is known in biological evolution variously as geographic, allopatric, or peripatric speciation; that is, potters became, for whatever reason, more isolated and their decorative styles evolved in independent directions, resulting in progressively more "types." Lipo et al. (1997) have recently indicated how the resulting local areas might be identified archaeologically.

Increasing localization of decorative styles may result as symbols of one's social identity become more important. Styles have a production cost, and thus are "driven by selection" (Lipo et al. 1997:318). Production of progressively more styles over a relatively large area, then, must have been selected for in some sense. Identifying those selective forces is beyond our scope here, though we note that some interesting hypotheses have been outlined that might prove applicable (e.g., Braun 1985b, 1991; Braun and Plog 1982). Our point here has been to identify the

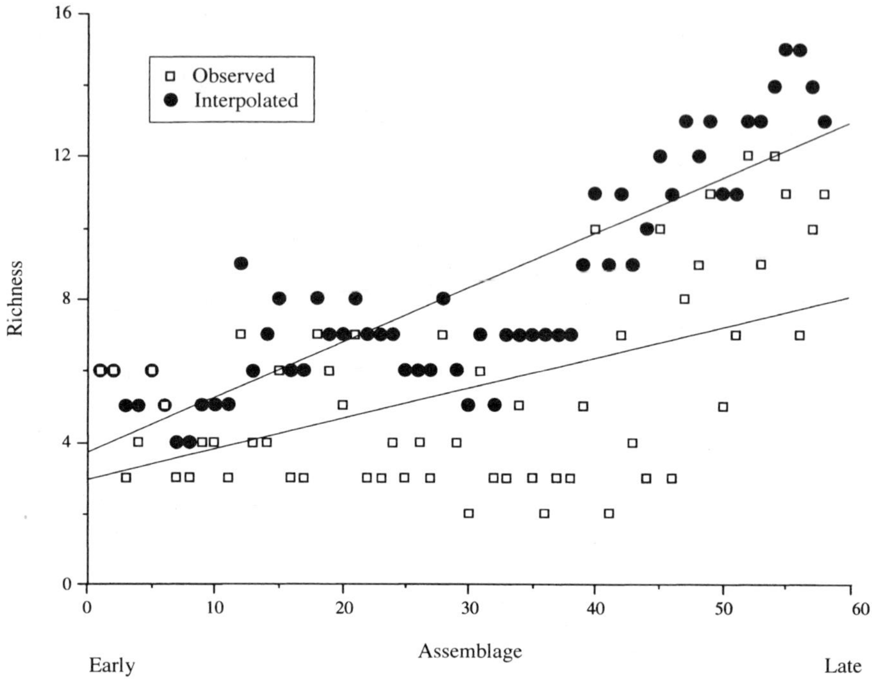

FIGURE 7.16. Change in richness of ceramic types across 58 assemblages from the St. Francis River area of the Lower Mississippi Alluvial Valley (data from Phillips et al. 1951). The lower, less-steep line is the simple best-fit regression line for the data plotted as squares (observed richness values). The upper, steeper line is the simple best-fit regression line for data plotted as dots (interpolated richness values).

historical shape of artifact-clade diversity as a springboard to detecting when such hypotheses warrant testing. We emphasize that although the examples we review concern dynamic histories of diversity, a history displaying stasis in diversity over time also requires explanation.

CLASSIFICATION AND CLADE DIVERSITY

The critical issue that has largely gone undiscussed is the role of classification in efforts to measure clade diversity, whether of biological organisms or of artifacts. Paleobiologists classify fossils as species; we suspect this is why such units often will not produce the lenticular distributions archaeologists have come to expect of historical types when their relative frequencies are plotted against time. Archaeologists interested in frequency seriation construct classifications of historical types such that the relative frequencies of specimens within each do fluctuate unimodally over time. The critical question is whether it is appropriate to construct clade-diversity diagrams for historical types of artifacts, that is, those whose frequencies can be seriated successfully. Perhaps it *is* appropriate if one wishes to have some measure of change in the relative amount of selectively neutral variation in artifact form. The seriation model guides the construction of classes to be used in a seriation: Each artifact type is constructed by the analyst—often by trial and error—explicitly to display a unimodal frequency distribution through time. Phillips et al.'s data comprise types of pottery that, based on their lenticular frequency distributions, comprise historical types. Furthermore, that the types display such lenticular frequency distributions through time *confirms* that they indeed reflect heritable continuity. Just as paleobiologists do with species conceived of as "individuals" that have historical beginnings and endings, one could plot the functional diversity of successfully seriated styles that have been reclassified on the basis of interspecimen variation measured with units related to use. Heritable continuity would be ensured within those units; seriated styles would comprise the replicators and when reclassified as functional types would comprise the interactors.

The classification system one uses to sort artifacts influences measures of richness and also the frequencies of specimens per class. However, theory is needed to *explain* both the fluctuating frequencies of specimens within historical types (classes) and the fluctuating diversity values. Seriable historical types may tell us something about changes in the diversity of selectively neutral variants over time, and they may reveal something about increases in adaptive plasticity of cultural phenotypes, given that styles of myriad scales may comprise a reservoir of alternatives that can be called into service as needed. Such would be exaptations in the sense of Gould and Vrba (1982). What about the units used to generate clade-diversity diagrams? Schiffer's (1996) data comprise "models" of portable radios and are proxies for models of vacuum-tube radios. Although Schiffer (1996) provides explanations for the bottom-heavy graph in Figure 7.10 and the top-heavy graph in Figure 7.11, we note that the particular historical contingencies of the events depicted in the graphs are known. Thus, he is able to suggest the operation of what he terms *stimulated variation*, defined by him as "bursts of variety-generation [caused by] changed conditions in selective contexts [that] stimulate an increase in inventive activities of behavioral components and ... foster the creation of new behavioral components" (Schiffer 1996:655). Although we view this as reducing the cause to the fulfillment of desires, some measure of diversity is required. That measure must comprise units that faithfully record the variable(s) of interest. When we view a clade-diversity diagram, we need to know *what* has been graphed in order to explain it. We could not explain weight as a result of color, or color as a result of length. Nor would a graph of typological diversity necessarily represent a clade-diversity diagram if it could not somehow be argued that the included types were all descendants of the same ancestor. In the next chapter, we turn to explaining lineage histories and there discuss further how various tempos and modes of change might be identified.

Chapter 8

Explaining Lineage Histories

> We work in a historical science and the narrative patterns of life's long-term history are as important as theories invoked to explain the regularities. These patterns emerge partly by operation of the theory through time, but also in large measure from the influence of contingent events, some of great magnitude, that have impacted the history of life. The fascination of scientific history (as opposed to mere description of events in temporal sequence) lies in its melding of general theory with recognition of the irreducible importance of particular events that are largely unpredictable (at least from the theory). (Gould 1995a:7–8)

> Biologists interested in culture are often struck by the absence of viable general theories in the social sciences. All of biology is united by the theory of biological evolution. Perhaps progress in the social sciences is impeded because there is no general theory of cultural evolution. (Sober 1992:30)

Sober is correct in noting that there is no general theory of cultural evolution, although we wonder just how many philosophers and biologists interested in culture have actually noticed this. It seems to us that most of them have written culture and cultural inheritance off as components of a Lamarckian system that is driven by human intent and necessity. Gould is correct in pointing out that narrative patterns of life's long-term history are just as important as theories that we invoke to explain the regularities. Interestingly, Gould (e.g., 1996) is one of numerous biologists—Ernst Mayr being another (e.g., Angier 1997; Mayr

1982a)—who have examined culture and found it to be everything Mother Nature is not. He and others of a similar bent have done what countless anthropologists have done for over a century: set humans aside as being something special because of their extrasomatic means of adaptation—culture. To us, this locates cause in the wrong place. Yes, the mode of transmission is different when culture is involved, although we view this more in quantitative rather than qualitative terms in light of what we know of animal behavior (e.g., Bonner 1980, 1988), and there can be no doubt that the tempo of cultural transmission differs significantly from that of genetic transmission. But do these differences lead to the inescapable conclusion that humans have stopped evolving—that they somehow are beyond the reach of selection? Do these differences indicate that other evolutionary processes, such as drift, play at best minimal roles in reshuffling *both* somatic and nonsomatic characters? In our opinion, the answer to both questions is an emphatic "no." Humans today are no more immune to evolutionary processes than they were ten thousand or fifty thousand years ago. The game of life has perhaps become more sophisticated, at least as seen through modern eyes, and there certainly are more players, but the rules of the game have not changed.

To us, evolutionary archaeology offers a systematic means of writing a comprehensible historical narrative of the human past. We do not need what philosopher Daniel Dennett (1995:74) playfully referred to as "skyhooks"—imaginary devices suspended from the sky that act to lift "unwieldy objects out of difficult circumstances"—to get the job done. Others are less convinced. After examining the evolutionary-archaeology agenda, critics such as Patty Jo Watson (1986) have claimed that the approach is little more than a sterile physics of artifacts—a methodologically sound but trivial analytical procedure for measuring "the form and distribution of behavioral by-products ... in a behavioral vacuum" (DeBoer and Lathrap 1979:103). In the minds of the critics, all that evolutionary archaeology is able to produce are lineages of artifacts and some engineering analyses of the mechanical properties of artifacts. Watson (1986:446) characterized archaeologists who adhere to this "vacuous" approach as "narrow empiricists," and in a similar vein, philosopher Alison Wylie (1995:206) saw in evolutionary archaeology "an uncompromising commitment to essentially positivist (specifically deductivist) ideals of security in inference." Such comments are in large part true, though we fail to see the sterility to which Watson alludes. Constructing and explaining historical artifact lineages may not be as glamorous as spinning interpretive stories about the past, and we admit it is tedious and time consuming, but it certainly is not sterile. It is, to put it bluntly, the only legitimate approach that we know of to use in trying to explain how and why the archaeological record takes the shape it does.

There should be no reason to suggest that evolutionary archaeology is anything but a deductivist strategy (but see Broughton and O'Connell 1999:158). Wylie (1995:206) argues that evolutionary archaeology inherited its deductivist component from processual archaeology, but this in incorrect. Darwinian evolutionism has *always* had a strongly deductivist aspect to it; Darwin's ideas on natural selection were deduced directly from his theory of descent with modification, and ascertaining whether artifacts or various of their attributes are functional and have adaptive value rests on hypothesis testing (e.g., Jaksic 1981). Hence, this component was not something evolutionary archaeologists happened to inherit from Lewis Binford and company's program for studying cultural processes. Wylie and Watson are correct in stating that evolutionary archaeology, in its bid to construct historical lineages, views behavioral reconstruction as being largely beyond its purview. In that sense, evolutionary archaeology *is* carried out in a behavioral vacuum, since there is no direct way, at least none that we see, to access the *historically contingent* behavioral past (Dunnell 1982; O'Brien and Holland 1995b; O'Brien et al. 1998). Again, this hardly makes evolutionary archaeology sterile or trivial. Rather, in our minds it lends scientific legitimacy to archaeology by jerking it out of the interpretive malaise that has come to grip the discipline over the past half century. Behavioralism is only one aspect of this malaise, but it is a key one. Americanist archaeology has always had a love affair with the study of prehistoric behavior—a liaison that grew from its close intellectual ties to anthropology (Lyman and O'Brien 1997).

Behavioral sequences (Schiffer and Skibo 1997)—histories of prehistoric behaviors that may or may not comprise historical lineages—are interpretations (inductions) about the past. They may be real, or sort of real, or not at all real; we simply have no way of knowing because of the shaky ground—or as Neff and Larson (1997:77) put it, the "lengthy, ad hoc inferential chains"—upon which they are constructed. In our opinion, there is little explanatory theory to help with the task, as so-called behavioral theory has yet to be spelled out as more than a list of empirical generalizations. Furthermore, behavioral theory and method contain no consideration of how the analyst is to ensure that historical lineages rather than mere sequences are constructed (O'Brien et al. 1998). Here, behavioralists might claim ownership of what we have characterized as historical continuity, but this does not ensure *heritable* continuity.

Behavioral archaeologists, a splinter group of the processual archaeologists, believe they *can* access the past through analogy—that is, by using the ethnographic present or the very recent past as a way of understanding how things were in the distant past. Some behavioral archaeologists (e.g., Schiffer 1996, 1999) view the emphasis on behavior as the means of rescuing evolutionary archaeology from "the 'artifacts' horn of the archaeologist's dilemma," but we have serious reservations about this position (O'Brien et al. 1998). The clarion call for what became known as behavioral archaeology was made in 1974 in a two-page article in *American Antiquity* that carried the ambitious title "Expanding Archaeology" (Reid et al. 1974; for background, see Reid 1995; Schiffer 1995b). The tone for and principles behind a behavioral archaeology were set the following year with the publication of two articles—"Archaeology as Behavioral Science" (Schiffer 1975) and "Behavioral Archaeology: Four Strategies" (Reid et al. 1975). The leaders of the behavioral movement—J. Jefferson Reid, Michael Schiffer, and William Rathje—defined archaeology as "the study of relationships between human behavior and material culture" (Reid et al. 1975:864), and they employed the term *behavioral archaeology* to refer to "the study of material objects regardless of time or space in order to describe and explain human behavior" (Reid et al. 1975:864). They also noted, "By virtue of years of research ... archaeologists now possess an expanding body of theory, method, and behavioral laws for the study of material objects and human behavior *regardless of time or space*" (Reid et al. 1975:866; emphasis added).

Despite the intervening years since those articles appeared, many archaeologists today still would subscribe to the notion that "behavioral archaeology is a synthesis of what archaeologists have done and aspire to do and that the essential interrelatedness among the strategies has roots deep in the progressive development of the discipline as a whole" (Reid et al. 1975:866). Schiffer certainly never changed his mind on the importance of studying behavior archaeologically, announcing recently that "constructing behavioral theory to explain variation and change in human behavior, conceived as people–artifact interactions, is archaeology's highest scientific calling" (1995a:35). Considerable disagreement exists in the archaeological literature over the most appropriate methods for meeting this calling—disagreements that run the gamut from philosophical to methodological (see various chapters in Skibo et al. 1995)—but these disagreements belie the fact that behavioral archaeology, by definition, has a central focus on the search for and application of behavioral *laws* and their corollaries (Schiffer 1996). Despite heated arguments over the proper role of ethnographic analogy in archaeological research (e.g., Binford 1981, 1985; R. A. Gould 1985; R. A. Gould and Watson 1982; Schiffer 1985; Watson 1982), many of the same principles that guided argument from analogy in the 1970s and 1980s still guide behaviorally oriented research today (O'Brien 1996d; O'Brien and Holland 1995b); that is, it is thought that within certain parameters, examination of contemporary material remains and behavior can guide and inform reconstructions of past behavior.

Some behavioral archaeologists, the most notable being Schiffer (1996), see considerable overlap between their approach and evolutionary archaeology, but some of the similarities are illusory (O'Brien et al. 1998). In our view, behavioral archaeologists are borrowing theoretical tidbits from evolutionism that can be used as a rationale for explaining apparent behavior and changes in behavior. When previous searches for general laws of behavior—behavioral archaeologists label them "principles"—turned up empty (O'Brien and Holland 1995b), Darwinian evolutionism became attractive. Nonetheless, some common ground exists between evolutionary archaeology and certain behavioral-archaeology strategies, specifically the analysis of prehistoric materials from the standpoint of engineering design (Cogswell and O'Brien 1997; O'Brien and Holland 1990, 1992, 1995a, 1995b; O'Brien et al. 1994). We have always agreed with behavioral archaeologists (e.g., Deal and Hagstrum 1995; Schiffer and Skibo 1987, 1997; Skibo and Schiffer 1995) that technological and functional analyses of how certain objects were used at specific times put one on a firmer foundation relative to inferring the *specific* behaviors, or *activities*, as Schiffer (1976) terms them, of prehistoric manufacturers and users of those objects (O'Brien and Holland 1995b; see Saunders 1990). Importantly, these inferences are derived *directly* from experimental evidence viewed against the archaeological context containing the materials being examined (e.g., Lyman et al. 1998a). This kind of engineering research agenda, one based on experimental evidence regarding immanent properties, will allow us to begin to understand not only the evolutionary trajectories of the humans responsible for the technological products but also the nature of selective regimens that affected those humans. In short, it will allow us to begin writing historical narratives (Lyman and O'Brien 1998; O'Brien and Lyman 2000a; O'Brien et al. 1998).

It is important to repeat, however, that our approach is *not* compatible with all aspects of behavioral archaeology, which itself has widely divergent goals (Walker et al. 1995). The common ground to which we refer has nothing to do with approaches grounded in the search for universal laws of behavior that some archaeologists (e.g., R. A. Gould 1978a, 1978b, 1980a, 1980b; Hole 1979; Schiffer 1978, 1996) think exist. In our view, such universalities not only *do not* exist, they *cannot* exist. If they did, constructing lineages would be greatly simplified: Everything would more or less be the same, the only exception being slightly changing boundary conditions that made humans slightly adjust their responses. After all, behavioral archaeologists do not claim that the present is exactly the same as the past; rather, they claim that humans will exhibit similar behaviors in the face of similar social and physical environmental situations (e.g., Schiffer and Skibo 1997). All we need do is approximate the environment and we can then predict the behavior. This strikes us as a form of possibilism at best and determinism at worst.

Our rejection of the basic tenets of behavioral archaeology does not imply that analogs to modern behavioral outputs do not occur in the archaeological record. Nor does it imply that some of the behaviors that create modern outputs are not similar to those that operated in the past. We object to the term *cultural universals*, which implies that we have at our disposal the necessary means of distinguishing between those behaviors that were the same and those that were different. This feat is difficult, and probably impossible, to accomplish. Importantly, we should ask here, what is at issue: similar *behaviors* or similar *processes*? Can we say that two behaviors are the same because two groups reduce stone in the same way? We would argue that there is a world of difference between the two. We would argue further, as O'Brien and Holland (1995b) have, that conflation of behaviors and processes has led to the belief that the archaeological record can be used to construct universal laws of behavior. Again, we view this as impossible. We *do* believe that the archaeological record can be used to examine (not reconstruct) *specific* human behaviors and that Darwinian evolutionary theory offers the best grounding for such an examination.

Some culture historians knew the difference between immanent and configurational properties, though they did not use those labels, but they employed the wrong kinds of units to build lineages. Some processualists, such as Kent Flannery, also knew the difference—hence his lament (1986) that although we want to know about adaptive plateaus and the like, we have only successions of phases with which to monitor change (Chapter 6). Although it would be ill-founded, the argument could be made that at present, not enough work has been done, and that what we need are more and more phases stacked up so that we approach a complete picture of the past. In the best tradition of archaeology in the lower Mississippi River valley (e.g., Phillips 1970), all we need to do is chop up time and space into finer and finer units. There is something to be said for obtaining more information, but even with the addition of more and still more phases, we are afraid all we would have would be a mishmash of variously useful and useless information. We might know something about assemblage similarities and differences, but we would know little, if anything, about change. And if the last quarter century of archaeology in the lower Mississippi Valley has taught us anything, it is that we are no better off for having carved up the record in supposedly finer and finer spatiotemporal units (Fox 1998; O'Brien 1995, 1996a; O'Brien and Dunnell 1998; O'Brien and Fox 1994b).

At this point, it seems we have to face the fact that many of the constructs in archaeology—whether those using phases as their basis or those employing ethnological units such as bands, tribes, chiefdoms, and states—sit on very weak foundations. Even if they somehow were sitting on fairly strong foundations, we need to admit that they are descriptive, interpretive constructs, not explanatory ones. Surely there is a better way of explaining the content and context of the archaeological record than by reducing it to a collection of units based on similarity or dissimilarity, or by holding it up against an ethnological analog. There *is* a better way, and it begins with the detailed piecing together of myriad artifact lineages. This requires, at a minimum, placing things in correct temporal order and observing the spatial and temporal locations of particular variants.

PUTTING THE PIECES TOGETHER

Theoretically, one could use superposition to construct lineages, in that it provides information on the relative chronological position of particular variants—information that perhaps can be supplemented with associated radiometric determinations to create a calendrical sequence. But as we have emphasized repeatedly, superposition may produce merely a sequence rather than a lineage. Any of several processes could create a sequence of items or organisms that are not linked directly through heredity. In biology, one species might spread out of its range and into that of another. Individuals of one species might resemble individuals of the other species, and the unwary biostratigrapher might assume that the superpositionally higher taxon evolved directly from the lower one. This kind of sequence is not unusual in nature, and if it were, punctuated equilibrium would be in big trouble given that stratigraphic discontinuity is one of its key tenets. Under the punctuationalist perspective, morphological discontinuity occurs because of peripatric speciation and migration. Specifically, a single taxon exists in one locale, but through time, daughter populations split off and inhabit new, geographically isolated locales. Gene flow ceases between or among the populations, and the daughter population evolves so that its members are genetically (taxonomically) and phenotypically distinct, though perhaps only in subtle ways, from the parental population. Then, descendants of the daughter population migrate back to the parental locale, and upon death, their remains accumulate with those of the other descendants of the parental population. Likewise, one human group could move into a locale and replace another group with which it was related only in the distant past, depositing its tools and by-products with, or on top of, those left by the group it replaced. Both of these scenarios are ones in which there is historical continuity but not direct heritable continuity at the scale being examined.

Paleobiologists and archaeologists have long been interested in how to distinguish between historical continuity and heritable continuity (Lyman and O'Brien 1999b), though they rarely, if ever, use those terms. As we briefly saw in Chapter 7, the normal procedure in archaeology is to examine the magnitude of a perceived difference as the result of innovation, diffusion, invasion, or a host of other possible causes (e.g., Lathrap 1956). This practice does not directly address the difference between historical and heritable continuity in any informed way. Paleobiologists also regularly use magnitude in morphological difference to infer taxonomic continuity or discontinuity, employing processes such as anagenesis or peripatric speciation as causes of a particular pattern.

To produce lineages as opposed to sequences requires first that we have some method of tightly controlling variation—intentionally created classes are our preferred units—and second that we have a means of ensuring that we actually have heritable continuity—seriation is our preferred method. Although creating a lineage of artifacts is important in its own right, the lineage does not really tell us *why* it took the shape it did. Why, for example, did projectile-point manufacturers stop putting concave bases on their points, only to start up again at a later time? Why did the temper of pottery change from grit to shell over large sections of the midwestern and southeastern United States? These and a host of similar questions can be answered only through a two-step analytical process: Construct a lineage, then attempt to explain the evidence of change in mechanical, engineering, and functional terms. Both are difficult procedures.

The number of examples that we might use to illustrate the various points we have made are few in number, largely because the units typically used to describe the archaeological record are not conducive to the analyses that must be performed under the Darwinian umbrella. We have chosen several examples that required minimum additional analyses on our part, and they should be taken as illustrating, if not exemplifying, what we take to be the major goal of a Darwinian archaeology: writing historical chronicles (O'Hara 1988)—constructing artifact lineages—and writing historical narratives (O'Hara 1988)—explaining those lineages. The first example deals with technological change in projectile points during the period from about 9250 B.C. to roughly 7900 B.C. In our opinion, there is no better example of some of the problems raised by focusing exclusively on essentialist types and ignoring the often subtle variation that individual specimens within the types exhibit. There also may be no clearer example of how some of the methods and techniques discussed throughout the book—especially seriation and cladistics—can be used to construct artifact lineages. Our emphasis in this chapter is on the explanation of artifact lineages, but it does not hurt to backtrack a bit and begin the discussion with how to construct the very lineages we hope to explain.

Evolutionary Change in Paleoindian-Period Projectile Points

It is commonplace in archaeology to employ essentialist units based on artifact style, morphology, or function when referring to groups of people. We do not need to cite long lists of references to make the point that statements such as "Clovis people hunted mammoths, whereas Folsom people hunted bison" are common occurrences in the literature. Such statements, it could be argued, are simply shorthand ways of saying that human groups at a certain time and place in the prehistoric past used a kind of stone weapon tip known archaeologically as a Clovis point and that at a later time human groups used what are commonly referred to as Folsom points. There is considerable truth in this argument, but regardless of its merits, it is undeniable that such statements are essentialistic. It also is undeniable that essentialist units can sometimes be used to examine evolutionary issues *if* we scale our analysis to examine the specimens placed in the units rather than focus solely on the units themselves.

Our use here of traditional projectile-point types—extensionally derived units—might at first glance appear to run counter to much of our argument thus far, but it really does not. We never said all extensional types are worthless units; to the contrary, many of the extensional units that have been created in archaeology are excellent historical types, meaning that they are useful for tracking the passage of time. Likewise, we are perfectly happy with the notion of species. Our position is that you can define a species any way you want—biologically, paleobiologically, stratophenetically, or whatever. Just do not get so wrapped up in such units that you lose sight of the real goal, which is measuring variation, tracking it through time and across space, and then explaining it. Our point has always been that with classes created through paradigmatic classification, one knows for sure what one is getting. With extensionally defined types, one is rolling the dice. Here, we roll the dice, admitting they are loaded in our favor, and incorporate several well-known projectile-point types in a phyletic seriation to determine if we can tease out information useful for building lineages. Phyletic seriation, as we will see, is only a preliminary step in lineage construction. Once we determine its merits, we turn attention to subsequent steps that are prefatory to actually explaining a lineage.

Phyletic Seriation and the Clovis–Dalton Lineage

One of the projectile-point types we examine is Dalton, the empirical specimens of which are common occurrences over much of the midwestern and southeastern United States. Carl Chapman, who coined the type name in honor of artifact collector Judge Samuel P. Dalton of Missouri, described Dalton points as being

> lanceolate or pentagonal in shape. The blade is essentially triangular. Edges of the blade are straight or slightly convex. Generally there are serrations on the blade edges, perhaps the result of resharpening the tool. The blade is often beveled by sharpening. Beveling may be steep or shallow. The length ranges from 4 cm to 17 cm. Most specimens are 5–7 cm. Width is 1.5 cm–3 cm. Proportions are 2 to 5 times as long as wide. Thickness is 5 mm–7 mm. Basal thinning is prominent and extends the same length as the grinding at the sides of the base. The base is usually the widest portion. It is relatively deeply concave with straight or slightly concave sides which are ground smooth. (Chapman 1975:245)

What Chapman's description omits is the exceptional workmanship that many Dalton points exhibit (Figure 8.1). Oftentimes, flake scars extend from one edge of a specimen to the other edge, created by the removal of continuous flakes across the face.

Archaeologists have long supposed that the dating of Dalton points, though somewhat imprecise, was adequate for many purposes. We, like Albert Goodyear (1982), were skeptical of many of the late dates for Dalton points, but we thought Goodyear was too conservative in his estimate that manufacture of Dalton points ended around 7900 B.C.[1] For one thing, one of us (MJO) had excavated what

[1]For simplicity, we use uncorrected radiocarbon ages throughout the chapter instead of calibrated (calendrical) ages. There may be as much as 2,000 years difference between the two for certain portions of the Paleoindian period (Fiedel 1999). Although the relative sequence of projectile-point ages *probably* is correct regardless of which kinds of dates are used, we stress that future analyses of the kind we present here should be done with calibrated dates to ensure that all age assessments are similarly based.

Figure 8.1. Dalton points from central Missouri. All specimens at same scale; specimen a is 8.0 centimeters long.

appeared at the time to be a Dalton-age horizon at the Pigeon Roost Creek site in northeastern Missouri that produced a calibrated radiocarbon date in the seventh millennium B.C. The fact that Dalton points were found alongside several points that were usually placed in the Early Archaic period (7500–5000 B.C.) did not really mean much at the time, since it seemed entirely possible that those Archaic-period forms began earlier than had previously been assumed. For another thing, we were not thinking in materialist terms at the time.

It now is obvious to us that what typically are referred to as Dalton points were first manufactured about 8900 B.C. and lasted no later than about 8500–7900 B.C. (O'Brien 1998; O'Brien and Wood 1998), probably closer to the former than the latter. We say "about" because we have no adequate means of dating the beginning or ending dates of Dalton-point manufacture other than on phyletic grounds, as we explain below. But this is not a weakness; in fact, the case for positioning Dalton points chronologically through seriation is stronger than if we had relied solely on a dozen radiocarbon dates, which, if we had them, would still require us to use the qualifier "about" when talking about the beginning and ending dates of manufacture. One end of our phyletic sequence—the early end—is anchored by a large suite of radiocarbon dates from contexts producing Clovis points (Figure 8.2). These are distinctive forms whose type would generally be agreed upon by knowledgeable archaeologists.

Clovis points occur across much of the United States, though the contexts that have been dated are restricted to the Plains and Southwest, where the points span a short period, from about 9250 B.C. to 8950 B.C. (Haynes 1991; Taylor et al. 1996). Given the widespread distribution, we find it inconceivable that these dates bracket the age of Clovis points in all parts of the country. In fact, four decades ago, Ronald J. Mason (1962) argued on morphological grounds that Clovis-point manufacture began in the Southeast and spread west—a proposition with which we happen to agree (O'Brien and Wood 1998). The "Clovis-first" hypothesis, which states that people using Clovis points colonized the Americas (or began making Clovis points soon after colonization), used to be commonly accepted, but the tide has begun to turn in the last several years as archaeologists have begun to realize that humans were in the Americas before Clovis points, at least those in the Plains and Southwest, were manufactured (Meltzer 1997). Regardless, Clovis is currently the earliest type of projectile point commonly recognized throughout much of the United States.

Clovis points are followed on the Plains and in the Southwest by Folsom points (Figure 8.3), a smaller point on average than Clovis but one that resembles it in several key ways. It, like Clovis, has a concave base, fluting on both faces, and grinding along the distal third or so of the lateral edges. Folsom points have been found stratigraphically above Clovis points, the former associated with bison remains and the latter with mammoth remains. The first reported occurrence of stratigraphic separation (Cotter 1937) was at Blackwater Draw in eastern New Mexico in the 1930s. This finding set the stage for the two-part subdivision that

FIGURE 8.2. Clovis points from central Missouri. Both specimens at same scale; specimen on the left is 8.6 centimeters long.

became archaeological canon: first Clovis, then Folsom. Subsequent chronometric dating of contexts producing Folsom points demonstrates that they indeed are younger than Clovis points. After closely examining stratochronological data from widely distributed sites in the Plains and Southwest, R. E. Taylor and his colleagues stated that the "latest North American Clovis occupation predates the earliest occurrence of Folsom" (Taylor et al. 1996:523). They also noted that the "transition from Clovis to Folsom may have occurred within a period of 100 years or less" (Taylor et al. 1996:524). Folsom points ceased to be manufactured by about 8550 B.C.

We later return to the issue of a "transition" from Clovis to Folsom, but here, our attention is on the Folsom "period" outside the Plains and Southwest. Despite the established sequence of Clovis followed by Folsom in the Southwest and Plains, it is not of much help in the Midwest, where Folsom points are rare occurrences, or in the Southeast, where Folsom points do not occur. For years, the absence of Folsom points in those regions was difficult to explain, as was the apparent gap of at least several hundred years between the terminal date for

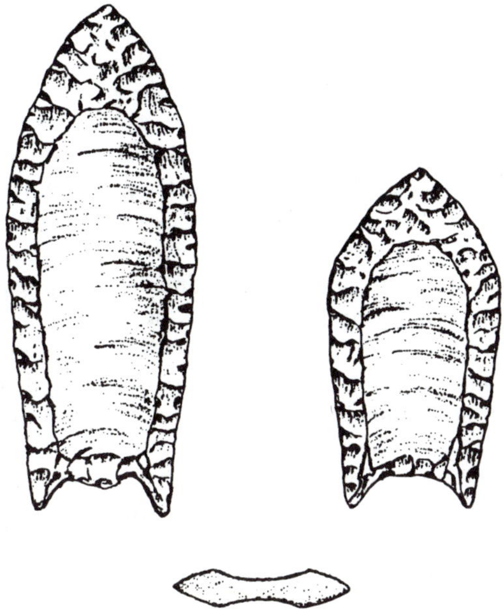

FIGURE 8.3. Folsom points from the Lindenmeier site, Larimer County, Colorado. Specimen on the left is 6.8 cm long (after Roberts 1935).

Clovis-point manufacture and the beginning date for the manufacture of Dalton points and a host of similar lanceolate points placed in other types. This gap, however, was more apparent than real. In fact, it was not real at all. The chronological relation between Clovis and Dalton points has been staring archaeologists in the face for years, but we never saw it. We had become so attached to essentialist units that they had taken on a reality of their own, at least in our minds. The "reality," however, is that heritable continuity appears to exist between what, in a sense akin to the way biologists refer to species, archaeologists refer to as Clovis points and Dalton points. To follow our reasoning, we examine Dalton points from the standpoint of technology, the best study of which is by lithic specialist Bruce Bradley (1997), who examined 146 complete specimens from the Sloan site in northeastern Arkansas (Morse 1997). One of Bradley's important findings concerns the morphological variation inherent in the Dalton-point assemblage from Sloan and the relation of that variation to point manufacture. Although his comments are aimed at points in a single assemblage, they are applicable to the majority of what typically are termed Dalton points:

> There is a wide range of morphological variation in the Sloan-site Dalton-point assemblage. This includes outline, size, margin form, basal treatment, and finish. There is, however, a single production technology present. It is possible to identify various stages and steps of a manufacturing process, and

> to classify each of the pieces into specific stages and steps of manufacture. Although this is possible, I view the technology as a continuous process, with different pieces representing different points of manufacture termination along a continuum. (Bradley 1997:53)

Bradley identified several distinct traits on the specimens: substantial preliminary pressure flaking; alternate serial pressure flaking in preparation for systematic reduction of point margins; pressure fluting; and distinctive serrations along the blade edges. The trick was to use these traits as general guidelines and to work backward through the manufacturing sequence—in other words, to reconstruct the manufacturing sequence through the investigation of immanent properties. Bradley eventually broke the sequence down into eight steps. First, a roughly lanceolate biface is produced by generalized percussion flaking. Second, the biface is modified by selective pressure flaking to produce regular faces and margins. This process also thins the biface. Bradley suggests that the biface might be fluted at this point. Although no evidence of such a procedure was noted on the Sloan specimens, fluting is not unknown on Dalton points (see below). Third, the biface is beveled continuously along opposite margins on opposing faces through pressure flaking, which produces a noticeable bibeveled appearance. Fourth, the abrupt margins serve as continuous platforms for carefully spaced and executed serial pressure-flake removals.

Fifth, biface edges are again continuously beveled on the same face as the edge platform bevel. This removes the small triangular areas between the flake scars and regularizes the edges. It also moves the edges in toward the center of the biface and accentuates the alternate beveling. Because the serial pressure flakes are removed from fairly steep platforms, in some cases steep enough that the flakes run past the midline, the negative bulbs are pronounced. The centers of the negative bulbs are used to produce the distinctive serrations found on many Dalton points. Sixth, at each point where there is a negative bulb scar adjacent to the edge, a shallow notch is produced by abrupt edge retouch on the same face as all previous edge beveling. Seventh, after a notch is created, a large flake is removed on the same face as the beveling to deepen the notch. Overlapping notching up and down the edges creates the serrated shape. Eighth, after the biface is serrated on both edges, the hafting area is thinned by removal of one or more fairly long flakes parallel to the long axis of the point, and the edges and base are ground.

Removal of these thinning flakes parallel to the long axis technically is fluting, but Bradley makes an important distinction between what he terms *technological fluting* and *morphological fluting*:

> Technological fluting is where basal thinning is done by the removal of one or more flakes that proportionally reduce the longitudinal thickness of the biface. This involves the removal of flakes that travel past the point of maximum longitudinal thickness. It makes no difference whatsoever if the resulting flake

> scars are retained in subsequent flaking. Morphological flutes, on the other hand, are simply basal flake scars that extend past the point of the hafting element and are visible on the finished object.
>
> With these definitions, it is possible to have points that were technologically fluted but are not morphologically fluted, if the channel scars are removed by subsequent flaking. It is also possible to have points that are morphologically fluted but whose basal flakes did not extend past the point of maximum longitudinal thickness. These did not technically thin the preform.
>
> Most Sloan Dalton points, and probably most other Dalton points, are not morphologically fluted. I suspect, however, that the majority were technologically fluted at some stage in their manufacture and probably more than once. (Bradley 1997:54–55)

If we had originally adopted a materialist outlook and thus paid more attention to formal variation the way Bradley did, we might quickly have seen where Dalton points fell chronologically. They exhibit numerous characteristics in common with several Paleoindian-period points from the Plains, including not only Clovis but also Goshen/Plainview—lanceolate points that, although they are unfluted, resemble both Clovis and Folsom. Most knowledgeable archaeologists and artifact collectors readily distinguish between Clovis points and Dalton points, but there are numerous specimens that are difficult to place in one type or the other. For example, look at the points shown in Figure 8.4, especially the one on the right. The photograph does not show it, but this 13-centimeter-long specimen is only about half a centimeter thick at its widest point—thinner than the majority of "typical" Clovis points but well within the range of "typical" Dalton points. The specimen also has the deeply concave base found on Dalton points, but it has morphological (Bradley's term) flutes like those on Clovis points. What does one do with such a specimen? Two obvious options come to mind: (1) create a new type in which to place it, as Carl Chapman (1975) did when he coined the term "fluted Dalton," or (2) place it in one of the two existing types—Clovis or Dalton. Of course, there is a third option: use paradigmatic classification in the first place and thus avoid worrying about what to do with such specimens. But we said at the beginning of this section that we were going to use established types in our discussion, essentialist and problematic though they are, so let us stick with them for a little while longer.

Regardless of what kind of units we employ, specimens such as those in Figure 8.4 scream out to be recognized for what they are—specimens morphologically (formally) between Clovis and Dalton. It is these so-called "transitional" forms that are often overlooked by archaeologists hell-bent on placing them in one type or another, without worrying too much about the significata (differentia) of those types. But what marvelous things such forms are because of their chronological significance and for what they can tell us about evolutionary change within a lineage—here, the lineage that includes Clovis and Dalton points.

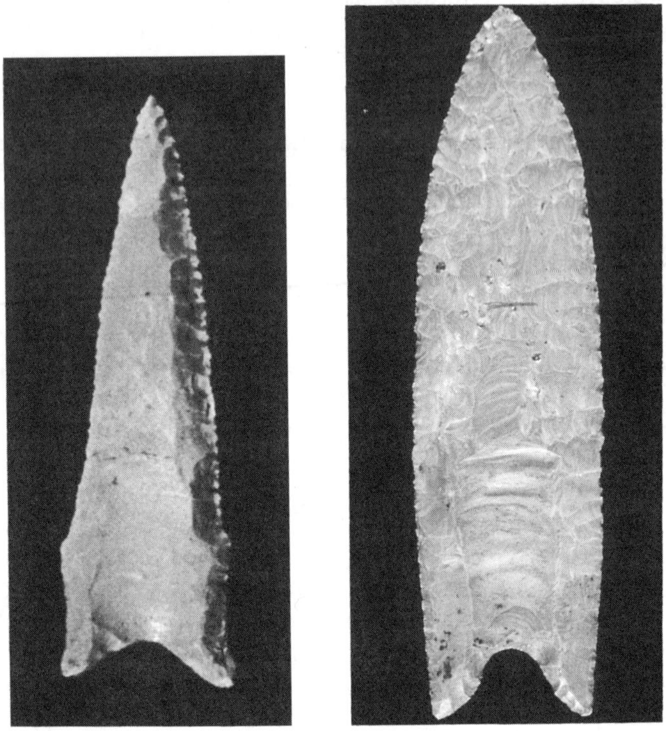

FIGURE 8.4. Clovis–Dalton "transitional" points from Arkansas and Missouri. Specimen on the left is 11.6 centimeters long.

No one seriously believes that sometime around 8950 B.C. there was an abrupt and complete end to the manufacture of Clovis points and the equally abrupt and complete shift to the manufacture of Dalton points, though all too often our actions and words make it appear as if this were the case. For example, we talk about Clovis people and Dalton people as if those were real (empirical) units, and we talk about the disappearance of Clovis points and the appearance of Dalton points, again, as if they were real units. We might talk about variation among, say, Clovis points, but in our minds there is an archetype of what such a point is supposed to look like, and it is that mental picture that guides what we do with newly discovered points. Minimal consideration is given to the significata of Clovis, Dalton, or other types that a particular specimen might resemble.

Why should we not expect that if Dalton points fell chronologically close on the heels of Clovis points in parts of the Midwest and Southeast, then Dalton points manufactured around 8950 B.C. would share attributes with "classic" Clovis points? Why, for example, if things are always in the state of becoming something else (the materialist position), should we not expect to see the bases of fluted points become more concave or the basal feet become slightly flared? Based on published criteria, it is perfectly acceptable for a Dalton point to have a deeply concave base and eared feet; it would seem that the only criterion that would exclude the points illustrated in Figure 8.4 from being placed in the Dalton type is that they contain what Bradley terms *morphological flutes*—a key characteristic of points placed in the Clovis type and, on the basis of available dating, manufactured before Dalton points.

Now let us look at the problem a little differently. Even if we did not know where in time Dalton points fall, specimens such as those in Figure 8.4 ought to give us a clue as to their age. If we were to line up three points—one that all archaeologists agree belongs in the Clovis type, another that all agree belongs in the Dalton type, and one of the points pictured in Figure 8.4, we should not have much trouble getting our experts to agree that the evolutionary sequence went either Clovis–"transitional"–Dalton or Dalton–"transitional"–Clovis. We might guess, and here we would be correct, that the evolution was fairly rapid—on the order, we suspect, of only a hundred years or so. Phyletic seriation provides the means to make an initial pass at establishing a lineage, and in some cases it can be done with essentialist categories, the standards of the industry. But to make use of phyletic seriation, we have to get beyond the types themselves and examine the specimens in terms of their singular variation. And we have to be careful that we are producing a lineage as opposed to merely a historical sequence of types. That is why we emphasize that phyletic seriation represents a means of making an *initial* pass at creating a lineage. There could actually be large chunks missing between the units in the seriation, and phyletic seriation might not show this. We can add to the strength of our seriation through superpositional evidence, realizing once again that chunks might be missing. However, in the case of Clovis and Dalton, to the best of our knowledge, no site in the Southeast or Midwest—the heartland of Dalton points—has produced such data. Interestingly, a site in the Southwest has.

Take a close look at the point shown in Figure 8.5. Now compare it to the specimens shown in Figure 8.1. The point in Figure 8.5 could have come from Missouri, Arkansas, or any of a half-dozen other states, and from a morphological standpoint, no archaeologist would find it the slightest bit odd. In all respects, it looks like a Dalton point, albeit a fluted one, but "fluted Daltons" are not particularly rare in the Midwest. The thing is, the point did not come from the Midwest. Instead, it came from Blackwater Draw, the Clovis-point type site in eastern New Mexico. The importance of Blackwater Draw lay not only in its association of Clovis points with mammoth bones but also in the fact that the

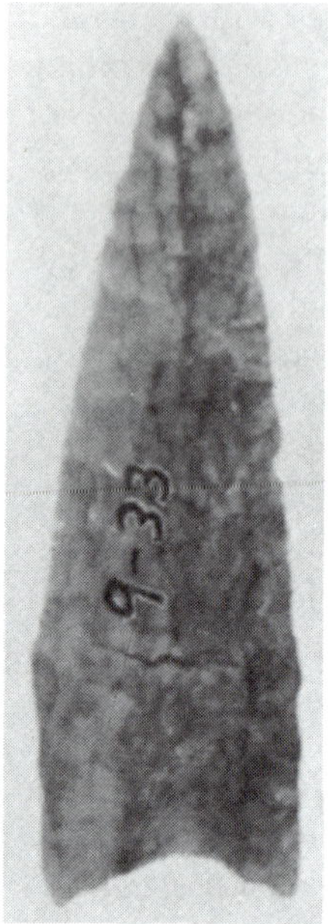

FIGURE 8.5. Dalton point from the Blackwater Draw locality, eastern New Mexico. The point was found in a level containing mammoth bones and Clovis points, all of which were stratigraphically beneath a level containing Folsom points and bison bones. Specimen is 5.5 centimeters long (from Cotter 1937).

association occurred stratigraphically below Folsom points and bison bones (Cotter 1937). The red chalcedony point shown in Figure 8.5 came out of the level that produced Clovis points and mammoth remains (Cotter 1937:9, 12).

Specimens such as the one from Blackwater Draw could prompt us to ask, "What in the world is a Dalton point doing in New Mexico, when the type range is restricted to the Midwest and Southeast?" Or it might prompt us to comment, "Well, if types are ideational units—classes—as opposed to empirical units, then there is no reason the distribution of the Dalton type *couldn't* include specimens from New Mexico." At that point, we might ask whether what we are looking at is a case of multiple development or, alternatively, a case of single development and eventual spread. In other words, did Dalton points develop out of Clovis points in a single location, or were there multiple centers of development? Unfortunately, not enough analysis has been done to answer this question satisfactorily, in large part because of our reliance on normative (essentialist) categories. The view that "a Dalton point is a Dalton point" precludes shifting attention to small-scale morphological variation that might have spatial significance.

It takes careful analysis to tease out engineering data, as Bradley did with the subtle difference between morphological and technological fluting, but this often is the only way to begin to make sense out of an archaeological record that is exponentially more complicated than we give it credit for being. It also is the only way to begin to get a handle on lineage histories, literally by piecing together tiny clues and deciding whether similarities are homologous, the result of common ancestry, or analogous, the result of functional convergence. Based on his examination of characteristics on literally thousands of points manufactured between roughly 9250 B.C. and 7500 B.C., Bradley (1997:57) stated, "I believe that the data currently point to an in situ technological development of Dalton points directly out of a Clovis technology." Thus, Bradley not only suggested heritable continuity between Clovis and Dalton but he also gave us the solution to the "missing" piece of the "what-happened-after-Clovis" puzzle for the Midwest and Southeast. If we buy Bradley's argument, and our previous discussion suggests there is every reason to do so, then there was no missing piece between Clovis points and Dalton points. The latter were the midwestern and southeastern *chronological* (and *only* chronological) equivalents of Folsom points in the Southwest.

Bradley's statement is written in evolutionary language. He is not speaking metaphorically when he says that Dalton points developed out of a Clovis technology, any more than we are speaking metaphorically when we say that one biological species evolved from another. Bradley's language highlights the phyletic relation between points normally placed in various types—a language based on careful examination of technological variation. Once we start thinking in terms of variation and begin measuring it for other than purely descriptive purposes, our attention has shifted to dimensions and attributes. What dimensions of variation should we monitor? The presence or absence of flutes is an obvious one, though we suggest that Bradley's (1997) distinction between technological fluting and morphological fluting is an important one to keep in mind. As the notion of heritable continuity suggests, the issue does not reduce to a simple presence or absence of a single dimension. Maximum length, maximum width, and maximum thickness might be examined, along with things such as the length of the haft area as designated by edge grinding, whether blade edges were beveled or serrated, whether the base (proximal end) was straight or concave, and if the latter, whether that concavity was V-shaped, U-shaped, or arc-shaped. The technology used to produce the point—percussion versus pressure flaking, the shape and extent of the resulting flake scars, and so forth—could also be noted. Once these and other potentially relevant variables are documented, one can examine their distribution across a series of points. Continuity of individual attribute states and particular combinations thereof over time would suggest heritable continuity, or its absence, in much greater detail than is currently possible.

Occurrence Seriation and Attribute States

Frequently seriation provides one means of testing heritable continuity at the tradition/lineage level, and occurrence seriation provides another means, as shown in Table 6.1. But it can also do it at the level of empirical units—in this case, artifacts. We can shift our attention from the presence or absence of ideational units—here, artifact types—in collections of things to the presence or absence of ideational units—attributes—in individual objects. Table 8.1 shows how this works. Ten objects are examined in terms of the presence or absence of six traits (top), and those data are used to produce the only logical solution as to the order of the objects (bottom). Note that this technique measures time continuously because of the overlap in trait presences. The result is strong supporting evidence that we are dealing with heritable continuity. Importantly, the technique does not require the use of extensional units: The only important things are the empirical specimens themselves and the ideational units we use to measure variation. Could we, at the end of the analysis, chop up the lineage shown in Table 8.1 into two pieces and call one, say, Clovis and the other Dalton? Certainly, just as we could call the two groups anything we like. The important point is that it does not really matter where we make the break; we are doing so strictly for the sake of either analytical or conversational reasons, *and* our steps in creating the groups and carving up the continuum are clearly traceable.

Table 8.1. An Example of an Occurrence-Seriation Procedure Using Traits of Objects

	Traits					
Assemblage	A	B	C	D	E	F
Unordered						
1	+	+		+		
2	+	+		+	+	
3	+				+	+
4	+	+	+			
5	+			+	+	
6		+	+			
7	+	+	+	+		
8	+				+	
9	+					+
10	+				+	+
Ordered						
6		+	+			
4	+	+	+			
7	+	+	+	+		
1	+	+		+		
2	+	+		+	+	
5	+			+	+	
8	+				+	
3/10	+				+	+
9	+					+

Time ↑

States						
1	⑥	6	①	4	3	6
1	4	6	4	4	③	6
1	4	⑥	4	4	2	⑥
1	4	5	④	4	2	4
1	④	5	3	④	2	4
1	3	5	③	3	②	4
1	3	⑤	5	3	1	4
1	3	4	5	③	1	④
1	③	4	5	2	①	3
1	2	4	⑤	2	4	3
1	2	④	2	②	4	3
1	2	1	2	5	4	3

Figure 8.6. Chronological arrangement of paradigmatic classes of specimens showing the evolution of attribute states through time. Circled attribute states denote a change from the immediately earlier state.

A similar kind of analysis is illustrated in Figure 8.6, but here we have switched to a dimensional analysis and recorded individual attribute states as opposed to presence–absence. This is identical to what is shown in Tables 5.1 and 5.2 relative to the paradigmatic classification of projectile points from the Pigeon Roost Creek site. In Figure 8.6, we show only seven dimensions (the columns), each of which has several possible attributes states. Each specimen is a unique row. In this fictitious example, we can track changes in individual dimensions through shifts in attribute states. Circled attribute states signify a change in state from the preceding specimen (presuming that time runs from bottom to top). For example, there are two changes in attribute state from the oldest specimen to the next oldest, one in dimension 3 and another in dimension 5. Importantly, all 12 specimens share either five or six attribute states with their immediate neighbor. Given the sequence as we constructed it, heritable continuity is evident from top to bottom.

It should be obvious that each specimen in Figure 8.6 is a member of a group extracted from the classes represented by the number strings. As such, they have unique time and space positions. As a further confirmation of heritable continuity, we could, if the groups contained multiple specimens from redundant locales, create a frequency seriation. We would simply arrange the groups across the top of the matrix, list the locales down the leftmost column, and tally the frequencies of each group at each locale. Provided we could exclude sampling error from consideration (perhaps because of small sample sizes), would we be guaranteed that our frequency seriation would be in agreement with the occurrence-based order shown in Figure 8.6? There actually should be very *close* agreement, because from the looks of the arrangement in Figure 8.6, the groups appear to be good historical types. Therefore, they should be amenable to frequency seriation—as long as they meet the provisions of the seriation model (Chapter 6).

Tempo and Mode of Change

Our emphasis to this point has been on heritable continuity, and we have not said much about either the tempo or the mode of projectile-point evolution. We admit that our discussion has made it sound as if projectile points evolve only through phyletic gradualism, and within it through anagenesis. There is, however, no reason to model all projectile-point evolution on gradualism, nor is there any reason to suspect that even in those instances when gradualism occurs that the only mode is anagenesis. In fact, we would argue that anagenesis probably plays a smaller role than cladogenesis in producing artifact lineages because of the rapidity with which selection works on variation. Thus, tempo and mode are highly interconnected here: Change occurs rapidly, and it does so by branching. Cladograms can be constructed to show the evolutionary relations among projectile-point lineages *if* we can isolate synapomorphies. This has rarely been attempted in archaeology.

Figure 8.7 compares an example of anagenesis (top) against a cladogram using paradigmatic classes and derived groups (bottom). In anagenesis group 134698 is the ancestral group at time *t*1. By time *t*2, the attribute state of dimension 4 has changed from 6 to 7, and by time *t*3, the attribute state of dimension 3 has changed from 4 to 3. The attribute states of dimensions 1, 2, 5, and 6 never change throughout the sequence. But maybe in reality, evolution proceeded by cladogenesis, as shown in the bottom half of Figure 8.7. Group 134698 is still the ancestral group, but instead of evolving directly into group 134798, it splits into two lines—the 134699 line, which we will ignore, and another one. After a time, that "other" line again splits, with one leading eventually to group 124798 and the other leading off in a different direction in morphospace. After a time, that new "other" line splits, with a line leading eventually to group 133798 and another line leading to group 134798. Production of the monophyletic groups shown in Figure 8.7 can be accomplished only by identifying derived characteristics (synapomorphies) and using them (and only them) in analysis. Mixing derived and nonderived (ancestral, or primitive) characteristics more than likely will lead to incorrect phylogeny.

If cladogenesis has occurred, is the change modeled best by phyletic gradualism or by punctuated equilibrium? So far as we know, punctuated equilibrium is an untested notion in archaeology, but there is no reason to suspect it is not applicable, nor is there reason to assume that phyletic gradualism is not an

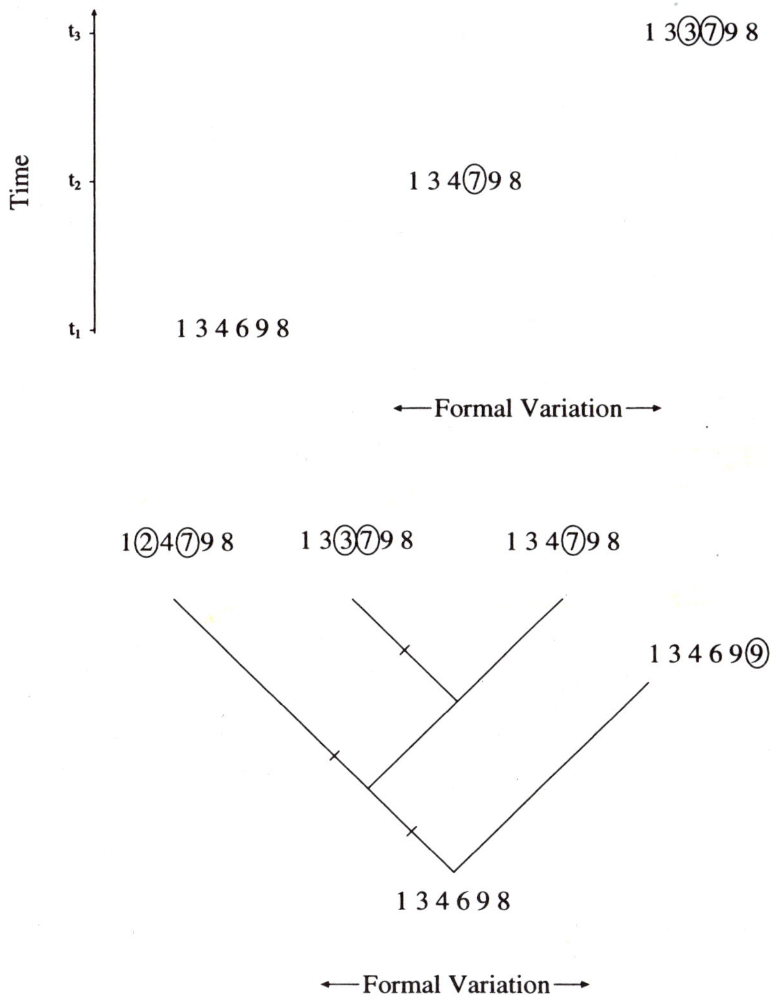

FIGURE 8.7. Examples of anagenesis (top) and cladogenesis (bottom) relative to the evolution of projectile-point groups. In anagenesis, groups evolve directly one into the other through changes in attribute states of dimensions. The same kinds of changes characterize evolution through cladogenesis, but here groups branch into daughter groups.

evolutionary mode when it comes to humans. In fact, one careful study—Anthony Barnosky's (1987) analysis of the Quaternary mammalian record—suggests that both play a major role in evolution. As yet we do not have the requisite data to examine the role punctuated equilibrium might play in human evolution, but we can at least outline one example that bears closer examination. That example is what was going in North America after roughly 9250 B.C. To make our point, we are again going to refer to established projectile-point types, using them in the same way that proponents of punctuated equilibrium use species.

We assume that Clovis points, and probably other as yet unidentified (and hence unnamed) types of points were derived from some ancestral type. Sometime around 9000 B.C., give or take a few hundred years, Dalton points evolved from Clovis points in some locale, probably in the Midwest or Southeast, and quickly spread out to other locales. Identifying *the* precise locale where Dalton points were first made is not only impossible but illogical from a materialist perspective, just as it is for identifying where a species first appeared. For the sake of argument, and to keep us from rehashing the discussion in Chapters 2 and 5, we will just accept that this happened somewhere at some time and that the technology spread. What really was happening, of course, is that there was an uninterrupted evolution of form, the cumulative changes in which become great enough at certain points that our eyes capture accumulated change in terms of types. In the Southeast, certainly, the change going on after 8950 B.C. or so was rapid and left a high visibility. Archaeologists have coined any number of type names—Suwannee, Simpson, Quad, Cumberland—to deal with the variation.

It has long been suspected that in the western United States, Folsom points evolved out of Clovis points—a logical proposition based on the facts that (1) John Cotter (1937) found them stratigraphically above Clovis points at Blackwater Draw; (2) Folsom dates are consistently later than Clovis dates (Taylor et al. 1996); and (3) points in the two types look similar. Archaeologists might make reference to "Folsom peoples" and "Clovis peoples" as if they were two separate groups, but this does not obscure the fact that we often make it sound as if whoever was around at the point that mammoths went extinct threw away their Clovis points and started making Folsom points with which to kill bison. But maybe this is not the way it happened. We base this statement on some of the findings Bruce Bradley and George Frison have recently made relative to projectile-point manufacture in the northwestern Plains. Admittedly, their comments are tentative, but they bear close scrutiny.

At issue is the chronological placement of Goshen points, which for all practical purposes are indistinguishable from Plainview points (Frison et al. 1996)—well-made, basally thinned, unfluted and unnotched points found throughout the Plains. In the southern Plains, where they were first found and described, Plainview points are believed to be of post–Folsom age, but matters are different in the northern Plains, where a growing number of radiocarbon dates suggest that Plainview (Goshen) points there are contemporaneous with—if not *earlier* than—Clovis points. Regardless, radiocarbon dates indicate Plainview/Goshen points in that region are older than Folsom points. Based on their analysis of Clovis, Folsom, and Goshen points in general, and especially of Goshen points from the Mill Iron site in southeastern Montana, Bradley and Frison (1996) stated that in their opinion, Goshen points are much closer technologically to Folsom points than they are to Clovis points. They certainly do not represent "a projectile point type that Paleoindian archaeologists ... would expect to see. They would expect something technologically and morphologically transitional between Clovis and Folsom" (Frison et al. 1996:206). This led Frison et al. (1996:208) to conclude that "the present interpretation is that fluting as it appears in Folsom on the Northern and Northwestern plains was a post–Goshen phenomenon since there is no problem in accepting Goshen as a technological precursor of Folsom based on the Mill Iron site evidence." In other words, Folsom evolved out of Goshen, not out of Clovis.

Detailed technological analysis might long ago have assisted in the chronological placement of Goshen points if it had not been for the received wisdom that Clovis produced Folsom. This is how Frison et al. (1996:209) put it: "Stratigraphic interpretations may be leading Paleoindian archaeologists into erroneous conclusions. When we look at a stratified site such as Hell Gap, Agate Basin, or Carter/Kerr-McGee, most of us are tempted to interpret the Paleoindian chronological record as one cultural complex evolving directly out of its stratigraphic antecedent without overlap in time, which does not necessarily have to be the case." Frison and colleagues are correct: Received wisdom among Paleoindian experts holds that cultures or "cultural complexes" evolve out of the ones stratigraphically below them—a view that grows directly out of archaeology's dependence on superposition not only as a way of keeping track of time (historical continuity) but also as a means of demonstrating heritable continuity. But maybe what we are witnessing on the northwestern Plains is a case of replacement instead of continuity. Continuing in our essentialist tone for the moment, perhaps the Goshen and Clovis traditions *were* contemporaries, with Folsom evolving out of Goshen. Perhaps later traditions—Agate Basin, Hell Gap, and so on—evolved out of Folsom. What about Clovis descendants? To the southeast, Clovis produced Dalton; what about the Plains and Southwest? We do not have an immediate answer except to note that the point type Meserve—usually thought of as a "western" point type—is so similar to Dalton in terms of morphological characteristics as to be indistinguishable. Once Dalton and Meserve are equated morphologically, the gap in the post–Clovis lineage in the West disappears.

If Clovis and Goshen were separate but contemporary traditions, did they have an immediate common ancestor? We would guess they probably did, but there currently are no data to answer that question. Bradley (1997:57) makes an interesting observation when he notes the resemblance between Dalton points and several "styles" from the Plains that postdate Folsom points: "Resemblances [of Dalton points] to post–Clovis Paleoindian styles in the High Plains are superficial and at most represent a common origin out of a Clovis predecessor." In other words, the pre–Clovis predecessor evolved into Clovis on the one hand—the line that produced Dalton points—and into (probably) Goshen on the other.

It should be obvious after this foray into early projectile-point morphology on the Plains that the entire issue is ripe for a long-term, detailed cladistical analysis. We have deliberately used essentialist language, employing type names in the same way a biologist employs essentialist terms to talk about the evolution of biological taxa. This in no way suggests that we do not believe that ultimately paradigmatic classification should be used to create the groups to be examined. Our discussion should also suggest that cladogenesis might have played a major role in the evolution of projectile-point technology. Further research, if structured along the lines suggested here, should focus on the extent of the role played by *both* cladogenesis and anagenesis in that evolution. To do so will require examination of specimens from the point of view of identifying synapomorphies, shared derived characteristics, so that cladograms are meaningful. This is going to be painstaking work—deciding which dimensions to use, identifying attribute states, and so on, but the result will be unparalleled in Americanist archaeology.

Such a foundation also might allow us to do two things. First, we might be able to examine what, if any, role punctuated equilibrium played in effecting the pattern evident in the archaeological record. If Frison and Bradley are correct that Folsom points and Clovis points have different histories, we would guess that some of the instances where Clovis points and Folsom points occur at the same locale might be explainable through punctuated equilibrium. This very well might be the case at Blackwater Draw, where Folsom points occur stratigraphically above Clovis points and the Dalton/Meserve point. Of considerable interest here is the aforementioned statement by R. E. Taylor and his colleagues that the "latest North American Clovis occupation predates the earliest occurrence of Folsom" (Taylor et al. 1996:523). Their "transition" from Clovis to Folsom, which they speculate could have been accomplished in a century, might not be a transition at all. Rather, it might reflect the amount of time it took for Folsom points to spread from their center of origin.

Second, such a foundation might allow us to examine the tempo of evolution. Certainly on the northwestern Plains the proposed Goshen–Folsom line evolved into a plethora of forms very rapidly. Again, if Frison and Bradley are correct, points such as Midland and Agate Basin appear to have evolved out of the Goshen–Folsom line rather than the Clovis line, and they did so rapidly. In the Midwest and Southeast, there was a literal explosion in forms out of the Clovis–Dalton line. At this point, the timing of the appearance of various forms is tied solely to a few radiocarbon dates. Hence, we know little about changes in evolutionary tempo.

Explaining Change in Projectile-Point Lineages

Let us assume that we have created the requisite foundation: We have examined thousands of projectile points and placed them into groups derived from paradigmatic classification, we have constructed what appear to be logical sequences and tested them to ensure heritable continuity, and we have made some preliminary assessments of evolutionary tempo and mode. This is a sizable contribution in itself, but it only gets us so far—admittedly much farther than we have ever gone before. But we still need to address the issue of why the lineages took the form they did. One set of clues might reside in the haft region of projectile points. We assume that haft-related variables are related to the manner in which projectile points were attached to shafts or foreshafts (recall our discussion in Chapter 7). Hafting itself is technological, but we assume that some morphological characteristics are related directly to function, since that is the reason why something is engineered in the first place. Form and function are independent—one does not change a flat tire with a soup ladle—but they are also correlated, if only loosely; that is, changes in form can lead to or result from changes in function. It is up to us to determine, through experimental means, how and why certain engineering designs were superior to others and the effects these changes might have had on the bearers of the traits—in this case, humans who depended on hunting as one means of making a living.

O'Brien and Wood (1998:114–117), in a review of hunting technology during the period 9000 B.C.–7000 B.C., suggest that the high incidence of edge beveling and the large size of Dalton points are results of the need for dart or spear stabilization during flight—itself the result of the fact that fletching had not yet been invented. Susan Hughes (1998), in an excellent study of projectile points from Mummy Cave in northwestern Wyoming, suggests that heavier projectile points compensated for the absence of fletching on early weapons. We believe the engineering changes that post–Clovis points went through—such as the addition of bevels to what are called Dalton points—were tied to the appearance of the atlatl, or throwing stick, as an addition to the hunting arsenal. Equally important, but for different functional reasons, was the reengineering of the proximal, or haft, ends of projectile points to coincide with improvements made in fastening the points to bone or wooden foreshafts or main shafts (for discussions and references on Paleoindian-period hafting, see Hughes 1998 and Lyman et al. 1998a).

In a cogent discussion of projectile-point hafting, Robert Musil (1988) tackled some of the important issues related to engineering design. Whether or not all his conclusions are correct is immaterial. What is important is that he lays out in clear language the kind of study needed to explain historical lineages. By titling his paper "Functional Efficiency and Technological Change: A Hafting Tradition Model for Prehistoric North America," he tied technological change—the ways in which points were hafted—to functional efficiency—how well the products of technology performed their intended tasks. Musil's emphasis on functional efficiency as the independent variable in the hafting equation differs significantly from approaches that have emphasized such things as ethnicity (different groups made different points), chronology (different points were made at different times), and economy (different points were used for different game). His explanation for the observed temporal sequence of projectile-point form in the Plains and Southwest, from what he termed the fluted/lanceolate tradition to the stemmed tradition to the notched tradition, is that

> each new hafting tradition and change in hafting technology rendered each successive design a more effective killing implement, allowed for more efficient reuse of broken projectile points and better prevented damage to the wooden shaft. These changes in point form usually involved the haft element, the feature of projectile points on which most type designations are based. In this view, then, the variety of forms represented among projectile points is not seen as derivative of the migration of peoples or as the result of different adaptations to a specific environment or hunting economy, but as technological change in one element of material culture—the projectile point. In this view, successive changes in projectile point form were made and adopted because they were functionally more efficient than designs of the preceding tradition. (Musil 1988:373)

Projectile points, Musil (1988:373) noted, are only one component of a complete weapon, the others being the shaft, which itself might be a compound piece, especially if it is a spear or dart, the method of binding, and the hafting technique. He distinguished among four dimensions of the design of a projectile point that allow it both to function as a piercing implement and to conserve the wooden shaft and the projectile point: (1) a sharp point that permits the projectile to pierce an animal's hide; (2) sharp blade edges to open a sizable wound; (3) a haft-element design that absorbs the force of the impact without splitting the shaft; and (4) an overall haft design that minimizes damage to the projectile point and allows the point to be reworked in case of breakage (see also Howard 1995). He identified three major hafting traditions (lineages), one of which he divided into two subtraditions. The fluted/lanceolate tradition includes points assigned to the Clovis, Folsom, Goshen, Midland, and Plainview types. The stemmed-point tradition consists of a parallel-sided subtradition of lanceolate points with shoulders and a stem added (including Alberta, Scottsbluff/Eden, Hardin barbed, and Windust types) and a contracting-sided subtradition of lanceolate points with thicker cross sections and stems (including Agate Basin, Hell Gap, and Haskett types). The notched-point tradition is the third major tradition.

Musil found the fluted/lanceolate tradition, which contains the oldest points, to be the least efficient of the three major hafting traditions because a considerable amount of lithic material is discarded when a point breaks, less opportunity exists for remanufacture, and greater possibility of shaft damage exists. Based on the width of their bases, Clovis, Folsom, and related lanceolate points, including Dalton points, apparently were hafted in split-end shafts (Figure 8.8a). On impact, most of the force travels through the projectile point—which, in the case of specimens that normally would be placed in those point types, already have thin bases—down into the wooden shaft. Frison and Todd (1986:123–128) noted, as did Musil (1988:376), that this arrangement invites damage to the shaft by further splitting it (see also Howard 1995). Also, based on the number of lanceolate-point bases that have been found with transverse breaks at the haft area, the method of hafting these specimens would have done little to extend the use-life of the points.

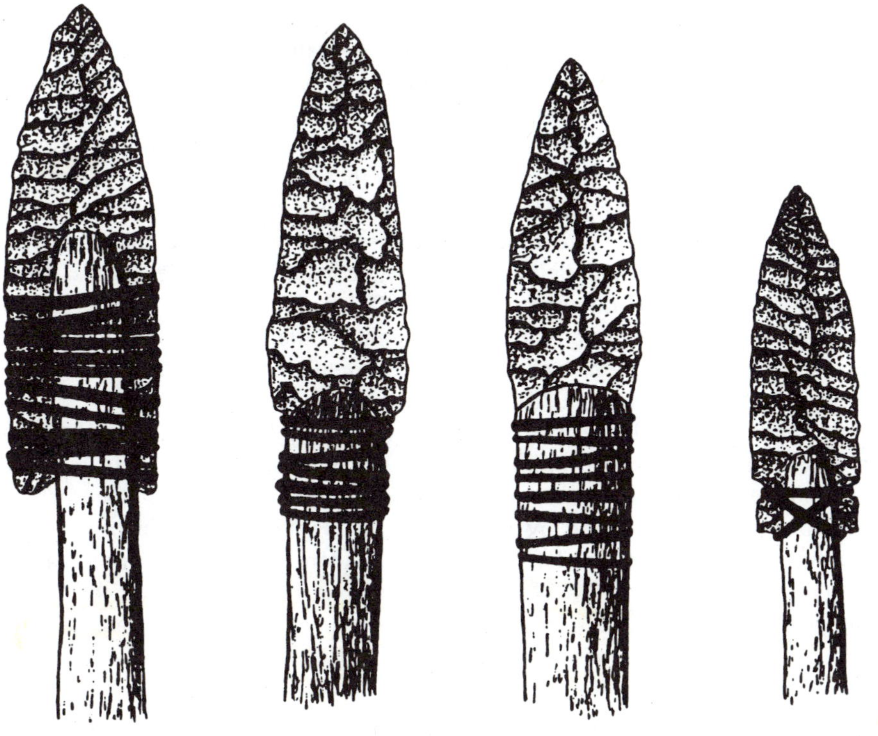

a b c d

FIGURE 8.8. Proposed techniques for hafting a projectile point to a shaft or foreshaft: a, split-shaft hafting of a parallel-edge lanceolate point; b, split-shaft hafting of a stemmed point; c, socketed hafting of a contracting-stem point; d, split-shaft hafting of a side-notched point (from Musil 1988).

Within the stemmed tradition, points belonging to the parallel-sided subtradition appear to have been more suited to a split shaft (Figure 8.8b), whereas points belonging to the contracting-sided tradition, because of their thick cross sections and rounded stem, probably were placed in socketed foreshafts (Figure 8.8c). Musil (1988:379) noted that "resting of the shoulders [of the parallel-sided points] on the bindings may have reduced the impact at the base/shaft juncture, thus reducing shaft damage more than in nonshouldered lanceolate points while also removing the bindings from the blade edges." If the purpose of the blade is to create a large wound, then the more blade that is exposed, the deeper the dart/spear point will penetrate (Howard 1995). The advantages of a socketed contracting-stem design over a fluted/lanceolate design are twofold: "The contracting of the stem removes the bindings from the blade edges, thus increasing the penetrating ability of the point; and ... the long thick contracting stem is in contact with the wooden shaft over the entire stem length and the blunt thick base absorbs more of the force of the thrust" (Musil 1988:379).

In terms of historical relationships, some stratigraphic associations in the Plains (such as at the Hell Gap site in Wyoming), coupled with radiocarbon dates, suggest that Agate Basin contracting-stem points are older than Hell Gap contracting-stem points (8500–8000 B.C. for Agate Basin and 8000–7500 B.C. for Hell Gap), and some associations suggest that they were contemporary forms but that Hell Gap outlasted Agate Basin (8500–8000 B.C. for Agate Basin and 8500–7500 B.C. for Hell Gap) (Agenbroad 1978; Frison 1991). It appears that both types of points are older than parallel-sided Alberta, Scottsbluff, and Eden points, which span portions of the millennium 7500–6500 B.C. (Frison 1991), though Musil (1988:382) suggested the possibility that points containing parallel-sided stems may have overlapped temporally with points containing contracting-sided stems on the Plains (Agenbroad 1978).

Regardless of whether there was a sequence that led from fluted/lanceolate points to contracting-sided points and then to parallel-sided points, or from the earlier fluted/lanceolate points to two parallel traditions of point manufacture, they eventually were supplanted by notched points (Figure 8.8d), which, as Musil (1988:382) noted, contained features that "rendered all of the earlier hafting techniques obsolete." Although it is a bit of an overstatement, the fact that notching persisted from ca. 7500 B.C. until historical times forces one to admit that it was "the most efficient hafting technique to have developed in prehistoric North America" (Musil 1988:382). There are several measures of its efficiency:

> The use of notches removes the bindings from the blade edges by inserting them into the notches. On a corner-notched or basally-notched point, the blade edges are completely unobstructed by the bindings or notches, making a clean cutting edge. The major advantage of notching is that the damage to the haft element usually occurs across the notches on impact, which means there is much less waste of lithic material when the basal fragment is discarded. The

> recovery of the remaining point and the addition of new notches farther up the blade edges allows the point to be rehafted more frequently and successfully than previous traditions. In this way it is possible to rehaft the same specimen after several breakage events. (Musil 1988:382)

The rapidity with which notched points replaced lanceolate points attests to their superiority; the fact that they appeared in many localities at about the same time suggests that the new innovation spread like wildfire or that there were multiple centers of appearance. Probably, both occurred. The appearance of side notching and the technological development of it have been documented at length elsewhere (for the West, see Beck 1995, 1998; for the Midwest, see O'Brien and Wood 1988). What we have done here is first to outline a historical narrative—an explanation—of that chronicle. The explanation is lodged in cause—selection: Functional traits of weapon-delivery systems were under continuous selection, which can be monitored through fine-scale analysis of components of the system. At this point, the only such analysis is that of Hughes (1998), but her work is a model for future studies. She convincingly demonstrates that engineering analysis undertaken within an evolutionary framework puts us well on the road to being able to explain the nature of change as opposed simply to interpreting it.

Frequency Seriation and Pottery Change in the Midwest

David Braun's (1983, 1985a, 1985b, 1987, 1991, 1995) extensive and detailed study of pottery change in the lower Illinois River valley is an excellent example of the kinds of analyses that might be performed and the insights to change that might result from application of Darwinian evolutionary theory to the archaeological record. Although we do not necessarily agree with all of Braun's views on style (e.g., Braun 1995), his articles provide cogent guidelines as to how an analysis focusing on lineage construction and explanation might progress from beginning to end. Some of the attributes Braun examined, the decorative ones, comprise (in our terms) style and thus can be used to measure the passage of time *and* interaction between human groups (e.g., Braun 1985b). Other attributes he measured, the technological or, in our terms, functional ones, might be used to measure the passage of time (e.g., Braun 1985a, 1987), but more importantly, they allow one to monitor the mechanical efficiency of artifacts as *tools* (Braun 1983).

Braun used a large set of sherd samples with associated radiocarbon dates to construct a chronology. Those specimens represented various established types, and Braun could have performed a frequency seriation or used superposition combined with relative frequencies of those types to produce a chronological ordering of his materials for the time period 200 B.C.–A.D. 800. Pottery produced in the lower Illinois River valley and adjacent regions between about 200 B.C. and A.D. 300 often contained elaborate decorations (Figure 8.9) that changed over fairly short periods of time. His analysis (Braun 1985b; Braun and Plog 1982)

FIGURE 8.9. Middle Woodland–period pottery from eastern Missouri. Pottery from eastern Missouri and western Illinois made between about 200 B.C. and A.D. 300 often contained elaborate decorations that changed over fairly short periods of time. Thus, sherds from those pots are amenable to such techniques as frequency seriation (from O'Brien and Wood 1998).

and related studies (e.g., O'Brien and Hoard 1996) demonstrated that there is heritable continuity within decorative types created to partition pottery assemblages from western Illinois and eastern Missouri.

In constructing his chronology Braun ignored decoration in favor of several technological and functional dimensions, one of which was the thickness of vessel walls. Braun (1985a, 1987) plotted the wall thicknesses of sherds from sixty-four well-dated contexts from thirty-two sites in west-central Illinois to determine both the direction and magnitude of change. His resulting time-series curve (Figure 8.10) illustrated that a thickening of vessel walls occurred between 200 B.C. and A.D. 50, after which time the walls became thinner. Specifically, mean wall thickness at around 200 B.C. was approximately 7.4 millimeters and increased to a maximum of approximately 9.0 millimeters by the start of the Christian era. After

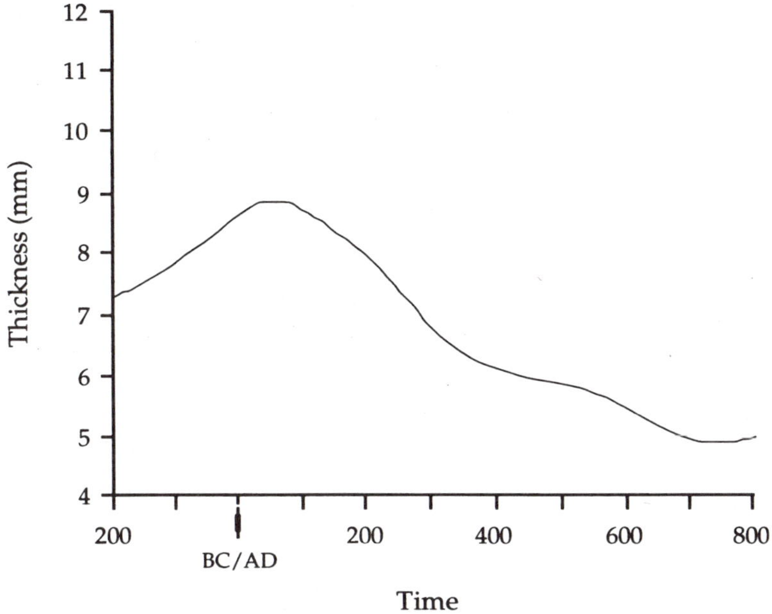

FIGURE 8.10. Curve of vessel-wall thickness from ca. 200 B.C. to A.D. 800 based on sherd samples from western Illinois (after Braun 1987).

approximately A.D. 50, wall thickness declined at a rapid rate until A.D. 300 (mean wall thickness of 6.1 millimeters), at which point it decreased at a slower rate until A.D. 550 (mean thickness of 5.8 millimeters), when the rate of decrease again accelerated, such that by A.D. 750 mean wall thickness was 5.3 millimeters. Braun (1985a:526) also found that after about A.D. 450 vessel diameter had less and less influence on wall thickness, such that by A.D. 800 mean thickness actually decreased with increasing vessel diameter—a phenomenon he interpreted as evidence of increasing demand for thermal conductivity and thermal-shock resistance.

Note an important feature of Figure 8.10: Wall thickness alone cannot be used to sort collections chronologically because several classes of wall thickness occur more than once in time, indicating it is a functional as opposed to a stylistic variable. Using data from Braun (1991:375), we constructed a centered-bar graph displaying the relative frequencies of each class of wall thickness (Figure 8.11).[2]

[2]Braun did not publish the raw data necessary to construct frequency seriations of pottery styles or centered bar graphs showing changes in frequencies of classes of wall thickness. Our graph is thus an approximation that may be incorrect in some details.

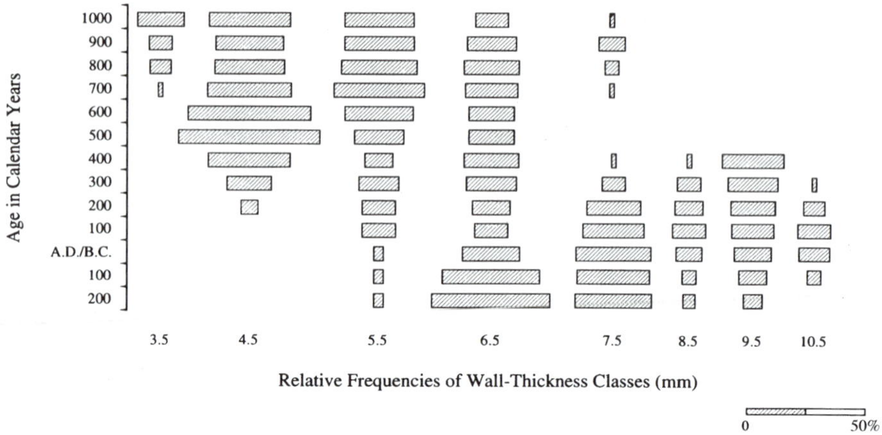

FIGURE 8.11. Centered-bar graph displaying the relative frequencies of each class of wall thickness based on sherds from western Illinois (data from Braun 1991).

The graph shows that changes in relative frequencies of wall-thickness classes are not unimodal and that some classes do not meet the continuity-through-time criterion of historical classes. We would expect this of function-related variables as opposed to the continuous distribution of stochastic (style-related) variables. We reiterate that functional features are not useless for keeping track of time. Certainly, the vast number of projectile-point types in existence, some of which are good historical types, are based on functional features (e.g., Beck 1995, 1998). But care must be exercised to sort out cases of evolutionary convergence—analogs—from those of divergence—homologs. We do not have to worry about convergence when dealing with stylistic traits; we just have to be able to identify them as such (O'Brien and Holland 1990, 1992).

The trend in wall thickness noted by Braun occurs in other ceramic assemblages, including those from the Saline Valley of southern Illinois (Hargrave 1981; Hargrave and Braun 1981) and that from the Burkemper site in eastern Missouri (Hoard 1992; O'Brien and Hoard 1996; O'Brien et al. 1994). Interestingly, Michael Hargrave's (1981) analysis of Saline Valley materials showed that, on average, vessel-wall thickness was 5 millimeters thicker than the averages Braun found for contemporary materials 60 kilometers away. Similarly, O'Brien and Robert Hoard (1996; see also Hoard 1992) demonstrated that Burkemper materials were on average 2 millimeters thicker than Braun's (1985a, 1987) averages for contemporary materials from the Illinois River valley.

What are we to make of these trends in Woodland-period cooking vessels? As Braun (1983) noted, it does us little good to be able to measure minute changes in things such as vessel walls and the size of temper particles if we cannot then explain the variation that we measured. What *performance characteristics* (Braun 1983; O'Brien et al. 1994; Schiffer and Skibo 1987) did Woodland potters look for when they manufactured their pots, and why were they looking for them? What explains the trend toward larger, thinner-walled pots through time? The possible answer to this intriguing question—one certainly well within the Darwinian framework—was provided by over a decade's worth of integrated work, primarily in the lower Illinois River valley, that produced considerable insights into the settlement and subsistence systems of Middle Woodland– and early Late Woodland–period peoples. Widespread evidence from midwestern riverine settings in Illinois and eastern Missouri (e.g., D. L. Asch and N. B. Asch 1978, 1985a, 1985b, 1985c; D. L. Asch et al. 1979; N. B. Asch and D. L. Asch 1980, 1981, 1986; O'Brien 1987; O'Brien and Pulliam 1996) indicates that around the beginning of the Christian era, Woodland groups increased their dependence on the seeds of native oily and starchy plants. There also is evidence that the frequency of juvenile dental caries increased significantly between A.D. 1 and A.D. 600 in west-central Illinois (Buikstra 1977; Cook 1979; Cook and Buikstra 1979), suggesting that seeds were being boiled until they were gelatinized and then fed to infants (Braun 1983).

Thus, it appears the trend in decreasing vessel thickness was tied to evolving food-preparation systems: "This parallel [between decreasing wall thickness and increasing appearance of starchy and oily seeds in the archaeobotanical record] strongly suggests that these ceramic changes all reflect increasing attention to the cooking of seed broths in meal preparation, increasing nutrient extraction and possibly improving palatability" (Braun 1987:164). Localized differences in vessel manufacture varied in relation to the demands placed on the vessels by the users. Braun (1985a:527) noted that the differences in wall thickness between his sample and the one from the Saline Valley paralleled archaebotanical evidence for a much lower importance of edible starchy and oily seeds in the Saline Valley diet (Lopinot 1982:804–806). Evidence of the role played by vessels in cooking seeds is provided by hundreds of sherds from Burkemper, which exhibit thick (often up to a millimeter) carbonized residues on the interiors (O'Brien and Hoard 1996).

The trend toward thinner cooking-vessel walls and the rise in the use of seeds as dietary staples led bioarchaeologist Jane Buikstra to propose that these changes led directly to changes in fertility of Woodland-period women. Infants were weaned at an earlier age and placed on a seed-based, carbohydrate-rich diet. The shortening of the lactation period in Woodland mothers, coupled with a corresponding resumption of ovulation, then led to a rise in fertility. As it turned out, Buikstra was correct (Buikstra et al. 1986; see also Holland 1989); evidence of increased fertility during the early Late Woodland period (ca. A.D. 400–750) was found in skeletal series from western Illinois.

Buikstra's proposition was clearly a result of using one evolutionary trend evident in the archaeological record to posit, in logical fashion, the existence of another. The work of Buikstra and Braun, when coupled with faunal analysis from localities in western Illinois (e.g., Styles 1981; Styles and Purdue 1986; Styles et al. 1985) and eastern Missouri (e.g., Purdue et al. 1989; Styles and Purdue 1996) produced a wealth of information on the strategies Woodland-period groups used to procure and process their food. Analysis of some of the physical properties of Woodland pottery provided a means not only of chronologically ordering depositional contexts (Braun's wall-thickness curve) but also of monitoring changes in several aspects of Woodland-period life.

Transmission and Interaction: Style and Nonselection-Driven Change

Our examples thus far comprise lineages the shape of which we explained in functional, or adaptational, terms because we focused on what we believe to be *functional* attributes; that is, we proposed in both cases that selection drove the changes documented. What about drift? That is, what about lineages or portions thereof that owe their form merely to the vagaries of transmission? Occurrence seriation and frequency seriation rest in part on the assumption of transmission, termed *persistence* if that transmission is through time and, generically, *diffusion* if over geographic space. Although most uses of frequency seriation have been aimed at monitoring persistence, a notable few have used it to monitor diffusion. As far as we can determine, it was first used, if simplistically and commonsensically, this way with some effect by George Quimby (1943) in the early 1940s to identify diffusion in the Great Lakes region (for discussion see Lyman et al. 1998b:252–253). Of course, the question is, was Quimby measuring persistence or diffusion? We suspect it was the latter, given that Quimby's four seriated collections could be arrayed in more or less a straight line geographically. Dave Davis (1981:57) was well attuned to the issue when he wrote, "The sizes of the critical temporal and geographic expanses within which a unit of artifact classification is 'effective' for purposes of chronological alignment [read *frequency seriation*] are controlled by various cultural variables. One such variable is the degree of community stability over time.... Thus, for a given region, nonregularity in the nature and rate of inter-community contact should render type-frequency seriation more difficult."

The underlying assumption to seriation as a method of monitoring transmission is that changes in types, varieties, or attributes are "products of changes in those aspects of cultural behavior which influence patterns of design transmission" (Davis 1981:70). Davis (1981:62) used the type–variety system of artifact classification (e.g., Gifford 1960; Phillips 1958; Wheat et al. 1958) to make his point that "varieties are assumed to represent the basic units of intra and intergenerational design transmission." Thus, in his view, types, because they comprise multiple varieties, "may well not follow the classic battleship [frequency] curve" through time, whereas varieties might (Davis 1981:63). He went on to note that attributes may sometimes display the unimodal frequency distributions predicted by the seriation model, whereas varieties may not. His notion that detecting transmission of either the persistence or the diffusion sort could vary in scale is a critical point. To underscore the importance of it, Davis (1981:73) quoted from Phillips, Ford, and Griffin (1951:61–62): "There is magic in names. Once let a hatful of miserable fragments of fourth-rate pottery be dignified by a 'Name,' and there will follow inevitably the tendency for the name to become an entity, particularly in the mind of him who gives it. Go a step further and publish a description and the type embarks on an independent existence of its own. At that point the classification ceases to be a 'tool,' and the archaeologist becomes one." We could not agree more. But what of transmission? How can Quimby's and Davis's insights be used?

Braun (1985b), following arguments developed earlier (Braun and Plog 1982), suggested that fluctuation in the richness of artifact styles over time suggests changes in intercommunity interaction. Braun largely ignored geographic variation within the area that produced the data he examined, but his point is a strong one. Changes in the decorative diversity of pottery in the lower Illinois River valley suggest changes in "the intensity of supralocal communication"; in short, what previously had been loosely termed "stylistic similarity" would correlate directly with what was also loosely termed "social proximity" (Braun 1985b:132). Braun (1985b:132) characterized much previous work as research in which "style is seen as socially passive" and contrasted that perspective with one in which "style is seen as socially active, an aspect of social display behavior in general, providing visual clues about social identity relationships among people who have opportunities to observe each other's products"; that is, he sought a sociofunctional reason for the spatiotemporal distributions of particular stylistic classes. Although this would incorporate much that has been written about artifact styles over the past two decades (e.g., Conkey and Hastorf 1990; Hegmon 1992; Plog 1983; and references therein), and we are sympathetic to such endeavors, we wonder how such efforts are to be tested in the absence of explanatory theory (see also Plog and Braun 1984; Saitta 1983).

The most recent and most methodologically sophisticated use of frequency seriation to identify interaction among groups is that presented by Carl Lipo and his colleagues (Lipo et al. 1997). Their analysis effectively begins with the notion that a "community or localized group of communities produces a distinctive style of pottery that is distinguished easily from the products of other centers" of communication and pottery production (Neff 1992:151). Using data from the Lower Mississippi Alluvial Valley generated by Phillips et al. (1951) and the principles of frequency seriation and the transmission model of Fraser Neiman (1995), Lipo et al. simulated trait transmission and mixture over time and across space. They found that perfect lenticular curves could be generated with sufficient control of space and time. Applying their findings to the archaeological data, they identified a history of prehistoric interaction and community structure. This innovative study is founded explicitly in the Darwinian tradition and in our knowledge of transmission. They were explicitly addressing Dunnell's (1995) question, "What is it that actually evolves?" using theoretically structured empirical investigation to answer the question and identify communities that interacted—that is, spatially distinct groups of humans that exchanged or transmitted pottery-making information. We suspect more such studies will appear in the future now that the basics of the requisite methods have been developed.

REPLICATORS AND INTERACTORS REVISITED

There are several important matters common to the examples we have discussed that require some elaboration. These concern the notions of scale and the identification of selection on the basis of engineering or mechanical properties. Both can be dealt with by returning to the notions of interactors and replicators that we introduced earlier. Recall that a *replicator* is "an entity that passes on its structure directly in replication" and an *interactor* is "an entity that directly interacts as a cohesive whole with its environment in such a way that replication is differential" (Hull 1980:318). Therefore, the two units represent "two quite distinct functions" (Hull 1981b:30). In the projectile-point example, the units of interaction comprise variation in the haft element, and in the example from the lower Illinois River valley, the units of interaction comprise variation in wall thickness of ceramic vessels. What are replicated (Leonard and Jones 1987) are, of course, projectile points and ceramic vessels comprising various combinations of attributes that are empirical manifestations—phenotypic expressions, if you will—of replicators. We suggest in both examples that it is the "structure of [the] replicators [that] is differentially perpetuated because of the relative success of the interactors of which the replicators are part" (Hull 1981b:41); that is, there is a scale shift from attributes of discrete objects to the discrete objects themselves when one conceptually shifts from replicator to interactor units.

The preceding addresses two significant discussions in the evolutionary literature. First, it is merely another way of expressing Elliott Sober's (1984a:100) distinction between "*selection of objects* and ... *selection for properties*.... 'Selection of' pertains to the *effects* of a selection process, whereas 'selection for' describes its *causes*. To say that there is selection for a given property means that having that property *causes* success in survival and reproduction. But to say that a given sort of object was selected is merely to say that the result of the selection process was to increase the representation of that kind of object" (emphasis in the original). Second, it expresses Elisabeth Vrba and Stephen Jay Gould's (1986:217) concept of *sorting* as a "a simple description of differential representation [that] contains, in itself, no statement about causes.'' The point here is that sorting ''can occur as an effect of causes [such as selection] at other levels'' within a hierarchy of *individuals*, whereas the latter can be genes, organisms, populations, species, or multispecies communities (Vrba and Gould 1986:219). In other words, scale again is critical to the distinction between replicators and interactors. Our examples focus on the scale of attributes of discrete objects (as proxies for replicators) and on differential persistence of classes of discrete objects (interactors). From the functional perspective, other scales are apparent when one considers that archaeologists often speak of ''tool kits'' as a set of particular classes of tools (perhaps in particular relative frequencies) that are functionally interrelated. Here, the scale of replicator would be a particular class of tool (and, again, perhaps its relative frequency), and the interactor would be the functionally integrated whole comprising the tool kit.

Lipo et al. (1997) focus on the scale of artifact types, particularly their frequencies, but examine stylistic variables, or more correctly, particular combinations of stylistic attributes. They also examine not only types but also more-inclusive types, which might be construed as wares, and succeed in identifying what they refer to as ''coarser,'' or geographically larger, evolutionary units (Lipo et al. 1997:328). Such is not unexpected, as the less-inclusive types they use tend mainly to comprise stylistic attributes, confirmed by the fact that the attribute combinations could be arrayed successfully in a frequency seriation. This suggests that those combinations were largely the result of transmission rather than selection. In short, if there is no apparent increase in engineering or mechanical efficiency of objects such as pottery, given changes in some of their particular attributes, then this would suggest the attributes that changed were not changing as a result of selection (O'Brien and Holland 1990). Frequency changes in types comprising these attribute combinations would display a stochastic distribution through time. One could, of course, perform the seriation first and, failing to find a lenticular frequency distribution, check first for spatial variation and second for functional explanations for the changes.

Chapter 9

Evolutionary Archaeology
An Epilogue

Tenacious, if not pig-headed, adherence to a research program is necessary, if the conceptual possibilities in the program are to be explored adequately. (Sober 1980:225)

[U]nderstanding evolution seems to necessitate the abandonment of common sense. (Hull 1980:330)

Over eighty years ago, A. L. Kroeber expressed in somewhat unique terminology what a number of anthropologists had believed for decades: Culture was, borrowing a term from Herbert Spencer, a "superorganic" entity (Kroeber 1917:188). Biological evolution involves heredity (genetic transmission), Kroeber noted, whereas cultural "development" involves social learning. The former is intergenerational and unidirectional, whereas the latter can be either inter- or intragenerational and multidirectional; thus, cultural development can occur more rapidly than biological evolution. Furthermore, this superorganic entity had, at some unidentified point in its history, "passed beyond natural selection" (Kroeber 1917:209). As a result, two "wholly disparate evolutions must be recognized" (Kroeber 1917:208)—one biological and the other cultural. Nearly five decades after Kroeber made his remarks, several of his key points were reiterated by David Kaplan: Whatever entities comprise culture, "they are transmitted across time and through space by non-biological mechanisms; and ... they do not appear to be explainable by an appeal to either genetic or panhuman psychic traits" (Kaplan 1965:960). Debate and discussion in the middle of the twentieth century over the constituent elements of culture seemed to miss a significant aspect of Kroeber's discussion. Kaplan (1965:962) suggested that Kroeber had identified, not defined, culture in such a way as to provide "a methodological superorganicism [that] might be used as a procedure in the study of cultural phenomena." Kroeber failed to pursue this line. As Kaplan implied, had anthropologists not focused on some of the other issues Kroeber dealt with, they might have realized that the concept of culture as superorganic demands not only a particular ontology but also its own particular concepts and "theoretical schemes" (Kaplan 1965:965).

We subscribe to the notion that anthropology in general and archaeology in particular comprise empirical and thus potentially scientific research. The subject matter might be construed as "culture," whatever that is, or its by-products, whatever those are, but as Kaplan (1965) argued, culture is what *we* decide it is; it is a concept invented by anthropologists in order to specify a field of observation and study. Thus, there is no immediate reason not to begin with one of the first, if not *the* first, definitions of culture available in the social-science literature, that by Edward Burnett Tylor (1871) in his classic *Primitive Culture*: "Culture ... taken in its wide ethnographic sense, is that complex whole which includes knowledge, belief, art, morals, law, custom, and any other capabilities and habits acquired by man as a member of society." Culture is "shared ideas" (Dunnell 1971a:121)—shared because the ideas can be transmitted. *Why* ideas are shared—in those instances where transmission is successful—or not—in those instances where transmission is unsuccessful—is a potential subject of historical investigation. Ideas, in other words, comprise replicators; their empirical manifestations—behaviors and artifacts—are interactors. Culture is the superorganic—or extrasomatic, in White's (1959b) terms—equivalent of genes. It is *what* is transmitted and also the *means* of transmission. More to the point, culture is the product of the interaction between human genotypes and their environment.

This comprises, in the briefest of forms, our perspective, and nothing in it goes against any of the basic tenets of Darwinian evolution. Nor, in our opinion, does it remotely suggest that we need to recognize two brands of evolution to begin to understand the hows and whys of human evolution. Kroeber was correct that biological evolution differs from cultural evolution in terms of tempo, as well as in particular mechanisms, but the correctness of such a claim rests solidly on dividing evolution into two separate kinds, one biological and the other cultural. We see little to be gained by doing this. The human phenotype is an amalgam of genetically as well as culturally transmitted features, which makes it both a remarkable and a complicated entity. But as O'Brien and Holland (1995a) emphasize, we do not need to complicate the issue further by establishing two theories to guide research—one based in biology to cover the genetic aspects and the other in anthropology to cover everything else. Dunnell (1995) makes this point equally well in his discussion of William Durham's (1991) formulation of a cultural evolution that is parallel to genetic evolution. There is no need for such a dual construction; one theory—Darwinian evolutionism—does the job quite well.

What we need are methods and techniques that enable us to understand the evolutionary histories of humans and the features they exhibit. Those histories are reflected in the material remains—the products and by-products of human behavior—left behind by the makers and users of the items that ended up in the archaeological record. Archaeologists, like paleontologists, have access to an incredible record of when and where certain phenotypic variants arose; the key to using archaeological remains to write chronicles and explanations lies in changing our perspective to include those remains as parts of past phenotypes.

Throughout this volume, we have borrowed heavily from the literature on biological evolutionism in general and on paleobiological evolutionism in particular. The latter provides much from which we, as archaeologists, can learn given similarities in analytical materials—fossils and artifacts, both by-products of organisms spanning tremendous time depths—and in research goals—understanding the history and evolution of variation within and between populations of organisms. We have devoted considerable space to drawing out these similarities and parallels, delving, where appropriate, into the history of each discipline to help identify the sources of concepts and ideas as well as the weaknesses in each. Some of this we have said previously in other venues (e.g., Lyman and O'Brien 1998; O'Brien and Lyman 2000a, 2000b; O'Brien et al. 1998), but much of this warrants repetition and more detail, as a great deal of it still seems to be misunderstood (e.g., Boone and Smith 1998; Preucel 1999; Rosenberg 1998; Schiffer 1996; Spencer 1997). In those other venues, we have bypassed what perhaps is the most critical issue in applying Darwinian evolutionary theory to the archaeological record—the systematics used to measure and record the variation displayed by that record. We have attempted in this volume to address that issue head-on.

HEREDITY AND TAXONOMY

Darwinian evolutionary theory focuses on documenting and explaining variation in biological phenomena. This sounds simple enough, and for good reason. Stripped to its essentials, Darwinian evolutionism is nothing more than a means of answering the question, "Why do organisms change over time?" In its simplest form, Darwinian evolutionism is not about tempo, or mode, or units, or dozens of other issues and concepts that we have touched on in this book, though these are vital to the logical conduct of evolutionism. Key though they may be, such topics and issues tend to cloud the simplicity of the question that Darwin answered so elegantly. As he figured out, and as biologists and paleobiologists have documented countless times since, organisms change because they differ in terms of fitness relative to one another. Or, as we now know, some are just plain luckier than others.

There are three critical aspects of Darwinian evolution: "1. *Variation* ... in phenotypic traits among members of a population. 2. *Inheritance*. These traits are heritable to some degree, meaning that offspring are more like their parents with respect to these traits than they are like the population mean. 3. *Differential reproductive success*. Different variants leave different numbers of offspring in succeeding generations" (Brandon 1990:7; see also Endler 1986a:4; Lewontin 1978; Sober 1993:9). Darwin did not understand inheritance the way we do today, but he understood the connection between inheritance and reproductive success. We prefer the term "replicative success" (Leonard and Jones 1987) rather than reproductive success, since it is portions of the human cultural phenotype—artifacts—that are evident in the archaeological record. The relation between the replicative success of artifacts, at least those that influence adaptive fitness, and the reproductive success of the organisms that bear them is clear: "The replicative success of artifact traits depends, at least in part, upon their contribution to the fitness of individuals possessing those traits as part of their phenotype but almost certainly also to the effectiveness of transmission mechanisms" (Jones et al. 1995:19).

What is important in Darwinian evolution is that variation, however it is generated, exists, and that transmission, however it is realized, takes place. Reproductive success among variant forms of organisms will result in large part (though not exclusively) from selection and drift. Similarly, replicative success among variant artifact forms will result from selection as well as from mere differences in transmission. Thus, we cannot agree with Stephen Jay Gould's (1996, 1997a) plea that the phenomenon of cultural evolution be labeled cultural development rather than "evolution" because the latter carries too many biological connotations. The Darwinian mechanisms of selection and transmission, when incorporated into an explanatory theory, provide *precisely* what culture historians were looking for and what processual archaeologists seem to view as unnecessary. They provide archaeologists with the tools to begin explaining cultural lineages, that is, to answering the *why* questions so fervently voiced by the processualists.

The key to why organisms change is founded in heredity and the way in which processes such as selection, drift, and recombination act to reshuffle genes in each succeeding generation. Here is where units begin to make an analytical appearance—discounting individual organisms, which are empirical units—and as we know, biologists and paleobiologists deal with any number of such units at varying scales of inclusiveness—populations, races, subspecies, species, genera, families, and so on. Most biologists see the modern version of Linnaean taxonomy as reflecting evolutionary descent relationships between and among those units. Phenetic disparity plays an important role, but it is most often used to aid evolutionary classification rather than being central to it (e.g., Mayr 1969a, 1981, 1995b). Certainly that can be said for the culture-history paradigm that was a focus of Americanist archaeology between about 1915 and 1960 (Lyman et al. 1997b).

That focus was replaced in the 1960s by one centered on artifact function—a focus that failed to explicitly incorporate roles for what we call style and heritable continuity. Our central point is that both style and function are important and both *must* be incorporated into any effort to write an explicable evolutionary history, whether it be the history of a cultural lineage or that of a biological lineage.

Paleobiologists and biologists seem intent on establishing biologically meaningful units, typically species. Thus, biologists regularly "revise" the taxonomy of an order or family of organisms and variously debate the taxonomic validity of particular sets of organisms for purposes of ecological study and wildlife management. An excellent example of the latter, occurring in the midwestern United States, where we live, concerns that group of organisms assigned to the folk taxon "red wolf" and to the scientific *Canis rufus* or *C. lupus rufus* (e.g., Nowak and Federoff 1998; Wayne et al. 1998). Here, the biological-species concept is necessary but insufficient to aid in resolution of whether red wolves should be recognized as a species or subspecies; instead, attention should be focused on the evolutionary history of these animals.

The wolf example is particularly good for illustrating some of the problems that arise when units are built without close attention to theoretical issues, but perhaps of more relevance here is an example drawn from paleobiology concerning fossil species. One might think that the taxonomy of the genus *Bison* would be straightforward and entail clear differentia, to use Eldredge and Gould's (1977) term, but the issue is not so simple. Newly discovered fossils are often "misidentified," and "incorrect" phylogenetic histories are still being written (Wyckoff and Dalquest 1997). In our view, efforts to identify species-like units—sometimes referred to as "nominal species" (Wyckoff and Dalquest 1997)—within the fossil record, because they are largely extensional efforts, have resulted in no end of problems. This applies to hominid paleontology as well (e.g., Tattersall 1986). The most serious problem is that those units tend to take on characteristics of essentialist units. More precisely, the *cause* of the problem is rooted in essentialism and the units that it spawns. When this occurs, any hopes of using Darwinian evolutionism to explain change is precluded for the simple reason that such units cannot evolve. This is why the distinction between theoretical and empirical units that we have stressed is so important. *Classes* of morphometrically distinct variants, regardless of whether they comprise real-world biological species or not, allow measurement of evolutionary change within biological lineages. Other kinds of units do not.

Similar problems plague the classification of artifacts, and they, too, revolve around heredity and taxonomy. An excellent and recently documented instance of the problems encountered when type descriptions and definitions are extensionally derived is found in Philippe LeTourneau's (1998) discussion of the "Folsom Problem." After 1927, all fluted points were placed in the type "Folsom."

What eventually became known as Clovis points were originally termed "Folsomoid" or "Folsom-like." The diagnostic criteria—differentia (significata)—separating what are today seldom-confused types were simply unknown and unspecifiable. According to LeTourneau (1998:60), it was E. H. "Sellards (1952) [who] eventually named as Clovis the larger fluted form that possessed proportionally shorter and narrower flutes than Folsom points." A few years later, Marie Wormington (1957:263) provided the following discussion:

> Clovis Points: Fluted lanceolate points with parallel or slightly convex sides and concave bases. They range in length from one and a half to five inches, but are usually some three inches or more in length and fairly heavy. The flutes sometimes extend almost the full length of the point but usually they extend no more than half way from the base to the tip. Normally, one face will have a longer flute than the other. The fluting is generally produced by the removal of multiple flakes. In most instances the edges of the basal portion show evidence of smoothing by grinding....
>
> Folsom Points: A more specialized type, of excellent workmanship, thought to be derived from the Clovis type. There is some overlap in size between Clovis and Folsom points, but the latter are lighter and usually smaller. They range in length from three quarters of an inch to three inches with an average of about two inches. They are lanceolate in outline and have concave bases usually marked by ear-like projections. There is frequently a small central nipple in the basal concavity. The points are fluted through the removal of longitudinal flakes. The flutes usually extend over most of the length of the point. In most cases one major channel flake was removed from each face but sometimes only one face was fluted. Most specimens have a fine marginal retouch. The lower edges usually bear evidence of grinding.

The modifiers "usually," "sometimes," "normally," "generally," "frequently," and "in most cases" preceding the specification of a character state, plus the specification of ranges and means, signify the extensional nature of the descriptions. Illustrations meant to supplement these descriptions were of limited utility because they showed archetypical specimens—an ideal or at best average example. Finally, which of the attributes listed were definitive, and which were merely descriptive? One could begin to sort out which was which only by comparing the descriptions. Edge grinding was, for example, descriptive because specimens of both Folsom and Clovis points display this, although apparently not all of them do. Similarly, the concave base seems to be descriptive; the nipple-like projection in the concave base of some Folsom points might be definitive, but this is unknown because not all Clovis points have been polled. The extent of the flute might be definitive, but the concept of fluting is operationalized for the identification of points of unknown type only with difficulty because it is not clear how large the flute has to be for a specimen of unknown type to be identified.

Have things improved much in the last forty years? Undoubtedly they have, some would argue, if for no other reason than we now rarely confuse Clovis and Folsom points (Chapter 8). Thus, any archaeologist or artifact collector who knows anything about early points can recognize the difference between the two. There is more than an element of truth in this statement, but it belies a subtle tautology that lurks just below the surface—one that we have tried to identify throughout the book. The tautology is this: Something is something simply because we say it is. This statement is a lot more complex than it sounds, and if we are not clear on exactly what we mean, it might appear as if we think anarchy is reigning in archaeology, which we clearly do not. Of course, archaeologists, like other taxonomists, assign objects to taxa (using the term broadly). Strictly speaking, something is something precisely because we *do* categorize it as such. That is why we said earlier that there is more than an element of truth to the statement that knowledgeable archaeologists and collectors can usually tell the difference between a Folsom point and a Clovis point. No tautology is involved if something is something because we *define* it that way, but the vast majority of units in archaeology, types being foremost in our minds, are not defined. Rather, they are described. At best, they are extensionally defined, but this is simply a nice way of saying that the type "definition" is nothing more than a composite description based on a closed array—the specimens examined originally and on which the description is based. Maybe the array opens at times, as when more specimens are included, but then what usually happens is that the definition is expanded to include the new variants. When it becomes too cumbersome, the type is split into two or more separate types, and the process starts all over again. Of course, the same might be said of species and other extensionally defined units.

Americanist archaeology has long been involved in the construction of types for various purposes. In the nineteenth century, many categories of artifacts were little more than commonsensical units, each with a functional name such as hammerstone, arrowhead, and the like. With the discovery that if types were constructed in particular ways, then time could be measured, attention began to center on historical types, often referred to as "styles." Once time was measured, whether by seriation or stratigraphic superpositioning (especially the latter), artifact styles were used to construct sequences of cultural periods. With the greater temporal resolution afforded by radiocarbon and other absolute-dating methods, superposition and seriation began to be forgotten as ways of building and testing suspected historically continuous sequences. The emergence of cultural reconstruction and processual archaeology in the 1950s and 1960s resulted in an emphasis on functional types, but these were now units founded in such things as use-wear, with less focus on analogous shapes documented ethnographically. Such units granted insights into functional variation and, no doubt, variation in adaptation, but they did not allow the study of heritable continuity or change. Monitoring functional variation within and between chronological periods was surely more "anthropological," but it neither modified the periods nor ensured heritable continuity between them.

IN DEFENSE OF SYSTEMATICS

Michael Schiffer (1996) once stated, we think facetiously, that "apparently ... evolutionary theory alone can solve archaeology's myriad explanatory problems." Our response to this comment (O'Brien et al. 1998) was no, it cannot, but it *can* solve archaeology's *evolutionary* problems. To that end, evolutionary archaeology has constructed concepts and principles to make Darwinian theory applicable to the archaeological record. Most fundamentally, artifacts, as well as the behaviors that created them, are viewed as being parts of the human phenotype—a point on which evolutionists and behavioralists agree. The concepts of replicative success, style, and function are necessary inclusions. Evolutionary archaeology seeks to write evolutionary history and to explain that history in terms of the mechanisms of natural selection and drift. Those are its goals (Lyman and O'Brien 1998), and in order to realize those goals, the distinction between theoretical and empirical units is mandatory. This is demanded by the materialist ontology of Darwinian evolutionary theory. Theoretical units are measurement units, nothing more, and they are used to record the variation displayed by empirical phenomena. Without accurate recording of variation, the very thing upon which evolutionary processes act, any explanation of change is meaningless.

Archaeologists do not like hearing that the units they have been using throughout their careers are inappropriate, just as they bristle when someone calls into question the validity of their findings. We cannot blame them. Critics of evolutionary archaeology maintain, and probably will continue to maintain, that much of the discussion of systematics that has been published over the last decade, especially when it focuses on the supposed misuse of standard archaeological units—phases and types, for example—is nothing more than "posturing" (Muller 1999). We take exception to this charge, and our answer is straightforward: We do not find it "posturing" to point out that *if* archaeology, as we define it, is a historical science, then we have to adopt or develop methods and procedures of *doing* historical science. This requires logical systematics that are applied uniformly time after time, not simply when it is convenient. We agree that it would be much easier were this not the case, but it is. The one discipline to which we can look for some guidance is paleobiology because of the nature of its subject matter. However, the issue is more than simply borrowing paleobiological methods and applying them to archaeological data. If that were all there were to it, we could have made this a very short book.

Archaeology needs its own systematics, and over the past century or so, it has created them. As we have stated repeatedly (e.g., Lyman and O'Brien 1999a; Lyman et al. 1997a, 1997b, 1998b; O'Brien and Lyman 1998, 1999a), various of these systematics were nothing short of remarkable in what they were and still are able to do. Frequency seriation, first used by Kroeber to order surface collections of pottery from New Mexico, was a brilliant breakthrough not only in terms of chronological ordering but also for what it revealed about stylistic change. Similarly, the Midwestern Taxonomic Method of William McKern and the later scheme of Gordon Willey and Philip Phillips showed significant insights. There is nothing inherently wrong with *any* of the units devised by archaeologists of the twentieth century. What is inherently wrong is how they often are misused, even though the misusers have the best of intentions. The pottery-classification schemes of Harold Colton and Harold Gladwin in the Southwest are cases in point. The biological basis of their schemes was groundbreaking; the naivete that went into discussions of what the classifications showed was heartbreaking.

Part of the backlash against evolutionary archaeology may reside in the fact that systematics is, well, kind of boring. One surefire way, we have found, of driving away colleagues at meetings is to begin talking about systematics. People will roll their eyes, clear their throats, and begin to ease away, usually with the excuse that they have to catch a paper that is about to be delivered. This is understandable. It takes considerable effort to learn even the basics of archaeological systematics, especially in light of the fact that they rarely are taught anymore. All archaeologists use the terms *type* and *phase*, for example, but in our opinion few have questioned what kind of units those really are and in which situations they are inappropriate. If it requires considerable effort to learn the basics of systematics, it requires even more effort to begin to break down the received wisdom regarding such things as types and phases and to start looking at things differently. There is more than a little truth in the quote from philosopher David Hull that we used to begin this chapter: Understanding evolution—and certainly the kind of systematics that evolutionism entails—*does* seem to necessitate abandoning common sense. It also necessitates, as Elliott Sober points out in the other quote used at the beginning of the chapter, tenacious, if not pig-headed, adherence to a research program in order to explore conceptual possibilities. Strict adherence to a research program is not posturing; rather, it is the result of believing in what you are doing.

As a final note on systematics, we should point out that traditional archaeological units, when used in a traditional manner, are not worthless constructs, despite the fact that many of them were extensionally derived. Type names, phase names, and the like, serve useful purposes, primarily as shorthand devices to facilitate communication. If we tell you that we found a Dalton point, that term immediately imparts considerable information. You know that the specimen is lanceolate shaped, has a concave base and a ground hafting element, and does not have notches that were chipped on the base, sides, or corners. Our critics might be surprised, and perhaps pleased, that we use the term *Dalton* ourselves. We do not, in normal speech, say, "I found a specimen of class 87324301." No one, including ourselves, would know what we were talking about. Types, even extensionally derived ones, convey information, and that, in and of itself, makes them useful. But it makes them useful only at a particular level. And that level often, though not always, is well below that which a historical science demands.

A CONCLUDING NOTE

As we said in Chapter 1, we would like to think that one day in the not-too-distant future Darwinian evolutionism will be commonplace in Americanist archaeology. As we have seen, numerous products of an evolutionary archaeology have appeared in the last decade, but despite these much-needed additions, it is important that those weighing the merits of the approach—as well as those who have embraced it wholeheartedly—understand precisely the basic tenets of evolutionary archaeology. Equally as important is understanding the systematics that underlie any attempt to explain change in Darwinian evolutionary terms. Maybe, as one of us once put it, "Ultimately it is not necessarily the role of science to explain phenomena—archaeological or otherwise—as much as it is to order observable phenomena into a system consistent with our perception of reality" (O'Brien and Holland 1995a:195).

Our perception of reality is, we freely admit, strikingly different from that of many of our colleagues. This is not to say that it is any better or worse than theirs, only different. Thinking about perceptions of reality brought to mind an interesting exchange that occurred between two characters in Michael Crichton's *The Lost World*. We chose it as a means of ending the book because it sums up our perception of reality beautifully. Mathematician Ian Malcolm is having a conversation with eleven-year-old Arby Benton (unfortunately, he did not make it into the movie version) as they and others are watching a group of parasaurs and apatosaurs in a mutually defensive posture against predators:

> "You realize this is all highly significant," Malcolm said. "For the question of extinction. Already it's obvious that the extinction of the dinosaurs is a far more complex problem than anyone has recognized."
>
> "It is?' Arby said.
>
> "Well, consider," Malcolm said. "All extinction theories are based on the fossil record. But the fossil record doesn't show the sort of behavior we're seeing here. It doesn't record the complexity of groups interacting."
>
> "Because fossils are just bones," Arby said.
>
> "Right. And bones are not behavior. When you think about it, the fossil record is like a series of photographs: frozen moments from what is really a moving, ongoing reality. Looking at the fossil record is like thumbing through

a family photo album. You know that the album isn't complete. You know life happens between the pictures. But you don't have any record of what happens in between, you only have the pictures. So you study them, and study them. And pretty soon, you begin to think of the album not as a series of moments, but as reality itself. And you begin to explain everything in terms of the album, and you forget the underlying reality." (Crichton 1995:192)

Glossary

adaptation Any feature of an organism that allows it to survive and reproduce in a given environment better than if it lacked the feature. As used here, there is the requirement that the feature was shaped by selection for the function it presently serves. Compare with *exaptation*.

adaptive peak Metaphor used by Sewall Wright (1932) to refer to a place on the adaptive landscape where a population has a local mean-fitness maximum measured by allelic frequencies.

allele Alternate state of a gene and the genotypic equivalent of a *character state*. For example, if there were a single gene for hair color, one allele would code for red, whereas another allele would code for brown.

allopatric speciation Speciation by means of geographically separated populations of the same species such that gene flow between or among populations is disrupted. Compare with *sympatric speciation* and *peripatric speciation*.

anagenesis Directional evolution of a character, combination of characters, or a species within a lineage without diversification, or branching, of the lineage. "Phyletic evolution" is sometimes used as a synonym. Compare with *cladogenesis*.

analog A similar character shared by a set of taxa that is not present in their common ancestor. The character does not occur in the set of taxa because of their shared ancestry but rather for other reasons. Compare with *homolog*.

attribute state A physical manifestation of a *dimension*. For example, under the dimension "shape," attribute state 1 might be "rounded," attribute state 2, "square," and so forth. Also referred to as a *character state* or simply as an "attribute."

Bauplan A basic structural design with a long history. For example, we can speak of a bilaterally symmetrical bauplan, a vertebrate bauplan, a mammalian bauplan, or a quadrupedal bauplan.

biface Any stone tool manufactured by removal of flakes from both faces.

character polarity The temporal order of appearance of one *character state* relative to another. Not used in *cladistics*, but such chronological data are used in *stratocladistics*.

character state The physical manifestation of a character, such as hair color, and the phenotypic equivalent of an *allele*. Red, brown, and blond are states of the character "hair." Also referred to as an attribute or *attribute state*.

chorospecies A segment of a suspected *lineage* occurring in the fossil record that is different enough morphologically from other suspected members of that lineage occurring in other geographic locales to be placed in a separate *species*. The focus is geographic variation; all specimens should be of the same age. Compare with *chronospecies* and *morphospecies*.

chronology The temporal relations between and among units or events.

chronospecies A segment of a suspected *lineage* occurring in the fossil record that is different enough morphologically from earlier and later segments of the lineage in a geographic locale to be placed in a separate species. The focus is temporal variation; all specimens should be from the same geographic area. Compare with *chorospecies* and *morphospecies*.

clade A set of taxa descended from a common ancestor, plus that ancestor. A clade comprises the ancestor and all of its descendants. Also known as a *monophyletic group*.

clade-diversity diagram A plot of the frequency of taxa within a particular evolutionary *lineage* over time, showing the multiplication and extinction of taxa within a *clade*.

cladistics A system of phylogenetic analysis that classifies things on the basis of the sequence by which they diverged from a common ancestor. Time is irrelevant except in a relative sense, and only shared derived characters, or *synapomorphies*, are used to make assignments.

cladogenesis The branching and diversification of *lineages* during evolution. It involves the multiplication of taxa within a lineage. Compare with *anagenesis*.

cladogram A branching diagram showing suspected phylogenetic relations among taxa based solely on shared derived characteristics, or *synapomorphies*. Compare with *phenogram*.

class An *ideational* unit that has been defined *intensionally* in terms of its characteristic *attribute states*; a definition that specifies the necessary and sufficient conditions—*significata*—that must be displayed by a specimen in order for it to be considered a member of the class. Unlike a *group*, a class has a distribution in time and space.

cline A character gradient, wherein a character such as length increases or decreases gradually and continuously. A cline distributed over geographic space is a "chorocline"; a cline distributed over time is a "chronocline." Compare with *chorospecies* and *chronospecies*.

component An assemblage of associated artifacts believed to represent a single occupation of a given locale by a group of people.

configurational properties A set of characteristic traits that a thing possesses and which depends on its position in time and space. Compare with *immanent properties*.

contingency An event that is in part predicated on what comes before it. Thus, we say that whatever happens at a particular point in time is contingent on everything that led up to that point; those previous points condition the outcome at the future point.

deme A local, potentially interbreeding, population.

denotata The empirical members of a *class*.

differentia Synonymous with *significata*.

diffusion The movement of a culture trait from one culture to another. Generally speaking, it may be realized by the movement of a concept of the trait, an empirical instance of the trait, or a person possessing the trait as a concept or empirical item. Compare with *transmission*.

dimension A set of mutually exclusive alternative features, or *attribute states*. For example, dimension 1 might be base shape, which contains an infinitely expandible set of mutually exclusive attribute states; dimension 2 might be notch angle, again, with an infinitely expandible set of attribute states; and so forth.

drift Random changes in trait frequency in a population resulting from the vagaries of *transmission*. In biology, it is typically defined as intergenerational sampling error of the available *alleles*. Compare with *selection* and *sorting*.

empirical unit An analytical unit that has an empirical existence, as opposed to an *ideational unit*.

epistemology The branch of philosophy that examines the origin and nature of knowledge. Compare with *metaphysics* and *ontology*.

essentialism A metaphysic that presumes the existence of discoverable, discrete kinds of things. Membership within a kind is based on a shared essence among all things placed in that kind. Compare with *materialism*.

evolution In a broad sense, the generation of entities that vary in terms of *character states* relative to their ancestors and changes in their proportional representation over time. Biological evolution is a shift over time in the proportion of organisms differing genetically in one or more traits. Cultural evolution has become a catchall term for any kind of long-term change exhibited by human groups over time.

exaptation A term coined by Stephen Jay Gould and Elisabeth Vrba (1982) to refer to characters that evolved for other functions (or no function at all) and later were coopted for their current function. Compare with *adaptation*.

extensional definition Establishes the necessary and sufficient conditions for membership in a unit. The conditions are derived by enumerating selected attributes of the members placed in a previously constructed set as opposed to being imposed on specimens to derive a set. Compare with *intensional definition*.

frequency seriation A technique for chronologically ordering collections of presumably historical types such that each type has a continuous distribution and a unimodal frequency distribution. The ordering is based on *type* frequencies.

function Usually a term synonymous with use. In evolutionary archaeology, the term has a restricted meaning and is used to refer to the adaptational aspects of an object or organism. Compare with *style*.

gene flow The mixing of genes from one gene pool of one population into the gene pool of another population.

genotype The set of genes that an organism possesses.

group A phenomenological unit comprising empirical things. Groups can be derived from *intensional* classification or from *extensional* clustering approaches. As opposed to a *class*, a group has a location in space.

heredity The result of *transmission* of information, whether in the form of genes or ideas, from one organism to another.

heritable continuity A genetic-like connection between things, produced by *transmission*, that results in an unbroken sequential ordering of things along a temporal continuum. *Sequences* are based on *overlapping*. Compare with *historical continuity*.

historical continuity A sequential ordering of things based on similarity and assumed to reflect the passage of time. *Heritable continuity* may be the implicit basis for the ordering, in which case *overlapping* is the principle of ordering, or it may be an inference derived from observed *overlapping*. Failure to distinguish explicitly between heritable continuity and historical continuity results in this awkward situation in archaeology. Compare with *sequence*.

homolog A character shared by a set of taxa that is present in their common ancestor. The character is shared among the taxa because of their shared ancestry. Compare with *analog*.

homomorphy Similarity in form; in and of itself, homomorphy says nothing about relatedness.

hybrid An individual created by mating between unlike forms.

ideational unit A conceptual unit that does not have an empirical existence, as opposed to an *empirical unit*. Ideational units can be either theoretical units—the kind most important in evolutionary archaeology—or descriptional units.

immanent properties A set of characteristic traits that a thing possesses regardless of where it exists in time and space. Compare with *configurational properties*.

index fossil Synonymous with *marker type*.

intensional definition Establishes the necessary and sufficient conditions for membership in a unit. The conditions are imposed on empirical specimens in order to sort them into sets as opposed to being extracted from previously constructed sets. Compare with *extensional definition*.

interactor An entity that directly interacts as a unified whole with its environment in such a manner that it is differentially replicated. An example would be a mammal's skin, endoskeleton, circulatory system, and so on. Compare with *replicator.*

isolating mechanisms Genetically (not environmentally) determined differences between or among populations that create barriers to gene flow.

Lamarckism The notion, derived from the work of French naturalist Jean-Baptiste Lamarck, that there is a direction to variation spawned by nature. Charles Darwin's notion of undirected variation is the antithesis of Lamarckism. Lamarckism historically has come to mean the inheritance of acquired characteristics, or soft inheritance—a notion with which Darwin toyed. This obscures the real difference between Lamarck's and Darwin's views—the nature of the directionality of variation.

lineage A temporal line of direct descent from a particular ancestor or ancestral group owing its existence to the *heredity* between ancestors and descendants produced by *transmission*. Compare with *tradition.*

marker type A kind of phenomenon used to denote a particular time period—the briefer the better—within a particular area. It allows the temporal correlation of geographically separate units; its function is the same as that of an index fossil.

materialism A metaphysic that holds that phenomena cannot exist as discrete kinds because they are constantly in the process of becoming something else. Compare with *essentialism.*

metaphysics The branch of philosophy that examines problems of ultimate reality. Compare with *epistemology* and *ontology.*

mode Patterns and processes of evolution. *Anagenesis* is an evolutionary mode, as is *cladogenesis*. Compare with *tempo.*

Modern Synthesis Term coined by Julian Huxley (1942) to highlight the reconciliation of Darwin's notion of *selection* with the empirical evidence of genetics that occurred between about 1935 and 1948. The product of that union often is referred to as neo–Darwinism.

monophyletic group A set of taxa, all of which are derived from a common ancestral taxon. Also known as a *clade*. Compare with *polyphyletic group.*

morphometric Related to the measurement of the size and form of an object.

morphospecies Species distinguished only on the basis of morphometric differences. Compare with *chorospecies* and *chronospecies.*

mutationism As used here, it refers to the belief in the late nineteenth and early twentieth century that speciation was the result of the sudden production of a variant—a mutant—that was discontinuous with its parent.

neontologists The unified group of population geneticists, naturalists, and experimentalists that grew out of the *Modern Synthesis* in the 1940s. Although there was general agreement between this group and the paleontologists on numerous points, the issue of the reality of *species* divided them deeply.

nomological Of or pertaining to laws.

novelty A new trait or form of trait that arises as a result of the imperfect fidelity of replication of its ancestor. This term can be used to characterize the result of "innovation" and "invention," but these two terms usually imply or signify intent.

occurrence seriation A technique for chronologically ordering collections of presumably historical types such that each type has a continuous distribution. The ordering is based on the types' presence–absence in the collections.

ontogenetic Pertaining to the development, or life history, of a thing.

ontology The branch of philosophy that deals with existence. Compare with *epistemology*.

outgroup A *taxon* that diverged from a group of other taxa prior to the point at which they diverged from each other. Outgroups are used in *cladistics* to identify ancestral and derived characteristics, the logic being that an ancestral characteristic will be found in the outgroup but not in the other, more closely related taxa.

overlapping The occurrence of a kind of phenomenon in multiple contiguous time periods. Multiple instances of overlapping serve to connect sets of material from different time periods, such as in *seriation*. Overlapping often serves as part of the basis for inferring *heritable continuity* between sets of material ordered chronologically.

paleobiology The study and explanation of the paleontological record using evolutionary theory.

paleontology The general study of fossils, especially their form and historical sequence.

paradigmatic classification A nonhierarchical classification procedure in which *classes* are created by the intersection of *dimensions*. Compare with *taxonomic classification*.

percentage stratigraphy Measuring the passage of time based on fluctuating percentages of kinds of objects over their geologically superposed distribution. This method often is confused with *seriation*, but the latter method does not employ superposition.

performance standards A set of measurable criteria used to gauge how poorly or how well something functions.

peripatric speciation A kind of *allopatric speciation* in which a new species is formed from a small population living in isolation at the edge of the geographic range of the parental population.

persistence The generic process resulting in what came to be known in archaeology as a *tradition*. Compare with *stasis*.

phenetics A kind of classification system based solely on phenotypic similarity.

phenogram A branching diagram showing suspected phylogenetic relations among taxa that is based on phenotypic similarities among the taxa. Compare with *cladogram*.

phenotype Properties of an organism created by the *genotype* interacting with the environment.

phyletic gradualism Term coined by Niles Eldredge and Stephen Jay Gould (1972) to characterize the view that *evolution* proceeds by the production of a continuous sequence of insensibly graded intermediate forms that link ancestors and descendants. The production of new forms is slow and takes place through *anagenesis*. Compare with *punctuated equilibrium*.

phyletic seriation Chronological ordering of objects based on similarity in appearance. For example, ceramic vessels could be ordered based on suspected change in form or decoration.

phylogenetic evolution *Evolution* characterized by branching of *lineages*. Compare with *reticulate evolution*.

phylogeny A reconstruction of evolutionary patterns and relationships. Also refers to a branching diagram that shows ancestral relations among taxa.

pleiotropy The phenotypic effect of a gene on more than one character.

polygeny The phenotypic effect of multiple genes influencing a single character.

polyphyletic group A set of taxa comprising members derived from more than one ancestral *taxon*. Compare with *monophyletic group*.

processual archaeology A paradigm that arose in the 1960s as archaeologists grew tired of chronological issues and cultural reconstruction and moved toward the identification of processes that produce culture change.

punctuated equilibrium Term coined by Niles Eldredge and Stephen Jay Gould (1972) to refer to the view that *evolution* proceeds primarily through abrupt *allopatric speciational* events that are separated by long periods of evolutionary *stasis*. New *species* arise only by the splitting of *lineages*, or *cladogenesis*, not by *anagenesis*. The new daughter species originate near the periphery of the range of the species and may spread back into the locale still occupied by the parent population. Thus, in a local geological section, there should be a clear morphological difference between ancestor and descendant.

replicative success Term coined by Robert Leonard and Tom Jones (1987) to denote the differential persistence of the traits of organisms from one generation to the next. It implies neither a mode of trait transmission nor that a selective advantage is attained by an organism as a result of trait transmission and retention.

replicator An entity that passes on its structure directly in replication—for example, genes. Compare with *interactor*.

reticulate evolution *Evolution* characterized by the branching and merging of *lineages*. Compare with *phylogenetic evolution*.

saltation A discontinuity of considerable magnitude in phenotypic traits between generations.

scale A particular degree of inclusiveness or rank. For example, one scale of analysis might be at the level of an attribute of a discrete object (e.g., neck width of a projectile point), a secord scale at the level of a discrete object (e.g., type of ceramic), and a third at the level of a set of discrete objects (e.g., tool kit).

selection Process by which certain forms in a population that are better adapted to a particular environment increase in proportion to less well-adapted forms. The contrast often is made between "natural selection" and "cultural selection," the former referring to instances where the physical environment is the selective agent and the latter only to instances where humans are the selective (choosing) agents. The distinction is immaterial given the outcome—differential representation of variant forms. Compare with *drift* and *sorting*.

sequence A temporal ordering of things. It may be based on chronological data independent of the things, or it may be based on an assumption of *historical continuity* between the things. An inference of *heritable continuity* might be derived from a sequence, but not all sequences measure heritable continuity.

seriation The arrangement of objects in presumed chronological order on the basis of some logical principle other than superposition. The order is based on intrinsic properties of the units being used rather than on extrinsic properties such as the geographic or geological positions of the things. The seriation method has several techniques housed within it, including *frequency seriation*, *occurrence seriation*, and *phyletic seriation*.

significata The necessary and sufficient conditions that must be displayed by a specimen in order for it to be considered a mumber of a class. Also referred to as *differentia*.

somatic Pertaining to bodily features.

sorting A general description of differential representation of traits in succeeding generations. Used by itself, it contains no statement about cause. *Selection* and *drift* are two sorting mechanisms.

spandrel An architectural term referring to the triangular space between the right or left exterior curve of an arch and the framework surrounding it. Stephen Jay Gould and Richard Lewontin (1979) introduced it into evolutionary biology as a metaphor for a *class* of forms and space arising as necessary by-products of another design decision and not as *adaptations*.

species A basic taxonomic category in which organisms are placed. A "biological species" is a group of populations, the members of which actually or potentially interbreed with one another under natural conditions. A "paleobiological species" is a group of fossil organisms, each member of which more resembles the other members of that group phenotypically than any does a member of another group. The latter can be *chorospecies*, *chronospecies*, or *morphospecies*.

stasis A state of static balance or equilibrium. The lack of change in *species* over long spans of time is a critical component of *punctuated equilibrium*. Compare with *tempo*.

stochastic Often used erroneously as a synonym for random, it correctly refers to "random but not an unlimited number of possibilities." The limit on the number of possible outcomes is set by historical *contingency*, what came before, or *heredity*. For example, without extensive testing, we cannot know phenotypically what a baby will look like before it is born, but the choices are not limitless. Its hair color might be black, or brown, or red, but it will not be green or purple.

stratigraphy The study of sedimentary units, the sediments contained within those units, and the vertical relations of the units and sediments. Often used colloquially to refer to a stratigraphic section (e.g., "the stratigraphy at the site was complex").

stratocladistics Method involving the use of temporal data, such as from superposition, to assist in constructing a *cladogram*.

stratum A three-dimensional unit of sediment that represents a depositional event and is distinguishable from other such units.

style Usually a catchall term used to denote a particular look or design of a thing, including such dimensions as form and decoration. In evolutionary archaeology, the term has a restricted meaning and is used to refer to the nonfunctional (nonadaptational) aspects of an object or organism, similar to the way in which biologists use the term "neutrality." For example, we could say that style A confers no fitness value on its possessor different than that conferred by style B. Compare with *function*.

superposition The geologically vertical relation between or among things, such as artifacts, fossils, or sedimentary units. The law of superposition, in simplified terms, states that things on the bottom were deposited before those on the top. Superposition does not refer to the age of the things themselves, only to the order in which they were deposited.

symplesiomorphies Ancestral characters or *character states* that two or more taxa share in common with their immediate ancestor and with preceding ancestors. Also known as "shared ancestral characteristics." Compare with *synapomorphies*.

sympatric speciation Speciation that occurs among populations with overlapping ranges, thus affording an opportunity to interbreed.

synapomorphies Derived characters or *character states* that two or more taxa share in common with their immediate ancestor but not with a preceding ancestor. Also known as "shared derived characteristics." *Cladistics* depends solely on the identification and use of synapomorphies. Compare with *symplesiomorphies*.

systematics Procedures for creating units derived logically for a specific analytical purpose. The units are used to characterize and measure similarities and differences between phenomena, and to sort them into sets such that like goes with like. Typically, things that are alike are thought to have a particular affinity. As used in this book, one critically important kind of affinity is evolutionary, or genetic-like, affinity.

taxon A named taxonomic unit of unspecified rank (e.g., genus, species, family) to which individual organisms or sets of organisms are assigned.

taxonomic classification A hierarchical classification made up of lower-level classes nested within higher-level classes, which themselves are nested in even higher-order classes, and so on. Compare with *paradigmatic classification*.

tempo The rate of evolutionary change. The rate can be zero, in which case the rate comprises *stasis*; it can be gradual; or it can be relatively rapid. Compare with *mode*.

theory An organized set of knowledge that is applied systematically to explain the nature of a set of phenomena.

tradition A temporal continuum represented by persistent configurations of related forms. Compare with *heredity*, *lineage*, and *persistence*.

tradition/lineage scale of heritable continuity Congeneric *species*—distinct species belonging to the same genus—denote *heritable continuity* given genetic transmission and evolutionary divergence from a common ancestor. The same applies for confamilial genera, and so on. Similarly, artifact taxa that are nested within higher taxa are assumed, on the basis of transmission, to denote heritable continuity. Compare with *type/species scale of heritable continuity*.

transmission The movement of information, whether genes or cultural ideas, from one organism to another. Thus, it is not necessarily intergenerational. This process produces inheritance and drives *heritable continuity*. Compare with *diffusion* and *heredity*.

type Synonymous with kind or category of unspecified derivation.

type/species scale of heritable continuity Each fossil specimen belonging to a particular *species* given genetic transmission between individuals within the species is assumed to be hereditarily continuous with every other fossil specimen identified as a member of that species. Similarly, each artifact placed in an archaeological type is assumed, on the basis of transmission, to be hereditarily continuous with every other specimen identified as a member of that type. Compare with *tradition/lineage scale of heritable continuity*.

References

Abbott, A. L., R. D. Leonard, and G. T. Jones
1996 Explaining the Change from Biface to Flake Technology: A Selectionist Application. In *Darwinian Archaeologies*, edited by H. D. G. Maschner, pp. 33–42. Plenum Press, New York.

Agenbroad, L. D.
1978 *The Hudson–Meng Site: An Alberta Bison Kill in the Nebraska High Plains*. University Press of America, Washington, DC.

Allen, G. E.
1980 The Evolutionary Synthesis: Morgan and Natural Selection Revisited. In *The Evolutionary Synthesis: Perspectives on the Unification of Biology*, edited by E. Mayr and W. B. Provine, pp. 356–382. Harvard University Press, Cambridge, MA.

Allen, W. L., and J. B. Richardson, III
1971 The Reconstruction of Kinship from Archaeological Data: The Concepts, the Methods. *American Antiquity* 36:41–53.

Amsden, C. A.
1931 Black-on-White Ware. In The Pottery of Pecos, Vol. 1, by A. V. Kidder, pp. 17–72. *Papers of the Southwestern Expedition, Phillips Academy* No. 5.

Anderson, E.
1970 Quaternary Evolution of the Genus *Martes* (Carnivora, Mustelidae). *Acta Zoologica Fennica* 130.
1994 Evolution, Prehistoric Distribution, and Systematics of *Martes*. In *Martens, Sables, and Fishers: Biology and Conservation*, edited by S. W. Buskirk, A. S. Harestad, M. G. Raphael, and R. A. Powell, pp. 1–25. Cornell University Press, Ithaca, NY.

Anderson, K. M.
1969 Ethnographic Analogy and Archaeological Interpretation. *Science* 163:133–138.

Angier, N.
1997 Ernst Mayr at 93. *Natural History* 106(4):8–11.

Antonovics, J.
1987 The Evolutionary Dys-Synthesis: Which Bottles for Which Wine? *American Naturalist* 129:321–331.

Arnold, A. J., and K. Fristrup
1982 The Theory of Evolution by Natural Selection: A Hierarchical Expansion. *Paleobiology* 8:113–129.

Arnold, M. L.
1997 *Natural Hybridization and Evolution*. Oxford University Press, New York.

Asch, D. L., and N. B. Asch
1978 The Economic Potential of *Iva annua* and Its Prehistoric Importance in the Lower Illinois River Valley. In The Nature and Status of Ethnobotany, edited by R. I. Ford, pp. 300–341. *University of Michigan, Museum of Anthropology, Anthropological Papers* No. 67.
1985a Prehistoric Plant Cultivation in West-Central Illinois. In Prehistoric Food Production in North America, edited by R. I. Ford, pp. 149–203. *University of Michigan, Museum of Anthropology, Anthropological Papers* No. 75.
1985b Archeobotany. In Smiling Dan: Structure and Function at a Middle Woodland Settlement in the Illinois Valley, edited by B. Stafford and M. Sant, pp. 327–401. *Kampsville Archeological Center, Center for American Archeology, Research Series* 2.
1985c Archeobotany. In Massey and Archie: A Study of Two Hopewellian Homesteads in the Western Illinois Uplands, edited by K. B. Farnsworth and A. L. Koski, pp. 162–220. *Kampsville Archeological Center, Center for American Archeology, Research Series* 3.

Asch, D. L., K. B. Farnsworth, and N. B. Asch
1979 Woodland Subsistence and Settlement in West Central Illinois. In *Hopewell Archaeology: The Chillicothe Conference*, edited by D. S. Brose and N. Greber, pp. 80–85. Kent State University Press, Kent, OH.

Asch, N. B., and D. L. Asch
1980 The Dickson Camp and Pond Sites: Middle Woodland Archaeobotany in Illinois. In Dickson Camp and Pond: Two Early Havana Tradition Sites in the Central Illinois Valley, by A. Cantwell, pp. 152–160. *Illinois State Museum, Reports of Investigation* No. 36.
1981 Archeobotany of Newbridge, Carlin, and Weitzer Sites—the White Hall Component. In Faunal Exploitation and Resource Selection, by B. W. Styles, pp. 275–291. *Northwestern University Archeological Program, Scientific Papers* No. 3.
1986 Woodland Period Archeobotany of the Napoleon Hollow Site. In Woodland Period Occupations of the Napoleon Hollow Site in the Lower Illinois Valley, edited by M. D. Wiant and C. R. McGimsey, pp. 427–512. *Kampsville Archeological Center, Center for American Archeology, Research Series* 6.

Ascher, R.
1961a Analogy in Archaeological Interpretation. *Southwestern Journal of Anthropology* 17:317–325.
1961b Experimental Archeology. *American Anthropology* 63:793–816.

Bargatzky, T.
1984 Culture, Environment, and the Ills of Adaptationism. *Current Anthropology* 25:399–415.

Barnosky, A. D.
1987 Punctuated Equilibrium and Phyletic Gradualism: Some Facts from the Quaternary Mammalian record. *Current Mammology* 1:109–147.

Barton, N. H. and B. Charlesworth
1984 Genetic Revolutions, Founder Effects, and Speciation. *Annual Review of Ecology and Systematics* 15:133–164.

Bateson, W.
1894 *Materials for the Study of Variation*. Macmillan, London.

Baum, D. A., and A. Larson
1991 Adaptation Reviewed: A Phylogenetic Methodology for Studying Character Macroevolution. *Systematic Zoology* 40:1–18.

Bayard, D. T.
1969 Science, Theory and Reality in the "New Archaeology." *American Antiquity* 34:376–384.

Beatty, J.
1995 The Evolutionary Contingency Thesis. In *Concepts, Theories, and Rationality in the Biological Sciences*, edited by G. Wolters and J. G. Lennox, pp. 45–81. University of Pittsburgh Press, Pittsburgh, PA.

Beck, C.
1995 Functional Analysis and the Differential Persistence of Great Basin Dart Forms. *Journal of California and Great Basin Anthropology* 17:222–243.
1998 Projectile Point Types as Valid Chronological Units. In *Unit Issues in Archaeology: Measuring Time, Space, and Material*, edited by A. F. Ramenofsky and A. Steffen, pp. 21–40. University of Utah Press, Salt Lake City.
Beck, C., and G. T. Jones
1989 Bias and Archaeological Classification. *American Antiquity* 54:244–261.
Behrensmeyer, A. K., and S. M. Kidwell
1985 Taphonomy's Contribution to Paleobiology. *Paleobiology* 11:105–119.
Behrensmeyer, A. K., and D. Schindel
1983 Resolving Time in Paleobiology. *Paleobiology* 9:1–8.
Bell, G.
1997 *The Basics of Selection*. Chapman and Hall, New York.
Bell, R. E.
1958 Guide to the Identification of Certain American Indian Projectile Points. *Oklahoma Anthropological Society, Special Bulletin* No. 1.
1960 Guide to the Identification of Certain American Indian Projectile Points. *Oklahoma Anthropological Society, Special Bulletin* No. 2.
Bell, W. C.
1950 Stratigraphy: A Factor in Paleontologic Taxonomy. *Journal of Paleontology* 24:492–496.
Bennett, J. W.
1976 Anticipation, Adaptation, and the Concept of Culture in Anthropology. *Science* 192:847–853.
Benton, M. J.
1995 Diversification and Extinction in the History of Life. *Science* 268:52–58.
Benton, M. J., and R. Hitchin
1997 Congruence between Phylogenetic and Stratigraphic Data on the History of Life. *Proceedings of the Royal Society of London: Biological Sciences* 264B:885–890.
Benton, M. J., and G. W. Storrs
1994 Testing the Quality of the Fossil Record: Paleontological Knowledge Is Improving. *Geology* 22:111–114.
Bettinger, R. L.
1991 *Hunter–Gatherers: Archaeological and Evolutionary Theory*. Plenum Press, New York.
Bettinger, R. L., J. F. O'Connell, and D. H. Thomas
1991 Projectile Points as Time Markers in the Great Basin. *American Anthropologist* 93:166–172.
Bettinger, R. L., and P. J. Richerson
1996 The State of Evolutionary Archaeology: Evolutionary Correctness, or the Search for the Common Ground. In *Darwinian Archaeologies*, edited by H. D. G. Maschner, pp. 221 –231. Plenum Press, New York.
Biers, W. R.
1992 *Art, Artefacts, and Chronology in Classical Archaeology*. Routledge, New York.
Binford, L. R.
1962 Archaeology as Anthropology. *American Antiquity* 28:217–225.
1963 "Red Ocher" Caches from the Michigan Area: A Possible Case of Cultural Drift. *Southwestern Journal of Anthropology* 19:89–108.
1964 A Consideration of Archaeological Research Design. *American Antiquity* 29:425–441.
1965 Archaeological Systematics and the Study of Culture Process. *American Antiquity* 31: 203–210.
1967 Smudge Pits and Hide Smoking: The Use of Analogy in Archaeological Reasoning. *American Antiquity* 32:1–12.
1968a Archeological Perspectives. In *New Perspectives in Archeology*, edited by S. R. Binford and L. R. Binford, pp. 5–32. Aldine, New York.

1968b Methodological Considerations of the Archeological Use of Ethnographic Data. In *Man the Hunter*, edited by R. B. Lee and I. DeVore, pp. 268–273. Aldine, Chicago.

1968c Some Comments on Historical versus Processual Archaeology. *Southwestern Journal of Anthropology* 24:267–275.

1969 Conceptual Problems in Dealing with Units and Rates of Cultural Evolution. *Anthropology U.C.L.A.* 1:27–35.

1972 [Untitled Introduction to Part I: Method and Archeological Reasoning.] In *An Archeological Perspective*, by L. R. Binford, pp. 17–19. Seminar Press, New York.

1977 General Introduction. In *For Theory Building in Archaeology*, edited by L. R. Binford, pp. 1–10. Academic Press, New York.

1981 Behavioral Archaeology and the Pompeii Premise. *Journal of Anthropological Research* 37:195–208.

1982 The Archaeology of Place. *Journal of Anthropological Archaeology* 1:5–31.

1983 *In Pursuit of the Past: Decoding the Archaeological Record.* Thames and Hudson, New York.

1985 "Brand X" versus the Recommended Product. *American Antiquity* 50:580–590.

Binford, L. R., and J. A. Sabloff

1982 Paradigms, Systematics, and Archaeology. *Journal of Anthropological Research* 38: 137–153.

Binford, S. R., and L. R. Binford (editors)

1968 *New Perspectives in Archeology.* Aldine, Chicago.

Boas, F.

1896 The Limitations of the Comparative Method of Anthropology. *Science* 4:901–908.

1902 Some Problems in North American Archaeology. *American Journal of Archaeology* 6:1–6.

1904 The History of Anthropology. *Science* 20:513–524.

1932 The Aims of Anthropological Research. *Science* 76:605–613.

Bock, W. J.

1977 Foundations and Methods of Evolutionary Classification. In *Major Patterns in Vertebrate Evolution*, edited by M. K. Hecht, P. C. Goody, and B. M. Hecht, pp. 851–895. Plenum Press, New York.

Bonner, J. T.

1980 *The Evolution of Culture in Animals.* Princeton University Press, Princeton, NJ.

1988 *The Evolution of Complexity.* Princeton University Press, Princeton, NJ.

Boone, J. L., and E. A. Smith

1998 Is It Evolution Yet? A Critique of Evolutionary Archaeology. *Current Anthropology* 39:S141–S173.

Boucot, A. J.

1982 Ecophenotypic or Genotypic? *Nature* 295:609–610.

Bowler, P. J.

1990 *Charles Darwin: The Man and His Influence.* Blackwell, London.

Boyd, R., and P. J. Richerson

1985 *Culture and the Evolutionary Process.* University of Chicago Press, Chicago.

1992 How Microevolutionary Processes Give Rise to History. In *History and Evolution*, edited by M. H. Nitecki and D. V. Nitecki, pp. 179–209. State University of New York Press, Albany.

Boyden, A.

1943 Homology and Analogy: A Century after the Definition of "Homologue" and "Analogue" of Richard Owen. *Quarterly Review of Biology* 18:228–241.

1947 Analogy and Homology. *American Midland Naturalist* 37:648–669.

Bradley, B. A.

1997 Sloan Site Biface and Projectile Point Technology. In *Sloan: A Paleoindian Dalton Cemetery in Arkansas*, by D. F. Morse, pp. 53–57. Smithsonian Institution Press, Washington, DC.

Bradley, B. A., and G. C. Frison
1996 Flaked-Stone and Worked-Bone Artifacts from the Mill Iron Site. In *The Mill Iron Site*, edited by G. C. Frison, pp. 43–69. University of New Mexico Press, Albuquerque.

Brady, R. H.
1985 On the Independence of Systematics. *Cladistics* 1:113–126.

Braidwood, R. J.
1959 Archeology and the Evolutionary Theory. In *Evolution and Anthropology: A Centennial Appraisal*, edited by B. J. Meggers, pp. 78–89. Anthropological Society of Washington, Washington, DC.

Brainerd, G. W.
1951a The Place of Chronological Ordering in Archaeological Analysis. *American Antiquity* 16: 301–313.
1951b The Use of Mathematical Formulations in Archaeological Analysis. In Essays on Archaeological Methods, edited by J. B. Griffin, pp. 117–127. *University of Michigan, Museum of Anthropology, Anthropological Papers* No. 8.

Brandon, R. N.
1990 *Adaptation and Environment*. Princeton University Press, Princeton, NJ.

Braun, D. P.
1983 Pots as Tools. In *Archaeological Hammers and Theories*, edited by J. A. Moore and A. S. Keene, pp. 107–134. Academic Press, New York.
1985a Absolute Seriation: A Time-Series Approach. In *For Concordance in Archaeological Analysis*, edited by C. Carr, pp. 509–539. Westport, Boulder, CO.
1985b Ceramic Decorative Diversity and Illinois Woodland Regional Integration. In *Decoding Prehistoric Ceramics*, edited by B. A. Nelson, pp. 128–153. Southern Illinois University Press, Carbondale.
1987 Coevolution of Sedentism, Pottery Technology, and Horticulture in the Central Midwest, 200 B.C.–A.D. 600. In Emergent Horticultural Economies of the Eastern Woodlands, edited by W. F. Keegan, pp. 153–181. *Southern Illinois University at Carbondale, Center for Archaeological Investigations, Occasional Paper* No. 7.
1991 Why Decorate a Pot? Midwestern Household Pottery, 200 B.C.–A.D. 600. *Journal of Anthropological Archaeology* 10:360–397.
1995 Style, Selection, and Historicity. In *Style, Society, and Person: Archaeological and Ethnological Perspectives*, edited by C. Carr and J. E. Neitzel, pp. 123–141. Plenum Press, New York.

Braun, D. P., and S. Plog
1982 Evolution of "Tribal" Social Networks: Theory and Prehistoric North American Evidence. *American Antiquity* 47:504–525.

Brew, J. O.
1946 Archaeology of Alkali Ridge, Southeastern Utah. *Peabody Museum of American Archaeology and Ethnology, Papers* 21.

Broughton, J. M., and J. F. O'Connell
1999 On Evolutionary Ecology, Selectionist Archaeology, and Behavioral Archaeology. *American Antiquity* 64:153–165.

Brower, J. C., and W. A. Burroughs
1982 A Simple Method for Quantitative Biostratigraphy. In *Quantitative Stratigraphic Correlation*, edited by J. M. Cubitt and R. A. Reyment, pp. 61–83. Wiley, Chichester, UK.

Brower, A. V. Z., and V. Schawaroch
1996 Three Steps in Homology Assessment. *Cladistics* 12:265–272.

Browman, D. L., and D. R. Givens
1996 Stratigraphic Excavation: The First "New Archaeology." *American Anthropologist* 98: 80–95.

Brown, W. L., Jr.

1987 Punctuated Equilibrium Excused: The Original Examples Fail to Support It. *Biological Journal of the Linnean Society* 31:383–404.

Buikstra, J. E.

1977 Biocultural Dimensions of Archaeological Study: A Regional Perspective. In Biocultural Adaptation in Prehistoric America, edited by R. L. Blakely, pp. 67–84. *Southern Anthropological Society, Proceedings* 11.

Buikstra, J. E., L. Koningsberg, and J. Bullington

1986 Fertility and the Development of Agriculture in the Prehistoric Midwest. *American Antiquity* 51:528–546.

Burgh, R. F.

1959 Ceramic Profiles in the Western Mound at Awatovi, Northeastern Arizona. *American Antiquity* 25:184–202.

Burian, R. M.

1988 Challenges to the Evolutionary Synthesis. *Evolutionary Biology* 23:247–269.

1992 Adaptation: Historical Perspectives. In *Keywords in Evolutionary Biology*, edited by E. F. Keller and E. A. Lloyd, pp. 7–12. Harvard University Press, Cambridge, MA.

Butler, B. R.

1973 Review of "An Archeological Perspective" by L. R. Binford. *American Antiquity* 39: 646–647.

Cain, J. A.

1992 Building a Temporal Biology: Simpson's Program for Paleontology during an American Expansion of Biology. *Earth Sciences History* 11:30–36.

1993 Common Problems and Cooperative Solutions: Organizational Activity in Evolutionary Studies, 1936–1947. *Isis* 84:1–25.

Caldwell, J. R.

1959 The New American Archaeology. *Science* 129:303–307.

Cambron, J. W., and D. C. Hulse

1960 An Excavation on the Quad Site. *Tennessee Archaeologist* 16(1):14–26.

1964 *Handbook of Alabama Archaeology, Part 1: Point Types.* Archaeological Research Association of Alabama, Moundville.

1975 *Handbook of Alabama Archaeology, Part 1: Point Types* (rev. ed.). Archaeological Research Association of Alabama, Tuscaloosa.

Carneiro, R. L.

1970 A Theory of the Origin of the State. *Science* 169:733–738.

1972 The Devolution of Evolution. *Social Biology* 19:248–258.

1973 Classical Evolution. In *Main Currents in Cultural Anthropology*, edited by R. Naroll and F. Naroll, pp. 57–121. Appleton–Century–Crofts, New York.

Carothers, S. W.

1978 Biology. *Plateau* 50(4):24–32.

Carr, C., and J. E. Neitzel

1995 Integrating Approaches to Material Style in Theory and Philosophy. In *Style, Society, and Person: Archaeological and Ethnological Perspectives*, edited by C. Carr and J. E. Neitzel, pp. 3–20. Plenum Press, New York.

Carroll, R. L.

1986 Physical and Biological Constraints on the Pattern of Vertebrate Evolution. *Geoscience Canada* 13:85–90.

Chang, K. C.

1967 *Rethinking Archaeology.* Random House, New York.

Chapman, C. H.

1975 *The Archaeology of Missouri, I.* University of Missouri Press, Columbia.

Chapman, C. H., and E. F. Chapman
1964 *Indians and Archaeology of Missouri*. University of Missouri Press, Columbia.
Charlesworth, B.
1982 Neo-Darwinism—the Plain Truth. *New Scientist* 94:133–137.
Charlesworth, B., and R. Lande
1982 Morphological Stasis and Developmental Constraint: No Problem for Neo–Darwinism. *Nature* 296:610.
Childe, V. G.
1951a *Social Evolution*. Fontana, London.
1951b *Man Makes Himself*. Watts, London.
Clarke, D. L.
1968 *Analytical Archaeology*. Methuen, London.
1978 *Analytical Archaeology* (2nd ed.). Columbia University Press, New York.
Clyde, W. C., and D. C. Fisher
1997 Comparing the Fit of Stratigraphic and Morphologic Data in Phylogenetic Analysis. *Paleobiology* 23:1–19.
Cogswell, J. W., and M. J. O'Brien
1997 A Comparison of Laboratory Results to Archaeological Data: Pottery Surface Treatments in Eastern Missouri. *Southeastern Archaeology* 16:169–174.
Cohen, A. S., and H. L. Schwartz
1983 Speciation in Molluscs from Turkana Basin. *Nature* 304:659–660.
Colton, H. S.
1939 Prehistoric Culture Units and Their Relationships in Northern Arizona. *Museum of Northern Arizona, Bulletin* 17.
1942 Archaeology and the Reconstruction of History. *American Antiquity* 8:33–40.
Colton, H. S., and L. L. Hargrave
1937 Handbook of Northern Arizona Pottery Wares. *Museum of Northern Arizona, Bulletin* 11.
Conkey, M., and C. Hastorf (editors)
1990 *The Uses of Style in Archaeology*. Cambridge University Press, Cambridge, UK.
Conway Morris, S.
1998 *The Crucible of Creation: The Burgess Shale and the Rise of Animals*. Oxford University Press, Oxford, UK.
Cook, D. C.
1979 Subsistence Base and Health in Prehistoric Illinois Valley: Evidence from the Human Skeleton. *Medical Anthropology* 3:109–124.
Cook, D. C., and J. E. Buikstra
1979 Health and Differential Survival in Prehistoric Populations: Prenatal Dental Defects. *American Journal of Physical Anthropology* 51:649–664.
Cotter, J. L.
1937 The Occurrence of Flints and Extinct Animals in Pluvial Deposits near Clovis, New Mexico. In Report on Excavation at the Gravel Pit, 1936 (Part 4). *Academy of Natural Sciences of Philadelphia, Proceedings* 89:1–16.
Courbin, P.
1962 *What Is Archaeology? An Essay on the Nature of Archaeological Research*. University of Chicago Press, Chicago.
Cowgill, G. L.
1963 Review of "A Quantitative Method for Deriving Cultural Chronology" by J. A. Ford. *American Anthropologist* 65:696–699.
1968 Review of "Computer Analysis of Chronological Seriation" by F. Hole and M. Shaw. *American Antiquity* 33:517–519.

1972 Models, Methods and Techniques for Seriation. In *Models in Archaeology*, edited by D. L. Clarke, pp. 381–424. Methuen, London.

1982 Clusters of Objects and Associations between Variables: Two Approaches to Archaeological Classification. In *Essays on Archaeological Typology*, edited by R. Whallon and J. A. Brown, pp. 30–55. Center for American Archeology Press, Evanston, IL.

Coyne, J. A., and B. Charlesworth

1997 On Punctuated Equilibria. *Science* 276:339–341.

Cracraft, J.

1979 Phylogenetic Analysis, Evolutionary Models, and Paleontology. In *Phylogenetic Analysis and Paleontology*, edited by J. Cracraft and N. Eldredge, pp. 7–39. Columbia University Press, New York.

1981a Pattern and Process in Paleobiology: The Role of Cladistic Analysis in Systematic Paleontology. *Paleobiology* 7:456–468.

1981b The Use of Functional and Adaptive Criteria in Phylogenetic Systematics. *American Zoologist* 21:21–36.

1982 A Non-Equilibrium Theory for the Rate-Control of Speciation and Extinction and the Origin of Macroevolutionary Patterns. *Systematic Zoology* 31:348–365.

1983 Species Concepts and Speciation Analysis. *Current Ornithology* 1:159–187.

1985 Species Selection, Macroevolutionary Analysis, and the "Hierarchical Theory" of Evolution. *Systematic Biology* 34:222–229.

1987 Species Concepts and the Ontology of Evolution. *Biology and Philosophy* 2:329–346.

1989a Speciation and Its Ontology: The Empirical Consequences of Alternative Species Concepts for Understanding Patterns and Processes of Differentiation. In *Speciation and Its Consequences*, edited by D. Otte and J. A. Endler, pp. 28–59. Sinauer, Sunderland, MA.

1989b Species as Entities of Biological Theory. In *What the Philosophy of Biology Is*, edited by M. Ruse, pp. 31–52. Kluwer, Dordrecht, Netherlands.

Crichton, M.

1995 *The Lost World*. Knopf, New York.

Crompton, A. W., and P. Parker

1978 Evolution of the Mammalian Masticatory Apparatus. *American Scientist* 66:192–201.

Cullen, B.

1993 The Darwinian Resurgence and the Cultural Virus Critique. *Cambridge Archaeological Journal* 3:179–202.

1996 Cultural Virus Theory and the Eusocial Pottery Assemblage. In *Darwinian Archaeologies*, edited by H. D. G. Maschner, pp. 43–59. Plenum Press, New York.

Curry, M., M. J. O'Brien, and M. K. Trimble

1985 The Classification of Pointed, Hafted Bifaces. In Archaeology of the Central Salt River Valley: An Overview of the Prehistoric Occupation, by M. J. O'Brien. *Missouri Archaeologist* 46:77–189.

Dall, W. H.

1877 On Succession in the Shell-Heaps of the Aleutian Islands. *Contributions to North American Ethnology* 1:41–91.

Daniel, G.

1975 *A Hundred and Fifty Years of Archaeology* (2nd ed.). Duckworth, London.

1981 *A Short History of Archaeology*. Thames and Hudson, London.

Darwin, C.

1859 *On the Origin of Species by Means of Natural Selection or the Preservation of Favored Races in the Struggle for Life*. Murray, London.

Darwin, F. (editor)

1893 *Charles Darwin, His Life Told in an Autobiographical Chapter and in a Selected Series of His Published Letters*. Appleton, New York.

Davis, D. D.
1981 Ceramic Classification and Temporal Discrimination: A Consideration of Later Prehistoric Stylistic Change in the Mississippi River Delta. *Mid-Continental Journal of Archaeology* 6:55–89.

Davis, J. I.
1996 Phylogenetics, Molecular Variation, and the Species Concept. *BioScience* 46:502–511.

Dawkins, R.
1976 *The Selfish Gene*. Oxford University Press, Oxford, UK.
1990 *The Extended Phenotype: The Long Reach of the Gene* (new ed.). Oxford University Press, Oxford, UK.

Deal, M., and M. B. Hagstrum
1995 Ceramic Reuse Behavior among the Maya and Wanka: Implications for Archaeology. In *Expanding Archaeology*, edited by J. M. Skibo, W. H. Walker, and A. E. Nielsen, pp. 111–125. University of Utah Press, Salt Lake City.

DeBoer, W. R., and D. W. Lathrap
1979 The Making and Breaking of Shipibo–Conibo Ceramics. In *Implications of Ethnography for Archaeology*, edited by C. Kramer, 102–138. Columbia University Press, New York.

Deetz, J.
1965 The Dynamics of Stylistic Change in Arikara Ceramics. *Illinois Studies in Anthropology* No. 4.
1968 The Inference of Residence and Descent Rules from Archaeological Data. In *New Perspectives in Archeology*, edited by S. R. Binford and L. R. Binford, pp. 41–48. Aldine, Chicago.
1970 Archeology as a Social Science. In Current Directions in Anthropology: A Special Issue, edited by A. Fischer, pp. 115–125. *American Anthropological Association, Bulletin* 3(3)(2).

Deetz, J., and E. Dethlefsen
1965 The Doppler Effect and Archaeology: A Consideration of the Spatial Aspects of Seriation. *Southwestern Journal of Anthropology* 21:196–206.
1971 Some Social Aspects of New England Colonial Mortuary Art. In Approaches to the Social Dimensions of Mortuary Practices, edited by J. A. Brown, pp. 30–38. *Society for American Archaeology, Memoirs* No. 25.

DeJarnette, D. L., and V. J. Knight, Jr.
1976 LaGrange. *Journal of Alabama Archaeology* 22:1–60.

Dempsey, P., and M. Baumhoff
1963 The Statistical Use of Artifact Distributions to Establish Chronological Sequence. *American Antiquity* 28:496–509.

Dennett, D. C.
1995 *Darwin's Dangerous Idea*. Simon and Schuster, New York.

Desmond, A.
1984 *Archetypes and Ancestors: Palaeontology in Victorian London, 1850–1875*. University of Chicago Press, Chicago.

Dethlefsen, E., and J. Deetz
1966 Death's Heads, Cherubs, and Willow Trees: Experimental Archaeology in Colonial Cemeteries. *American Antiquity* 31:502–510.

De Vries, H.
1901 *Die Mutationstheorie. Versuche und Beobachtungen über die Entstehung der Arten im Pflanzenreich* (Vol. 1). *Die Entstehung der Arten durch Mutation*. Veit, Leipzig. (English translation, J. B. Farmer and A. D. Darbishire. 1909. Open Court, Chicago)

Dingus, L., and P. M. Sadler
1982 The Effects of Stratigraphic Completeness on Estimates of Evolutionary Rates. *Systematic Zoology* 31:400–412.

Dixon, R. B.
1913 Some Aspects of North American Archeology. *American Anthropologist* 15:549–577.

Dobzhansky, T.
1935 A Critique of the Species Concept in Biology. *Philosophy of Science* 2:344–355.
1937a *Genetics and the Origin of Species*. Columbia University Press, New York.
1937b The Genetic Nature of Species Differences. *American Naturalist* 71:404–420.
1951 Mendelian Populations and Their Evolution. In *Genetics in the 20th Century*, edited by L. C. Dunn, pp. 573–589. Macmillan, New York.

Dole, G.
1973 Foundations of Contemporary Evolutionism. In *Main Currents in Cultural Anthropology*, edited by R. Naroll and F. Naroll, pp. 247–279. Appleton–Century–Crofts, New York.

Donoghue, M. J.
1992 Homology. In *Keywords in Evolutionary Biology*, edited by E. F. Keller and E. A. Lloyd, pp. 170–179. Harvard University Press, Cambridge, MA.

Donoghue, M. J., J. A. Doyle, J. Gauthier, A. G. Kluge, and T. Rowe
1989 The Importance of Fossils in Phylogeny Reconstruction. *Annual Review of Ecology and Systematics* 20:431–460.

Doran, J. E., and F. R. Hodson
1975 *Mathematics and Computers in Archaeology*. Harvard University Press, Cambridge, MA.

Driver, H. E., and A. L. Kroeber
1932 Quantitative Expression of Cultural Relationships. *University of California, Publications in American Archaeology and Ethnology* 31:211–256.

Dumond, D. E.
1977 Science in Archaeology: The Saints Go Marching In. *American Antiquity* 42:330–349.

Dunnell, R. C.
1970 Seriation Method and Its Evaluation. *American Antiquity* 35:305–319.
1971a *Systematics in Prehistory*. Free Press, New York.
1971b *Anthropological and Scientific Models of Function in Archaeology*. Paper presented at the 70th Annual Meeting of the American Anthropological Association, New York.
1978a *Natural Selection, Scale, and Cultural Evolution: Some Preliminary Considerations*. Paper presented at the 77th Annual Meeting of the American Anthropological Association, Los Angeles. [published in O'Brien 1996b]
1978b Archaeological Potential of Anthropological and Scientific Models of Function. In *Archaeological Essays in Honor of Irving B. Rouse*, edited by R. C. Dunnell and E. S. Hall, Jr., pp. 41–73. Mouton, The Hague.
1978c Style and Function: A Fundamental Dichotomy. *American Antiquity* 43:192–202.
1980 Evolutionary Theory and Archaeology. In *Advances in Archaeological Method and Theory*, Vol. 3, edited by M. B. Schiffer, pp. 35–99. Academic Press, New York.
1981 Seriation, Groups, and Measurements. In *Manejo de Datos y Metodos Matematicos de Arqueologia*, compiled by G. L. Cowgill, R. Whallon, and B. S. Ottaway, pp. 67–90. Union Internacional de Ciencias Prehistoricas y Protohistoricas, Mexico D.F.
1982 Science, Social Science, and Common Sense: The Agonizing Dilemma of Modern Archaeology. *Journal of Anthropological Research* 38:1–25.
1985a Archaeological Survey in the Lower Mississippi Alluvial Valley, 1940–1947: A Landmark Study in American Archaeology. *American Antiquity* 50:297–300.
1985b Methodological Issues in Contemporary Americanist Archaeology. In *Proceedings of the 1984 Biennial Meeting of the Philosophy of Science Association*, Vol. 2, edited by P. D. Asquith and P. Kitcher, pp. 717–744.
1986 Methodological Issues in Americanist Artifact Classification. In *Advances in Archaeological Method and Theory*, Vol. 9, edited by M. B. Schiffer, pp. 149–207. Academic Press, Orlando, FL.
1988a *Archaeology and Evolutionary Theory*. Paper presented at the University of Missouri–Columbia.

1988b The Concept of Progress in Cultural Evolution. In *Evolutionary Progress*, edited by M. H. Nitecki, pp. 169–194. University of Chicago Press, Chicago.
1989 Aspects of the Application of Evolutionary Theory in Archaeology. In *Archaeological Thought in America*, edited by C. C. Lamberg-Karlovsky, pp. 35–49. Cambridge University Press, Cambridge, UK.
1992a Archaeology and Evolutionary Science. In Quandaries and Quests: Visions of Archaeology's Future, edited by L. Wandsnider, pp. 209–224. *Southern Illinois University at Carbondale, Center for Archaeological Investigations, Occasional Paper* No. 20.
1992b Is a Scientific Archaeology Possible? In *Metaarchaeology*, edited by L. Embree, pp. 75–97. Kluwer, Dordrecht, Netherlands.
1994 *Problems in the Application of Evolutionary Theory in Archaeology*. Paper presented at Arizona State University, Tempe.
1995 What Is It That Actually Evolves? In *Evolutionary Archaeology: Methodological Issues*, edited by P. A. Teltser, pp. 33–50. University of Arizona Press, Tucson.
1996 Foreword. In *Evolutionary Archaeology: Theory and Application*, edited by M. J. O'Brien, pp. vii–xii. University of Utah Press, Salt Lake City.

Dunnell, R. C., and J. K. Feathers
1991 Late Woodland Manifestations of the Malden Plain, Southeast Missouri. In *Stability, Transformation, and Variation: The Late Woodland Southeast*, edited by M. S. Nassaney and C. R. Cobb, pp. 21–45. Plenum Press, New York.

Dunnell, R. C., and R. J. Wenke
1980a Cultural and Scientific Evolution: Some Comments on "The Decline and Rise of Mesopotamian Civilization." *American Antiquity* 45:605–609.
1980b If You've Nothing Better to Do, Honk. *American Antiquity* 45:612–613.

Durham, J. W.
1967 The Incompleteness of Our Knowledge of the Fossil Record. *Journal of Paleontology* 41:559–565.

Durham, W. H.
1991 *Coevolution: Genes, Culture, and Human Diversity*. Stanford University Press, Palo Alto, CA.

Edson, M. M., T. C. Foin, and C. M. Knapp
1981 "Emergent Properties" and Ecological Research. *American Naturalist* 118:593–596.

Ehrich, R. W.
1963 Further Reflections on Archaeological Interpretation. *American Anthropologist* 65:16–31.
1965 Ceramics and Man: A Cultural Perspective. In Ceramics and Man, edited by F. R. Matson, pp. 1–19. *Viking Fund Publications in Anthropology* No. 41.

Eldredge, N.
1971 The Allopatric Model and Phylogeny in Paleozoic Invertebrates. *Evolution* 25:156–167.
1972 Systematics and Evolution of *Phacops rana* (Green, 1832) and *Phacops iowensis* Delo, 1935 (Trilobita) in the Middle Devonian of North America. *American Museum of Natural History, Bulletin* 47:45–114.
1974 Testing Evolutionary Hypotheses in Paleontology: A Comment on Makurath and Anderson (1973). *Evolution* 28:479–481.
1979 Alternative Approaches to Evolutionary Theory. In Models and Methodologies in Evolutionary Theory, edited by J. H. Schwartz and H. B. Rollins, pp. 7–19. *Carnegie Museum of Natural History, Bulletin* 13.
1982 Phenomenological Levels and Evolutionary Rates. *Systematic Zoology* 31:338–347.
1985 *Unfinished Synthesis: Biological Hierarchies and Modern Evolutionary Thought*. Oxford University Press, Oxford, UK.
1989 Punctuated Equilibria, Rates of Change, and Large-Scale Entities in Evolutionary Systems. *Journal of Biological and Social Structures* 12:173–184.

1990 Hierarchy and Macroevolution. In *Paleobiology: A Synthesis*, edited by D. E. G. Briggs and P. R. Crowther, pp. 124–129. Blackwell, Oxford, UK.

1992 Marjorie Grene, "Two Evolutionary Theories" and Modern Evolutionary Theory. *Synthese* 92:135–149.

1995a *Reinventing Darwin: The Great Debate at the High Table of Evolutionary Theory*. Wiley, New York.

1995b Species, Speciation, and the Context of Adaptive Change in Evolution. In *New Approaches to Speciation in the Fossil Record*, edited by D. H. Erwin and R. L. Anstey, pp. 39–63. Columbia University Press, New York.

1996 Hierarchies in Macroevolution. In *Evolutionary Paleobiology*, edited by D. Jablonski, D. H. Erwin, and J. H. Lipps, pp. 42–61. University of Chicago Press, Chicago.

1999 *The Pattern of Evolution*. Freeman, New York.

Eldredge, N., and S. J. Gould

1972 Punctuated Equilibria: An Alternative to Phyletic Gradualism. In *Models in Paleobiology*, edited by T. J. M. Schopf, pp. 82–115. Freeman, Cooper, San Francisco.

1974 Reply to M. K. Hecht (1974). *Evolutionary Biology* 7:303–308.

1977 Evolutionary Models and Biostratigraphic Strategies. In *Concepts and Methods of Biostratigraphy*, edited by E. G. Kauffman and J. E. Hazel, pp. 25–40. Dowden, Hutchinson and Ross, Stroudsburg, PA.

1988 Punctuated Equilibrium Prevails. *Nature* 332:211–212.

1997 On Punctuated Equilibria. *Science* 276:338–339.

Eldredge, N., and M. J. Novacek

1985 Systematics and Paleobiology. *Paleobiology* 11:65–74.

Endler, J. A.

1986a *Natural Selection in the Wild*. Princeton University Press, Princeton, NJ.

1986b The Newer Synthesis? Some Conceptual Problems in Evolutionary Biology. *Oxford Surveys in Evolutionary Biology* 3:224–243.

1998 The Place of Hybridization in Evolution. *Evolution* 52:640–644.

Endler, J. A., and T. McLellan

1988 The Processes of Evolution: Toward a Newer Synthesis. *Annual Review of Ecology and Systematics* 19:395–421.

Ereshefsky, M.

1989 Where's the Species? Comments on the Phylogenetic Species Concepts. *Biology and Philosophy* 4:89–96.

1992 The Historical Nature of Evolutionary Theory. In *History and Evolution*, edited by M. H. Nitecki and D. V. Nitecki, pp. 81–99. State University of New York Press, Albany.

1994 Some Problems with the Linnaean Hierarchy. *Philosophy of Science* 61:186–205.

Evans, J.

1850 On the Date of British Coins. *The Numismatic Chronicle and Journal of the Numismatic Society* 12(4):127–137.

1875 On the Coinage of the Ancient Britons and Natural Selection. *Royal Institution of Great Britain, Proceedings* 7:24–32.

Fiedel, S. J.

1999 Older Than We Thought: Implications of Corrected Dates for Paleoindians. *American Antiquity* 64:95–115.

Figgins, J. D.

1927 The Antiquity of Man in America. *Natural History* 27:229–239.

Fisher, A. K.

1997 Origins of the Midwestern Taxonomic Method. *Mid-Continental Journal of Archaeology* 22:117–122.

Fisher, D. C.

1994 Stratocladistics: Morphological and Temporal Patterns and Their Relation to Phylogenetic

Process. In *Interpreting the Hierarchy of Nature*, edited by L. Grande and O. Rieppel, pp. 133–171. Academic Press, San Diego.

Fisher, R. A.
1930 *The Genetical Theory of Natural Selection*. Clarendon, Oxford, UK.
1936 Has Mendel's Work Been Rediscovered? *Annals of Science* 1:115–137.

Flannery, K. V.
1967 Culture History v. Cultural Process: A Debate in American Archaeology. *Scientific American* 217(2):119–122.
1972 The Cultural Evolution of Civilizations. *Annual Review of Ecology and Systematics* 3:399–426.
1973 The Origins of Agriculture. *Annual Review of Anthropology* 2:271–310.
1986 Adaptation, Evolution, and Archaeological Phases: Some Implications of Reynolds' Simulation. In *Guilá Naquitz: Archaic Foraging and Early Agriculture in Oaxaca, Mexico*, edited by K. V. Flannery, pp. 501–507. Academic Press, Orlando, FL.

Flenniken, J. J., and P. J. Wilke
1989 Typology, Technology, and Chronology of Great Basin Dart Points. *American Anthropologist* 91:149–185.

Foote, M.
1991 Morphological and Taxonomic Diversity in a Clade's History: The Blastoid Record and Stochastic Simulations. *University of Michigan, Contributions from the Museum of Paleontology* 28:101–140.
1993a Contributions of Individual Taxa to Overall Morphological Disparity. *Paleobiology* 19:403–419.
1993b Discordance and Concordance between Morphological and Taxonomic Diversity. *Paleobiology* 19:185–205.
1994 Morphological Disparity in Ordovician–Devonian Crinoids and the Early Saturation of Morphological Space. *Paleobiology* 20:320–344.
1996 Perspective: Evolutionary Patterns in the Fossil Record. *Evolution* 50:1–11.
1997a Estimating Taxonomic Durations and Preservation Probability. *Paleobiology* 23:278–300.
1997b Sampling, Taxonomic Description, and Our Evolving Knowledge of Morphological Diversity. *Paleobiology* 23:181–206.

Foote, M., and D. M. Raup
1996 Fossil Preservation and the Stratigraphic Ranges of Taxa. *Paleobiology* 22:121–140.

Ford, J. A.
1935a An Introduction to Louisiana Archaeology. *Louisiana Conservation Review* 4(5):8–11.
1935b Outline of Louisiana and Mississippi Pottery Horizons. *Louisiana Conservation Review* 4(6):33–38.
1935c Ceramic Decoration Sequence at an Old Indian Village Site near Sicily Island, Louisiana. *Louisiana Department of Conservation, Anthropological Study* No. 1.
1936a Analysis of Indian Village Site Collections from Louisiana and Mississippi. *Louisiana Department of Conservation, Anthropological Study* No. 2.
1936b Archaeological Methods Applicable to Louisiana. *Proceedings of the Louisiana Academy of Sciences* 3:102–105.
1938 An Examination of Some Theories and Methods of Ceramic Analysis. M.A. thesis, University of Michigan.
1949 Cultural Dating of Prehistoric Sites in Virú Valley, Peru. *American Museum of Natural History, Anthropological Papers* 43(1):29–89.
1951 Greenhouse: A Troyville–Coles Creek Period Site in Avoyelles Parish, Louisiana. *American Museum of Natural History, Anthropological Papers* 44(1):1–132.
1952 Measurements of Some Prehistoric Design Developments in the Southeastern States. *American Museum of Natural History, Anthropological Papers* 44(3):313–384.
1954a Comment on A. C. Spaulding's "Statistical Techniques for the Discovery of Artifact Types." *American Antiquity* 19:390–391.

1954b Spaulding's Review of Ford. *American Anthropologist* 56:109–112.
1954c On the Concept of Types: The Type Concept Revisited. *American Anthropologist* 56:42–53.
1962 A Quantitative Method for Deriving Cultural Chronology. *Pan American Union, Technical Bulletin* No. 1.

Ford, J. A., and Griffin, J. B.
1937 [A proposal for a] Conference on Pottery Nomenclature for the Southeastern United States. Mimeographed. [reprinted in *Newsletter of the Southeastern Archaeological Conference* 7(1):5–9]
1938 Report of the Conference on Southeastern Pottery Typology. Mimeographed. [reprinted in *Newsletter of the Southeastern Archaeological Conference* 7(1):10–22, and in Lyman et al. 1997a (editors)].

Ford, J. A., Phillips, P., and Haag, W. G.
1955 The Jaketown Site in West-Central Mississippi. *American Museum of Natural History, Anthropological Papers* 46(1).

Ford, J. A., and Quimby, G. I., Jr.
1945 The Tchefuncte Culture, an Early Occupation of the Lower Mississippi Valley. *Society for American Archaeology, Memoirs* No. 2.

Ford, J. A., and Webb, C. H.
1956 Poverty Point, a Late Archaic Site in Louisiana. *American Museum of Natural History, Anthropological Papers* 46(1).

Forey, P. L.
1990 Cladistics. In *Palaeobiology: A Synthesis*, edited by D. E. G. Briggs and P. R. Crowther, pp. 430–434. Blackwell, Oxford, UK.
1992 Fossils and Cladistic Analysis. In *Cladistics: A Practical Course in Systematics*, edited by P. L. Forey, C. J. Humphries, I. J. Kitching, R. W. Scotland, D. J. Siebert, and D. M. Williams, pp. 124–136. Clarendon Press, Oxford, UK.

Fortey, R. A.
1985 Gradualism and Punctuated Equilibria as Competing and Complementary Theories. In Evolutionary Case Histories from the Fossil Record, edited by J. C. W. Cope and P. W. Skelton, pp. 17–28. *Palaeontological Association, Special Papers in Palaeontology* No. 33.
1988 Seeing is Believing: Gradualism and Punctuated Equilibria in the Fossil Record. *Science Progress* 72:1–19.

Fowke, G.
1896 Stone Art. *Bureau of American Ethnology, Annual Report* 13:57–178.
1922 Archaeological Investigations. *Bureau of American Ethnology, Bulletin* 76.

Fox, G. L.
1992 *A Critical Evaluation of the Interpretive Framework of the Mississippi Period in Southeast Missouri*. Ph.D. dissertation, Department of Anthropology, University of Missouri–Columbia.
1998 An Examination of Mississippian-Period Phases in Southeastern Missouri. In *Changing Perspectives on the Archaeology of the Central Mississippi Valley*, edited by M. J. O'Brien and R. C. Dunnell, pp. 31–58. University of Alabama Press, Tuscaloosa.

Fox, R. C.
1986 Species in Paleontology. *Geoscience Canada* 13:73–84.

Frison, G. C.
1968 A Functional Analysis of Certain Chipped Stone Tools. *American Antiquity* 33: 149–155.
1991 *Prehistoric Hunters of the High Plains* (2nd ed.). Academic Press, New York.

Frison, G. C., C. V. Haynes, Jr., and M. L. Larson
1996 Discussion and Conclusions. In *The Mill Iron Site*, edited by G. Frison, pp. 205–216. University of New Mexico Press, Albuquerque.

Frison, G. C., and L. C. Todd
1986 *The Colby Mammoth Site: Taphonomy and Archaeology of a Clovis Kill in Northern Wyoming*. University of New Mexico Press, Albuquerque.
Fritz, J. M.
1972 Archaeological Systems from Indirect Observation of the Past. In *Contemporary Archaeology: A Guide to Theory and Contributions*, edited by M. P. Leone, pp. 135–157. Southern Illinois University Press, Carbondale.
Fritz, J. M., and F. T. Plog
1970 The Nature of Archaeological Explanation. *American Antiquity* 35:405–412.
Fryer, G., P. H. Greenwood, and J. F. Peake
1983 Punctuated Equilibria, Morphological Stasis and the Palaeontological Documentation of Speciation: A Biological Appraisal of a Case History in an African Lake. *Biological Journal of the Linnean Society* 20:195–205.
1985 The Demonstration of Speciation in Fossil Molluscs and Living Fishes. *Biological Journal of the Linnean Society* 26:325–336.
Futuyma, D. J.
1986 *Evolutionary Biology* (2nd ed.). Sinauer, Sunderland, MA.
Futuyma, D. J., R. C. Lewontin, G. C. Mayer, J. Seger, and J. W. Stubblefield, III
1981 Macroevolution Conference. *Science* 211:770.
Gallagher, M.
1978 Anthropology. *Plateau* 50(4):18–23.
Gamio, M.
1913 Arqueologia de Atzcapotzalco, D.F., Mexico. *Eighteenth International Congress of Americanists, Proceedings*, pp. 180–187.
Gans, C.
1979 Momentarily Excessive Construction as the Basis for Protoadaptation. *Evolution* 33:227–233.
1988 Adaptation and the Form–Function Relation. *American Zoologist* 28:681–697.
1993 On the Merits of Adequacy. *American Journal of Science* 293A:391–406.
Ghiselin, M. T.
1966 On Psychologism in the Logic of Taxonomic Controversies. *Systematic Zoology* 15:207–215.
1974a A Radical Solution to the Species Problem. *Systematic Zoology* 25:536–544.
1974b *The Economy of Nature and the Evolution of Sex*. University of California Press, Berkeley.
1981 Categories, Life and Thinking. *Behavioral and Brain Sciences* 4:269–283.
1987 Species Concepts, Individuality, and Objectivity. *Biology and Philosophy* 2:127–143.
Gibbon, G.
1984 *Anthropological Archaeology*. Columbia University Press, New York.
1989 *Explanation in Archaeology*. Blackwell, Oxford, UK.
Gifford, J. C.
1960 The Type–Variety Method of Ceramic Classification as an Indicator of Cultural Phenomena. *American Antiquity* 25:341–347.
Gilinsky, N. L., and R. K. Bambach
1986 The Evolutionary Bootstrap: A New Approach to the Study of Taxonomic Diversity. *Paleobiology* 12:251–268.
Gilinsky, N. L., S. J. Gould, and R. Z. German
1989 Asymmetries of Clade Shape and the Direction of Evolutionary Time. *Science* 243:1613–1614.
Gingerich, P. D.
1976a Cranial Anatomy and Evolution of Early Tertiary Plesiadapidae. *University of Michigan Papers in Paleontology* 15:1–140.
1976b Paleontology and Phylogeny: Patterns of Evolution at the Species Level in Early Tertiary Mammals. *American Journal of Science* 276:1–28.

1979a Paleontology, Phylogeny, and Classification: An Example from the Mammalian Fossil Record. *Systematic Zoology* 28:451–464.

1979b The Stratophenetic Approach to Phylogeny Reconstruction in Vertebrate Paleontology. In *Phylogenetic Analysis and Paleontology*, edited by J. Cracraft and N. Eldredge, pp. 41–77. Columbia University Press, New York.

1980 Evolutionary Patterns in Early Cenozoic Mammals. *Annual Review of Earth and Planetary Sciences* 8:407–424.

1983 Rates of Evolution: Effects of Time and Temporal Setting. *Science* 222:159–161.

1984a Darwin's Gradualism and Empiricism. *Nature* 309:116.

1984b Punctuated Equilibria—Where Is the Evidence? *Systematic Zoology* 33:335–338.

1984c Reply to Gould 1984. *Science* 226:995–996.

1985 Species in the Fossil Record: Concepts, Trends, and Transitions. *Paleobiology* 11:27–41.

Gingerich, P. D., and E. L. Simons

1977 Systematics, Phylogeny, and Evolution of Early Eocene Adapidae (Mammalia, Primates) in North America. *University of Michigan, Museum of Paleontology, Contributions* 24:245–279.

Ginzburg, L. R., and J. D. Rost

1982 Are "Punctuations" Artifacts of Time-Scales? *Nature* 296:610–611.

Givens, D. R.

1992 *Alfred Vincent Kidder and the Development of Americanist Archaeology*. University of New Mexico Press, Albuquerque.

Gladwin, H. S.

1936 Editorials: Methodology in the Southwest. *American Antiquity* 1:256–259.

1937 Conclusions. In Excavations at Snaketown, I: Material Culture, by H. S. Gladwin, E. W. Haury, E. B. Sayles, and N. Gladwin, pp. 247–269. *Medallion Papers* No. 25.

Gladwin, W., and H. S. Gladwin

1930 A Method for the Designation of Southwestern Pottery Types. *Medallion Papers* No. 7.

1934 A Method for the Designation of Cultures and Their Variations. *Medallion Papers* No. 15.

Gleason, H. A.

1923 Review of "Age and Area, a Study in Geographical Distribution and Origin of Species" by J. C. Willis. *Ecology* 4:196–201.

Goldschmidt, R.

1940 *The Material Basis of Evolution*. Yale University Press, New Haven, CT.

Goodfriend, G. A., and S. J. Gould

1996 Paleontology and Chronology of Two Evolutionary Transitions by Hybridization in the Bahamian Land Snail *Cerion*. *Science* 274:1894–1897.

Goodyear, A. C.

1977 Review of "Behavioral Archeology" by M. B. Schiffer. *American Antiquity* 42:668–671.

1982 The Chronological Position of the Dalton Horizon in the Southeastern United States. *American Antiquity* 47:382–395.

Gould, R. A.

1978a From Tasmania to Tucson: New Directions in Ethnoarchaeology. In *Explorations in Ethnoarchaeology*, edited by R. A. Gould, pp. 1–10. University of New Mexico Press, Albuquerque.

1978b Beyond Analogy in Ethnoarchaeology. In *Explorations in Ethnoarchaeology*, edited by R. A. Gould, pp. 249–293. University of New Mexico Press, Albuquerque.

1980a *Living Archaeology*. Cambridge University Press, New York.

1980b Review of *Ethnoarchaeology: Implications of Ethnography for Archaeology*, edited by C. Cramer. *Archaeology* 33:65–66.

1985 The Empiricist Strikes Back. *American Antiquity* 50:639–644.

Gould, R. A., and P. J. Watson

1982 A Dialogue on the Meaning and Use of Analogy in Ethnoarchaeological Reasoning. *Journal of Anthropological Archaeology* 1:355–381.

Gould, S. J.
1970 Evolutionary Paleontology and the Science of Form. *Earth-Science Reviews* 6:77–119.
1976 The Genomic Metronome as a Null Hypothesis. *Paleobiology* 2:177–179.
1977 The Return of Hopeful Monsters. *Natural History* 86(6):22–30.
1980a Is a New and General Theory of Evolution Emerging? *Paleobiology* 6:119–130.
1980b G. G. Simpson, Paleontology, and the Modern Synthesis. In *The Evolutionary Synthesis: Perspectives on the Unification of Biology*, edited by E. Mayr and W. B. Provine, pp. 153–172. Harvard University Press, Cambridge, MA.
1980c The Evolutionary Biology of Constraint. *Daedalus* 109(2):39–52.
1982a Darwinism and the Expansion of Evolutionary Theory. *Science* 216:380–387.
1982b The Meaning of Punctuated Equilibrium and Its Role in Validating a Hierarchical Approach to Macroevolution. In *Perspectives in Evolution*, edited by R. Milkman, pp. 83–104. Sinauer, Sunderland, MA.
1982c Punctuated Equilibrium—A Different Way of Seeing. *New Scientist* 94:137–141.
1983a Punctuated Equilibrium and the Fossil Record. *Science* 219:439–440.
1983b The Hardening of the Modern Synthesis. In *Dimensions of Darwinism*, edited by M. Grene, pp. 71–93. Cambridge University Press, Cambridge, UK.
1983c Irrelevance, Submission, and Partnership: The Changing Role of Palaeontology in Darwin's Three Centennials, and a Modest Proposal for Macroevolution. In *Evolution from Molecules to Men*, edited by D. S. Bendall, pp. 347–366. Cambridge University Press, Cambridge, UK.
1984a Smooth Curve of Evolutionary Rates: A Psychological and Mathematical Artifact. *Science* 226:994–995.
1984b Toward the Vindication of Punctuational Change. In *Catastrophes and Earth History: The New Uniformitarianism*, edited by W. A. Berggren and J. A. VanCouvering, pp. 9–34. Princeton University Press, Princeton, NJ.
1985 The Paradox of the First Tier: An Agenda for Paleobiology *Paleobiology* 11:2–12.
1986 Evolution and the Triumph of Homology, or Why History Matters. *American Scientist* 74:60–69.
1987a The Panda's Thumb of Technology. *Natural History* 96(1):14–23.
1987b *Time's Arrow, Time's Cycle: Myth and Metaphor in the Discovery of Geological Time*. Harvard University Press, Cambridge, MA.
1989a *Wonderful Life: The Burgess Shale and the Nature of History*. Norton, New York.
1989b Punctuated Equilibrium in Fact and Theory. *Journal of Biological and Social Structures* 12:117–136.
1989c A Developmental Constraint in *Cerion*, with Comments on the Definition and Interpretation of Constraint in Evolution. *Evolution* 43:516–539.
1990 Speciation and Sorting as the Source of Evolutionary Trends, or "Things are Seldom What They Seem." In *Evolutionary Trends*, edited by K. J. McNamara, pp. 3–27. University of Arizona Press, Tucson.
1991a Exaptation: A Crucial Term for an Evolutionary Psychology. *Journal of Social Issues* 47(3):43–65.
1991b The Disparity of the Burgess Shale Arthropod Fauna and the Limits of Cladistic Analysis: Why We Must Strive to Quantify Morphospace. *Paleobiology* 17:411–423.
1993 How to Analyze the Burgess Shale Disparity—A Reply to Ridley. *Paleobiology* 19: 522–523.
1995a A Task for Paleobiology at the Threshold of the Majority. *Paleobiology* 21:1–14.
1995b Tempo and Mode in the Macroevolutionary Reconstruction of Darwinism. In *Tempo and Mode in Evolution*, edited by W. M. Fitch and F. J. Ayala, pp. 125–144. National Academy Press, Washington, DC.
1996 *Full House: The Spread of Excellence from Plato to Darwin*. Harmony, New York.
1997a A Tale of Two Worksites. *Natural History* 106(9):18–22, 29, 62–68.

1997b The Exaptive Excellence of Spandrels as a Term and Prototype. *National Academy of Sciences, Proceedings* 94:10750–10755.

1997c The Paradox of the Visibly Irrelevant. *Natural History* 106(11):12–18, 60–66.

Gould, S. J., and N. Eldredge

1977 Punctuated Equilibria: The Tempo and Mode of Evolution Reconsidered. *Paleobiology* 3:115–151.

1983 Darwin's Gradualism. *Systemic Zoology* 32:444–445.

1986 Punctuated Equilibrium at the Third Stage. *Systematic Zoology* 35:143–148.

1988a Punctuated Equilibrium Prevails. *Nature* 332:211–212.

1988b Species Selection: Its Range and Power. *Nature* 334:19.

1993 Punctuated Equilibrium Comes of Age. *Nature* 366:223–227.

Gould, S. J., N. L. Gilinsky, and R. Z. German

1987 Asymmetry of Lineages and the Direction of Evolutionary Time. *Science* 236:1437–1441.

Gould, S. J., and R. C. Lewontin

1979 The Spandrels of San Marco and the Panglossian Paradigm: A Critique of the Adaptationist Programme. *Proceedings of the Royal Society of London Series B* 205:581–598.

Gould, S. J., D. M. Raup, J. J. Sepkoski, Jr., T. J. M. Schopf, and D. S. Simberloff

1977 The Shape of Evolution: A Comparison of Real and Random Clades. *Paleobiology* 3:23–40.

Gould, S. J., and E. S. Vrba

1982 Exaptation—A Missing Term in the Science of Form. *Paleobiology* 8:4–15.

Graham, R. W., and M. A. Graham

1994 Late Quaternary Distribution of *Martes* in North America. In *Martens, Sables, and Fishers: Biology and Conservation*, edited by S. W. Buskirk, A. S. Harestad, M. G. Raphael, and R. A. Powell, pp. 26–58. Cornell University Press, Ithaca, NY.

Grantham, T. A.

1995 Hierarchical Approaches to Macroevolution: Recent Work on Species Selection and the "Effect Hypothesis." *Annual Review of Ecology and Systematics* 26:301–321.

Grayson, D. K.

1993 *The Desert's Past: A Natural Prehistory of the Great Basin*. Smithsonian Institution Press, Washington, DC.

Gumerman, G. J., and D. A. Phillips, Jr.

1978 Archaeology beyond Anthropology. *American Antiquity* 43:184–191.

Haag, W. G.

1959 The Status of Evolutionary Theory in American History. In *Evolution and Anthropology: A Centennial Appraisal*, edited by B. J. Meggers, pp. 90–105. Anthropological Society of Washington, Washington, DC.

1961 Review of "The Evolution of Man: Man, Culture and Society," edited by S. Tax. *American Antiquity* 26:439–441.

Haas, O., and G. G. Simpson

1946 Analysis of Some Phylogenetic Terms, with Attempts at Redefinition. *American Philosophical Society, Proceedings* 90:319–349.

Haldane, J. B. S.

1932 *The Causes of Evolution*. Longmans, Green, New York.

Hall, B. K.

1996 *Baupläne*, Phylotypic Stages, and Constraint: Why There Are So Few Types of Animals. *Evolutionary Biology* 29:251–261.

Hall, E. R.

1926 A New Marten from the Pleistocene Cave Deposits of California. *Journal of Mammalogy* 7:127–130.

1936 Mustelid Mammals from the Pleistocene of North America with Systematic Notes on Some Recent Members of the Genera *Mustela*, *Taxidea* and *Mephitis*. *Carnegie Institute of Washington, Publication* 73:41–119.

Hancock, J. M.
1977 The Historic Development of Concepts of Biostratigraphic Correlation. In *Concepts and Methods of Biostratigraphy*, edited by E. G. Kauffman and J. E. Hazel, pp. 3–22. Dowden, Hutchinson and Ross, Stroudsburg, PA.

Hargrave, M. L.
1981 *Woodland Ceramic Chronometry and Occupational Intensity at the Carrier Mills Archaeological District, Saline County, Illinois*. M.A. thesis, Department of Anthropology, University of Southern Illinois, Carbondale.

Hargrave, M. L., and D. P. Braun
1981 Chronology of Mechanical Performance Characteristics of Woodland Ceramics: Methods, Results, Applications. Paper presented at the 46th Annual Meeting of the Society for American Archaeology, San Diego.

Harper, C. W., Jr.
1975 Origin of Species in Geologic Time: Alternatives to the Eldredge–Gould Model. *Science* 190:47–48.
1976 Stability of Species in Geologic Time. *Science* 192:269.

Harrington, M. R.
1909a Ancient Shell Heaps near New York City. In The Indians of Greater New York and the Lower Hudson, edited by C. Wissler. *American Museum of Natural History, Anthropological Papers* 3:167–179.
1909b The Rock-Shelters of Armonk, New York. In The Indians of Greater New York and the Lower Hudson, edited by C. Wissler. *American Museum of Natural History, Anthropological Papers* 3:123–138.

Harris, M.
1979 *Cultural Materialism: The Struggle for a Science of Culture*. Random House, New York.

Haslam, N. O.
1998 Natural Kinds, Human Kinds, and Essentialism. *Social Research* 65:291–314.

Hayden, B. (editor)
1979 *Lithic Use-Wear Analysis*. Academic Press, New York.

Haynes, C. V.
1991 Geoarchaeological and Paleohydrological Evidence for a Clovis-Age Drought in North America and Its Bearing on Extinction. *Quaternary Research* 35:438–450.

Hazard, E. B.
1998 Teaching about "Intermediate Forms." *American Biology Teacher* 60:359–361.

Hegmon, M.
1992 Archaeological Research on Style. *Annual Review of Anthropology* 21:517–536.

Hempel, C. G.
1965 *Aspects of Scientific Explanation, and Other Essays in the Philosophy of Science*. Free Press, New York.

Herschel, J.
1850 Quetelet on Probabilities. *Edinburgh Review* (July).

Heyes, C. M., and H. C. Plotkin
1989 Replicators and Interactors in Cultural Evolution. In *What the Philosophy of Biology Is*, edited by M. Ruse, pp. 139–162. Kluwer, Dordrecht, Netherlands.

Hill, J. N.
1966 A Prehistoric Community in Eastern Arizona. *Southwestern Journal of Anthropology* 22:9–30.
1970 Broken K Pueblo: Prehistoric Social Organization in the American Southwest. *University of Arizona, Anthropological Papers* No. 18.
1972 The Methodological Debate in Contemporary Archaeology: A Model. In *Models in Archaeology*, edited by D. L. Clarke, pp. 61–107. Methuen, London.
1985 Style: A Conceptual Evolutionary Framework. In *Decoding Prehistoric Ceramics*, edited by B. A. Nelson, pp. 362–385. Southern Illinois University Press, Carbondale.

Hilts, V. L.
1973 Statistics and Social Science. In *Foundations of Scientific Method: The Nineteenth Century*, edited by R. N. Giere and R. S. Westfall, pp. 206–233. Indiana University Press, Bloomington.

Hoard, R. J.
1992 Woodland-Period Cooking Vessels from Missouri. Ph.D. dissertation, Department of Anthropology, University of Missouri–Columbia.

Hodson, F. R., P. H. A. Sneath, and J. E. Doran
1966 Some Experiments in the Numerical Analysis of Archaeological Data. *Biometrika* 53: 311–324.

Hoffman, A.
1982 Punctuated versus Gradual Mode of Evolution: A Reconsideration. *Evolutionary Biology* 15:411–436.

Hole, F.
1979 Rediscovering the Past in the Present: Ethnoarchaeology in Luristan, Iran. In *Ethnoarchaeology: Implications of Ethnography for Archaeology*, edited by C. Cramer, pp. 192–218. Columbia University Press, New York.

Hole, F., and M. Shaw
1967 Computer Analysis of Chronological Seriation. *Rice University Studies* 53(3).

Holland, T. D.
1989 Fertility in the Prehistoric Midwest: A Critique of Unifactorial Models. *American Antiquity* 54:389–426.

Holmes, W. H.
1886a Pottery of the Ancient Pueblos. *Bureau of Ethnology, Annual Report* 4:257–360.
1886b Ancient Pottery of the Mississippi Valley. *Bureau of Ethnology, Annual Report* 4:361–436.
1886c Origin and Development of Form and Ornamentation in Ceramic Art. *Bureau of Ethnology, Annual Report* 4:437–465.
1892 Modern Quarry Refuse and the Paleolithic Theory. *Science* 20:295–297.
1897 Stone Implements of the Potomac–Chesapeake Tidewater Province. *Bureau of American Ethnology, Annual Report* 15:13–152.
1903 Aboriginal Pottery of the Eastern United States. *Bureau of American Ethnology, Annual Report* 20:1–201.

Howard, C. D.
1995 Projectile Point and Hafting Design Review. *North American Archaeologist* 16:291–301.

Hubbs, C. L.
1944 Concepts of Homology and Analogy. *American Naturalist* 78:289–307.

Huelsenbeck, J. P., and B. Rannala
1997 Maximum Likelihood Estimation of Phylogeny Using Stratigraphic Data. *Paleobiology* 23:174–180.

Hughes, N. C., and C. C. Labandeira
1995 The Stability of Species in Taxonomy. *Paleobiology* 21:401–403.

Hughes, S. S.
1998 Getting to the Point: Evolutionary Change in Prehistoric Weaponry. *Journal of Archaeological Method and Theory* 5:345–408.

Hull, D.
1965 The Effect of Essentialism on Taxonomy: 2000 Years of Stasis. *British Journal for the Philosophy of Science* 15:314–326; 16:1–18.
1970 Contemporary Systematic Philosophies. *Annual Review of Ecology and Systematics* 1:19–54.
1976 Are Species Really Individuals? *Systematic Zoology* 25:174–191.
1978 A Matter of Individuality. *Philosophy of Science* 45:335–360.
1980 Individuality and Selection. *Annual Review of Ecology and Systematics* 11:311–332.

1981a Metaphysics and Common Usage. *Behavioral and Brain Sciences* 4:290–291.

1981b Units of Evolution: A Metaphysical Essay. In *The Philosophy of Evolution*, edited by U. J. Jenson and R. Harré, pp. 23–44. St. Martin's Press, New York.

1982 The Naked Meme. In *Learning, Development, and Culture*, edited by H. C. Plotkin, pp. 273–327. Wiley, New York.

1984 Can Kripke Alone Save Essentialism? A Reply to Kitts. *Systematic Zoology* 33:110–112.

1988 Interactors versus Vehicles. In *The Role of Behavior in Evolution*, edited by H. C. Plotkin, pp. 19–50. MIT Press, Cambridge, MA.

1992 Biological Species: An Inductivist's Nightmare. In *How Classification Works*, edited by M. Douglas and D. Hull, pp. 42–68. Edinburgh University Press, Edinburgh.

1997 The Ideal Species Concept—and Why We Can't Get It. In *Species: The Units of Biodiversity*, edited by M. F. Claridge, H. A. Dawah, and M. R. Wilson, pp. 357–380. Chapman and Hall, London.

1998 Species, Subspecies, and Races. *Social Research* 65:351–367.

Huxley, A.

1982 Anniversary Address 1981. *Proceedings of the Royal Society* A379:ix–xx.

Huxley, J. S.

1942 *Evolution, the Modern Synthesis*. Allen and Unwin, London.

Ihde, A.

1964 *The Development of Modern Chemistry*. Harper and Row, New York.

Jablonski, D.

1986 Background and Mass Extinctions: Alternation of Macroevolutionary Regimes. *Science* 231:129–133.

Jacob, F.

1977 Evolution and Tinkering. *Science* 196:1161–1166.

Jaksic, F. M.

1981 Recognition of Morphological Adaptations in Animals: The Hypothetico-Deductive Method. *BioScience* 31:667–670.

Jennings, Jesse D.

1947 Review of "Hiwassee Island" by T. M. N. Lewis and M. Kneberg. *American Antiquity* 12:191–193.

Jernvall, J., J. P. Hunter, and M. Fortelius

1996 Molar Tooth Diversity, Disparity, and Ecology in Cenozoic Ungulate Radiations. *Science* 274:1489–1492.

Johnson, J. G.

1975 Allopatric Speciation in Fossil Brachiopods. *Journal of Paleontology* 49:646–661.

1982 Occurrence of Phyletic Gradualism and Punctuated Equilibria through Geologic Time. *Journal of Paleontology* 56:1329–1331.

Johnson, L., Jr.

1972a Introduction to Imaginary Models for Archaeological Scaling and Clustering. In *Models in Archaeology*, edited by D. L. Clarke, pp. 309–379. Methuen, London.

1972b Problems in "Avant Garde" Archaeology. *American Anthropologist* 74:366–377.

Jones, G. T., R. D. Leonard, and A. L. Abbott

1995 The Structure of Selectionist Explanations in Archaeology. In *Evolutionary Archaeology: Methodological Issues*, edited by P. A. Teltser, pp. 13–32. University of Arizona Press, Tucson.

Judd, N. M.

1929 The Present Status of Archaeology in the United States. *American Anthropologist* 31:401–418.

Justice, N. D.

1987 *Stone Age Spear and Arrow Points of the Midcontinental and Eastern United States*. Indiana University Press, Bloomington.

Kaplan, D.
1965 The Superorganic: Science or Metaphysics? *American Anthropologist* 67:956–976.
Kat, P. W., and G. M. Davis
1983 Speciation in Molluscs from Turkana Basin. *Nature* 304:660–661.
Keesing, R. M.
1974 Theories of Culture. *Annual Review of Anthropology* 3:73–97.
Kelley, P. H.
1983 Evolutionary Patterns of Eight Chesapeake Group Molluscs: Evidence for the Model of Punctuated Equilibria. *Journal of Paleontology* 57:581–598.
Kidder, A. V.
1915 Pottery of the Pajarito Plateau and of Some Adjacent Regions in New Mexico. *American Anthropological Association, Memoir* 2:407–462.
1916 Archeological Explorations at Pecos, New Mexico. *National Academy of Sciences, Proceedings* 2:119–123.
1917 A Design-Sequence from New Mexico. *National Academy of Sciences, Proceedings* 3: 369–370.
1919 Review of "An Outline for a Chronology of Zuñi Ruins; Notes on Some Little Colorado Ruins; Ruins in the White Mountains, Arizona" by L. Spier. *American Anthropologist* 21:296–301.
1924 An Introduction to the Study of Southwestern Archaeology, with a Preliminary Account of the Excavations at Pecos. *Papers of the Southwestern Expedition, Phillips Academy*, No. 1.
1931 The Pottery of Pecos, Vol. 1. *Papers of the Southwestern Expedition, Phillips Academy* No. 5.
1932 The Artifacts of Pecos. *Papers of the Southwestern Expedition, Phillips Academy* No. 6.
1936a Discussion. In The Pottery of Pecos Vol. II, by A. V. Kidder and A. O. Shepard, pp. 589–628. *Papers of the Southwestern Expedition, Phillips Academy* No. 7.
1936b Introduction. In The Pottery of Pecos Vol. II, by A. V. Kidder and A. O. Shepard, pp. xvii–xxxi. *Papers of the Southwestern Expedition, Phillips Academy* No. 7.
Kidder, M. A., and A. V. Kidder
1917 Notes on the Pottery of Pecos. *American Anthropologist* 19:325–360.
Kimura, M.
1979 The Neutral Theory of Molecular Evolution. *Scientific American* 241(5):98–126.
1983 *The Neutral Theory of Molecular Evolution*. Cambridge University Press, Cambridge, UK.
1991 Neutral Evolution. In *Evolution of Life: Fossils, Molecules, and Culture*, edited by S. Osawa and T. Honjo, pp. 67–78. Springer-Verlag, Tokyo.
King, J. L., and T. H. Jukes
1969 Non–Darwinian Evolution. *Science* 164:788–798.
Kintigh, K. W.
1984 Measuring Archaeological Diversity by Comparison with Simulated Assemblages. *American Antiquity* 49:44–54.
Kirchner, J. W., and A. Weil
1998 No Fractals in Fossil Extinction Statistics. *Nature* 395:337–338.
Kitchell, J. A., and N. MacLeod
1988 Macroevolutionary Interpretations of Symmetry and Synchroneity in the Fossil Record. *Science* 240:1190–1193.
1989 Asymmetries of Clade Shape and the Direction of Evolutionary Time. *Science* 243:1614–1615.
Kitchell, J. A., and D. Pena
1984 Periodicity of Extinctions in the Geologic Past: Deterministic versus Stochastic Explanations. *Science* 226:689–692.
Kitcher, P.
1984a Species. *Philosophy of Science* 51:308–333.

1984b Against the Monism of the Moment: A Reply to Elliott Sober. *Philosophy of Science* 51: 616–630.

Kitts, D. B.

1974 Paleontology and Evolutionary Theory. *Evolution* 28:458–472.

1984 The Names of Species: A Reply to Hull. *Systematic Zoology* 33:112–115.

Kitts, D. B., and D. J. Kitts

1979 Biological Species as Natural Kinds. *Philosophy of Science* 46:613–622.

Klejn, L. S.

1973 On Major Aspects of the Interrelationship of Archeology and Ethnology. *Current Anthropology* 14:311–320.

1977 A Panorama of Theoretical Archaeology. *Current Anthropology* 18:1–47.

Kluckhohn, C.

1939a Discussion. In Preliminary Report on the 1937 Excavations, Bc 50–51. Chaco Canyon, New Mexico, edited by C. Kluckhohn and P. Reiter, pp. 151–162. *University of New Mexico, Bulletin* 345.

1939b The Place of Theory in Anthropological Studies. *Philosophy of Science* 6:328–344.

1960 The Use of Typology in Anthropological Theory. In *Selected Papers of the Fifth International Congress of Anthropological and Ethnological Sciences*, edited by A. F. C. Wallace, pp. 134–140. University of Pennsylvania Press, Philadelphia.

Kneberg, M.

1956 Some Important Projectile Point Types Found in the Tennessee Area. *Tennessee Archaeologist* 12(1):17–28.

Knudson, R.

1979 Inference and Imposition in Lithic Analysis. In *Lithic Use-Wear Analysis*, edited by B. Hayden, pp. 269–281. Academic Press, New York.

Koch, P. L.

1986 Clinal Geographic Variation in Mammals: Implications for the Study of Chronoclines. *Paleobiology* 12:269–281.

Kramer, C. (editor)

1979 *Ethnoarchaeology: Implications of Ethnography for Archaeology*. Columbia University Press, New York.

Krieger, A. D.

1944 The Typological Concept. *American Antiquity* 9:271–288.

Kroeber, A. L.

1909 The Archaeology of California. In *Putnam Anniversary Volume*, edited by F. Boas, pp. 1–42. Stechert, New York.

1916a Zuñi Culture Sequences. *National Academy of Sciences, Proceedings* 2:42–45.

1916b Zuñi Potsherds. *American Museum of Natural History, Anthropological Papers* 18(1):1–37.

1917 The Superorganic. *American Anthropologist* 19:163–213.

1919 On the Principle of Order in Civilization as Exemplified by Changes of Fashion. *American Anthropologist* 21:235–263.

1923a *Anthropology*. Harcourt, Brace, New York.

1923b The History of Native Culture in California. *University of California, Publications in American Archaeology and Ethnology* 20:123–142.

1925 Handbook of the Indians of California. *Bureau of American Ethnology, Bulletin* 78.

1931a Historical Reconstruction of Culture Growths and Organic Evolution. *American Anthropologist* 33:149–156.

1931b The Culture-Area and Age–Area Concepts of Clark Wissler. In *Methods in Social Science*, edited by S. A. Rice, pp. 248–265. University of Chicago Press, Chicago.

1939 Cultural and Natural Areas of Native North America. *University of California, Publications in American Archaeology and Ethnology* 38:1–242.

1940 Statistical Classification. *American Antiquity* 6:29–44.

1942 Tapajó Pottery. *American Antiquity* 7:403–405.

1943 Structure, Function and Pattern in Biology and Anthropology. *Scientific Monthly* 56: 105–113.

1946 History and Evolution. *Southwestern Journal of Anthropology* 2:1–15.

1948 *Anthropology* (rev. ed.). Harcourt, Brace, New York.

Kurtén, B.

1960 Rates of Evolution in Fossil Mammals. *Cold Spring Harbor Symposia on Quantitative Biology* 24:205–215.

1965 Evolution in Geological Time. In *Ideas in Modern Biology*, edited by J. A. Moore, pp. 329–354. Natural History Press, Garden City, NY.

Kushner, G.

1970 A Consideration of Some Processual Designs for Archaeology as Anthropology. *American Antiquity* 35:125–132.

Lande, R.

1980 Genetic Variation and Phenotypic Evolution during Allopatric Speciation. *American Naturalist* 116:463–479.

LaPorte, J.

1997 Essential Membership. *Philosophy of Science* 64:96–112.

Laporte, L. F.

1983 Simpson's *Tempo and Mode in Evolution* Revisited. *American Philosophical Society, Proceedings* 127:365–417.

1991 George G. Simpson, Paleontology, and the Expansion of Biology. In *The Expansion of American Biology*, edited by K. R. Benson, J. Maienschein, and R. Rainger, pp. 80–106. Rutgers University Press, New Brunswick, NJ.

1994 Simpson on Species. *Journal of the History of Biology* 27:141–159.

Lathrap, D. W. (editor)

1956 An Archaeological Classification of Culture Contact Situations. In Seminars in Archaeology: 1955. *Society for American Archaeology, Memoirs* No. 11:1–30.

Laudan, R.

1992 What's So Special about the Past? In *History and Evolution*, edited by M. H. Nitecki and D. V. Nitecki, pp. 55–67. State University of New York Press, Albany.

Leaf, M. L.

1979 *Man, Mind, and Science: A History of Anthropology.* Columbia University Press, New York.

LeBlanc, S. A.

1973 Two Points of Logic Concerning Data, Hypotheses, General Laws and Systems. In *Research and Theory in Current Archaeology*, edited by C. L. Redman, pp. 199–214, Wiley, New York.

1975 Micro-Seriation: A Method for Fine Chronologic Differentiation. *American Antiquity* 40:22–38.

Leonard, R. D., and G. T. Jones

1987 Elements of an Inclusive Evolutionary Model for Archaeology. *Journal of Anthropological Archaeology* 6:199–219.

1989 (Editors) *Quantifying Diversity in Archaeology*. Cambridge University Press, Cambridge, UK.

Leonard, R. D., and H. E. Reed

1993 Population Aggregation in the Prehistoric American Southwest: A Selectionist Model. *American Antiquity* 58:648–661.

Leone, M. P.

1971 Review of "New Perspectives in Archeology," edited by S. R. Binford and L. R. Binford. *American Antiquity* 36:220–222.

Leroi, A. M., M. R. Rose, and G. V. Lauder
1994 What Does the Comparative Method Reveal about Adaptation? *American Naturalist* 143:381–402.

LeTourneau, P. D.
1998 The "Folsom Problem." In *Unit Issues in Archaeology: Measuring Time, Space, and Material*, edited by A. F. Ramenofsky and A. Steffen, pp. 52–73. University of Utah Press, Salt Lake City.

Levin, M. E.
1973 On Explanation in Archaeology: A Rebuttal to Fritz and Plog. *American Antiquity* 38: 387–395.
1976 On the Ascription of Functions to Objects, with Special Reference to Inference in Archaeology. *Philosophy of the Social Sciences* 6:227–234.

Levinton, J. S.
1983 Stasis in Progress: The Empirical Basis of Macroevolution. *Annual Review of Ecology and Systematics* 14:103–137.
1986 Punctuated Equilibrium. *Science* 231:1490.

Levinton, J. S., and D. J. Futuyma
1982 Macroevolution: Pattern and Process, Introduction and Background. *Evolution* 36:425–426.

Levinton, J. S., and C. M. Simon
1980 A Critique of the Punctuated Equilibria Model and Implications for the Detection of Speciation in the Fossil Record. *Systematic Zoology* 29:130–142.

Lewin, R.
1980 Evolutionary Theory under Fire. *Science* 210:883–887.
1982 *Thread of Life: The Smithsonian Looks at Evolution*. Smithsonian Institution Press, Washington, DC.
1986 Punctuated Equilibrium Is Now Old Hat. *Science* 231:672–673.

Lewontin, R. C.
1970 The Units of Selection. *Annual Review of Ecology and Systematics* 1:1–18.
1974a *The Genetic Basis of Evolutionary Change*. Columbia University Press, New York.
1974b Darwin and Mendel—The Materialist Revolution. In *The Heritage of Copernicus: Theories Pleasing to the Mind*, edited by J. Neyman, pp. 166–183. MIT Press, Cambridge, MA.
1978 Adaptation. *Scientific American* 239(9):156–169.
1983 The Organism as the Subject and Object of Evolution. *Scientia* 118:65–82.

Lewontin, R. C., and L. C. Birch
1966 Hybridization as a Source of Variation for Adaptation to New Environments. *Evolution* 20:315–336.

Lindsay, D. W.
1982 Punctuated Equilibria and Punctuated Environments. *Nature* 296:611.

Lipe, W. D.
1964 Comment on Dempsey and Baumhoff's "The Statistical Use of Artifact Distributions to Establish Chronological Sequence." *American Antiquity* 30:103–104.

Lipo, C. P., M. E. Madsen, R. C. Dunnell, and T. Hunt
1997 Population Structure, Cultural Transmission, and Frequency Seriation. *Journal of Anthropological Archaeology* 16:301–334.

Lister, A.
1984 Evolutionary Case Histories from the Fossil Record. *Nature* 309:114–115.

Logan, W. D.
1952 Graham Cave, an Archaic Site in Montgomery County, Missouri. *Missouri Archaeological Society, Memoir* No. 2.

Longacre, W. A.
1964 Archaeology as Anthropology: A Case Study. *Science* 144:1454–1455.

1966 Changing Patterns of Social Integration: A Prehistoric Example from the American Southwest. *American Anthropologist* 68:94–102.

1968 Some Aspects of Prehistoric Society in East-Central Arizona. In *New Perspectives in Archeology*, edited by S. R. Binford and L. R. Binford, pp. 89–102. Aldine, Chicago.

1970 Archaeology as Anthropology: A Case Study. *University of Arizona, Anthropological Papers* 17.

Loomis, Frederic B., and D. B. Young

1912 On the Shell Heaps of Maine. *American Journal of Science* 34:17–42.

Lopinot, N. H.

1982 Summary of the Botanical Analysis. In The Carrier Mills Archaeological Project: Human Adaptation in the Saline Valley, Illinois, edited by R. W. Jefferies and B. M. Butler, pp. 797–806. *Southern Illinois University at Carbondale, Center for Archaeological Investigations, Research Paper* No. 33.

Lowe, J. W. G., and R. J. Barth

1980 Systems in Archaeology: A Comment on Salmon. *American Antiquity* 45:568–575.

Lyman, R. L., and M. J. O'Brien

1997 The Concept of Evolution in Early Twentieth-Century Americanist Archeology. In Rediscovering Darwin: Evolutionary Theory in Archeological Explanation, edited by C. M. Barton and G. A. Clark, pp. 21–48. *American Anthropological Association, Archeological Papers* No. 7. Washington, DC.

1998 The Goals of Evolutionary Archaeology: History and Explanation. *Current Anthropology* 39:615–652.

1999a Americanist Stratigraphic Excavation and the Measurement of Culture Change. *Journal of Archaeological Method and Theory* 6:55–108.

1999b Measuring and Explaining Change in Artifact Variation with Clade-Diversity Diagrams. *Journal of Anthropological Archaeology*. In press.

n.d.a Graphing Change and Testing Lineages with Seriation. Unpublished manuscript in possession of the authors.

n.d.b Chronometers and Units in Early Archaeology and Paleontology. *American Antiquity*. In press.

Lyman, R. L., M. J. O'Brien, and R. C. Dunnell

1997a Editors. *Americanist Culture History: Fundamentals of Time, Space, and Form*. Plenum Press, New York.

1997b *The Rise and Fall of Culture History*. Plenum Press, New York.

Lyman, R. L., M. J. O'Brien, and V. Hayes

1998a A Mechanical and Functional Study of Bone Rods from the Richey–Roberts Clovis Cache, Washington, U.S.A. *Journal of Archaeological Science* 25:887–906.

Lyman, R. L., S. Wolverton, and M. J. O'Brien

1998b Seriation, Superposition, and Interdigitation: A History of Americanist Graphic Depictions of Culture Change. *American Antiquity* 63:239–261.

Mace, R., and M. Pagel

1994 The Comparative Method in Anthropology *Current Anthropology* 35:549–564.

MacFadden, B. J.

1992 *Fossil Horses: Systematics, Paleobiology, and Evolution of the Family Equidae*. Cambridge University Press, Cambridge, UK.

MacLeod, N.

1991 Punctuated Anagenesis and the Importance of Stratigraphy to Paleobiology. *Paleobiology* 17:167–188.

Maglio, V. J.

1971 The Nomenclature of Intermediate Forms: An Opinion. *Systematic Zoology* 20:370–373.

Mallory, V. S.

1970 Biostratigraphy—A Major Basis for Paleontologic Correlation. In *Proceedings of the North*

American Paleontological Convention, edited by E. L. Yochelson, pp. 553–566. Allen Press, Lawrence, KS.

Malmgren, B. A., W. A. Berggren, and G. P. Lohmann

1983 Evidence for Punctuated Gradualism in the Late Neogene *Globorotalia tumida* Lineage of Planktonic Foraminifera. *Paleobiology* 9:377–389.

1984 Species Formation through Punctuated Gradualism in Planktonic Foraminifera. *Science* 225:317–319.

Marquardt, W. H.

1978 Advances in Archaeological Seriation. In *Advances in Archaeological Method and Theory*, Vol. 1, edited by M. B. Schiffer, pp. 257–314. Academic Press, New York.

Martin, P. S.

1936 Lowry Ruin in Southwestern Colorado. *Fieldiana: Anthropology* 23(1):1–216.

1938 Archaeological Work in the Ackmen–Lowry Area, Southwestern Colorado, 1937. *Fieldiana: Anthropology* 23(2):217–304.

1939 Modified Basket Maker Sites, Ackmen–Lowry Area, Southwestern Colorado, 1938. *Fieldiana: Anthropology* 23(3):305–499.

1971 The Revolution in Archaeology. *American Antiquity* 36:1–8.

Martin, R. A., and A. D. Barnosky

1993 Quaternary Mammals and Evolutionary Theory: Introductory Remarks and Historical Perspective. In *Morphological Change in Quaternary Mammals of North America*, edited by R. A. Martin and A. D. Barnosky, pp. 1–12. Cambridge University Press, Cambridge, UK.

Maschner, H. D. G.

1996 Editor. *Evolutionary Archaeologies*. Plenum Press, New York.

1998 Review of *Evolutionary Archaeology: Theory and Application*, edited by M. J. O'Brien. *Journal of the Royal Anthropological Institute* 4:354–355.

Mason, O. T.

1896 Influence of Environment upon Human Industries or Arts. *Smithsonian Institution, Annual Report* (1894):639–665.

1905 Environment. In Handbook of American Indians, edited by F. W. Hodge, pp. 427–430. *Bureau of American Ethnology, Bulletin* 30.

Mason, R. J.

1962 The Paleo-Indian Tradition in Eastern North America. *Current Anthropology* 3:227–246.

Mayden, R. L.

1997 A Hierarchy of Species Concepts: The Denouement in the Saga of the Species Problem. In *Species: The Units of Biodiversity*, edited by M. F. Claridge, H. A. Dawah, and M. R. Wilson, pp. 381–424. Chapman and Hall, London.

Mayr, E.

1942 *Systematics and the Origin of Species*. Columbia University Press, New York.

1943 Criteria for Subspecies, Species and Genera in Ornithology. *Annals of the New York Academy of Science* 44:133-140.

1947 Foreword. *Evolution* 1:i–ii.

1949 The Species Concept: Semantics versus Semantics. *Evolution* 3:371–372.

1954 Change of Genetic Environment and Evolution. In *Evolution as a Process*, edited by J. Huxley, A. C. Hardy, and E. B. Ford, pp. 157–180. Allen and Unwin, London.

1957 Species Concepts and Definitions. In *The Species Problem*, edited by E. Mayr, pp. 1–22. American Association for the Advancement of Science, Washington, DC.

1959a Darwin and the Evolutionary Theory in Biology. In *Evolution and Anthropology: A Centennial Appraisal*, edited by B. J. Meggers, pp. 409–412. Anthropological Society of Washington, Washington, DC.

1959b Isolation as an Evolutionary Factor. *American Philosophical Society, Proceedings* 103: 221–230.

1961 Cause and Effect in Biology. *Science* 134:1501–1506.

1962 Accident or Design: The Paradox of Evolution. In *The Evolution of Living Organisms*, edited by G. W. Leeper, pp. 1–14. Melbourne University Press, Melbourne.
1963 *Animal Species and Evolution*. Harvard University Press, Cambridge, MA.
1968 The Role of Systematics in Biology. *Science* 159:595–599.
1969a *Principles of Systematic Zoology*. McGraw-Hill, New York.
1969b The Biological Meaning of Species. *Biological Journal of the Linnean Society* 1:311–320.
1970 *Populations, Species, and Evolution*. Harvard University Press, Cambridge, MA.
1972 The Nature of the Darwinian Revolution. *Science* 176:981–989.
1976a *Evolution and the Diversity of Life*. Harvard University Press, Cambridge, MA.
1976b Typological versus Population Thinking. In *Evolution and the Diversity of Life*, by E. Mayr, pp. 26–29. Harvard University Press, Cambridge, MA.
1977 Darwin and Natural Selection. *American Scientist* 65:321–337.
1980 Prologue: Some Thoughts on the History of the Evolutionary Synthesis. In *The Evolutionary Synthesis: Perspectives on the Unification of Biology*, edited by E. Mayr and W. B. Provine, pp. 1–48. Harvard University Press, Cambridge, MA.
1981 Biological Classification: Toward a Synthesis of Opposing Methodologies. *Science* 214:510–516.
1982a *The Growth of Biological Thought: Diversity, Evolution, and Inheritance*. Harvard University Press, Cambridge, MA.
1982b Speciation and Macroevolution. *Evolution* 36:1119–1132.
1982c Process of Speciation in Animals. In *Mechanisms of Speciation*, edited by C. Barigozzi, pp. 1–19. Wiley–Liss, New York.
1983 How to Carry Out the Adaptationist Program? *American Naturalist* 121:324–334.
1987 The Ontological Status of Species: Scientific Progress and Philosophical Terminology. *Biology and Philosophy* 2:145–166.
1988 *Toward a New Philosophy of Biology: Observations of an Evolutionist*. Harvard University Press, Cambridge, MA.
1989 Speciational Evolution or Punctuated Equilibria. *Journal of Social and Biological Structures* 12:137–158.
1991a An Overview of Current Evolutionary Biology. In *New Perspectives on Evolution*, edited by L. Warren and H. Koprowski, pp. 1–14. Wiley–Liss, New York.
1991b *One Long Argument: Charles Darwin and the Genesis of Modern Evolutionary Thought*. Harvard University Press, Cambridge, MA.
1993 Fifty Years of Progress in Research on Species and Speciation. *California Academy of Sciences, Proceedings* 48:131–140.
1994 Ordering Systems. *Science* 266:715–716.
1995a Darwin's Impact on Modern Thought. *American Philosophical Society, Proceedings* 139:317–325.
1995b Systems of Ordering Data. *Biology and Philosophy* 10:419–434.
1996 What Is a Species, and What Is Not? *Philosophy of Science* 63:262–277.
1997 *This Is Biology: The Science of the Living World*. Harvard University Press, Cambridge, MA.

McCartney, P. H., and M. F. Glass
1990 Simulation Models and the Interpretation of Archaeological Diversity. *American Antiquity* 55:521–536.

McGregor, J. C.
1941 *Southwestern Archaeology*. Wiley, New York.

McKee, J. K., J. F. Thackeray, and L. R. Berger
1995 Faunal Assemblage Seriation of Southern African Pliocene and Pleistocene Fossil Deposits. *American Journal of Physical Anthropology* 96:235–250.

McKern, W. C.
1934 *Certain Culture Classification Problems in Middle Western Archaeology*. National Research Council, Committee on State Archaeological Surveys, Washington, DC.

1937 Certain Culture Classification Problems in Middle Western Archaeology. In The Indianapolis Archaeological Conference, pp. 70–82. *National Research Council, Committee on State Archaeological Surveys, Circular* No. 17.

1939 The Midwestern Taxonomic Method as an Aid to Archaeological Culture Study. *American Antiquity* 4:301–313.

1940 Application of the Midwestern Taxonomic Method. *Bulletin of the Archaeological Society of Delaware* 3:18–21.

1942 Taxonomy and the Direct Historical Approach. *American Antiquity* 8:170–172.

1943 Regarding Midwestern Archaeological Taxonomy. *American Anthropologist* 45:313–315.

1944 An Inaccurate Description of Midwestern Taxonomy. *American Antiquity* 9:445–556.

McKitrick, M. C.

1994 On Homology and the Ontological Relationship of Parts. *Systematic Biology* 43:1–10.

McShea, D. W.

1993 Arguments, Tests, and the Burgess Shale—A Commentary on the Debate. *Paleobiology* 19:399–402.

Meggers, B. J.

1955 The Coming of Age of American Archaeology. In *New Interpretations of Aboriginal American Culture History*, edited by B. J. Meggers and C. Evans, pp. 116–129. Anthropological Society of Washington, Washington, DC.

1959 Editor. *Evolution and Anthropology: A Centennial Appraisal.* Anthropological Society of Washington, Washington, DC.

Meighan, C. W.

1959 A New Method for the Seriation of Archaeological Collections. *American Antiquity* 25: 203–211.

1977 Recognition of Short Time Periods through Seriation. *American Antiquity* 42:628–629.

Meltzer, D. J.

1979 Paradigms and the Nature of Change in American Archaeology. *American Antiquity* 44: 644–657.

1981a A Study of Style and Function in a Class of Tools. *Journal of Field Archaeology* 8:313–326.

1981b Paradigms Lost—Paradigms Found? *American Antiquity* 46:662–665.

1983 The Antiquity of Man and the Development of American Archaeology. In *Advances in Archaeological Method and Theory*, Vol. 6, edited by M. B. Schiffer, pp. 1–51. Academic Press, New York.

1985 North American Archaeology and Archaeologists 1879–1934. *American Antiquity* 50:249–260.

1991 On "Paradigms" and "Paradigm Bias" in Controversies over Human Antiquity in America. In *The First Americans: Search and Research*, edited by T. Dillehay and D. J. Meltzer, pp. 13–49. CRC Press, Boca Raton, FL.

1997 Monte Verde and the Pleistocene Peopling of the Americas. *Science* 276:754–755.

Meltzer, D. J., and R. C. Dunnell (editors)

1992 *The Archaeology of William Henry Holmes.* Smithsonian Institution Press, Washington, DC.

Miller, A. I., and J. J. Sepkoski, Jr.

1988 Modeling Bivalve Diversification: The Effect of Interaction on a Macroevolutionary System. *Paleobiology* 14:364–369.

Miller, R.

1982 Pseudo-Tools Created by Livestock from Halawa, Syria. *Journal of Field Archaeology* 9:281–283.

Mishler, B., and R. Brandon

1987 Individuality, Pluralism, and the Phylogenetic Species Concept. *Biology and Philosophy* 2:37–54.

Mishler, B., and M. Donoghue

1982 Species Concepts: A Case for Pluralism. *Systematic Zoology* 31:491–503.

Mitchell, S. D.
1987 Competing Units of Selection?: A Case of Symbiosis. *Philosophy of Science* 54:351–367.
Moment, G. B.
1945 The Relationship between Serial and Special Homology and Organic Similarity. *American Naturalist* 79:445–455.
Morgan, C. G.
1973 Archaeology and Explanation. *World Archaeology* 4:259–276.
1974 Explanation and Scientific Archaeology. *World Archaeology* 6:133–137.
Morgan, L. H.
1877 *Ancient Society*. Holt, New York.
Morgan, T. H.
1923 The Bearing of Mendelism on the Origin of Species. *Scientific Monthly* 16:237–246.
1932 *The Scientific Basis of Evolution*. Norton, New York.
Morgan, T. H., A. H. Sturtevant, H. J. Muller, and C. B. Bridges
1915 *The Mechanism of Mendelian Heredity*. Holt, New York.
Morse, D. F.
1997 *Sloan: A Paleoindian Dalton Cemetery in Arkansas*. Smithsonian Institution Press, Washington, DC.
Morse, D. F., and P. A. Morse
1983 *Archaeology of the Central Mississippi Valley*. Academic Press, New York.
Muller, H. J.
1949 The Darwinian and Modern Conceptions of Natural Selection. *American Philosophical Society, Proceedings* 93:459–470.
Muller, J.
1973 Review of "Broken K. Pueblo: Prehistoric Social Organization in the American Southwest" by J. N. Hill. *American Antiquity* 38:249–251.
1999 Review of "Changing Perspectives on the Archaeology of the Central Mississippi Valley," edited by M. J. O'Brien and R. C. Dunnell. *American Anthropologist* 101:440–441.
Munson, P. J.
1969 Comments on Binford's "Smudge Pits and Hide Smoking: The Use of Analogy in Archaeological Reasoning." *American Antiquity* 34:83–85.
Murdock, G. P.
1948 Clark Wissler, 1870–1947. *American Anthropologist* 50:292–304.
Murray, T., and M. J. Walker
1988 Like WHAT? A Practical Question of Analogical Inference and Archaeological Meaningfulness. *Journal of Anthropological Archaeology* 7:248–287.
Musil, R. R.
1988 Functional Efficiency and Technological Change: A Hafting Tradition Model for Prehistoric North America. In Early Human Occupation in Far Western North America: The Clovis–Archaic Interface, edited by J. A. Willig, C. M. Aikens, and J. L. Fagan, pp. 373–387. *Nevada State Museum, Anthropological Papers* No. 21.
Nagel, E.
1961 *The Structure of Science*. Harcourt, Brace, and World, New York.
Neff, H.
1992 Ceramics and Evolution. In *Archaeological Method and Theory*, Vol. 4, edited by M. B. Schiffer, pp. 141–193. University of Arizona Press, Tucson.
1993 Theory, Sampling, and Analytical Technique in the Archaeological Study of Prehistoric Ceramics. *American Antiquity* 58:23–44.
1998 *Differential Persistence of What? The Scale of Selection Issue in Evolutionary Archaeology*. Paper presented at the 63rd Annual Meeting of the Society for American Archaeology, Seattle.

Neff, H., and D. O. Larson

1997 Methodology of Comparison in Evolutionary Archeology. In Rediscovering Darwin: Evolutionary Theory and Archeological Explanation, edited by C. M. Barton and G. A. Clark, pp. 75–94. *American Anthropological Association, Archeological Papers* No. 7. Washington, DC.

Neiman, F. D.

1995 Stylistic Variation in Evolutionary Perspective: Implications for Middle Woodland Ceramic Diversity. *American Antiquity* 60:7–36.

Nelson, N. C.

1906 Excavation of the Emeryville Shellmound [Nels C. Nelson's Final Report, transcribed by J. M. Broughton, 1996]. *University of California, Archaeological Research Facility, Contributions* No. 54.

1910 The Ellis Landing Shellmound. *University of California, Publications in American Archaeology and Ethnology* 7(5):357–426.

1913 Ruins of Prehistoric New Mexico. *American Museum Journal* 13:63–81.

1914 Pueblo Ruins of the Galisteo Basin, New Mexico. *American Museum of Natural History, Anthropological Papers* 15(1):1–124.

1915 Ancient Cities of New Mexico. *American Museum Journal* 15:389–394.

1916 Chronology of the Tano Ruins, New Mexico. *American Anthropologist* 18:159–180.

1919a Human Culture. *Natural History* 19:131–140.

1919b The Archaeology of the Southwest: A Preliminary Report. *National Academy of Sciences, Proceedings* 5:114–120.

1919c The Southwest Problem. *El Palacio* 6:132–135.

1932 The Origin and Development of Material Culture. *Sigma Xi Quarterly* 20:102–123.

1937 Prehistoric Archeology, Past, Present and Future. *Science* 85:81–89.

Newell, N. D.

1959 Adequacy of the Fossil Record. *Journal of Paleontology* 33:488–499.

Nickles, T.

1977 On the Independence of Singular Causal Explanation in Social Science: Archaeology. *Philosophy of the Social Sciences* 7:163–187.

Nitecki, M. H., and D. V. Nitecki (editors)

1992 *History and Evolution*. State University of New York Press, Albany.

Norrell, M. A.

1993 Tree-Based Approaches to Understanding History: Comments on Ranks, Rules, and the Quality of the Fossil Record. *American Journal of Science* 293A:407–417.

Norell, M. A., and M. J. Novacek

1992 The Fossil Record and Evolution: Comparing Cladistic and Paleontologic Evidence for Vertebrate History. *Science* 255:1690–1693.

Novacek, M. J.

1982 Introduction to the Symposium: Phylogeny and Rates of Evolution. *Systematic Zoology* 31:337.

1992 Fossils as Critical Data for Phylogeny. In *Extinction and Phylogeny*, edited by M. J. Novacek and Q. D. Wheeler, pp. 46–88. Columbia University Press, New York.

1996 Paleontological Data and the Study of Adaptation. In *Adaptation*, edited by M. R. Rose and G. V. Lauder, pp. 311–359. Academic Press, New York.

Nowak, R. M., and N. E. Federoff

1998 Validity of the Red Wolf: Response to Roy et al. *Conservation Biology* 12:722–725.

O'Brien, M. J.

1985 Radiocarbon Dates. In Archaeology of the Central Salt River Valley: An Overview of the Prehistoric Occupation, by M. J. O'Brien. *The Missouri Archaeologist* 46:67–75.

1987 Sedentism, Population Growth, and Resource Selection in the Woodland Midwest: A Review of Coevolutionary Developments. *Current Anthropology* 28:177–197.

1995 Archaeological Research in the Central Mississippi Valley: Culture History Gone Awry. *The Review of Archaeology* 16:23–36.

1996a *Paradigms of the Past: The Story of Missouri Archaeology*. University of Missouri Press, Columbia.

1996b Editor. *Evolutionary Archaeology: Theory and Application*. University of Utah Press, Salt Lake City.

1996c Evolutionary Theory: An Introduction. In *Evolutionary Archaeology: Theory and Application*, edited by M. J. O'Brien, pp. 1–15. University of Utah Press, Salt Lake City.

1996d The Historical Development of an Evolutionary Archaeology. In *Darwinian Archaeologies*, edited by H. D. G. Maschner, pp. 17–32. Plenum Press, New York.

1998 Sloan: Dalton-Age Occupation of Northeastern Missouri. *The Review of Archaeology* 19: 16–30.

O'Brien, M. J., and R. C. Dunnell

1998 A Brief Introduction to the Archaeology of the Central Mississippi River Valley. In *Changing Perspectives on the Archaeology of the Central Mississippi Valley*, edited by M. J. O'Brien and R. C. Dunnell, pp. 1–30. University of Alabama Press, Tuscaloosa.

O'Brien, M. J., and G. L. Fox

1994a Sorting Artifacts in Space and Time. In *Cat Monsters and Head Pots: The Archaeology of Missouri's Pemiscot Bayou*, by M. J. O'Brien, pp. 25–60. University of Missouri Press, Columbia.

1994b Assemblage Similarities and Differences. In *Cat Monsters and Head Pots: The Archaeology of Missouri's Pemiscot Bayou*, by M. J. O'Brien, pp. 61–93. University of Missouri Press, Columbia.

O'Brien, M. J., and R. J. Hoard

1996 Ceramic Vessels. In Middle and Late Woodland Subsistence and Ceramic Technology in the Central Mississippi River Valley: Selected Studies from the Burkemper Site, Lincoln County, Missouri, by M. J. O'Brien, pp. 91–144. *Illinois State Museum, Reports of Investigations* No. 52.

O'Brien, M. J., and T. D. Holland

1990 Variation, Selection, and the Archaeological Record. In *Archaeological Method and Theory*, Vol. 2, edited by M. B. Schiffer, pp. 31–79. University of Arizona Press, Tucson.

1992 The Role of Adaptation in Archaeological Explanation. *American Antiquity* 57:3–59.

1995a The Nature and Premise of a Selection-Based Archaeology. In *Evolutionary Archaeology: Methodological Issues*, edited by P. A. Teltser, pp. 175–200. University of Arizona Press, Tucson.

1995b Behavioral Archaeology and the Extended Phenotype. In *Expanding Archaeology*, edited by J. M. Skibo, W. H. Walker, and A. E. Nielsen, pp. 143–161. University of Utah Press, Salt Lake City.

O'Brien, M. J., T. D. Holland, R. J. Hoard, and G. L. Fox

1994 Evolutionary Implications of Design and Performance Characteristics of Prehistoric Pottery. *Journal of Archaeological Method and Theory* 1:259–304.

O'Brien, M. J., and R. L. Lyman

1997 *The Bureau of American Archaeology and Its Legacy to Southeastern Archaeology*. Paper presented at the 62nd Annual Meeting of the Society for American Archaeology, Nashville.

1998 *James A. Ford and the Growth of Americanist Archaeology*. University of Missouri Press, Columbia.

1999a *Seriation, Stratigraphy, and Index Fossils: The Backbone of Archaeological Dating*. Kluwer Academic/Plenum Publishers, New York.

1999b *Measuring the Flow of Time: The Works of James A. Ford, 1935–1941*. University of Alabama Press, Tuscaloosa.

2000a Evolutionary Archaeology: Reconstructing and Explaining Historical Lineages. In *Explora-*

tions in Social Theory, edited by M. B. Schiffer, in press. University of Utah Press, Salt Lake City.
2000b Darwinian Evolutionism Is Applicable to Historical Archaeology. *International Journal of Historical Archaeology*. In press.

O'Brien, M. J., R. L. Lyman, and R. D. Leonard
1998 Basic Incompatibilities between Evolutionary and Behavioral Archaeology. *American Antiquity* 63:485–498.

O'Brien, M. J., and C. B. Pulliam
1996 Exploitation of Floral Resources. In Middle and Late Woodland Subsistence and Ceramic Technology in the Central Mississippi River Valley: Selected Studies from the Burkemper Site, by M. J. O'Brien, pp. 177–201. *Illinois State Museum, Reports of Investigations* No. 52.

O'Brien, M. J., and R. E. Warren
1982 Chronology of the Preceramic Period. In *The Cannon Reservoir Human Ecology Project: An Archaeological Study of Cultural Adaptations in the Southern Prairie Peninsula*, edited by M. J. O'Brien, R. E. Warren, and D. E. Lewarch, pp. 103–116. Academic Press, New York.
1983 An Archaic Projectile Point Sequence from the Southern Prairie Peninsula: The Pigeon Roost Creek Site. In *Archaic Hunters and Gatherers in the American Midwest*, edited by J. L. Phillips and J. A. Brown, pp. 71–98. Academic Press, New York.
1985 Stratigraphy and Chronology at Pigeon Roost Creek. In Archaeology of the Central Salt River Valley: An Overview of the Prehistoric Occupation, by M. J. O'Brien. *The Missouri Archaeologist* 46:203–225.

O'Brien, M. J., and H. C. Wilson
1988 A Paradigmatic Shift in the Search for the Origin of Agriculture. *American Anthropologist* 90:958–965.

O'Brien, M. J., and W. R. Wood
1998 *The Prehistory of Missouri*. University of Missouri Press, Columbia.

O'Connell, J. F.
1995 Ethnoarchaeology Needs a General Theory of Behavior. *Journal of Archaeological Research* 3:205–255.

O'Connell, J. F., and C. M. Inoway
1994 Surprise Valley Projectile Points and Their Chronological Implications. *Journal of California and Great Basin Anthropology* 16:162–198.

O'Hara, R. J.
1988 Homage to Clio, or, toward an Historical Philosophy of Evolutionary Biology. *Systematic Zoology* 37:142–155.
1993 Systematic Generalization, Historical Facts, and the Species Problem. *Systematic Biology* 42:231–246.

Olson, A. P.
1962 A History of the Phase Concept in the Southwest. *American Antiquity* 27:457–472.

O'Meara, J. T.
1989 Anthropology as Empirical Science. *American Anthropologist* 91:354–369.

Orel, V.
1996 *Gregor Mendel: The First Geneticist*. Oxford University Press, New York.

Padian, K.
1989 The Whole Real Guts of Evolution? *Paleobiology* 15:73–78.
1994 Ordering Organisms. *Science* 265:1017.
1997 The Rehabilitation of Sir Richard Owen. *BioScience* 47:446–453.

Palmer, A. R.
1985 Quantum Changes in Gastropod Shell Morphology Need Not Reflect Speciation. *Evolution* 39:699–705.

Panchen, A. L.

1994 Richard Owen and the Concept of Homology. In *Homology: The Hierarchical Basis of Comparative Biology*, edited by B. K. Hall, pp. 21–62. Academic Press, San Diego.

Patterson, C.

1981 Significance of Fossils in Determining Evolutionary Relationships. *Annual Review of Ecology and Systematics* 12:195–223.

1982 Morphological Characters and Homology. In *Problems of Phylogenetic Reconstruction*, edited by K. A. Joysey and A. E. Friday, pp. 21–74. Academic Press, New York.

1988 Homology in Classical and Molecular Biology. *Molecular Biology and Evolution* 5: 603–625.

Patterson, T. C.

1963 Contemporaneity and Cross-Dating in Archaeological Interpretation. *American Antiquity* 28:389–392.

Peabody, C.

1904 Explorations of Mounds, Coahoma Co., Mississippi. *Peabody Museum, Papers* 3(2):23–66.

Peabody, C., and W. K. Moorehead

1904 The Exploration of Jacobs Cavern. *Phillips Academy, Department of Archaeology, Bulletin* 1.

Pearson, P. N.

1992 Survivorship Analysis of Fossil Taxa When Real-Time Extinction Rates Vary: The Paleogene Planktonic Foraminifera. *Paleobiology* 18:115–131.

1998a Evolutionary Concepts in Biostratigraphy. In *Unlocking the Stratigraphical Record: Advances in Modern Stratigraphy*, edited by P. Doyle and M. R. Bennett, pp. 123–144. Wiley, Chichester, UK.

1998b Speciation and Extinction: Asymmetries in Paleontological Phylogenies: Evidence for Evolutionary Progress? *Paleobiology* 24:305–335.

Penny, D.

1983 Charles Darwin, Gradualism and Punctuated Equilibrium. *Systematic Zoology* 32:72–74.

Perino, G.

1963 Tentative Classification of Two Projectile Points and One Knife from West-Central Illinois. *Central States Archaeological Journal* 10:95–100.

1968 Guide to the Identification of Certain American Indian Projectile Points. *Oklahoma Anthropological Society, Special Bulletin* No. 3.

1971 Guide to the Identification of Certain American Indian Projectile Points. *Oklahoma Anthropological Society, Special Bulletin* No. 4.

1985 *Selected Preforms, Points and Knives of the North American Indians*, Vol. 1. Points and Barbs Press, Idabel, OK.

1991 *Selected Preforms, Points and Knives of the North American Indians*, Vol. 2. Points and Barbs Press, Idabel, OK.

Peterson, C. H.

1977 The Paleoecological Significance of Undetected Short-Term Temporal Variability. *Journal of Paleontology* 51:976–981.

Petrie, W. M. F.

1899 Sequences in Prehistoric Remains. *Journal of the Royal Anthropological Institute of Great Britain and Ireland* 29:295–301.

1901 Diospolis Parva. *Egypt Exploration Fund, Memoir* No. 20.

Phillips, P.

1942 Review of "An Archaeological Survey of the Pickwick Basin in the Adjacent Portions of the States of Alabama, Mississippi, and Tennessee" by W. S. Webb and D. L. DeJarnette. *American Antiquity* 8:197–201.

1955 American Archaeology and General Anthropological Theory. *Southwestern Journal of Anthropology* 11:246–250.

1958 Application of the Wheat–Gifford–Wasley Taxonomy to Eastern Ceramics. *American Antiquity* 24:117–125.
1970 Archaeological Survey in the Lower Yazoo Basin, 1949–1955. *Peabody Museum of Archaeology and Ethnology, Papers* 60.

Phillips, P., J. A. Ford, and J. B. Griffin
1951 Archaeological Survey in the Lower Mississippi Valley, 1940–1947. *Peabody Museum of Archaeology and Ethnology, Papers* 25.

Phillips, P., and G. R. Willey
1953 Method and Theory in American Archaeology: An Operational Basis for Culture-Historical Integration. *American Anthropologist* 55:615–633.

Pitt-Rivers, A. L.
1870 Primitive Warfare, Sec. III; On the Resemblance of the Weapons of Early Races, Their Variations, Continuity and Development of Form—Metal Period. *Royal United Service Institute, Journal* 13:509–539.
1875a On the Principles of Classification Adopted in the Arrangement of His Anthropological Collection, Now Exhibited in the Bethnal Green Museum. *Anthropological Institute of Great Britain and Ireland, Journal* 4:293–308.
1875b The Evolution of Culture. *Royal Institution of Great Britain, Proceedings* 7:496–520.

Platnick, N. I., and D. Cameron
1977 Cladistic Methods in Textual, Linguistic, and Phylogenetic Analysis. *Systematic Zoology* 26:380–385.

Plog, F. T.
1974 *The Study of Prehistoric Change*. Academic Press, New York.
1975 Systems Theory in Archeological Research. *Annual Review of Anthropology* 4:207–224.

Plog, S.
1983 Analyses of Style in Artifacts. *Annual Review of Anthropology* 12:125–142.

Plog, S., and D. P. Braun
1984 Some Issues in the Archaeology of "Tribal" Social Systems. *American Antiquity* 49:619–625.

Praetzellis, A.
1993 The Limits of Arbitrary Excavation. In *Practices of Archaeological Stratigraphy*, edited by E. C. Harris, M. R. Brown, III, and G. J. Brown, pp. 68–86. Academic Press, London.

Preucel, R. W.
1999 Review of "Evolutionary Archaeology: Theory and Application," edited by M. J. O'Brien. *Journal of Field Archaeology* 26:93–99.

Prothero, D. R.
1985 North American Mammalian Diversity and Eocene–Oligocene Extinctions. *Paleobiology* 11:389–405.

Purdue, J. R., B. W. Styles, and M. C. Masulis
1989 Faunal Remains and White-Tailed Deer Exploitation from a Late Woodland Upland Encampment: The Boschert Site (23SC609), St. Charles County, Missouri. *Midcontinental Journal of Archaeology* 14:146–163.

Quetelet, A.
1835 *Sur L'Homme et le Developpement de se Facultes, ou Essai de Physique Sociale*. Bachelier, Paris.

Quimby, G. I.
1943 The Ceramic Sequence within the Goodall Focus. *Michigan Academy of Science, Arts and Letters, Papers* 28:543–548.

Rafferty, J.
1985 The Archaeological Record on Sedentariness: Recognition, Development, and Implications. In *Advances in Archaeological Method and Theory*, Vol. 5, edited by M. B. Schiffer, pp. 113–156. Academic Press, New York.

1994 Gradual or Step-Wise Change: The Development of Sedentary Settlement Patterns in Northeast Mississippi. *American Antiquity* 59:405–425.

Rau, C.

1876 The Archaeological Collections of the United States National Museum in Charge of the Smithsonian. *Smithsonian Contributions to Knowledge* 22(4).

Raup, D. M.

1972 Taxonomic Diversity during the Phanerozoic. *Science* 177:1065–1071.

1976a Species Diversity in the Phanerozoic: A Tabulation. *Paleobiology* 2:279–288.

1976b Species Diversity in the Phanerozoic: An Interpretation. *Paleobiology* 2:289–297.

1977 Stochastic Models of Evolutionary Paleontology. In *Patterns of Evolution as Illustrated by the Fossil Record*, edited by A. Hallam, pp. 59–78. Elsevier Scientific, Amsterdam.

1986 Biological Extinction in Earth History. *Science* 231:1528–1533.

Raup, D. M., and G. E. Boyajian

1988 Patterns of Genetic Extinction in the Fossil Record. *Paleobiology* 14:109–125.

Raup, D. M., and S. J. Gould

1974 Stochastic Simulation and Evolution of Morphology—Towards a Nomothetic Paleontology. *Systematic Zoology* 23:305–322.

Raup, D. M., S. J. Gould, T. J. M. Schopf, and D. S. Simberloff

1973 Stochastic Models of Phylogeny and the Evolution of Diversity. *Journal of Geology* 81: 525–542.

Raup, D. M., and J. J. Sepkoski, Jr.

1982 Mass Extinctions in the Fossil Record. *Science* 215:1501–1503.

1984 Periodicity of Extinctions in the Geologic Past. *Proceedings of the National Academy of Science USA* 81:801–805.

1986 Periodic Extinction of Families and Genera. *Science* 231:833–836.

Raup, D. M., and S. M. Stanley

1978 *Principles of Paleontology* (2nd ed.). Freeman, San Francisco.

Read, D. W.

1979 The Effective Use of Radiocarbon Dates in the Seriation of Archaeological Sites. In *Radiocarbon Dating*, edited by R. Berger and H. E. Suess, pp. 89–94. University of California Press, Berkeley.

Read, D. W., and S. A. LeBlanc

1978 Descriptive Statements, Covering Laws, and Theories in Archaeology *Current Anthropology*. 19:307–335.

Redman, C. L.

1973 Editor. *Research and Theory in Current Archeology*. Wiley, New York.

1991 In Defense of the Seventies—The Adolescence of New Archeology. *American Anthropologist* 93:295–307.

Reed, E. K.

1940 Review of "Prehistoric Culture Units and Their Relationships in Northern Arizona" by H. S. Colton. *American Antiquity* 6:189–192.

Reid, J. J.

1995 Four Strategies after Twenty Years. In *Expanding Archaeology*, edited by J. M. Skibo, W. H. Walker, and A. E. Nielsen, pp. 15–21. University of Utah Press, Salt Lake City.

Reid, J. J., W. L. Rathje, and M. B. Schiffer

1974 Expanding Archaeology. *American Antiquity* 39:125–126.

Reid, J. J., M. B. Schiffer, and W. L. Rathje

1975 Behavioral Archaeology: Four Strategies. *American Anthropologist* 77:864–869.

Renfrew, C. (editor)

1973 *The Explanation of Culture Change: Models in Prehistory*. Duckworth, London.

Renfrew, C., M. J. Rowlands, and B. A. Segraves (editors)
1982 *Theory and Explanation in Archaeology: The Southampton Conference*. Academic Press, New York.

Renner, S.
1994 Ordering Systems. *Science* 266:716.

Rhode, D.
1988 Measurement of Archaeological Diversity and the Sample-Size Effect. *American Antiquity* 53:708–716.

Rhodes, F. H. T.
1983 Gradualism, Punctuated Equilibrium and the *Origin of Species*. *Nature* 305:369–372.
1984 Darwin's Gradualism and Empiricism: Reply to Gingerich. *Nature* 309:116.

Ridley, M.
1980 Evolution and Gaps in the Fossil Record. *Nature* 286:444–445.
1985 *The Problems of Evolution*. Oxford University Press, Oxford, UK.

Rieppel, O.
1994 Homology, Topology, and Typology: The History of Modern Debates. In *Homology: The Hierarchical Basis of Comparative Biology*, edited by B. K. Hall, pp. 63–100. Academic Press, San Diego.

Rindos, D.
1980 Symbiosis, Instability, and the Origins and Spread of Agriculture: A New Model. *Current Anthropology* 21:751–772.
1984 *The Origins of Agriculture: An Evolutionary Perspective*. Academic Press, New York.
1985 Darwinian Selection, Symbolic Variation, and the Evolution of Culture. *Current Anthropology* 26:65–88.
1989 Undirected Variation and the Darwinian Explanation of Cultural Change. In *Archaeological Method and Theory*, Vol. 1, edited by M. B. Schiffer, pp. 1–45. University of Arizona Press, Tucson.

Ritchie, W. A.
1971 A Typology and Nomenclature for New York Projectile Points (rev. ed.). *New York State Museum and Science Service, Bulletin* No. 384.

Roberts, F. H. H., Jr.
1935 A Survey of Southwestern Archaeology. *American Anthropologist* 37:1–35.

Rose, K. D., and T. M. Bown
1986 Gradual Evolution and Species Discrimination in the Fossil Record. In Vertebrates, Phylogeny, and Philosophy, edited by K. M. Flanagan and J. A. Lillegraven, pp. 119–130. *University of Wyoming, Contributions to Geology, Special Paper* No. 3.

Rosenberg, M.
1990 The Mother of Invention: Evolutionary Theory, Territoriality, and the Origins of Agriculture. *American Anthropologist* 92:399–415.
1991 Population Pressure, Locational Constraints, and the Evolution of Culture: A Reply to Graber. *American Anthropologist* 93:695–697.
1994 Pattern, Process, and Hierarchy in the Evolution of Culture. *Journal of Anthropological Archaeology* 13:307–340.
1998 Comment on "The Goals of Evolutionary Archaeology: History and Explanation" by R. L. Lyman and M. J. O'Brien. *Current Anthropology* 39:639–640.

Rouse, I. B.
1939 Prehistory in Haiti: A Study in Method. *Yale University Publications in Anthropology* No. 21.
1953 The Strategy of Culture History. In *Anthropology Today*, edited by A. L. Kroeber, pp. 57–76. University of Chicago Press, Chicago.

1954 Review of "Measurements of Some Prehistoric Design Developments in the Southeastern States" by J. A. Ford. *American Antiquity* 19:296–297.
1955 On the Correlation of Phases of Culture. *American Anthropologist* 57:713–722.
1967 Seriation in Archaeology. In *American Historical Anthropology: Essays in Honor of Leslie Spier*, edited by C. L. Riley and W. W. Taylor, pp. 153–195. Southern Illinois University Press, Carbondale.

Rowe, J. H.
1954 Max Uhle, 1856–1944: A Memoir of the Father of Peruvian Archaeology. *University of California, Publications in American Archaeology and Ethnology* 46(1):1–134.
1959 Archaeological Dating and Cultural Process. *Southwestern Journal of Anthropology* 15:317–324.
1961 Stratigraphy and Seriation. *American Antiquity* 26:324–330.
1962 Alfred Louis Kroeber. *American Antiquity* 27:395–415.
1975 Review of "A History of American Archaeology" by G. R. Willey and J. A. Sabloff. *Antiquity* 49:156–158.

Rudwick, M. J. S.
1964 The Inference of Function from Structure in Fossils. *British Journal of the Philosophy of Science* 15:27–40.
1976 *The Meaning of Fossils* (rev. ed.). Neale Watson, New York.
1996 Cuvier and Brongniart, William Smith, and the Reconstruction of Geohistory. *Earth Sciences History* 15:25–36.

Sabloff, J. A., T. W. Beale, and A. M. Kurland, Jr.
1973 Recent Developments in Archaeology. *American Academy of Political and Social Science, Annals* 408:103–118.

Sabloff, J. A., L. R. Binford, and P. A. McAnany
1987 Understanding the Archaeological Record. *Antiquity* 61:203–209.

Sadler, P. M.
1981 Sediment Accumulation Rates and the Completeness of Stratigraphic Sections. *Journal of Geology* 89:569–584.

Sahlins, M. D., and E. R. Service (editors)
1960 *Evolution and Culture*. University of Michigan Press, Ann Arbor.

Saitta, D. J.
1983 On the Evolution of "Tribal" Social Networks. *American Antiquity* 48:820–824.

Salmon, M. H.
1975 Confirmation and Explanation in Archaeology. *American Antiquity* 40:459–464.
1978 What Can Systems Theory Do for Archaeology? *American Antiquity* 43:174–183.
1980 Reply to Lowe and Barth. *American Antiquity* 45:575–579.
1981 Ascribing Functions to Archaeological Objects. *Philosophy of the Social Sciences* 11:19–26.
1982 *Philosophy and Archaeology*. Academic Press, New York.

Salmon, M. H., and W. C. Salmon
1979 Alternative Models of Scientific Explanation. *American Anthropologist* 81:61–74.

Salt, G. W.
1979 Comment on the Use of the Term *Emergent Properties*. *American Naturalist* 113:145–148.

Sapir, E.
1916 Time Perspective in Aboriginal American Culture, a Study in Method. *Canada Department of Mines, Geological Survey, Memoir* 90.

Saunders, J.
1990 The Discovery of America's Past. In Megafauna and Man: Discovery of America's Heartland, edited by L. D. Agenbroad, J. I. Mead, and L. W. Nelson, pp. 136–143. *Mammoth Site of Hot Springs, South Dakota, Scientific Papers* 1.

Schaeffer, B., M. K. Hecht, and N. Eldredge
1972 Phylogeny and Paleontology. *Evolutionary Biology* 6:31–46.

Schankler, D. M.
1981 Local Extinction and Ecological Re-Entry of Early Eocene Mammals. *Nature* 293:135–138.
Scharrer, E.
1946 Anatomy and the Concept of Analogy. *Science* 103:578–579.
Schiffer, M. B.
1972 Archaeological Context and Systemic Context. *American Antiquity* 37:156–165.
1975 Archaeology as Behavioral Science. *American Anthropologist* 77:836–848.
1976 *Behavioral Archeology.* Academic Press, New York.
1978 Methodological Issues in Ethnoarchaeology. In *Explorations in Ethnoarchaeology*, edited by R. A. Gould, pp. 229–247. University of New Mexico Press, Albuquerque.
1985 Is There a "Pompeii Premise" in Archaeology? *Journal of Anthropological Research* 41: 18–41.
1987 *Formation Processes of the Archaeological Record.* University of New Mexico Press, Albuquerque.
1995a Social Theory and History in Behavioral Archaeology. In *Expanding Archaeology*, edited by J. M. Skibo, W. H. Walker, and A. E. Nielsen, pp. 22–35. University of Utah Press, Salt Lake City.
1995b A Personal History of Behavioral Archaeology. In *Behavioral Archaeology: First Principles*, by M. B. Schiffer, pp. 1–24. University of Utah Press, Salt Lake City.
1996 Some Relationships between Behavioral and Evolutionary Archaeologies. *American Antiquity* 61:643–662.
1999 Behavioral Archaeology: Some Clarification. *American Antiquity* 64:166–168.
Schiffer, M. B., and J. M. Skibo
1987 Theory and Experiment in the Study of Technological Change. *Current Anthropology* 28:595–622.
1997 The Explanation of Artifact Variability. *American Antiquity* 62:27–50.
Schindel, D. E.
1980 Microstratigraphic Sampling and the Limits of Paleontologic Resolution. *Paleobiology* 6:408–426.
1982 Resolution Analysis: A New Approach to the Gaps in the Fossil Record. *Paleobiology* 8:340–353.
Schoch, R. M.
1983 Evolution Debate. *Science* 220:360.
Schopf, T. J. M.
1980 Macroevolution: The Fifth Dimension? *Paleobiology* 6:380–382.
1981 Punctuated Equilibrium and Evolutionary Stasis. *Paleobiology* 7:156–166.
1982 A Critical Assessment of Punctuated Equilibria: I. Duration of Taxa. *Evolution* 36:1144–1157.
Schopf, T. J. M., and A. Hoffman
1983 Punctuated Equilibrium and the Fossil Record. *Science* 219:438–439.
Schwartz, S. P.
1981 Natural Kinds. *Behavioral and Brain Sciences* 4:301–302.
Scully, E. G.
1951 *Some Central Mississippi Valley Projectile Point Types.* University of Michigan, Museum of Anthropology, Ann Arbor.
Sellards, E. H.
1952 *Early Man in America.* University of Texas Press, Austin.
Sepkoski, J. J., Jr.
1975 Stratigraphic Biases in the Analysis of Taxonomic Survivorship. *Paleobiology* 1:343–355.
1981 A Factor Analytic Description of the Phanerozoic Marine Fossil Record. *Paleobiology* 7: 36–53.

Sepkoski, J. J., Jr., R. K. Bambach, D. M. Raup, and J. W. Valentine

1981 Phanerozoic Marine Diversity and the Fossil Record. *Nature* 293:435–437.

Sepkoski, J. J., Jr., and P. R. Crane

1985 Tenth Anniversary Issue of *Paleobiology*. *Paleobiology* 11:i.

Shennan, S. J.

1989 Cultural Transmission and Cultural Change. In *What's New? A Closer Look at the Process of Innovation*, edited by S. E. van der Leeuw and R. Torrence, pp. 330–346. Unwin Hyman, London.

1991 Tradition, Rationality, and Cultural Transmission. In Processual and Postprocessual Archaeologies: Multiple Ways of Knowing the Past, edited by R. W. Preucel, pp. 197–208. *Southern Illinois University Press, Carbondale, Center for Archaeological Investigations, Occasional Paper* No. 10.

Signor, P. W.

1982 Species Richness in the Phanerozoic: Compensating for Sampling Bias. *Geology* 10:625–628.

1990 The Geologic History of Diversity. *Annual Review of Ecology and Systematics* 21:509–539.

Simpson, G. G.

1937a Notes on the Clark Fork, Upper Paleocene, Fauna. *American Museum Novitates* 954.

1937b Patterns of Phyletic Evolution. *Geological Society of America Bulletin* 48:303–314.

1937c Super-Specific Variation in Nature and in Classification from the View-Point of Paleontology. *American Naturalist* 71:236–267.

1940 Types in Modern Taxonomy. *American Journal of Science* 238:413–431.

1943 Criteria for Genera, Species, and Subspecies in Zoology and Paleozoology. *New York Academy of Sciences, Annals* 44:145–178.

1944 *Tempo and Mode in Evolution*. Columbia University Press, New York.

1949 *The Meaning of Evolution*. Yale University Press, New Haven, CT.

1951 The Species Concept. *Evolution* 5:285–298.

1953 *The Major Features of Evolution*. Columbia University Press, New York.

1961 *Principles of Animal Taxonomy*. Columbia University Press, New York.

1963 Historical Science. In *The Fabric of Geology*, edited by C. C. Albritton, Jr., pp. 24–48. Freeman, Cooper, Stanford, CA.

1970 Uniformitarianism: An Inquiry into Principle, Theory, and Method in Geohistory and Biohistory. In *Essays in Evolution and Genetics in Honor of Theodosius Dobzhansky*, edited by M. K. Hecht and W. C. Steere, pp. 43–96. Appleton, New York.

1975 Recent Advances in Methods of Phylogenetic Inference. In *Phylogeny of Primates: A Multidisciplinary Approach*, edited by W. P. Luckett and F. S. Szalay, pp. 3–19. Plenum Press, New York.

1984 Introduction: Twenty Years Later. In *Tempo and Mode in Evolution* (rev. ed.), by G. G. Simpson, pp. xiii–xxx. Columbia University Press, New York.

Skibo, J. M., and M. B. Schiffer

1995 The Clay Cooking Pot: An Exploration of Women's Technology. In *Expanding Archaeology*, edited by J. M. Skibo, W. H. Walker, and A. E. Nielsen, pp. 80–91. University of Utah Press, Salt Lake City.

Skibo, J. M., W. H. Walker, and A. E. Nielsen (editors)

1995 *Expanding Archaeology*. University of Utah Press, Salt Lake City.

Sluys, R.

1996 The Notion of Homology in Current Comparative Biology. *Journal of Zoological Systematics and Evolutionary Research* 34:145–152.

Smith, A. B.

1994 *Systematics and the Fossil Record: Documenting Evolutionary Patterns*. Blackwell, London.

Smith, B. D.

1977 Archaeological Inference and Inductive Confirmation. *American Anthropologist* 79:598–617.

Smith, E. A., and B. Winterhalder (editors)
1992 *Evolutionary Ecology and Human Behavior*. Aldine, New York.

Smith, H. M.
1967 Biological Similarities and Homologies. *Systematic Zoology* 16:101–102.

Sober, E.
1980 Evolution, Population Thinking, and Essentialism. *Philosophy of Science* 47:350–383.
1984a *The Nature of Selection: Evolutionary Theory in Philosophical Focus*. MIT Press, Cambridge, MA.
1984b Sets, Species, and Evolution: Comments on Philip Kitcher's "Species." *Philosophy of Science* 51:334–341.
1988 *Reconstructing the Past: Parsimony, Evolution, and Inference*. MIT Press, Cambridge, MA.
1992 Models of Cultural Evolution. In *Trees of Life: Essays in Philosophy of Biology*, edited by P. Griffiths, pp. 17–39. Kluwer, Dordrecht, Netherlands.
1993 *Philosophy of Biology*. Westview Press, Boulder, CO.

Sober, E., and D. S. Wilson
1994 A Critical Review of Philosophical Work on the Units of Selection Problem. *Philosophy of Science* 61:534–555.

Sokal, R. R., and T. Crovello
1970 The Biological Species Concept: A Critical Evaluation. *American Naturalist* 104:127–153.

Sokal, R. R., and P. H. A. Sneath
1963 *Principles of Numerical Taxonomy*. Freeman, San Francisco.

Solow, A. R., and W. Smith
1997 On Fossil Preservation and the Stratigraphic Ranges of Taxa. *Paleobiology* 23:271–277.

Spaulding, A. C.
1949 Cultural and Chronological Classification in the Plains Area. *Plains Archaeological Conference Newsletter* 2(2):3–5.
1953a Review of "Measurements of Some Prehistoric Design Developments in the Southeastern States" by J. A. Ford. *American Anthropologist* 55:588–591.
1953b Statistical Techniques for the Discovery of Artifact Types. *American Antiquity* 18:305–313.
1954a Reply (to Ford). *American Anthropologist* 56:112–114.
1954b Reply to Ford. *American Antiquity* 19:391–393.
1954c Prehistoric Cultural Development in the Eastern United States. In *New Interpretations of Aboriginal American Culture History*, edited by B. J. Meggers and C. Evans, pp. 12–27. Anthropological Society of Washington, Washington, DC.
1957 Review of "Method and Theory in American Archeology: An Operational Basis for Cultural-Historical Integration" by P. Phillips and G. R. Willey, and "Method and Theory in American Archeology II: Historical–Developmental Interpretation" by G. R. Willey and P. Phillips. *American Antiquity* 23:85–87.
1960 The Dimensions of Archaeology. In *Essays in the Science of Culture in Honor of Leslie A. White*, edited by G. E. Dole and R. L. Carneiro, pp. 437–456. Crowell, New York.

Spencer, C. S.
1992 Homology, Analogy, and Comparative Research in Archaeology. *Behavior Science Research* 26:1–4.
1997 Evolutionary Approaches in Archaeology. *Journal of Archaeological Research* 5: 209–264.
1998 Comment on "The Goals of Evolutionary Archaeology: History and Explanation" by R. L. Lyman and M. J. O'Brien. *Current Anthropology* 39:641–642.

Spier, L.
1917a An Outline for a Chronology of Zuñi Ruins. *American Museum of Natural History, Anthropological Papers* 18(3):207–331.
1917b Zuñi Chronology. *National Academy of Sciences, Proceedings* 3:280–283.

Springer, K. B., and M. A. Murphy
1994 Punctuated Stasis and Collateral Evolution in the Devonian Lineage of *Monograptus hercynicus*. *Lethaia* 27:119–128.
Stahl, A. B.
1993 Concepts of Time and Approaches to Analogical Reasoning in Historical Perspective. *American Antiquity* 58:235–260.
Stanislawski, M. B.
1973 Review of "Archaeology as Anthropology: A Case Study" by W. A. Longacre. *American Antiquity* 38:117–121.
Stanley, S. M.
1975 A Theory of Evolution above the Species Level. *National Academy of Sciences, Proceedings* 72:646–650.
1976 Stability of Species in Geologic Time. *Science* 192:267–269.
1979 *Macroevolution: Pattern and Process*. Freeman, San Francisco.
1981 *The New Evolutionary Time Table: Fossils, Genes, and the Origin of Species*. Basic Books, New York.
1982 Macroevolution and the Fossil Record. *Evolution* 36:460–473.
1989 The Empirical Case for the Punctuational Model of Evolution. *Journal of Biological and Social Structures* 12:159–172.
Stanley, S. M., P. W. Signor III, S. Lidgard, and A. F. Karr
1981 Natural Clades Differ from "Random" Clades: Simulations and Analyses. *Paleobiology* 7:115–127.
Stearns, S. C.
1986 Natural Selection and Fitness, Adaptation and Constraint. In *Patterns and Processes in the History of Life*, edited by D. M. Raup and D. Jablonski, pp. 23–44. Springer-Verlag, Berlin.
Stebbins, G. L., and F. J. Ayala
1981 Is a New Evolutionary Synthesis Necessary? *Science* 213:967–971.
Stein, E., and P. Lipton
1989 Where Guesses Come from: Evolutionary Epistemology and the Anomaly of Guided Variation. *Biology and Philosophy* 4:33–56.
Sterns, F. H.
1915 A Stratification of Cultures in Nebraska. *American Anthropologist* 17:121–127.
Stevens, P. F.
1980 Evolutionary Polarity of Character States. *Annual Review of Ecology and Systematics* 11:333–358.
Stevens, S. S.
1946 On the Theory of Scales of Measurement. *Science* 103:677–680.
Steward, J. H.
1929 Diffusion and Independent Invention: A Critique of Logic. *American Anthropologist* 31: 491–495.
1941 Review of "Prehistoric Culture Units and Their Relationships in Northern Arizona" by H. S. Colton. *American Antiquity* 6:366–367.
1942 The Direct Historical Approach to Archaeology. *American Antiquity* 7:337–343.
1944 Re: Archaeological Tools and Jobs. *American Antiquity* 10:99–100.
1955 *Theory of Culture Change: The Methodology of Multilinear Evolution*. University of Illinois Press, Urbana.
Stiebing, W. H., Jr.
1993 *Uncovering the Past: A History of Archaeology*. Prometheus Books, Buffalo, New York.
Striedter, G. F., and R. G. Northcutt
1991 Biological Hierarchies and the Concept of Homology. *Brain, Behavior and Evolution* 38:177–189.

Strong, W. D.
1952 The Value of Archeology in the Training of Professional Anthropologists. *American Anthropologist* 54:318–321.
Styles, B. W.
1981 Early Late Woodland Subsistence in the Lower Illinois Valley. *Northwestern University Archeological Program, Scientific Papers* No. 3.
Styles, B. W., and J. R. Purdue
1986 Middle Woodland Faunal Exploitation. In The Woodland Period Occupations of the Napoleon Hollow Site in the Lower Illinois Valley, edited by M. D. Wiant and C. R. McGimsey, pp. 513–526. *Kampsville Archeological Center, Center for American Archeology, Research Series* 6.
1996 Animal Exploitation. In Middle and Late Woodland Subsistence and Ceramic Technology in the Central Mississippi River Valley: Selected Studies from the Burkemper Site, Lincoln County, Missouri, by M. J. O'Brien, pp. 145–176. *Illinois State Museum, Reports of Investigations* No. 52.
Styles, B. W., J R. Purdue, and M. L. Colburn
1985 Faunal Exploitation at the Smiling Dan Site. In Smiling Dan: Structure and Function at a Middle Woodland Settlement in the Illinois Valley, edited by B. D. Stafford and M. B. Sant, pp. 402–446. *Kampsville Archeological Center, Center for American Archeology, Research Series* 2.
Suhm, D. A., and E. B. Jelks
1962 Handbook of Texas Archeology: Type Descriptions. *Texas Archeological Society, Special Publication* No. 1.
Suhm, D. A., A. D. Krieger, and E. B. Jelks
1954 An Introductory Handbook of Texas Archeology. *Texas Archeological Society, Bulletin* 25.
Sylvester-Bradley, P. C. (editor)
1956 *The Species Concept in Palaeontology*. The Systematics Association, London.
Szalay, F. S.
1994 *Evolutionary History of the Marsupials and an Analysis of Osteological Characters*. Cambridge University Press, Cambridge, UK.
Szalay, F. S., and W. J. Bock
1991 Evolutionary Theory and Systematics: Relationships between Process and Patterns. *Zeitschrift für Zoologische Systematik und Evolutionsforschung* 29:1–39.
Szarski, H.
1949 The Concept of Homology in the Light of the Comparative Anatomy of Vertebrates. *Quarterly Review of Biology* 24:124–131.
Tattersall, I.
1986 Species Recognition in Human Paleontology. *Journal of Human Evolution* 15:165–175.
Taylor, P. J.
1987 Historical versus Selectionist Explanations in Evolutionary Biology. *Cladistics* 3:1–13.
Taylor, R. E., C. V. Haynes, Jr., and M. Stuiver
1996 Clovis and Folsom Age Estimates: Stratigraphic Context and Radiocarbon Calibration. *Antiquity* 70:515–525.
Taylor, W. W.
1948 A Study of Archeology. *American Anthropological Association, Memoir* 69.
Teltser, P. A.
1995a Editor. *Evolutionary Archaeology: Methodological Issues*. University of Arizona Press, Tucson.
1995b Introduction: The Methodological Challenge of Evolutionary Theory. In *Evolutionary Archaeology: Methodological Issues*, edited by P. A. Teltser, pp. 1–11. University of Arizona Press, Tucson.

1995c Culture History, Evolutionary Theory, and Frequency Seriation. In *Evolutionary Archaeology: Methodological Issues*, edited by P. A. Teltser, pp. 51–68. University of Arizona Press, Tucson.

Templeton, A. R.

1980 The Theory of Speciation via the Founder Principle. *Genetics* 94:1011–1038.

1981 Mechanisms of Speciation—A Population Genetic Approach. *Annual Review of Ecology and Systematics* 12:23–48.

Terrell, J.

1986 Causal Pathways and Causal Processes: Studying the Evolutionary Prehistory of Human Diversity in Language, Customs, and Biology. *Journal of Anthropological Archaeology* 5:187–198.

Thomas, C.

1891 Catalogue of Prehistoric Works East of the Rocky Mountains. *Bureau of Ethnology, Bulletin* 12.

1894 Report on the Mound Explorations of the Bureau of Ethnology. *Bureau of Ethnology, Annual Report* 12:3–742.

Thomas, D. H.

1981 How to Classify the Projectile Points from Monitor Valley, Nevada. *Journal of California and Great Basin Anthropology* 3:7–43.

1983a Cultural Stratigraphy at Gatecliff Shelter. In The Archaeology of Monitor Valley 2: Gatecliff Shelter, by D. H. Thomas, pp. 172–176. *American Museum of Natural History, Anthropological Papers* 59.

1983b Some Theoretical and Methodological Considerations. In The Archaeology of Monitor Valley 2: Gatecliff Shelter, by D. H. Thomas, pp. 419–440. *American Museum of Natural History, Anthropological Papers* 59.

Thomas, D. H., and S. L. Bierwirth

1983 Material Culture of Gatecliff Shelter: Projectile Points. In The Archaeology of Monitor Valley 2: Gatecliff Shelter, by D. H. Thomas, pp. 177–211. *American Museum of Natural History, Anthropological Papers* 59.

Thompson, P.

1983 Tempo and Mode in Evolution: Punctuated Equilibria and the Modern Synthetic Theory. *Philosophy of Science* 50:432–452.

Thompson, R. H.

1956a Editor. An Archaeological Approach to the Study of Cultural Stability. In Seminars in Archaeology: 1955. *Society for American Archaeology, Memoirs* No. 11:31–57.

1956b The Subjective Element in Archaeological Inference. *Southwestern Journal of Anthropology* 12:327–332.

1958a Editor. Migrations in New World Culture History. *University of Arizona, Social Science Bulletin* 27.

1958b Modern Yucatecan Maya Pottery Making. *Society for American Archaeology, Memoirs* No. 15.

Trigger, B. G.

1968 *Beyond History: The Methods of Prehistory*. Holt, Rinehart and Winston, New York.

1989 *A History of Archaeological Thought*. Cambridge University Press, Cambridge, UK.

1998 Review of "The Rise and Fall of Culture History," by R. L. Lyman, M. J. O'Brien, and R. C. Dunnell. *Journal of Field Archaeology* 25:363–366.

Trueman, E. R.

1979 Species Concept. In *The Encyclopedia of Paleontology*, edited by R. W. Fairbridge and D. Jablonski, pp. 764–767. Dowden, Hutchinson and Ross, Stroudsburg, PA.

Tschauner, H.

1994 Archaeological Systematics and Cultural Evolution: Retrieving the Honour of Culture History. *Man* 29:77–93.

Tuggle, H. D., A. H. Townsend, and T. J. Riley
1972 Laws, Systems and Research Designs: A Discussion of Explanation in Archaeology. *American Antiquity* 37:3–12.
Turner, J. R. G.
1986 The Genetics of Adaptive Radiation: A Neo–Darwinian Theory of Punctuational Evolution. In *Patterns and Processes in the History of Life*, edited by D. M. Raup and D. Jablonski, pp. 183–207. Springer-Verlag, Berlin.
Tylor, E. G.
1871 *Primitive Culture*. Murray, London.
Uhen, M. D.
1996 An Evaluation of Clade-Shape Statistics Using Simulations and Extinct Families of Mammals. *Paleobiology* 22:8–22.
Uhle, M.
1903 *Pachacamac*. University of Pennsylvania Press, Philadelphia.
1907 The Emeryville Shellmound. *University of California, Publications in American Archaeology and Ethnology* 7:1–107.
Valentine, J. W., and C. L. May
1996 Hierarchies in Biology and Paleontology. *Paleobiology* 22:23–33.
Van Riper, A. B.
1993 *Men among the Mammoths: Victorian Science and the Discovery of Human Prehistory*. University of Chicago Press, Chicago.
Vaupel Klein, J. C. von
1994 Punctuated Equilibria and Phyletic Gradualism: Even Partners Can Be Good Friends. *Acta Biotheoretica* 42:15–48.
Vrba, E. S.
1980 Evolution, Species and Fossils: How Does Life Evolve? *South African Journal of Science* 76:61–84.
1983 Macroevolutionary Trends: New Perspectives on the Roles of Adaptation and Incidental Effect. *Science* 221:387–389.
1984 What Is Species Selection? *Systematic Zoology* 33:318–328.
1985 Environment and Evolution: Alternative Causes of the Temporal Distribution of Evolutionary Events. *South African Journal of Science* 81:229–236.
1989 Levels of Selection and Sorting with Special Reference to the Species Level. *Oxford Surveys in Evolutionary Biology* 6:111–168.
Vrba, E. S., and N. Eldredge
1984 Individuals, Hierarchies and Processes: Towards a More Complete Evolutionary Theory. *Paleobiology* 10:146–171.
Vrba, E. S., and S. J. Gould
1986 The Hierarchical Expansion of Sorting and Selection: Sorting and Selection Cannot Be Equated. *Paleobiology* 12:217–228.
Wade, M. J., and S. Kalisz
1990 The Causes of Natural Selection. *Evolution* 44:1947–1955.
Wagner, P. J.
1995 Stratigraphic Tests of Cladistic Hypotheses. *Paleobiology* 21:153–178.
Walker, W. H., J. M. Skibo, and A. E. Nielsen
1995 Introduction: Expanding Archaeology. In *Expanding Archaeology*, edited by J. M. Skibo, W. H. Walker, and A. E. Nielsen, pp. 1–12. University of Utah Press, Salt Lake City.
Watson, P. J.
1982 Review of "Living Archaeology" by R. A. Gould. *American Antiquity* 47:445–448.
1986 Archaeological Interpretation, 1985. In *American Archaeology Past and Future: A Celebration of the Society for American Archaeology, 1935–1985*, edited by D. J. Meltzer, D. D. Fowler, and J. A. Sabloff, pp. 439–457. Smithsonian Institution Press, Washington, DC.

Watson, P. J., S. A. LeBlanc, and C. L. Redman
1971 *Explanation in Archeology: An Explicitly Scientific Approach*. Columbia University Press, New York.
1974 The Covering Law Model in Archaeology: Practical Uses and Formal Interpretations. *World Archaeology* 6:125–132.
1984 *Archeological Explanation: The Scientific Method in Archeology*. Columbia University Press, New York.

Watson, R. A.
1966 Is Geology Different: A Critical Discussion of "The Fabric of Geology." *Philosophy of Science* 33:172–185.
1976 Inference in Archaeology. *American Antiquity* 41:58–66.

Wauchope, R.
1965 Alfred Vincent Kidder, 1885–1963. *American Antiquity* 31:149–171.

Wayne, R. K., M. S. Roy, and J. L. Gittleman
1998 Origin of the Red Wolf: Response to Nowak and Federoff and Gardener. *Conservation Biology* 12:726–729.

Weller, J. M.
1949 Paleontologic Classification. *Journal of Paleontology* 23:680–690.
1961 The Species Problem. *Journal of Paleontology* 35:1181–1192.

Whallon, R.
1982 Editorial Introduction. *Journal of Anthropological Archaeology* 1:1–4.

Wheat, J. B., J. C. Gifford, and W. W. Wasley
1958 Ceramic Variety, Type Cluster, and Ceramic System in Southwestern Pottery Analysis. *American Antiquity* 24:34–47.

White, L. A.
1943 Energy and the Evolution of Culture. *American Anthropologist* 45:335–356.
1945 History, Evolution, and Functionalism: Three Types of Interpretation of Culture. *Southwestern Journal of Anthropology* 1:221–248.
1949 *The Science of Culture: A Study of Man and Civilization*. Grove Press, New York.
1959a The Concept of Evolution in Anthropology. In *Evolution and Anthropology: A Centennial Appraisal*, edited by B. J. Meggers, pp. 106–124. Anthropological Society of Washington, Washington, DC.
1959b *The Evolution of Culture: The Development of Civilization to the Fall of Rome*. McGraw-Hill, New York.

Whittaker, J. C., D. Caulkins, and K. A. Kamp
1998 Evaluating Consistency in Typology and Classification. *Journal of Archaeological Method and Theory* 5:129–164.

Willey, G. R.
1938 Time Studies: Pottery and Trees in Georgia. *Society for Georgia Archaeology, Proceedings* 1:15–22.
1939 Ceramic Stratigraphy in a Georgia Village Site. *American Antiquity* 5:140–147.
1940 Review of "Prehistory in Haiti: A Study in Method" by I. B. Rouse. *American Anthropologist* 42:673–675.
1949 Archeology of the Florida Gulf Coast. *Smithsonian Miscellaneous Collections* 113.
1953 Archaeological Theories and Interpretation: New World. In *Anthropology Today*, edited by A. L. Kroeber, pp. 361–385. University of Chicago Press, Chicago.
1968 One Hundred Years of American Archaeology. In *One Hundred Years of Anthropology*, edited by J. O. Brew, pp. 26–53. Harvard University Press, Cambridge, MA.
1988 *Portraits in American Archaeology: Remembrances of Some Distinguished Americanists*. University of New Mexico Press, Albuquerque.

Willey, G. R., and P. Phillips

1955 Method and Theory in American Archaeology, II: Historical–Developmental Interpretation. *American Anthropologist* 57:723–819.

1958 *Method and Theory in American Archaeology.* University of Chicago Press, Chicago.

Willey, G. R., and J. A. Sabloff

1993 *A History of American Archaeology* (3rd ed.). Freeman, New York.

Willey, G. R., and R. B. Woodbury

1942 A Chronological Outline for the Northwest Florida Coast. *American Antiquity* 7:232–254.

Williams, G. C.

1966 *Adaptation and Natural Selection.* Princeton University Press, Princeton, NJ.

Williamson, P. G.

1981 Palaeontological Documentation of Speciation in Cenozoic Molluscs from Turkana Basin. *Nature* 293:437–443.

1982 Williamson Replies. *Nature* 296:611–612.

1983 Williamson Replies. *Nature* 304:661–663.

1985a Punctuated Equilibrium, Morphological Stasis and the Palaeontological Documentation of Speciation: A Reply to Fryer, Greenwood and Peake's Critique of the Turkana Basin Mollusc Sequence. *Biological Journal of the Linnean Society* 26:307–324.

1985b In Reply to Fryer, Greenwood and Peake. *Biological Journal of the Linnean Society* 26: 337–340.

Willis, J. C.

1922 *Age and Area.* Cambridge University Press, Cambridge, UK.

Wilmsen, E. N.

1968a Functional Analysis of Flaked Stone Artifacts. *American Antiquity* 33:156–161.

1968b Lithic Analysis in Paleoanthropology. *Science* 161:982–987.

Wilson, B. E.

1996 Changing Conceptions of Species. *Biology and Philosophy* 11:405–420.

Wissler, C.

1909 Introduction. In The Indians of Greater New York and the Lower Hudson, edited by C. Wissler. *American Museum of Natural History, Anthropological Papers* 3:xiii–xv.

1914 Material Cultures of the North American Indians. *American Anthropologist* 16:447–505.

1915 Explorations in the Southwest by the American Museum. *American Museum Journal* 15:395–398.

1916a Correlations between Archeological and Culture Areas in the American Continents. In *Holmes Anniversary Volume: Anthropological Essays*, edited by F. W. Hodge, pp. 481–490. Washington, DC.

1916b The Application of Statistical Methods to the Data on the Trenton Argillite Culture. *American Anthropologist* 18:190–197.

1916c The Genetic Relations of Certain Forms in American Aboriginal Art. *National Academy of Sciences, Proceedings* 2:224–226.

1917a *The American Indian.* McMurtrie, New York.

1917b The New Archaeology. *American Museum Journal* 17:100–101.

1919 General Introduction. *American Museum of Natural History, Anthropological Papers* 18: iii–ix.

1923 *Man and Culture.* Crowell, New York.

1924 The Relation of Nature to Man as Illustrated by the North American Indian. *Ecology* 5:311–318.

Wobst, H. M.

1977 Stylistic Behavior and Information Exchange. In For the Director: Research Essays in Honor of James B. Griffin, edited by C. E. Cleland, pp. 317– 342. *University of Michigan, Museum of Anthropology, Anthropological Papers* No. 61.

1978 The Archaeo-Ethnology of Hunter–Gatherers or the Tyranny of the Ethnographic Record in Archaeology. *American Antiquity* 43:303–309.

Woodbury, R. B.

1960a Nels C. Nelson and Chronological Archaeology. *American Antiquity* 25:400–401.

1960b Nelson's Stratigraphy. *American Antiquity* 26:98–99.

1973 *Alfred V. Kidder*. Columbia University Press, New York.

Wormington, H. M.

1957 Ancient Man in North America (4th ed.). *Denver Museum of Natural History, Popular Series* No. 4.

Wright, C. W.

1950 Paleontologic Classification. *Journal of Paleontology* 24:746–748.

Wright, S.

1930 The Genetical Theory of Natural Selection: A Review. *Journal of Heredity* 21:349–356.

1931 Evolution in Mendelian Populations. *Genetics* 16:97–159.

1932 The Roles of Mutation, Inbreeding, Crossbreeding, and Selection in Evolution. *Proceedings of the 6th International Congress on Genetics* 1:356–366.

Wyckoff, D. G., and W. W. Dalquest

1997 From Whence They Came: The Paleontology of Southern Plains Bison. In Southern Plains Bison Procurement and Utilization from Paleoindian to Historic, edited by L. C. Bement and K. J. Buehler, pp. 5–32. *Plains Anthropologist Memoir* 29.

Wylie, A.

1982 An Analogy by Any Other Name Is Just as Analogical: A Commentary on the Gould–Watson Dialogue. *Journal of Anthropological Archaeology* 1:382–401.

1985 The Reaction against Analogy. In *Advances in Archaeological Method and Theory*, Vol. 5, edited by M. B. Schiffer, pp. 63–111. Academic Press, New York.

1995 An Expanded Behavioral Archaeology: Transformation and Redefinition. In *Expanding Archaeology*, edited by J. M. Skibo, W. H. Walker, and A. E. Nielsen, pp. 198–209. University of Utah Press, Salt Lake City.

Wyman, J.

1868 An Account of Some Kjoekkenmoeddings, or Shell-Heaps, in Maine and Massachusetts. *The American Naturalist* 1:561–584.

1875 Fresh-Water Shell Mounds of the St. John's River, Florida. *Peabody Academy of Science, Memoir* 4.

Yoffee, N.

1979 The Decline and Rise of Mesopotamian Civilization: An Ethnoarchaeological Perspective on the Evolution of Social Complexity. *American Antiquity* 44:5–35.

1980 Honk If You Know Darwin: Brief Reply to Dunnell and Wenke. *American Antiquity* 45: 610–612.

Youngman, P. M., and F. W. Schueler

1991 *Martes nobilis* Is a Synonym of *Martes americana*, Not an Extinct Pleistocene–Holocene Species. *Journal of Mammalogy* 72:567–577.

Index